A First Course in Logic

A First Course in Logic

An introduction to model theory, proof theory, computability, and complexity

---

SHAWN HEDMAN
Department of Mathematics, Florida Southern College

OXFORD
UNIVERSITY PRESS

*This book has been printed digitally and produced in a standard specification
in order to ensure its continuing availability*

OXFORD
UNIVERSITY PRESS

Great Clarendon Street, Oxford OX2 6DP

Oxford University Press is a department of the University of Oxford.
It furthers the University's objective of excellence in research, scholarship,
and education by publishing worldwide in

Oxford New York

Auckland   Cape Town   Dar es Salaam   Hong Kong   Karachi
Kuala Lumpur   Madrid   Melbourne   Mexico City   Nairobi
New Delhi   Shanghai   Taipei   Toronto

With offices in

Argentina   Austria   Brazil   Chile   Czech Republic   France   Greece
Guatemala   Hungary   Italy   Japan   South Korea   Poland   Portugal
Singapore   Switzerland   Thailand   Turkey   Ukraine   Vietnam

Oxford is a registered trade mark of Oxford University Press
in the UK and in certain other countries

Published in the United States
by Oxford University Press Inc., New York

© Oxford University Press 2004

The moral rights of the author have been asserted

Database right Oxford University Press (maker)

Reprinted 2008

All rights reserved. No part of this publication may be reproduced,
stored in a retrieval system, or transmitted, in any form or by any means,
without the prior permission in writing of Oxford University Press,
or as expressly permitted by law, or under terms agreed with the appropriate
reprographics rights organization. Enquiries concerning reproduction
outside the scope of the above should be sent to the Rights Department,
Oxford University Press, at the address above

You must not circulate this book in any other binding or cover
And you must impose this same condition on any acquirer

ISBN 978-0-19-852981-1

To Julia

Acknowledgments

Florida Southern College provided a most pleasant and hospitable setting for the writing of this book. Thanks to all of my friends and colleagues at the college. In particular, I thank colleague David Rose and student Biljana Cokovic for reading portions of the manuscript and offering helpful feedback. I thank my colleague Mike Way for much needed technological assistance. This book began as lecture notes for a course I taught at the University of Maryland. I thank my students and colleagues in Maryland for their encouragement in beginning this project.

The manuscript was prepared using the MikTex Latex system with a GNU Emacs editor. For the few diagrams that were not produced using Latex, the Gimp was used (the GNU Image Manipulation Program). I would like to thank the producers of this software for making it freely available.

I cannot adequately acknowledge all those who have shaped the subject and my understanding of the subject contained within these pages. For the many names of logicians and mathematicians mentioned in the book, I fear there are many deserving names that I have left out. My apologies to those I have slighted in this respect. Many people, through books and personal interaction, have influenced my presentation of the subject. The books are included in the bibliography. Of my teachers, two merit special mention. I thank John Baldwin and David Marker at the University of Illinois at Chicago from whom I learned so much not so long ago. It is my hope that this book should lead readers to their outstanding books on Stability Theory and Model Theory.

Most importantly, I must acknowledge my wife Julia and our young children Max and Sabrina. From Sabrina's perspective, this book has been a life-long project. To Julia and Max, it may have seemed like a lifetime. It is to Julia that I owe the greatest debt of gratitude. Without Julia's enduring patience, effort, and support, this book certainly would not exist.

Contents

1	**Propositional logic**		**1**
	1.1 What is propositional logic?		1
	1.2 Validity, satisfiability, and contradiction		7
	1.3 Consequence and equivalence		9
	1.4 Formal proofs		12
	1.5 Proof by induction		22
		1.5.1 Mathematical induction	23
		1.5.2 Induction on the complexity of formulas	25
	1.6 Normal forms		27
	1.7 Horn formulas		32
	1.8 Resolution		37
		1.8.1 Clauses	37
		1.8.2 Resolvents	38
		1.8.3 Completeness of resolution	40
	1.9 Completeness and compactness		44
2	**Structures and first-order logic**		**53**
	2.1 The language of first-order logic		53
	2.2 The syntax of first-order logic		54
	2.3 Semantics and structures		57
	2.4 Examples of structures		66
		2.4.1 Graphs	66
		2.4.2 Relational databases	69
		2.4.3 Linear orders	70
		2.4.4 Number systems	72
	2.5 The size of a structure		73
	2.6 Relations between structures		79
		2.6.1 Embeddings	80
		2.6.2 Substructures	83
		2.6.3 Diagrams	86
	2.7 Theories and models		89

3 Proof theory — 99

- 3.1 Formal proofs — 100
- 3.2 Normal forms — 109
  - 3.2.1 Conjunctive prenex normal form — 109
  - 3.2.2 Skolem normal form — 111
- 3.3 Herbrand theory — 113
  - 3.3.1 Herbrand structures — 113
  - 3.3.2 Dealing with equality — 116
  - 3.3.3 The Herbrand method — 118
- 3.4 Resolution for first-order logic — 120
  - 3.4.1 Unification — 121
  - 3.4.2 Resolution — 124
- 3.5 SLD-resolution — 128
- 3.6 Prolog — 137

4 Properties of first-order logic — 147

- 4.1 The countable case — 147
- 4.2 Cardinal knowledge — 152
  - 4.2.1 Ordinal numbers — 153
  - 4.2.2 Cardinal arithmetic — 156
  - 4.2.3 Continuum hypotheses — 161
- 4.3 Four theorems of first-order logic — 163
- 4.4 Amalgamation of structures — 170
- 4.5 Preservation of formulas — 174
  - 4.5.1 Supermodels and submodels — 175
  - 4.5.2 Unions of chains — 179
- 4.6 Amalgamation of vocabularies — 183
- 4.7 The expressive power of first-order logic — 189

5 First-order theories — 198

- 5.1 Completeness and decidability — 199
- 5.2 Categoricity — 205
- 5.3 Countably categorical theories — 211
  - 5.3.1 Dense linear orders — 211
  - 5.3.2 Ryll-Nardzewski et al. — 214
- 5.4 The Random graph and 0–1 laws — 216
- 5.5 Quantifier elimination — 221
  - 5.5.1 Finite relational vocabularies — 222
  - 5.5.2 The general case — 228
- 5.6 Model-completeness — 233
- 5.7 Minimal theories — 239

	5.8	Fields and vector spaces	247
	5.9	Some algebraic geometry	257
6	**Models of countable theories**		**267**
	6.1	Types	267
	6.2	Isolated types	271
	6.3	Small models of small theories	275
		6.3.1 Atomic models	276
		6.3.2 Homogeneity	277
		6.3.3 Prime models	279
	6.4	Big models of small theories	280
		6.4.1 Countable saturated models	281
		6.4.2 Monster models	285
	6.5	Theories with many types	286
	6.6	The number of nonisomorphic models	289
	6.7	A touch of stability	290
7	**Computability and complexity**		**299**
	7.1	Computable functions and Church's thesis	301
		7.1.1 Primitive recursive functions	302
		7.1.2 The Ackermann function	307
		7.1.3 Recursive functions	309
	7.2	Computable sets and relations	312
	7.3	Computing machines	316
	7.4	Codes	320
	7.5	Semi-decidable decision problems	327
	7.6	Undecidable decision problems	332
		7.6.1 Nonrecursive sets	332
		7.6.2 The arithmetic hierarchy	335
	7.7	Decidable decision problems	337
		7.7.1 Examples	338
		7.7.2 Time and space	344
		7.7.3 Nondeterministic polynomial-time	347
	7.8	**NP**-completeness	348
8	**The incompleteness theorems**		**357**
	8.1	Axioms for first-order number theory	358
	8.2	The expressive power of first-order number theory	362
	8.3	Gödel's First Incompleteness theorem	370
	8.4	Gödel codes	374
	8.5	Gödel's Second Incompleteness theorem	380
	8.6	Goodstein sequences	383

9 Beyond first-order logic — 388
- 9.1 Second-order logic — 388
- 9.2 Infinitary logics — 392
- 9.3 Fixed-point logics — 395
- 9.4 Lindström's theorem — 400

10 Finite model theory — 408
- 10.1 Finite-variable logics — 408
- 10.2 Classical failures — 412
- 10.3 Descriptive complexity — 417
- 10.4 Logic and the $\mathbf{P} = \mathbf{NP}$ problem — 423

Bibliography — 426

Index — 428

Preliminaries

What is a logic?

A logic is a language equipped with rules for deducing the truth of one sentence from that of another. Unlike natural languages such as English, Finnish, and Cantonese, a logic is an artificial language having a precisely defined syntax. One purpose for such artificial languages is to avoid the ambiguities and paradoxes that arise in natural languages. Consider the following English sentence.

> Let n be the smallest natural number that cannot be defined in fewer than 20 words.

Since this sentence itself contains fewer than 20 words, it is paradoxical. A logic avoids such pitfalls and streamlines the reasoning process. The above sentence cannot be expressed in the logics we study. This demonstrates the fundamental tradeoff in using logics as opposed to natural languages: to gain precision we necessarily sacrifice expressive power.

In this book, we consider classical logics: primarily first-order logic but also propositional logic, second-order logic and variations of these three logics. Each logic has a notion of *atomic formula*. Every sentence and formula can be constructed from atomic formulas following precise rules. One way that the three logics differ is that, as we proceed from propositional logic to first-order logic to second-order logic, there is an increasing number of rules that allow us to construct increasingly complex formulas from atomic formulas. We are able to express more concepts in each successive logic.

We begin our study with propositional logic in Chapter 1. In the present section, we provide background and prerequisites for our study.

What is logic?

Logic is defined as the study of the principles of reasoning. The study of logics (as defined above) is the part of this study known as symbolic logic. Symbolic logic is a branch of mathematics. Like other areas of mathematics, symbolic logic flourished during the past century. A century ago, the primary aim of symbolic logic was to provide a foundation for mathematics. Today, foundational studies are just one part of symbolic logic. We do not discuss foundational issues in this book, but rather focus on other areas such as model theory, proof theory, and computability theory. Our goal is to introduce the fundamentals and prepare the reader for further study in any of these related areas of symbolic logic. Symbolic logic views mathematics and computer science from a unique perspective and supplies distinct tools and techniques for the solution of certain problems. We highlight many of the landmark results in logic achieved during the past century.

Symbolic logic is exclusively the subject of this book. Henceforth, when we refer to "logic" we always mean "symbolic logic."

Time complexity

Logic and computer science share a symbiotic relationship. Computers provide a concrete setting for the implementation of logic. Logic provides language and methods for the study of theoretical computer science. The subject of complexity theory demonstrates this relationship well. Complexity theory is the branch of theoretical computer science that classifies problems according to how difficult they are to solve. For example, consider the following problem:

The Sum 10 Problem: Given a finite set of integers, does some subset add up to 10?

This is an example of a decision problem. Given input as specified (in this case, a finite set of integers) a decision problem asks a question to be answered with a "yes" or "no." Suppose, for example, that we are given the following set as input:

$$\{-26, -16, -12, -8, -4, -2, 7, 8, 27\}.$$

The problem is to decide whether or not this set contains a subset of numbers that add up to 10. One way to resolve this problem is to check every subset. Since 10 is not in our set, such a subset must contain more than one number. We can check to see if the sum of any two numbers is 10. We can then check to see if the sum of any three numbers is 10, and so forth. This method will eventually provide the correct answer to the question, but it is not efficient. We have $2^9 = 512$ subsets to check. In general, if the input contains n integers, then there are 2^n subsets to check. If the input set is large, then this is not feasible. If the set contains 23 numbers, then there are more than 8 million subsets to check. Although this is a lot of subsets, this is a relatively simple task for a computer. If, however, there are more than, say, 100 numbers in the input set, then, even for the fastest computer, the time required to check each subset exceeds the lifespan of earth.

Time complexity is concerned with the amount of time it takes to answer a problem. To answer a decision problem, one must produce an algorithm that, given any suitable input, will result in the correct answer of "yes" or "no." An algorithm is a step-by-step procedure. The "amount of time" is measured by how many steps it takes to reach the answer. Of course, the bigger the input, the longer it will take to reach a conclusion. An algorithm is said to be *polynomial-time* if there is some number k so that, given any input of size n, the algorithm reaches its conclusion in fewer than n^k steps. The class of all decision problems that can be solved by a polynomial-time algorithm is denoted by **P**. We said that complexity theory classifies problems according to how difficult they are to solve. The complexity class **P** contains problems that are relatively easy to solve.

To answer the Sum 10 Problem, we gave the following algorithm: check every subset. If some subset adds up to 10, then output "yes." Otherwise, output "no." This algorithm is not polynomial-time. Given input of size n, it takes at least 2^n steps for the algorithm to reach a conclusion and, for any k, $2^n > n^k$ for sufficiently large n. So this decision problem is not necessarily in **P**. It is in another complexity class known as **NP** (nondeterministic polynomial-time). Essentially, a decision problem is in **NP** if a "yes" answer can be obtained in polynomial-time by guessing. For example, suppose we somehow guess that the subset $\{-26, -4, -2, 7, 8, 27\}$ sums up to 10. It is easy to check that this guess is indeed correct. So we quickly obtain the correct output of "yes."

So the Sum 10 Problem is in **NP**. It is not known whether it is in **P**. The algorithm we gave is not polynomial-time, but perhaps there exists a better algorithm for this problem. In fact, maybe every problem in **NP** is in **P**. The question of whether **P** = **NP** is not only one of the big questions of complexity theory, it is one of the most famous unanswered questions of mathematics. The Clay Institute of Mathematics has chosen this as one of its seven Millennium Problems. The Clay Institute has put a bounty of one million dollars on the solution for each of these problems.

What does this have to do with logic? Complexity theory will be a recurring theme throughout this book. From the outset, we will see decision problems that naturally arise in the study of logic. For example, we may ask if a given sentence of propositional logic is always true. Likewise, we may ask if the sentence is sometimes true or never true. These related decision problems are in **NP**. It is not known whether they are in **P**. In Chapter 7, we show that these problems are **NP**-complete. This means that if one of these problems is in **P**, then so is every problem in **NP**. So if we can find a polynomial-time algorithm for determining whether or not a given sentence of propositional logic is always true, or if we can show that no such algorithm exists, then we will resolve the **P** = **NP** problem.

In Chapter 10, we turn this relationship between complexity and logic on its head. We show that, in a certain setting (namely, graph theory) the complexity classes of **P** and **NP** (and others) can be defined as logics. For example, Fagin's Theorem states that (for graphs) **NP** contains precisely those decision problems that can be expressed in second-order existential logic. So the **P** = **NP** problem and related questions can be rephrased as questions of whether or not two logics are equivalent.

From the point of view of a mathematician, this makes the **P** = **NP** problem more precise. Our above definitions of **P** and **NP** may seem hazy. After all, our definition of these complexity classes depends on the notion of a "step" of an algorithm. Although we could (and will) precisely define what constitutes a "step," we utterly avoid this issue by defining these classes as logics. From the point of view of a computer scientist, on the other hand, the relationship between logics and complexity classes justifies the study of logics. The fact that the familiar complexity classes arise from these logics is evidence that these logics are natural objects to study.

Clearly, we are getting ahead of ourselves. Fagin's Theorem is not mentioned until the final chapter. In fact, no prior knowledge of complexity theory is assumed in this book. Some prior knowledge of algorithms may be helpful, but is not required. We do assume that the reader is familiar with sets, relations, and functions. Before beginning our study, we briefly review these topics.

Sets and structures

We assume that the reader is familiar with the fundamental notion of a *set*. We use standard set notation:

$x \in A$ means x is an element of set A,
$x \notin A$ means x is not an element of A,
$\emptyset$ denotes the unique set containing no elements,
$A \subset B$ means every element of set A is also an element of set B,
$A \cup B$ denotes the union of sets A and B,
$A \cap B$ denotes the intersection of sets A and B, and
$A \times B$ denotes the Cartesian product of sets A and B.

Recall that the *union* $A \cup B$ of A and B is the set of elements that are in A or B (including those in both A and B), whereas the *intersection* $A \cap B$ is the set of only those elements that are in both A and B. The Cartesian product $A \times B$ of A and B is the set of ordered pairs (a, b) with $a \in A$ and $b \in B$. We simply write A^2 for $A \times A$. Likewise, for $n > 2$, A^n denotes the Cartesian product of A^{n-1} and A. This is the set of n-tuples $(a_1, a_2, \ldots, a_n)$ with each $a_i \in A$. For convenience, A^1 (the set of 1-tuples) is an alternative notation for A itself.

Example 1 Let $A = \{\alpha, \beta, \gamma\}$ and let $B = \{\beta, \delta, \epsilon\}$. Then

$$A \cup B = \{\alpha, \beta, \gamma, \delta, \epsilon\},$$
$$A \cap B = \{\beta\},$$
$$A \times B = \{(\alpha, \beta), (\alpha, \delta), (\alpha, \epsilon), (\beta, \beta), (\beta, \delta), (\beta, \epsilon), (\gamma, \beta), (\gamma, \delta), (\gamma, \epsilon)\},$$
and $B^2 = \{(\beta, \beta), (\beta, \delta), (\beta, \epsilon), (\delta, \beta), (\delta, \delta), (\delta, \epsilon), (\epsilon, \beta), (\epsilon, \delta), (\epsilon, \epsilon)\}.$

Two sets are equal if and only if they contain the same elements. Put another way, $A = B$ if and only if both $A \subset B$ and $B \subset A$. In particular, the order and repetition of elements within a set do not matter. For example,

$$A = \{\alpha, \beta, \gamma\} = \{\gamma, \beta, \alpha\} = \{\beta, \beta, \alpha, \gamma\} = \{\gamma, \alpha, \beta, \beta, \alpha\}.$$

Note that $A \subset B$ includes the possibility that $A = B$. We say that A is a *proper* subset of B if $A \subset B$ and $A \neq B$ and $A \neq \emptyset$.

A set is essentially a database that has no structure. For an example of a database, suppose that we have a phone book listing 1000 names in alphabetical order along with addresses and phone numbers. Let T be the set containing these names, addresses, and phone numbers. As a set, T is a collection of 3000 elements having no particular order or other relationships. As a database, our phone book is more than merely a set with 3000 entries. The database is a *structure*: a set together with certain relations.

Definition 2 Let A be a set. A **relation** R on A is a subset of A^n (for some natural number n). If $n = 1, 2$, or 3, then the relation R is called **unary**, **binary**, or **ternary** respectively. If n is bigger than 3, then we refer to R as an n-ary relation. The number n is called the **arity** of R.

As a database, our phone book has several relations. There are three types of entries in T: names, numbers, and addresses. Each of these forms a subset of T, and so can be viewed as a unary relation on T. Let N be the set of names in T, P be the set of phone numbers in T, and A be the set of addresses in T. Since a "relation" typically occurs between two or more objects, the phrase "unary relation" is somewhat of an oxymoron. We continue to use this terminology, but point out that a "unary relation" should be viewed as a predicate or an adjective describing elements of the set.

We assume that each name in the phone book corresponds to exactly one phone number and one address. This describes another relation between the elements of T. Let R be the ternary relation consisting of all 3-tuples (x, y, z) of elements in T^3 such that x is a name having phone number y and address z. Yet another relation is the order of the names. The phone book, unlike the set T, is in alphabetical order. Let the symbol $<$ represent this order. If x and y

are elements of N (that is, if they are names in T), then $x < y$ means that x precedes y alphabetically. This order is a binary relation on T. It can be viewed as the subset of T^2 consisting of all ordered pairs (x, y) with $x < y$.

Structures play a primary role in the study of first-order logic (and other logics). They provide a context for determining whether a given sentence of the logic is true or false. First-order structures are formally introduced in Chapter 2. In the previous paragraphs, we have seen our first example of a structure: a phone book. Let D denote the database we have defined. We have

$$D = (T|N, P, A, <, R).$$

The above notation expresses that D is the structure having set T and the five relations N, P, A, $<$, and R on T.

Although a phone book may not seem relevant to mathematics, the objects of mathematical inquiry often can be viewed as structures such as D. Number systems provide familiar examples of infinite structures studied in mathematics. Consider the following sets:

$\mathbb{N}$ denotes the set of natural numbers: $\mathbb{N} = \{1, 2, 3, ...\}$,
$\mathbb{Z}$ denotes the set of integers: $\mathbb{Z} = \{..., -3, -2, -1, 0, 1, 2, 3, ...\}$, and
$\mathbb{Q}$ denotes the set of rational numbers: $\mathbb{Q} = \{a/b | a, b \in \mathbb{Z}\}$.
$\mathbb{R}$ denotes the set of real numbers: $\mathbb{R}$ is the set of all decimal expansions of the form $z.a_1 a_2 a_3 \cdots$ where z and each a_i are integers and $0 \leq a_i \leq 9$.
$\mathbb{C}$ denotes the set of complex numbers: $\mathbb{C} = \{a+bi | a, b \in \mathbb{R}\}$ where $i = \sqrt{-1}$.

Note that $\mathbb{N}$, $\mathbb{Z}$, $\mathbb{Q}$, $\mathbb{R}$, and $\mathbb{C}$ each represents a set. These number systems, however, are more than sets. They have much structure. The structure includes relations (such as $<$ for less than) and functions (such as $+$ for addition). Depending on what our interests are, we may consider these sets with any number of various functions and relations.

The interplay between mathematical structures and formal languages is the subject of model theory. First-order logic, containing various relations and functions, is the primary language of model theory. We study model theory in Chapters 4–6. As we shall see, the perspective of model theory sheds new light on familiar structures such as the real and complex numbers.

Functions

The notation $f : A \to B$ expresses that f is a function from set A to set B. This means that, given any $a \in A$ as input, f yields at most one output $f(a) \in B$. It is possible that, given $a \in A$, f yields no output. In this case we say that $f(a)$ is undefined. The set of all $a \in A$ for which f does produce an output is called the *domain* of f. The *range* of f is the set of all $b \in B$ such that $b = f(a)$ for some $a \in A$. If the range of f is all of B, then the function is said to be *onto* B.

The *graph* of $f\colon A \to B$ is the subset of $A \times B$ consisting of all ordered pairs (a, b) with $f(a) = b$. If A happens to be B^n for some $n \in \mathbb{N}$, then we say that f is a function *on* B and n is the *arity* of f. In this case, the graph of f is an $(n+1)$-ary relation on B. The *inverse graph* of $f\colon A \to B$ is obtained by reversing each ordered pair in the graph of f. That is, (b, a) is in the inverse graph of f if and only if (a, b) is in the graph of f. The inverse graph does not necessarily determine a function. If it does determine a function $f^{-1}\colon B \to A$ (defined by $f^{-1}(b) = a$ if and only if (b, a) is in the inverse graph of f) then f^{-1} is called the *inverse function* of f and f is said to be *one-to-one*.

The concept of a function should be quite familiar to anyone who has completed a course in calculus. As an example, consider the function from $\mathbb{R}$ to $\mathbb{R}$ defined by $h(x) = 3x^2 + 1$. This function is defined by a rule. Put into words, this rule states that, given input x, h squares x, then multiplies it by 3, and then adds 1. This rule allows us to compute $h(0) = 1$, $h(3) = 28$, $h(72) = 15553$, and so forth. In addition to this rule, we must be given two sets. In this example, the real numbers serve as both sets. So h is a unary function on the real numbers. The domain of h is all of $\mathbb{R}$ since, given any x in $\mathbb{R}$, $3x^2 + 1$ is also in $\mathbb{R}$. The function h is not one-to-one since $h(x)$ and $h(-x)$ both equal the same number for any x. Nor is h onto $\mathbb{R}$ since, given $x \in \mathbb{R}$, $h(x) = 3x^2 + 1$ cannot be less than 1.

Other examples of functions are provided by various buttons on any calculator. Scientific calculators have buttons $\boxed{x^2}$, $\boxed{\log x}$, $\boxed{\sin x}$, and so forth. When you put a number into the calculator and then push one of these buttons, the calculator outputs at most one number. This is exactly what is meant by a "function." The key phrase is "at most one." As a nonexample, consider the square root. Given input 4, there are two outputs: 2 and -2. This is not a function. If we restrict the output to the positive square root (as most calculators do), then we do have a function. It is possible to get less than one output: you may get an ERROR message (say you input -35 and then push the $\boxed{\log x}$ button). The domain of a function is the set of inputs for which an ERROR does not occur. We can imagine a calculator that has a button $\boxed{h}$ for the function h defined in the previous paragraph. When you input 2 and then push $\boxed{h}$, the output is 13. This is what $\boxed{h}$ does: it squares 2, multiplies it by 3, and adds 1. Indeed, if we have a programmable calculator we could easily make it compute h at the push of a button.

Intuitively, any function behaves like a calculator button. However, this analogy must not be taken literally. Although calculators provide many examples of familiar functions, most functions cannot be programmed into a calculator.

Definition 3 A function f is **computable** if there exists a computer program that, given input x,

- outputs $f(x)$ if x is in the domain of f, and
- yields no output if x is not in the domain of f.

As we will see in Section 2.5, most functions are not computable. However, it is hard to effectively demonstrate a function that is not computable. How can we uniquely describe a particular function without providing a means for its computation? As we will see, logic provides many examples. Computability theory is a subject of Chapter 7. Odd as it may seem, computability theory studies things that cannot be done by a computer. This subject arose from Gödel's proof of his famous Incompleteness theorems. Proved in 1931, Gödel's theorems rank among the great mathematical achievements of the past century. They imply that there is no computer algorithm to determine whether or not a given statement of arithmetic is true or false. Again, we are getting way ahead of ourselves. We will come to Gödel's theorems (as well as Fagin's theorem) and state them precisely in due time. Let us now end our preliminary ramblings and begin our study of logic.

1 Propositional logic

1.1 What is propositional logic?

In propositional logic, atomic formulas are propositions. Any assertion will do. For example,

A = "Aristotle is dead,"

B = "Barcelona is on the Seine," and

C = "Courtney Love is tall"

are atomic formulas. Atomic formulas are the building blocks used to construct sentences. In any logic, a sentence is regarded as a particular type of formula. In propositional logic, there is no distinction between these two terms. We use "formula" and "sentence" interchangeably.

In propositional logic, as with all logics we study, each sentence is either true or false. A *truth value* of 1 or 0 is assigned to the sentence accordingly. In the above example, we may assign truth value 1 to formula A and truth value 0 to formula B. If we take proposition C literally, then its truth is debatable. Perhaps it would make more sense to allow truth values between 0 and 1. We could assign 0.75 to statement C if Miss Love is taller than 75% of American women. Fuzzy logic allows such truth values, but the classical logics we study do not.

In fact, the content of the propositions is not relevant to propositional logic. Henceforth, atomic formulas are denoted only by the capital letters A, B, C,... (possibly with subscripts) without referring to what these propositions actually say. The veracity of these formulas does not concern us. Propositional logic is not the study of truth, but of the relationship between the truth of one statement and that of another.

The language of propositional logic contains words for "not," "and," "or," "implies," and "if and only if." These words are represented by symbols:

$\neg$ for "not," $\land$ for "and," $\lor$ for "or,"

$\rightarrow$ for "implies," and $\leftrightarrow$ for "if and only if."

As is always the case when translating one language into another, this correspondence is not exact. Unlike their English counterparts, these symbols represent concepts that are precise and invariable. The meaning of an English word, on the other hand, always depends on the context. For example, $\land$ represents a concept that is similar but not identical to "and." For atomic formulas A and B, $A \land B$ always means the same as $B \land A$. This is not always true of the word "and." The sentence

She became violently sick and she went to the doctor.

does not have the same meaning as

She went to the doctor and she became violently sick.

Likewise $\lor$ differs from "or." Conversationally, the use of "A or B" often precludes the possibility of both A and B. In propositional logic $A \lor B$ always means either A or B or both A and B.

We must precisely define the symbols $\neg$, $\land$, $\lor$, $\rightarrow$, and $\leftrightarrow$. We are confronted with the conundrum of how to define the first word of a language (having recourse to no other words!). For this reason, we take the symbols $\neg$ and $\land$ as *primitives*. We define the other symbols in terms of these two symbols. Although we do not define $\neg$ and $\land$ in terms of the other symbols, we do describe the semantics of these symbols in an unambiguous manner.

Before describing the semantics of the language, we discuss the syntax. Whereas the semantics regards the meaning, or interpretation, of sentences in the language, the syntax regards the grammar of the language. The syntax of propositional logic tells us which strings of symbols are permissible as formulas. Naturally, any atomic formula is a formula. We also have the following two rules.

(R1) If F is a formula, then $\neg F$ is a formula.
(R2) If F and G are formulas, then $(F \land G)$ is a formula.

Definition 1.1 The formula $\neg F$ is the *negation* of F
and the formula $(F \land G)$ is the *conjunction* of F and G.

Definition 1.2 A finite string of symbols is a *formula* of propositional logic if and only if it is built up from atomic formulas by repeated application of rules (R1) and (R2).

Example 1.3 $\neg(\neg(A \land B) \land \neg C)$ is a formula and $((A\neg \land)B(C\neg$ is not.

Note that we have restricted the definition of formula to the primitive symbols $\neg$ and $\land$. If we were describing the syntax of propositional logic to a computer, then this definition of formula would suffice. However, to make formulas more palatable to humans, we include the other symbols ($\lor$, $\rightarrow$, and $\leftrightarrow$) to be defined later. We may regard formulas involving these symbols as abbreviations for more complicated formulas involving only $\neg$ and $\land$. The inclusion of these symbols make the formulas easier (for us humans) to read.

Also toward the aim of readability, we employ certain conventions. The use of these abbreviations and conventions alters our notion of "formula" somewhat. One of these conventions is the following:

(C1) If F or (F) is a formula, then we view F and (F) as the same formula.

That is, we may drop the outermost parentheses. This extends our definition of formula. Technically, by the above definition, $A \land B$ is not a formula. However, using convention (C1), we do not distinguish $A \land B$ from the formula $(A \land B)$.

The use of convention (C1) leads to some ambiguities that we presently address. Suppose that, in (R1), F denotes $A \land B$ (which, by (C1) is a formula). Then $\neg F$ does not represent the formula $\neg A \land B$. Rather, $\neg F$ denotes the formula $\neg(A \land B)$. As we shall see, $\neg A \land B$ and $\neg(A \land B)$ do not mean the same thing. Likewise, $F \land G$ denotes the formula $(F) \land (G)$.

The use of (C1) also requires care in defining the notion of "subformula." A subformula of a formula F (viewed as a string of symbols) is a substring of F that is itself a formula. However, because of (C1), not every such substring is a subformula. So we do not want to take this property as the definition of "subformula." Instead, we define "subformula" as follows.

Definition 1.4 The following rules define the *subformulas* of a formula.

Any formula is a subformula of itself.
Any subformula of F is also a subformula of $\neg F$.
Any subformula of F or G is also a subformula of $(F \land G)$.

Example 1.5 Let A and B be atomic and let F be the formula $\neg(\neg A \land \neg B)$.

The formula $A \land \neg B$ occurs as a substring of F, but it is not a subformula of F. There is no way to build the formula F from the formula $A \land \neg B$. The subformulas of F are A, B, $\neg A$, $\neg B$, $(\neg A \land \neg B)$, and $\neg(\neg A \land \neg B)$.

Having described the syntax of propositional logic, we now describe the semantics. That is, we say how to interpret the formulas. Not only must we describe the semantics for the symbols "$\land$" and "$\neg$," but we must also say how to interpret formulas in which these symbols occur together. For this we state the *order of operations*. It is the role of parentheses to dictate which subformulas are to be considered first when interpreting a formula. If no parentheses are present, then we use the following rule:

$$\neg \text{ has priority over } \land.$$

For example, the formula $\neg(A \land B)$ means "not both A and B." The parentheses tell us that the "$\land$" in this formula has priority over "$\neg$." The formula $\neg A \land B$, on the other hand, has a different interpretation. In the absence of parentheses, we use the rule that $\neg$ has priority over $\wedge$. So this formula means "both not A and B."

The semantics of propositional logic is defined by this rule along with Tables 1.1 and 1.2.

These are examples of *truth tables*. Each row of these tables assigns truth values to atomic formulas and gives resulting truth values for more complex formulas. For example, the third row of Table 1.1 tells us that if A is true and B is false, then $(A \wedge B)$ is false. We see that $(A \wedge B)$ has truth value 1 only if both A and B have truth value 1, corresponding to our notion of "and." Likewise, Table 1.2 tells us that $\neg A$ has the opposite truth value of A, corresponding to our notion of negation.

Using these two truth tables, we can find truth tables for any formula. This is because every formula is built from atomic formulas via rules (R1) and (R2). Suppose, for example, we want to find a truth table for the formula $\neg(\neg A \wedge \neg B)$. Given truth values for A and B, we can use Table 1.2 to find the truth values of $\neg A$ and $\neg B$. Given truth values for $\neg A$ and $\neg B$, we can then use Table 1.1 to find the truth value of $(\neg A \wedge \neg B)$. Finally, we can refer again to Table 1.2 to find the truth value of $\neg(\neg A \wedge \neg B)$. The resulting truth table for this formula is shown in Table 1.3.

Note that the formulas listed across the top of Table 1.3 are precisely the subformulas from Example 1.5. From this table we see that the formula $\neg(\neg A \wedge \neg B)$ has truth value 1 if and only if A or B has truth value 1. This formula corresponds to the notion of "or" discussed earlier. The symbol $\vee$ is used to denote this useful notion.

Table 1.1 Truth table for $A \wedge B$

A	B	$(A \wedge B)$
0	0	0
0	1	0
1	0	0
1	1	1

Table 1.2 Truth table for $\neg A$

A	$\neg A$
0	1
1	0

Propositional logic 5

Table 1.3 Truth table for $(A \vee B)$

A	B	$\neg A$	$\neg B$	$(\neg A \wedge \neg B)$	$\neg(\neg A \wedge \neg B)$
0	0	1	1	1	0
0	1	1	0	0	1
1	0	0	1	0	1
1	1	0	0	0	1

Table 1.4 Truth table for $(A \to B)$

A	B	$\neg A$	$(B \vee \neg A)$	$(A \to B)$
0	0	1	1	1
0	1	1	1	1
1	0	0	0	0
1	1	0	1	1

Definition 1.6 The symbol $\vee$ is defined as follows: for any formulas F and G, $(F \vee G)$ is an abbreviation for $\neg(\neg F \wedge \neg G)$. The formula $(F \vee G)$ is called the *disjunction* of F and G.

Two other abbreviations that are convenient are the following.

Definition 1.7 The symbols $\to$ and $\leftrightarrow$ are defined as follows:

$(F \to G)$ abbreviates $(G \vee \neg F)$, and
$(F \leftrightarrow G)$ abbreviates $((F \to G) \wedge (G \to F))$.

We previously remarked that the symbol $\to$ corresponds to the English word "implies" and the symbol $\leftrightarrow$ corresponds to the phrase "if and only if." Again, these correspondences are merely mnemonic devices for the semantics of the symbols. For example $(A \leftrightarrow B)$ is true if A and B have the same truth values and is otherwise false. So $\leftrightarrow$ behaves exactly like the phrase "if and only if." The relationship between $\to$ and "implies" is a bit tenuous. Consider the truth table for $(A \to B)$ (Table 1.4).

We see that $(A \to B)$ is true unless A is true and B is false. In particular, $(A \to B)$ is true whenever A is false. Thus, in logic, a false statement implies anything. This differs from the colloquial use of the word "implies." We would not say "Barcelona is on the Seine implies Aristotle is dead" or, even more egregious, "Barcelona is on the Seine implies Barcelona is not on the Seine." However, $(A \to B)$ and $(A \to \neg A)$ are true statements of propositional logic whenever A is false.

Having introduced new symbols, we must determine the order of operation for these symbols. When evaluating the truth value of a formula, we must know the order in which to proceed. Rather than ranking all of the symbols in a hierarchy, we state just one rule:

$\neg$ has priority over $\wedge$, $\vee$, $\rightarrow$, and $\leftrightarrow$.

Beyond this, the parentheses dictate the order in which to proceed.

Example 1.8 Consider the formula $((\neg A \rightarrow B) \wedge C) \vee \neg(A \wedge D)$. Call this formula F. Suppose we know that the truth values for A, B, C, and D are 1, 0, 1, and 0, respectively. To evaluate the truth value for F we begin with the subformula $\neg A$ (using Table 1.2) since $\neg$ has priority. We next evaluate the subformula $(\neg A \rightarrow B)$ (Table 1.4) which is in the innermost set of parentheses. We next evaluate the truth values for $((\neg A \rightarrow B) \wedge C)$ and $(A \wedge D)$ (Table 1.1). We then find the truth values for $\neg(A \wedge D)$ (Table 1.2) and, finally, for F (Table 1.3). We obtain the following truth values.

A	B	C	D	$\neg A$	$(\neg A \rightarrow B)$	$((\neg A \rightarrow B) \wedge C)$	$(A \wedge D)$	$\neg(A \wedge D)$	F
1	0	1	0	0	1	1	0	1	1

This is just one row of a truth table for this formula. We could choose other truth values for A, B, C, and D other than 1, 0, 1, and 0. Since there are two possible values for each of these four atomic formulas, there are $2^4 = 16$ ways to assign truth values to A, B, C, and D. So the full table has 16 rows. The completion of this table is left as Exercise 1.3(d).

The role of parentheses is not only to determine the order of operations, but also to make formulas more readable. Toward this aim, we omit parentheses when they are not necessary. We have already discussed convention (C1) that allows us to drop the outermost parentheses from the formula (F). We also use the following convention:

(C2) For any formulas F, G, and H,
we view $F \wedge G \wedge H$ as the same formula as $(F \wedge G) \wedge H$
and $F \vee G \vee H$ as the same formula as $(F \vee G) \vee H$.

Since the formulas $(F \wedge G) \wedge H$ and $F \wedge (G \wedge H)$ have the same truth tables, there is no ambiguity in dropping the parentheses and simply writing $F \wedge G \wedge H$. In contrast, $F \wedge G \vee H$ is ambiguous and is not permitted as a formula of propositional logic. The formulas $(F \wedge G) \vee H$ and $F \wedge (G \vee H)$ do not have the same truth tables.

We have now completely defined propositional logic.

In summary, propositional logic, like any logic, is a language. Its dictionary contains the words ¬, ∧, ∨, →, and ↔. (The symbols "(" and ")" are used only as punctuation.) The words ∨, →, and ↔ are defined in terms of ¬ and ∧. The words ¬ and ∧ are considered primitive and are listed in our hypothetical dictionary without definition. The dictionary also contains infinitely many atomic formulas that are merely listed as capital letters (with subscripts, perhaps). The grammar of this language consists of the rules (R1) and (R2) along with conventions (C1) and (C2) regarding parentheses.

Propositional logic, like any logic, also has rules for deduction. These rules follow from the semantics of the logic. The semantics of propositional logic are summarized by Tables 1.1 and 1.2 and the definitions of the symbols ∨, →, and ↔. The semantics and the rules for deduction that follow from the semantics are implicit in the words "not," "and," "or," "implies," and "if and only if" (although this correspondence is not exact). For example, if $(A \land B)$ is true (truth value 1), then we can deduce that both A and B are true. And if $A \to B$ and A both have truth value 1, then it follows that B also has truth value 1. We discuss these and other rules for deduction in Section 1.5.

1.2 Validity, satisfiability, and contradiction

Let $S = \{A_1, \ldots, A_n\}$ be a set of atomic formulas. Let $\mathcal{F}(S)$ be the set of all formulas that can be built from the atomic formulas in S.

Definition 1.9 An *assignment* of S is a function $\mathcal{A}: S \to \{0, 1\}$.

That is, an assignment of S assigns truth values to each atomic formula in S. An assignment $\mathcal{A}$ of S naturally extends to all of $\mathcal{F}(S)$. Given any formula F in $\mathcal{F}(S)$, an assignment $\mathcal{A}$ of S corresponds to a unique row of the truth table for F. We define $\mathcal{A}(F)$ to be the truth value of F in this row.

An assignment $\mathcal{A}$ of S also extends to certain formulas not in $\mathcal{F}(S)$. Suppose F_0 is a formula that is not in $\mathcal{F}(S)$. Let S_0 be the set of atomic subformulas of F_0. If every extension of $\mathcal{A}$ to $S \cup S_0$ has the same value for F_0, then we define $\mathcal{A}(F_0)$ to be this value.

Example 1.10 Let A and B be atomic formulas. Let $\mathcal{A}$ be the assignment of $\{A, B\}$ defined by $\mathcal{A}(A) = 1$ and $\mathcal{A}(B) = 0$. Then

$$\mathcal{A}(A \land B) = 0,$$
$$\mathcal{A}(A \lor B) = 1,$$
$$\mathcal{A}(A \land (C \lor \neg C)) = 1, \text{ and}$$
$$\mathcal{A}(B \lor (C \land \neg C)) = 0.$$

The reason $\mathcal{A}(A \land (C \lor \neg C)) = 1$ is that $\mathcal{A}(A) = 1$ and, no matter what truth value we assign to C, $(C \lor \neg C)$ has truth value 1. Likewise $\mathcal{A}(B \lor (C \land \neg C)) = 0$ because both B and $(C \land \neg C)$ have truth value 0, regardless of the truth value of C.

Let $\mathcal{A}$ be an assignment of $\mathcal{S}$ and let F be a formula. If $\mathcal{A}(F) = 1$, then we say F *holds* under assignment $\mathcal{A}$. Equivalently, we say $\mathcal{A}$ *models* F. We write $\mathcal{A} \models F$ to denote this concept.

Definition 1.11 A formula is *valid* if it holds under every assignment. We use $\models F$ to denote this. A valid formula is called a *tautology*.

Example 1.12 The formula $(C \lor \neg C)$ from the previous example is a tautology.

Definition 1.13 A formula is *satisfiable* if it holds under some assignment.

Definition 1.14 A formula is *unsatisfiable* if it holds under no assignment. An unsatisfiable formula is called a *contradiction*.

Example 1.15 The formula $(C \land \neg C)$ is a contradiction.

Suppose that we want to determine whether or not a given formula is valid. This is an example of a *decision problem*. A decision problem is any problem that, given certain input, asks a question to be answered with a "yes" or a "no." Given formula F as input, we may ask "Is F valid?" We refer to this as the *validity problem*. Likewise, we may ask "Is F satisfiable?," and refer to this the *satisfiability problem*. For propositional logic, truth tables provide a systematic approach for resolving such decision problems. If all of the truth values for F are 1s, then F is valid. If some truth value is 1, then F is satisfiable. Otherwise, if no truth values are 1s, F is unsatisfiable.

Example 1.16 Consider the formula $(A \land (A \to B)) \to B$. To determine whether this formula is satisfiable, we compute the following truth table.

A	B	$A \to B$	$A \land (A \to B)$	$(A \land (A \to B)) \to B$
0	0	1	0	1
0	1	1	0	1
1	0	0	0	1
1	1	1	1	1

We see that $(A \land (A \to B)) \to B$ has truth value 1 under any assignment. So not only is this formula satisfiable, it is valid.

Example 1.17 Consider now the formula $((A \to B) \to A) \land \neg A$. Suppose we want to determine whether this formula is satisfiable or not. Again we compute a truth table.

A	B	$(A \to B)$	$((A \to B) \to A)$	$\neg A$	$((A \to B) \to A) \land \neg A$
0	0	1	0	1	0
0	1	1	0	1	0
1	0	0	1	0	0
1	1	1	1	0	0

This formula is unsatisfiable. It is a contradiction.

Theoretically, we can determine whether any formula F is valid, satisfiable or unsatisfiable by looking at a truth table. Unfortunately, this is not always an efficient method. If F contains n atomic formulas, then there are 2^n rows to compute in the truth table for F. So if F happens to have, say, 23 atomic formulas, then computing a truth table is not feasible. One of our aims in this chapter is to find alternative methods for resolving the validity and satisfiability problems that avoid truth tables. More generally, our aim is to contrive various ways of determining whether or not a given formula is a consequence of a given set of formulas. This is a central problem of any logic.

1.3 Consequence and equivalence

We now introduce the fundamental notion of consequence. First, we define what it means for one formula to be a consequence of another. Later in this section, we similarly define what it means for a formula to be a consequence of a *set* of formulas.

Definition 1.18 Formula G is a *consequence* of formula F if for every assignment $\mathcal{A}$, if $\mathcal{A} \models F$ then $\mathcal{A} \models G$. We denote this by $F \models G$.

Note that the symbol $\models$ is used in a variety of ways. There is always a formula to the right of this symbol. When we write $\_\_ \models F$, the interpretation of "$\models$" depends on how we fill in the blank. The blank may either be filled with an assignment $\mathcal{A}$, a formula G, or not filled with the empty set. The three corresponding interpretations for $\models$ are as follows:

- $\mathcal{A} \models F$ means that $\mathcal{A}(F) = 1$. We read this as "$\mathcal{A}$ models F."
- $G \models F$ means every assignment that models G also models F. That is, F is a consequence of G.

- $\models F$ means every assignment models F. That is, F is a tautology.

So although $\models$ has multiple interpretations, in context it is not ambiguous.

The notion of consequence is closely related to the notion of "implies" discussed in Section 1.1. A formula G is a consequence of a formula F if and only if "F implies G" is always true. We restate this as the following proposition.

Proposition 1.19 For any formulas F and G, G is a consequence of F if and only if $F \to G$ is a tautology.

Proof We show that $F \to G$ is *not* a tautology if and only if G is *not* a consequence of F.

By the definition of "tautology," $F \to G$ is not a tautology if and only if there exists an assignment $\mathcal{A}$ such that $\mathcal{A} \models \neg(F \to G)$.

By the definition of "$\to$," $\mathcal{A} \models \neg(F \to G)$ if and only if $\mathcal{A} \models \neg(\neg F \vee G)$.

By the semantics of propositional logic, $\mathcal{A} \models \neg(\neg F \vee G)$ if and only if both $\mathcal{A} \models F$ and $\mathcal{A} \models \neg G$.

Finally, by the definition of "consequence," there exists an assignment $\mathcal{A}$ such that $\mathcal{A} \models F$ and $\mathcal{A} \models \neg G$ if and only if G is not a consequence of F. $\square$

Suppose we want to determine whether or not formula G is a consequence of a formula F. We refer to this as the *consequence problem*. By Proposition 1.19 this can be rephrased as a validity problem (since G is a consequence of F if and only if $G \to F$ is valid). Such problems can be resolved by computing a truth table. If the truth values for $F \to G$ are all 1s, then G is a consequence of F. Otherwise, it is not. In particular, if F is a contradiction, then G is a consequence of F regardless of G.

Example 1.20 Let F and G be formulas. Each of the following can easily be verified by computing a truth table.

$$(F \wedge G) \models F$$
$$F \models (F \vee G)$$
$$(F \wedge \neg F) \models G$$

Definition 1.21 If both G is a consequence of F and F is a consequence of G, then we say F and G are *equivalent*. We denote this by $F \equiv G$.

It follows from Proposition 1.19 that two formulas F and G are equivalent if and only if $F \leftrightarrow G$ is a tautology. So we can determine whether two formulas F and G are equivalent by computing a truth table. Each of the equivalences in the following examples can easily be verified in this manner.

Example 1.22 For all formulas F and G, $(F \wedge G) \equiv (G \wedge F)$ *and* $(F \vee G) \equiv (G \vee F)$.

Example 1.23 For any formula F and any tautology T, $(F \wedge T) \equiv F$ and $(F \vee T) \equiv T$.

Example 1.24 For any formula F and any contradiction $\bot$, $(F \wedge \bot) \equiv \bot$ and $(F \vee \bot) \equiv F$.

Example 1.25 (Distributivity rules) The following two equivalences exhibit the distributivity rules for $\wedge$ and $\vee$. For all formulas F, G, and H,

$$(F \wedge (G \vee H)) \equiv ((F \wedge G) \vee (F \wedge H)) \text{ and}$$
$$(F \vee (G \wedge H)) \equiv ((F \vee G) \wedge (F \vee H)).$$

Example 1.26 (DeMorgan's rules) For all formulas F and G

$$\neg(F \wedge G) \equiv (\neg F \vee \neg G), \text{ and}$$
$$\neg(F \vee G) \equiv (\neg F \wedge \neg G).$$

The equivalences in the previous examples are basic. Note that we refer to some of these equivalences as "rules." Each of these holds true for arbitrary formulas. From these basic equivalences, more elaborate equivalences can be created.

Example 1.27 Using the equivalences in the previous examples, we show that $((C \wedge D) \vee A) \wedge ((C \wedge D) \vee B) \wedge (E \vee \neg E) \equiv (A \wedge B) \vee (C \wedge D)$. Let L denote the formula on the left in this equivalence. Note that $(E \vee \neg E)$ is a tautology. By Example 1.23, L is equivalent to $((C \wedge D) \vee A) \wedge ((C \wedge D) \vee B)$. According to the second distributivity rule in Example 1.25, this is equivalent to $(C \wedge D) \vee (A \wedge B)$ (viewing $(C \wedge D)$ as the formula F in that rule).

By Example 1.22, this is equivalent to $(A \wedge B) \vee (C \wedge D)$ which is the formula on the right in our equivalence.

Using the basic rules in Examples 1.22–1.26, we were able to verify that $((C \wedge D) \vee A) \wedge ((C \wedge D) \vee B) \wedge (E \vee \neg E) \equiv (A \wedge B) \vee (C \wedge D)$. This is itself a rule, holding for any formulas A, B, C, D, and E. Alternatively, we could have verified this equivalence by computing a truth table. Such a truth table would have had $2^5 = 32$ rows. The previously established rules provided a more efficient method of verification.

Likewise, we could state "rules for consequence" that would allow us to show that one formula is a consequence of another without having to compute truth tables. In the next section, we exploit this idea and introduce the notion of formal proof. Formal proofs allow us to "derive" formulas from sets of formulas. The following definition extends the notion of consequence to this setting.

Propositional logic

Definition 1.28 Let $\mathcal{F} = \{F_1, F_2, F_3, \ldots\}$ be a set of formulas.

For any assignment $\mathcal{A}$, we say $\mathcal{A}$ *models* $\mathcal{F}$, denoted $\mathcal{A} \models \mathcal{F}$ if $\mathcal{A} \models F_i$ for each formula F_i in $\mathcal{F}$.

We say a formula G is a *consequence* of $\mathcal{F}$, and write $\mathcal{F} \models G$, if $\mathcal{A} \models \mathcal{F}$ implies $\mathcal{A} \models G$ for every assignment $\mathcal{A}$.

Suppose that we want to determine whether a formula G is a consequence of a set of formulas $\mathcal{F}$. If $\mathcal{F}$ is finite, then we could consider the conjunction $\bigwedge \mathcal{F}$ of all formulas in $\mathcal{F}$ and compute a truth table for $\bigwedge \mathcal{F} \rightarrow G$. This method would certainly produce an answer. However, if the set $\mathcal{F}$ is large, then computing such a truth table is neither an efficient, nor a pleasant, thing to do. If $\mathcal{F}$ is infinite, then this method does not work at all. Another approach is to *derive* G from $\mathcal{F}$. Consider the following example.

Example 1.29 Let $\mathcal{F}$ be the following set of formulas

$$\{A, (A \rightarrow B), (B \rightarrow C), (C \rightarrow D), (D \rightarrow E), (E \rightarrow F), (F \rightarrow G)\}.$$

Suppose each of the seven formulas in $\mathcal{F}$ is true. Then, in particular, A and $A \rightarrow B$ are true. It follows that B must also be true. Likewise, since B and $B \rightarrow C$ are true, then C must also be true, and so forth. If each formula in $\mathcal{F}$ is true, then A, B, C, D, E, F, and G are true. Each of these formulas is a consequence of $\mathcal{F}$. We do not need a truth table to see this.

Let $\bigwedge \mathcal{F}$ be the conjunction of all formulas in $\mathcal{F}$. That is,

$$\bigwedge \mathcal{F} = A \wedge (A \rightarrow B) \wedge (B \rightarrow C) \wedge (C \rightarrow D) \wedge (D \rightarrow E) \wedge (E \rightarrow F) \wedge (F \rightarrow G).$$

The truth table for $\bigwedge \mathcal{F} \rightarrow G$ comprises 128 rows. Without computing a single row, we can see that each row will have truth value 1. The formula $\bigwedge \mathcal{F} \rightarrow G$ is a tautology and, equivalently, G is a consequence of $\mathcal{F}$.

In the previous example, we repeatedly used the fact that if X and $X \rightarrow Y$ are both true, then Y is also true. That is, we used the fact that Y is a consequence of $X \wedge (X \rightarrow Y)$. This follows from the truth table we computed in Example 1.16. Rather than compute another truth table (having 128 rows), we used a truth table we have already computed (having only four rows) to deduce that G is a consequence of $\mathcal{F}$. We *derived* G from $\mathcal{F}$ using a previously validated *rule*.

1.4 Formal proofs

A logic, by definition, has rules for deducing the truth of one sentence from that of another. These rules yield a system of formal proof. In this section, we describe such a proof system for propositional logic.

A *proof system* consists of a set of basic rules for derivations. These rules allow us to deduce formulas from sets of formulas. It may take several steps to derive a given formula G from a set of formulas $\mathcal{F}$, where each "step" is an application of one of the basic rules. The list of these steps forms a *formal proof* of G from $\mathcal{F}$.

Of particular interest is the relationship between the notion of formal proof and the notion of consequence. We want a proof system that is *sound*. That is, we want the following property to hold.

(Soundness) If a formula G can be derived from a set of formulas $\mathcal{F}$, then G is a consequence of $\mathcal{F}$.

If a proof system is sound, then it provides an alternative to truth tables for determining whether a formula G is a consequence of a set of formulas $\mathcal{F}$. In this section, we present a proof system and prove that it is sound. We use this proof system in several examples. The proof system we introduce is intended to be user-friendly. To construct a proof deriving G from $\mathcal{F}$, one may consider the question: *why* is G a consequence of $\mathcal{F}$? The object is then to translate one's reasoning into a formal proof. Although this process is necessarily pedantic, the large yet coherent set of basic rules we provide is intended to aid the translation of thought into formal proof.

The aim of a formal proof system is to make the thought process infallible. Ideally, the proof system could replace the thought process. Instead of thinking, we could blindly follow a set of rules. If two people disagree about whether G truly is a consequence of $\mathcal{F}$, there would be no need for debate. Both parties could perform a computation to see whether or not G is indeed a consequence of $\mathcal{F}$. The reason that "not thinking" is ideal is that we could program a computer to perform this task. In Section 1.8, we introduce another proof system known as *resolution*. Resolution is a pared down proof system that takes the thinking out of formal proofs. Whereas resolution is intended for the mechanization of proofs, the proof system we describe in the present section is intended for human use.

We now list the basic rules for our proof system. The veracity of many of these rules is self-evident. For example, if F and G can be derived from $\mathcal{F}$, then $F \wedge G$ can also be derived from $\mathcal{F}$. We call this rule "$\wedge$-Introduction." We use the following notation: we write $\mathcal{F} \vdash G$ to abbreviate "G can be derived from $\mathcal{F}$." Using this notation, $\wedge$-Introduction is written as:

if $\mathcal{F} \vdash F$ and $\mathcal{F} \vdash G$ then $\mathcal{F} \vdash (F \wedge G)$.

The table below lists this and other basic rules for derivations.

There are a lot of rules. Note the organization of the list. It begins with a couple of rules that are quite intuitive: Assumption and Monotonicity. There follow a few similarly named rules for various symbols of propositional logic. The four rules that conclude the list reflect conventions (C1) and (C2). In addition to the rules in Table 1.5, we have the rules in Table 1.6 regarding the definitions of $\vee$, $\to$, and $\leftrightarrow$.

Our list of rules is both too big and too small. It is too big in the sense that some of these rules are redundant. For example, since $\vee$ can be expressed in terms of $\neg$ and $\wedge$, $\vee$-Symmetry follows from $\wedge$-Symmetry (see Exercise 1.13). We could pare these redundant rules from our list. In fact, we really need only

Table 1.5 Basic rules for derivations

Premise	Conclusion	Name
G is in $\mathcal{F}$	$\mathcal{F} \vdash G$	Assumption
$\mathcal{F} \vdash G$ and $\mathcal{F} \subset \mathcal{F}'$	$\mathcal{F}' \vdash G$	Monotonicity
$\mathcal{F} \vdash G$	$\mathcal{F} \vdash \neg\neg G$	Double negation
$\mathcal{F} \vdash F$, $\mathcal{F} \vdash G$	$\mathcal{F} \vdash (F \wedge G)$	$\wedge$-Introduction
$\mathcal{F} \vdash (F \wedge G)$	$\mathcal{F} \vdash F$	$\wedge$-Elimination
$\mathcal{F} \vdash (F \wedge G)$	$\mathcal{F} \vdash (G \wedge F)$	$\wedge$-Symmetry
$\mathcal{F} \vdash F$	$\mathcal{F} \vdash (F \vee G)$	$\vee$-Introduction
$\mathcal{F} \vdash (F \vee G)$, $\mathcal{F} \cup \{F\} \vdash H$, $\mathcal{F} \cup \{G\} \vdash H$	$\mathcal{F} \vdash H$	$\vee$-Elimination
$\mathcal{F} \vdash (F \vee G)$	$\mathcal{F} \vdash (G \vee F)$	$\vee$-Symmetry
$\mathcal{F} \cup \{F\} \vdash G$	$\mathcal{F} \vdash (F \to G)$	$\to$-Introduction
$\mathcal{F} \vdash (F \to G)$, $\mathcal{F} \vdash F$	$\mathcal{F} \vdash G$	$\to$-Elimination
$\mathcal{F} \vdash F$	$\mathcal{F} \vdash (F)$	$(,)$-Introduction
$\mathcal{F} \vdash (F)$	$\mathcal{F} \vdash F$	$(,)$-Elimination
$\mathcal{F} \vdash ((F \wedge G) \wedge H)$	$\mathcal{F} \vdash (F \wedge G \wedge H)$	$\wedge$-Parentheses rule
$\mathcal{F} \vdash ((F \vee G) \vee H)$	$\mathcal{F} \vdash (F \vee G \vee H)$	$\vee$-Parentheses rule

Table 1.6 More rules for derivations

Rules	Name
$\mathcal{F} \vdash (F \vee G)$ if and only if $\mathcal{F} \vdash \neg(\neg F \wedge \neg G)$	$\vee$-Definition
$\mathcal{F} \vdash (F \to G)$ if and only if $\mathcal{F} \vdash (\neg F \vee G)$	$\to$-Definition
$\mathcal{F} \vdash (F \leftrightarrow G)$ if and only if both $\mathcal{F} \vdash (F \to G)$ and $\mathcal{F} \vdash (G \to F)$	$\leftrightarrow$-Definition

Propositional logic

three rules! These rules must be phrased within the proper context and are the topic of Section 1.8. Our present concern is not economy, but utility. These rules allow us to derive formulas from other formulas. The more rules at our disposal, the better. In this sense, the above list is too small. As we shall see, there is no end to the rules that can be derived from those stated above.

Having listed these rules for derivations, we now define "formal proof."

Definition 1.30 A *formal proof* in propositional logic is a finite sequence of statements of the form "$\mathcal{X} \vdash Y$" (where $\mathcal{X}$ is a set of formulas and Y is a formula) each of which follows from the previous statements by one of the rules in Table 1.5 or Table 1.6. We say that G can be *derived* from $\mathcal{F}$ if there is a formal proof concluding with the statement $\mathcal{F} \vdash G$.

A formal proof can always be put into two-column form. The best way to describe formal proofs is to give an example.

Example 1.31 Let $\mathcal{H} = \{(\neg A \vee B), (\neg A \vee C), (A \vee \neg D)\}$.
We derive the formula $D \to (A \wedge B \wedge C)$ from $\mathcal{H}$.

Statement	Justification
1. $\mathcal{H} \cup \{D\} \vdash D$	Assumption
2. $\mathcal{H} \cup \{D\} \vdash (A \vee \neg D)$	Assumption
3. $\mathcal{H} \cup \{D\} \vdash (\neg D \vee A)$	∨-Symmetry applied to 2
4. $\mathcal{H} \cup \{D\} \vdash (D \to A)$	→-Definition applied to 3
5. $\mathcal{H} \cup \{D\} \vdash A$	→-Elimination applied to 4 and 1
6. $\mathcal{H} \cup \{D\} \vdash (\neg A \vee B)$	Assumption
7. $\mathcal{H} \cup \{D\} \vdash (A \to B)$	→-Definition applied to 6
8. $\mathcal{H} \cup \{D\} \vdash B$	→-Elimination applied to 7 and 5
9. $\mathcal{H} \cup \{D\} \vdash (\neg A \vee C)$	Assumption
10. $\mathcal{H} \cup \{D\} \vdash (A \to C)$	→-Definition applied to 9
11. $\mathcal{H} \cup \{D\} \vdash C$	→-Elimination applied to 10 and 5
12. $\mathcal{H} \cup \{D\} \vdash (A \wedge B)$	∧-Introduction applied to 5 and 8
13. $\mathcal{H} \cup \{D\} \vdash ((A \wedge B) \wedge C)$	∧-Introduction applied to 12 and 11
14. $\mathcal{H} \cup \{D\} \vdash (A \wedge B \wedge C)$	∧-Parenthesis rule applied to 13
15. $\mathcal{H} \vdash (D \to (A \wedge B \wedge C))$	→-Introduction applied to 14
16. $\mathcal{H} \vdash D \to (A \wedge B \wedge C)$	(,)-Elimination

At first glance, the above proof looks like a complicated way of demonstrating a fact that is not so complicated. However, reading the proof line-by-line we see that each line asserts a simple truth that is easy to verify. Proofs can be made more succinct by using additional rules. For example, note that in the previous proof we repeatedly introduced the symbol $\rightarrow$ only to eliminate it. Instead, we could have separately proved the following rule.

Premise: $\mathcal{F} \vdash (\neg F \vee G)$, $\mathcal{F} \vdash F$
Conclusion: $\mathcal{F} \vdash G$

Statement	Justification
1. $\mathcal{F} \vdash F$	Premise
2. $\mathcal{F} \vdash (\neg F \vee G)$	Premise
3. $\mathcal{F} \vdash (F \rightarrow G)$	$\rightarrow$-Definition applied to 2
4. $\mathcal{F} \vdash G$	$\rightarrow$-Elimination applied to 3 and 1

The proof in Example 1.31 essentially repeats the above argument three times. We can treat this four-line proof as a subroutine. Had we established this rule prior to Example 1.31, we could have referred to it three times to make the proof more concise. For lack of a better name, we christen this rule $\vee$-*Modus Ponens*. "Modus ponens" is a standard name for the rule we call $\rightarrow$-Elimination. As the archaic name suggests, this rule has been around for a while. It is found in what can be considered the origin of proofs: Euclid's *Elements*. The next five examples establish some other rules that facilitate the construction of proofs.

Example 1.32 (Tautology rule) This rule states that, for any formula G, $(\neg G \vee G)$ can be derived from any set of formulas.

Premise: None
Conclusion: $\mathcal{F} \vdash (\neg G \vee G)$

Statement	Justification
1. $\mathcal{F} \cup \{G\} \vdash G$	Assumption
2. $\mathcal{F} \vdash (G \rightarrow G)$	$\rightarrow$-Introduction applied to 1
3. $\mathcal{F} \vdash (\neg G \vee G)$	$\rightarrow$-Definition applied to 2

Example 1.33 (Contradiction rule) This rule states that any formula G can be derived from the contradiction $F \wedge \neg F$.

Premise: $\mathcal{F} \vdash (F \wedge \neg F)$
Conclusion: $\mathcal{F} \vdash G$

Propositional logic

Statement	Justification
1. $\mathcal{F} \vdash (F \wedge \neg F)$	Premise
2. $\mathcal{F} \vdash (\neg F \wedge F)$	∧-Symmetry applied to 1
3. $\mathcal{F} \vdash \neg F$	∧-Elimination applied to 2
4. $\mathcal{F} \vdash (\neg F \vee G)$	∨-Introduction applied to 3
5. $\mathcal{F} \vdash F$	∧-Elimination applied to 1
6. $\mathcal{F} \vdash G$	∨-Modus Ponens applied to 4 and 5

Example 1.34 (Contrapositive) This is the rule of logic that states that if p implies q, then $\neg q$ implies $\neg p$.

Premise: $\mathcal{F} \cup \{F\} \vdash G$
Conclusion: $\mathcal{F} \cup \{\neg G\} \vdash \neg F$

Statement	Justification
1. $\mathcal{F} \cup \{F\} \vdash G$	Premise
2. $\mathcal{F} \cup \{F\} \vdash \neg\neg G$	Double negation applied to 1
3. $\mathcal{F} \vdash (F \rightarrow \neg\neg G)$	→-Introduction applied to 2
4. $\mathcal{F} \vdash (\neg F \vee \neg\neg G)$	→-Definition applied to 3
5. $\mathcal{F} \vdash (\neg\neg G \vee \neg F)$	∨-Symmetry applied to 4
6. $\mathcal{F} \vdash (\neg G \rightarrow \neg F)$	→-Definition applied to 5
7. $\mathcal{F} \cup \{\neg G\} \vdash (\neg G \rightarrow \neg F)$	Monotonicity applied to 6
8. $\mathcal{F} \cup \{\neg G\} \vdash \neg G$	Assumption
9. $\mathcal{F} \cup \{\neg G\} \vdash \neg F$	→-Elimination applied to 6 and 8

Example 1.35 (Proof by cases) This rule provides a useful way to structure a proof.

Premise: $\mathcal{F} \cup \{F\} \vdash G$, $\mathcal{F} \cup \{\neg F\} \vdash G$
Conclusion: $\mathcal{F} \vdash G$

Statement	Justification
1. $\mathcal{F} \cup \{F\} \vdash G$	Assumption
2. $\mathcal{F} \cup \{\neg F\} \vdash G$	Assumption
3. $\mathcal{F} \vdash (\neg F \vee F)$	Tautology rule
4. $\mathcal{F} \vdash G$	∨-Elimination applied to 3, 2, and 1

Example 1.36 (Proof by contradiction) Another way to structure a proof is by contradiction. As the following proof indicates, proof by contradiction is (in this context) essentially the contrapositive of proof by cases.

Premise: $\mathcal{F} \cup \{F\} \vdash G$, $\mathcal{F} \cup \{F\} \vdash \neg G$
Conclusion: $\mathcal{F} \vdash \neg F$

Statement	Justification
1. $\mathcal{F} \cup \{F\} \vdash G$	Premise
2. $\mathcal{F} \cup \{\neg G\} \vdash \neg F$	Contrapositive applied to 1
3. $\mathcal{F} \cup \{F\} \vdash \neg G$	Premise
4. $\mathcal{F} \cup \{\neg\neg G\} \vdash \neg F$	Contrapositive applied to 3
5. $\mathcal{F} \vdash \neg F$	Proof by cases applied to 2 and 4

These are now established rules that may henceforth be used to justify statements in proofs. By "established" we mean that these rules are true, provided that the rules in Table 1.5 are true. We must prove that this is the case. We must prove that our proof system is sound: that if G can be derived from $\mathcal{F}$, then G is in fact a consequence of $\mathcal{F}$.

Theorem 1.37 (Soundness) If $\mathcal{F} \vdash G$, then $\mathcal{F} \models G$.

Proof If $\mathcal{F} \vdash G$, then there is a formal proof concluding with $\mathcal{F} \vdash G$. Each line of the proof contains a statement of the form $\mathcal{X} \vdash Y$ which is justified by one of the rules in Tables 1.5 or 1.6. We want to show that for each line of the proof, if $\mathcal{X} \vdash Y$, then $\mathcal{X} \models Y$. This can be accomplished by verifying each rule in the tables one-by-one. We demonstrate this by verifying three rules: Assumption, $\wedge$-Elimination, and $\rightarrow$-Introduction. The verification of the remaining rules are left as an exercise.

We begin with the first rule of Table 1.5: Assumption. The conclusion of this rule is that $\mathcal{F} \vdash G$. We must show, under the premise of this rule, that $\mathcal{F} \models G$. But this is clear since the premise states that G is in $\mathcal{F}$ (if $\mathcal{A}$ models $\mathcal{F}$ then $\mathcal{A}$ must model G).

Refer next to $\wedge$-Elimination. This rule states that if $\mathcal{F} \vdash (F \wedge G)$, then $\mathcal{F} \vdash F$. We must show that if $\mathcal{F} \models (F \wedge G)$, then $\mathcal{F} \models F$. That is, we must show that F is a consequence of $(F \wedge G)$. This is verified by the following truth table.

F	G	$(F \wedge G)$	$((F \wedge G) \rightarrow F)$
0	0	0	1
0	1	0	1
1	0	0	1
1	1	1	1

Propositional logic

Now consider →-Introduction. This rule states that if $\mathcal{F} \cup \{F\} \vdash G$ then $\mathcal{F} \vdash (F \to G)$. To verify this, we must show that if $\mathcal{F} \cup \{F\} \models G$ then $\mathcal{F} \models (F \to G)$. Assuming that $\mathcal{F} \cup \{F\} \models G$, we want to show that, for any assignment $\mathcal{A}$, if $\mathcal{A} \models \mathcal{F}$, then $\mathcal{A} \models (F \to G)$.

So suppose that $\mathcal{A} \models \mathcal{F}$ and $\mathcal{A}(F)$ is defined. If $\mathcal{A}(F) = 0$, then $\mathcal{A} \models (F \to G)$ regardless of the value of $\mathcal{A}(G)$. If, on the other hand, $\mathcal{A}(F) = 1$, then $\mathcal{A} \models \mathcal{F} \cup \{F\}$. By our assumption, $\mathcal{A} \models G$. In any case, we see that $\mathcal{A} \models (F \to G)$. Since $\mathcal{A}$ was an arbitrary assignment modeling $\mathcal{F}$, we conclude that $\mathcal{F} \models (F \to G)$ as was required.

Essentially, we must verify that each rule is true when $\vdash$ is replaced by $\models$. For most of the rules, like ∧-Elimination, this can be accomplished by computing a small truth table. For the last four rules of Table 1.5, there is really nothing to prove. These four rules hold by conventions (C1) and (C2). Also, each rule in Table 1.6 is sound by virtue of the definitions of ∨, →, and ↔. We leave the verification of the remaining rules in Table 1.5 as Exercise 1.22. □

Formal proofs provide a method for showing that a formula is a consequence of other formulas. The following Corollaries state that formal proofs can also show that a formula is valid or unsatisfiable.

Corollary 1.38 If G can be derived from the empty set, then G is a tautology.

Proof If $\emptyset \vdash G$, then, by Monotonicity, $\mathcal{F} \vdash G$ for every set of formulas $\mathcal{F}$. By Theorem 1.37, $\mathcal{F} \models G$ for every set of formulas $\mathcal{F}$. It follows that $\mathcal{A} \models G$ for any assignment $\mathcal{A}$ and G is a tautology. □

Corollary 1.39 If $\neg G$ can be derived from the empty set, then G is a contradiction.

Proof This is immediate from the previous Corollary and the definition of "contradiction." □

Example 1.40 The following formal proof shows that $((A \to B) \lor A)$ is a tautology.

Statement	Justification
1. $\{\neg A\} \vdash \neg A$	Assumption
2. $\{\neg A\} \vdash (\neg A \lor B)$	∨-Introduction applied to 1
3. $\{\neg A\} \vdash (A \to B)$	→-Definition applied to 2
4. $\{\neg A\} \vdash ((A \to B) \lor A)$	∨-Introduction applied to 3
5. $\{A\} \vdash A$	Assumption
6. $\{A\} \vdash (A \lor (A \to B))$	∨-Introduction applied to 5
7. $\{A\} \vdash ((A \to B) \lor A)$	∨-Symmetry applied to 6
8. $\emptyset \vdash ((A \to B) \lor A)$	Proof by cases applied to 4 and 7

Formal proofs can also show that two formulas are equivalent.

Definition 1.41 Formulas F and G are *provably equivalent* if both $\{F\} \vdash G$ and $\{G\} \vdash F$.

Corollary 1.42 If F and G are provably equivalent, then they are equivalent.

Proof This follows immediately from Theorem 1.37. □

Consider now the converses of Theorem 1.37 and its Corollaries. Theorem 1.37 states that if G can be derived from $\mathcal{F}$, then G is a consequence of $\mathcal{F}$. Is the opposite true? Can we derive from $\mathcal{F}$ every consequence of $\mathcal{F}$? Can every tautology be given a formal proof as in Example 1.40? If two formulas are equivalent, does this mean we can prove that they are equivalent? We claim that the answer to each of these questions is "yes." We claim that every rule that is true in propositional logic, all infinitely many of them, can be derived from the rules in Tables 1.5 and 1.6. This is not obvious.

Example 1.43 It may seem that our list of rules is incomplete. For example, the formulas F and $\neg\neg F$ are clearly equivalent. So if we can derive the formula $\neg\neg F$ from a set of formulas $\mathcal{F}$, then we should also be able to derive F from $\mathcal{F}$. However this is not one of our rules. Double negation states that if $\mathcal{F} \vdash F$, then $\mathcal{F} \vdash \neg\neg F$. We now show that the converse of Double negation, although not stated as a rule, can be derived from our rules.

Premise: $\mathcal{F} \vdash \neg\neg F$
Conclusion: $\mathcal{F} \vdash F$

Statement	Justification
1. $\mathcal{F} \vdash \neg\neg F$	Premise
2. $\mathcal{F} \cup \{\neg F\} \vdash \neg\neg F$	Monotonicity applied to 1
3. $\mathcal{F} \cup \{\neg F\} \vdash \neg F$	Assumption
4. $\mathcal{F} \cup \{\neg F\} \vdash (\neg F \wedge \neg\neg F)$	$\wedge$-Introduction applied to 3 and 2
5. $\mathcal{F} \cup \{\neg F\} \vdash F$	Contradiction rule (1.33) applied to 4
6. $\mathcal{F} \vdash \{F\} \vdash F$	Assumption
7. $\mathcal{F} \vdash F$	Proof by cases applied to 5 and 6

So not only are F and $\neg\neg F$ equivalent formulas, we can formally prove that they are equivalent formulas. We claim that each of the equivalences in the previous section are actually provably equivalent. In particular, we show that the Distributivity rules from Example 1.25 and DeMorgan's rules from Example 1.26 can be given formal derivations.

Proposition 1.44 (DeMorgan's rules) The equivalent pairs of formulas in Example 1.26 are each provably equivalent.

Proof We prove this for the second of DeMorgan's rules. We demonstrate formal proofs for each of the following:

$\{\neg(F \vee G)\} \vdash (\neg F \wedge \neg G)$, and
$\{(\neg F \wedge \neg G)\} \vdash \neg(F \vee G)$.

Statement	Justification
1. $\{\neg(\neg F \wedge \neg G)\} \vdash (F \vee G)$	$\vee$-Introduction
2. $\{\neg(F \vee G)\} \vdash \neg\neg(\neg F \wedge \neg G)$	Contrapositive
3. $\{\neg(F \vee G)\} \vdash (\neg F \wedge \neg G)$	Double negation

Statement	Justification
1. $\{(\neg F \wedge \neg G)\} \cup \{(F \vee G)\} \vdash (F \vee G)$	Assumption
2. $\{(\neg F \wedge \neg G)\} \cup \{(F \vee G)\} \vdash (\neg F \wedge \neg G)$	Assumption
3. $\{(\neg F \wedge \neg G)\} \cup \{(F \vee G)\} \vdash \neg F$	$\wedge$-Elimination applied to 2
4. $\{(\neg F \wedge \neg G)\} \cup \{(F \vee G)\} \vdash G$	$\vee$-Elimination applied to 1 and 3
5. $\{(\neg F \wedge \neg G)\} \cup \{(F \vee G)\} \vdash (\neg G \wedge \neg F)$	$\wedge$-Symmetry applied to 2
6. $\{(\neg F \wedge \neg G)\} \cup \{(F \vee G)\} \vdash \neg G$	$\wedge$-Elimination applied to 5
7. $\{(\neg F \wedge \neg G)\} \vdash \neg(F \vee G)$	Proof by contradiction applied to 4 and 6

We have demonstrated that $\neg(F \vee G)$ and $(\neg F \wedge \neg G)$ are provably equivalent. The verification of DeMorgan's first rule is left as Exercise 1.23. □

Proposition 1.45 ($\wedge$-Distributivity) For any formulas F, G, and H, the formulas $(F \wedge (G \vee H))$ and $((F \wedge G) \vee (F \wedge H))$ are provably equivalent.

Proof To prove this, we must derive each formula from the other. Instead of providing formal proofs, we outline the derivations and leave the details to the reader. First we show that $(F \wedge G) \vee (F \wedge H)$ can be derived from $F \wedge (G \vee H)$.

Premise: $\mathcal{F} \vdash F \wedge (G \vee H)$.
Conclusion: $\mathcal{F} \vdash (F \wedge G) \vee (F \wedge H)$

We sketch a formal proof using Proof by cases. Assuming the premise, we show that $(F \wedge G) \vee (F \wedge H)$ can be derived from both $\mathcal{F} \cup \{G\}$ and $\mathcal{F} \cup \{\neg G\}$.

From the premise, we see that $\mathcal{F} \cup \{G\} \vdash F$. It follows that $(F \wedge G)$ can be derived from $\mathcal{F} \cup \{G\}$. We then obtain $\mathcal{F} \cup \{G\} \vdash (F \wedge G) \vee (F \wedge H)$ by $\vee$-Introduction.

Next we show that $\mathcal{F} \cup \{\neg G\} \vdash (F \wedge G) \vee (F \wedge H)$. From the premise we see that both F and $(G \vee H)$ can be derived from $\mathcal{F} \cup \{\neg G\}$. Since, $\mathcal{F} \cup \{\neg G\} \vdash \neg G$, we obtain $\mathcal{F} \cup \{\neg G\} \vdash H$ from $(G \vee H)$ by $\vee$-Modus Ponens. It follows that $\mathcal{F} \cup \{\neg G\} \vdash (F \wedge H)$. Finally, we get $\mathcal{F} \cup \{\neg G\} \vdash (F \wedge G) \vee (F \wedge H)$ by $\vee$-Introduction.

We must also show that the converse holds.

Premise: $\mathcal{F} \vdash (F \wedge G) \vee (F \wedge H)$
Conclusion: $\mathcal{F} \vdash F \wedge (G \vee H)$

We prove this by twice applying $\vee$-Elimination. Since $(G \vee H)$ can be derived from both $(F \wedge G)$ and $(F \wedge H)$, we obtain $\mathcal{F} \vdash (G \vee H)$ by applying $\vee$-Elimination to the premise. We obtain $\mathcal{F} \vdash F$ in the same manner. The conclusion then follows by $\wedge$-Introduction.

These arguments can be arranged as formal two-column proofs. We leave this as Exercise 1.24. □

Proposition 1.46 ($\vee$-Distributivity) For any formulas F, G, and H, the formulas $(F \vee (G \wedge H))$ and $((F \vee G) \wedge (F \vee H))$ are provably equivalent.

Proof Exercise 1.25. □

Of course, we do not need formal proofs to verify these equivalences. We could use truth tables. In the case of the Distributivity rules and DeMorgan's rules, truth tables provide a more efficient method of verification than formal proofs. For now, the importance of Propositions 1.44, 1.45, and 1.46 is that they lend credence to our earlier claim that we can formally prove anything that is true in propositional logic. Later, these propositions will help us prove this claim.

At the outset of this section, we said we would be interested in the relationship between the notion of formal proof and the notion of consequence. We proved in Theorem 1.37 that if G can be formally proved from F then G is a consequence of F. We stated, without proof, that the opposite of this is also true: if $\mathcal{F} \models G$ then $\mathcal{F} \vdash G$. So the symbol $\models$ introduced in the previous section and the symbol $\vdash$ introduced in the present section mean the same thing in propositional logic. This is the Completeness theorem for propositional logic, the proof of which will be given at the conclusion of this chapter.

1.5 Proof by induction

There are two types of proofs that must be distinguished. We have discussed and given several examples of *formal proofs*. This type of proof arises from the rules of the logic. Such proofs are said to take place *within* the logic, and we refer to them as *internal proofs*. Formal proofs have a limited scope. They can prove only sentences that can be written in the logic. In contrast, we may want to prove something about the logic itself. We may want to prove, say, that every sentence in the logic has a certain property. Such statements that refer to the logic itself generally can neither be stated nor proved within the logic. We give *external* proofs for such statements. External proofs are sometimes called meta-mathematical. However, this terminology belies the fact that external proofs are often more mathematical in nature than formal proofs.

Induction is a method of external proof that is used repeatedly in this book. Suppose that we want to prove that some property holds for every formula of propositional logic. For example, in the next section we show that each formula of propositional logic is equivalent to some formula in conjunctive normal form. We will define "conjunctive normal form" later. Our present concern is the question of how can we prove such a thing for *all* formulas. We need a systematic way to check each and every formula F. We do this by *induction on the complexity* of F. Induction on the complexity of F is analogous to mathematical induction.

1.5.1 Mathematical induction. Recall that mathematical induction is a method of proof that allows us to prove something for *all* natural numbers. For example, suppose we want to prove that for all natural numbers n, the number $11^n - 4^n$ is divisible by 7. Using mathematical induction, we can do this in two steps. First, we show that the statement is true for $n = 1$. This is easy. Second, we show that if the statement holds for $n = m$ for some m, then it also holds for $n = m+1$. This is the inductive step. In our example, we can do this by observing that $11^{m+1} - 4^{m+1} = 11^{m+1} - 11 \cdot 4^m + 7 \cdot 4^m = 11(11^m - 4^m) + 7 \cdot 4^m$. It follows that if $11^m - 4^m$ is divisible by 7, then so is $11^{m+1} - 4^{m+1}$. This completes the proof. It's like the domino effect. It is true for $n = 1$, and so, by the second step of the proof, it must also be true for $n = 2$, and therefore $n = 3$, and $n = 4$, and so forth. We conclude that for every natural number n, $11^n - 4^n$ is divisible by 7.

An example of mathematical induction that is more relevant to propositional logic is provided by the proof of Proposition 1.47. This proposition is a generalization of DeMorgan's rules. First, we introduce some notation.

Notation 1 Let $F_1, \ldots, F_n$ be formulas. We write

$$\bigwedge_{i=1}^{n} F_i \text{ to abbreviate } F_1 \wedge F_2 \wedge \ldots \wedge F_n, \text{ and}$$

$$\bigvee_{i=1}^{n} F_i \text{ to abbreviate } F_1 \vee F_2 \vee \ldots \vee F_n.$$

Proposition 1.47 Let $\{F_1, \ldots, F_n\}$ be a finite set of formulas. Then both

$$\neg \left(\bigwedge_{i=1}^{n} F_i \right) \equiv \left(\bigvee_{i=1}^{n} \neg F_i \right) \text{ and } \neg \left(\bigvee_{i=1}^{n} F_i \right) \equiv \left(\bigwedge_{i=1}^{n} \neg F_i \right).$$

Proof We show that $\neg(\bigwedge_{i=1}^{n} F_i) \equiv (\bigvee_{i=1}^{n} \neg F_i)$ by induction on n.

First, suppose $n = 1$. We need to show that $\neg(\bigwedge_{i=1}^{1} F_i) \equiv (\bigvee_{i=1}^{1} \neg F_i)$. By the definitions of "$\bigwedge$" and "$\bigvee$," this is the same as $\neg(F_1) \equiv (\neg F_1)$, which is true by convention (C1).

Our induction hypothesis is that, for some $m \geq 1$ and any formulas $F_1, \ldots, F_m$, we have

$$\neg\left(\bigwedge_{i=1}^{m} F_i\right) \equiv \left(\bigvee_{i=1}^{m} \neg F_i\right).$$

We want to show that

$$\neg\left(\bigwedge_{i=1}^{m+1} F_i\right) \equiv \left(\bigvee_{i=1}^{m+1} \neg F_i\right).$$

By the definition of $\bigwedge$ we have

$$\neg\left(\bigwedge_{i=1}^{m+1} F_i\right) \equiv \neg\left(\left(\bigwedge_{i=1}^{m} F_i\right) \wedge F_{m+1}\right).$$

By DeMorgan's rule we get

$$(1)\ \neg\left(\bigwedge_{i=1}^{m+1} F_i\right) \equiv \left(\neg\left(\bigwedge_{i=1}^{m} F_i\right) \vee \neg F_{m+1}\right).$$

By our induction hypothesis,

$$\neg\left(\bigwedge_{i=1}^{m} F_i\right) \equiv \left(\bigvee_{i=1}^{m} \neg F_i\right).$$

(†) Substituting this into (1) yields

$$\neg\left(\bigwedge_{i=1}^{m+1} F_i\right) \equiv \left(\left(\bigvee_{i=1}^{m} \neg F_i\right) \vee \neg F_{m+1}\right).$$

Finally, by the definition of $\bigvee$ we arrive at

$$\neg\left(\bigwedge_{i=1}^{m+1} F_i\right) \equiv \bigvee_{i=1}^{m+1} \neg F_i.$$

We have shown that $\neg(\bigwedge_{i=1}^{m+1} F_i) \equiv (\bigvee_{i=1}^{m+1} \neg F_i)$ as was required. We conclude that $\neg(\bigwedge_{i=1}^{n} F_i) \equiv (\bigvee_{i=1}^{n} \neg F_i)$ for any n.

The second equivalence of the proposition follows from the first. Since $(\bigvee_{i=1}^{n+1} \neg F_i) \equiv \neg(\bigwedge_{i=1}^{n+1} F_i)$ holds for any formulas F_i, it holds when each F_i is replaced by $\neg F_i$:

$$\left(\bigvee_{i=1}^{n+1} \neg\neg F_i\right) \equiv \neg\left(\bigwedge_{i=1}^{n+1} \neg F_i\right).$$

Since these two formulas are equivalent, their negations are also equivalent:
$$\neg \left(\bigvee_{i=1}^{n+1} \neg\neg F_i \right) \equiv \neg\neg \left(\bigwedge_{i=1}^{n+1} \neg F_i \right).$$

Now $\neg(\bigvee_{i=1}^{n} F_i) \equiv (\bigwedge_{i=1}^{n} \neg F_i)$ by double negation. □

Likewise, we can generalize the distributivity rules as follows.

Proposition 1.48 Let $\{F_1, \ldots, F_n\}$ and $\{G_1, \ldots, G_m\}$ be finite sets of formulas. The following equivalences hold:

$$\left(\left(\bigwedge_{i=1}^{n} F_i \right) \vee \left(\bigwedge_{j=1}^{m} G_j \right) \right) \equiv \left(\bigwedge_{i=1}^{n} \left(\bigwedge_{j=1}^{m} (F_i \vee G_j) \right) \right)$$

$$\left(\left(\bigvee_{i=1}^{n} F_i \right) \wedge \left(\bigvee_{j=1}^{m} G_j \right) \right) \equiv \left(\bigvee_{i=1}^{n} \left(\bigvee_{j=1}^{m} (F_i \wedge G_j) \right) \right)$$

Proof Exercise 1.27. □

There is one unjustified step in the proof of Proposition 1.47. In the step labeled with (†), we essentially said that if $G' \equiv G$, then $(G \vee F) \equiv (G' \vee F)$. Although this substitution makes intuitive sense, we have not yet established this as a rule we may use. We validate this step in Theorem 1.49. We prove this theorem by induction on the complexity of formulas. We now describe this method of proof.

1.5.2 Induction on the complexity of formulas. Suppose we want to show that property $\mathcal{P}$ holds for every formula F. We can do this by *induction on the complexity of F* follows. First we show that every atomic formula possesses property $\mathcal{P}$. This corresponds to verifying case $n = 1$ in mathematical induction. The atomic case is our *induction basis*. We then assume that property $\mathcal{P}$ holds for formulas G and H. This is our *induction hypothesis*. Our aim is to show that property $\mathcal{P}$ necessarily holds for $\neg G$, $G \wedge H$, $G \vee H$, $G \to H$, and $G \leftrightarrow H$. If we succeed at this, then we can rightly conclude that $\mathcal{P}$ holds for all formulas. This completes the proof.

Theorem 1.49 (Substitution theorem) Suppose $F \equiv G$. Let H be a formula that contains F as a subformula. Let H' be the formula obtained by replacing some occurrence of F in H with G. Then $H \equiv H'$.

Proof We prove this by induction on the complexity of H.

First suppose H is atomic. Then the only subformula of H is H itself. So $F = H$. It follows that $H' = G$ and, since $F \equiv G$, we have $H \equiv H'$.

Our induction hypothesis is that the conclusion of the theorem holds for formulas H_1 and H_2 each of which contains an occurrence of F as a subformula. That is, $H_1 \equiv H_1'$ and $H_2 \equiv H_2'$ whenever H_1' and H_2' are formulas obtained from H_1 and H_2 by replacing an occurrence of F with G.

Suppose $H = \neg H_1$. Then $H' = \neg H_1'$. Since $H_1 \equiv H_1'$, we have $\neg H_1 \equiv \neg H_1'$. It follows that $H \equiv H'$ as was required.

Suppose H is one of the following formulas: $H_1 \wedge H_2$, $H_1 \vee H_2$, $H_1 \to H_2$, or $H_1 \leftrightarrow H_2$. Since F is a subformula of H, F is a subformula of H_1, a subformula of H_2, or is H itself. If $F = H$, then we have $H = F \equiv G = H'$ as in the atomic case. So we may assume that the occurrence of F that is to be replaced by G occurs either in H_1 or H_2. With no loss of generality, we may assume that it occurs in H_1.

If $H = H_1 \wedge H_2$ then $H' = H_1' \wedge H_2$. In this case we have:

$H_1 \wedge H_2$ is true if and only if
both H_1 and H_2 are true if and only if
both H_1' and H_2 are true (since $H_1 \equiv H_1'$) if and only if
$H_1' \wedge H_2$ is true.

That is, $H_1 \wedge H_2 \equiv H_1' \wedge H_2$. Since $H \equiv H_1 \wedge H_2$, we have $H \equiv H'$.

If $H = H_1 \vee H_2$, then $H' = H_1' \vee H_2$. By the definition of $\vee$, we have $H \equiv \neg(\neg H_1 \wedge \neg H_2)$ and $H' \equiv \neg(\neg H_1' \wedge H_2)$. It follows from the previous cases (corresponding to $\neg$ and $\wedge$) that $H \equiv H'$.

If $H = H_1 \to H_2$, then $H' = H_1' \to H_2$. By the definition of $\to$, $H \equiv (\neg H_1 \vee H_2$ and $H' \equiv (\neg H_1' \vee H_2)$. It follows from the previous cases (corresponding to $\neg$ and $\vee$) that $H \equiv H'$.

If $H = H_1 \leftrightarrow H_2$, then $H' = H_1' \leftrightarrow H_2$. By the definition of $\leftrightarrow$, $H \equiv (H_1 \to H_2) \wedge (H_2 \to H_1)$ and $H' \equiv (H_1' \to H_2) \wedge (H_2 \to H_1)$. It follows from the previous cases (corresponding to $\wedge$ and $\to$) that $H \equiv H'$.

We conclude that for any formula H that contains F as a subformula, $H \equiv H'$. □

In fact, this theorem remains true when "$\equiv$" is replaced by "provably equivalent."

Theorem 1.50 Suppose that F and G are provably equivalent. Let H be a formula that contains F as a subformula. Let H' be the formula obtained by replacing some occurrence of F in H with G. Then H and H' are provably equivalent.

Proof The proof is similar to the proof of Theorem 1.49. Proceed by induction on the complexity of H. The induction hypothesis is that both

H_1 and H_1' are provably equivalent, and
H_2 and H_2' are provably equivalent

where H_1' and H_2' are formulas obtained from H_1 and H_2 by replacing an occurrence of F with G. We want to verify in each of the five cases that H and H' are provably equivalent. To do this, we refer to the rules in Tables 1.5 and 1.6 (whereas in the proof of Theorem 1.49 we referred to the semantics of propositional logic). We leave the details of this proof as Exercise 1.28. □

The word "induction" indicates that we are reasoning from a particular case to the general case. Proofs by induction involve two steps and conclude that some statement holds in general *for all* natural numbers or *for all* formulas. These two steps are called the "base step" and the "induction step." In mathematical induction, the base step is the step where we show that the statement is true for $n = 1$. If we are using induction on the complexity of formulas, then the base step is the step where we verify the statement holds for all atomic formulas.

The induction step for mathematical induction is the step where we show that, if the statement is true for $n = m$, then it is also true for $n = m+1$. The induction step for induction on the complexity of formulas comprises five cases corresponding to $\neg$, $\wedge$, $\vee$, $\rightarrow$, and $\leftrightarrow$. Note that, in the proof of Theorem 1.49, the cases corresponding to $\vee$, $\rightarrow$, and $\leftrightarrow$ followed quickly from the cases regarding $\neg$ and $\wedge$. This is because $\vee$, $\rightarrow$, and $\leftrightarrow$ were defined in terms of $\neg$ and $\wedge$. This suggests an alternative form for the induction step which we now describe.

Suppose we want to show that some property P holds for all formulas of propositional logic. To do this by induction on the complexity of formulas, we first show that P holds for all atomic formulas (the base step). For the induction step, instead of verifying the five cases as above, we can sometimes do just three cases. First we show that P is preserved under equivalence. That is, we show that if $F \equiv G$ and G possess property P, then so does F. If this is true, then we only need to consider the cases corresponding to $\neg$ and $\wedge$. This suffices because every formula of propositional logic is equivalent to a formula that uses only $\neg$ and $\wedge$ (and neither $\vee$, $\rightarrow$, nor $\leftrightarrow$). We demonstrate this version of the induction step in the next section where we prove that every formula in propositional logic is equivalent to a formula that is in conjunctive normal form.

1.6 Normal forms

In Example 1.27 we showed that the formula $((C \wedge D) \vee A) \wedge ((C \wedge D) \vee B) \wedge (E \vee \neg E)$ is equivalent to the formula $(A \wedge B) \vee (C \wedge D)$ which is a disjunction of two conjunctions. In this section we show that there is nothing special about $((C \wedge D) \vee A) \wedge ((C \wedge D) \vee B) \wedge (E \vee \neg E)$. Every formula of propositional logic is equivalent to a formula that is a disjunction of conjunctions. We begin with some definitions.

Definition 1.51 A *literal* is an atomic formula or the negation of an atomic formula, and we refer to these as being *positive* or *negative*, respectively.

Example 1.52 If A is an atomic formula, then A is a positive literal and $\neg A$ is a negative literal.

Definition 1.53 A formula F is in *conjunctive normal form* (CNF) if it is a conjunction of disjunctions of literals. That is,

$$F = \bigwedge_{i=1}^{n} \left(\bigvee_{j=1}^{m} L_{i,j} \right)$$

where each $L_{i,j}$ is either atomic or a negated atomic formula.

Definition 1.54 A formula F is in *disjunctive normal form* (DNF) if it is a disjunction of conjunctions of literals. That is,

$$F = \bigvee_{i=1}^{n} \left(\bigwedge_{j=1}^{m} L_{i,j} \right)$$

where each $L_{i,j}$ is either atomic or a negated atomic formula .

Example 1.55

$(A \vee B) \wedge (C \vee D) \wedge (\neg A \vee \neg B \vee \neg D)$ *is in CNF*,

$(\neg A \wedge B) \vee C \vee (B \wedge \neg C \wedge D)$ *is in DNF, and*

$(A \vee B) \wedge ((A \wedge C) \vee (B \wedge D))$ *is neither CNF nor DNF*.

Lemma 1.56 Let F be a formula in CNF and G be a formula in DNF. Then $\neg F$ is equivalent to a formula in DNF and $\neg G$ is equivalent to a formula in CNF.

Proof If F is in CNF, then F is the formula

$$\bigwedge_{i=1}^{n} \left(\bigvee_{j=1}^{m} L_{i,j} \right)$$

for some literals $L_{i,j}$. The negation of this formula

$$\neg F = \neg \bigwedge_{i=1}^{n} \left(\bigvee_{j=1}^{m} L_{i,j} \right)$$

Propositional logic

is equivalent to

$$\bigvee_{i=1}^{n} \neg \left(\bigvee_{j=1}^{m} L_{i,j} \right)$$

by Proposition 1.47. Likewise, by the same proposition, this is equivalent to

$$\bigvee_{i=1}^{n} \left(\bigwedge_{j=1}^{m} \neg L_{i,j} \right).$$

This formula is in DNF and is equivalent to $\neg F$.
Similarly, using Proposition 1.47 twice, we can prove that $\neg G$ is equivalent to a formula in CNF. □

Theorem 1.57 Every formula F is equivalent to some formula F_1 in CNF and some formula F_2 in DNF.

Proof We prove this by induction on the complexity of F.

First suppose F is atomic. Then F is already both CNF and DNF. So we can take $F_1 = F_2 = F$.

Our induction hypothesis is that the conclusion of the theorem holds for formulas G and H. That is, we suppose there exist formulas H_1 and G_1 in CNF and H_2 and G_2 in DNF such that $H \equiv H_1 \equiv H_2$ and $G \equiv G_1 \equiv G_2$.

The property of being equivalent to formulas in CNF and DNF is clearly preserved under equivalence. If $F \equiv G$, then, by our induction hypothesis, we can just take $F_1 = G_1$ and $F_2 = G_2$. It therefore suffices to verify only two more cases corresponding to $\neg$ and $\wedge$.

Suppose first that F has the form $\neg G$. Then $F \equiv \neg G_1 \equiv \neg G_2$. Since G_1 is in CNF, $\neg G_1$ is equivalent to a formula G_3 in DNF by Lemma 1.56. Likewise, $\neg G_2$ is equivalent to a formula G_4 in CNF. So we can take $F_1 = G_4$ and $F_2 = G_3$.

Now suppose F has the form $G \wedge H$. Then $F \equiv G_1 \wedge H_1$ by substitution (Theorem 1.49). Since G_1 and H_1 are both in CNF, so is their conjunction.

It remains to be shown that $F = G \wedge H$ is equivalent to a formula in DNF. Again using Theorem 1.49, $F \equiv G_2 \wedge H_2$. Since each of these formulas is in DNF, they can be written as follows:

$$G_2 = \bigvee_i M_i \text{ and } H_2 = \bigvee_j N_j$$

where each M_i and N_i is a conjunction of literals. We then have

$$F \equiv \left(\bigvee_i M_i \right) \wedge \left(\bigvee_j N_j \right).$$

Using the second equivalence of Proposition 1.48, we have

$$F \equiv \bigvee_i \left(\bigvee_j (M_i \wedge N_j) \right)$$

which is a disjunction of conjunctions of literals as was required. □

Given a formula F, the previous theorem guarantees the existence of a formula in DNF that is equivalent to F. Suppose we want to find such a formula. One way to do this is to compute a truth table for F. For example, suppose F has the following truth table.

A	B	F
0	0	1
0	1	0
1	0	1
1	1	0

Then F is true under assignment $\mathcal{A}$ if and only if $\mathcal{A}$ corresponds to row 1 or 3 of the table. This leads to a formula in DNF. F is true if and only if either A and B are both false (row 1) OR A is true and B is false (row 3). So F is equivalent to $(\neg A \wedge \neg B) \vee (A \wedge \neg B)$, which is in DNF.

Likewise, by considering the rows in which F is false, we can find an equivalent formula in CNF. F is true if and only if we are not in row 2 AND we are not in row 4. That is, F is true if and only if A or $\neg B$ holds (NOT row 2) AND $\neg A$ or $\neg B$ holds (NOT row 4). So F is equivalent to $(A \vee \neg B) \wedge (\neg A \vee \neg B)$ which is in CNF.

This actually provides an alternative proof of Theorem 1.57. Given any formula F, we can use a truth table to find equivalent formulas in CNF and DNF. An alternative way to find a formula in CNF equivalent to F is provided by the following algorithm. This algorithm is often, but not always, more efficient than computing a truth table.

CNF Algorithm

Step 1: Replace all subformulas of the form $F \to G$ with $(\neg F \vee G)$ and all subformulas of the form $F \leftrightarrow G$ with $(\neg F \vee G) \wedge (\neg G \vee F)$. When there are no occurrences of $\to$ or $\leftrightarrow$, proceed to Step 2.

Step 2: Get rid of all double negations and apply DeMorgan's rules wherever possible. That is, replace all subformulas of the form

$$\neg\neg G \text{ with } G,$$
$$\neg(G \wedge H) \text{ with } (\neg G \vee \neg H), \text{ and}$$
$$\neg(G \vee H) \text{ with } (\neg G \wedge \neg H).$$

When there are no subformulas having these forms, proceed to Step 3.

Step 3: Apply the distributivity rule for $\vee$ wherever possible. That is, replace all subformulas of the form

$$(G \vee (H \wedge K)) \text{ or } ((H \wedge K) \vee G) \text{ with } ((G \vee H) \wedge (G \vee K)).$$

If we rid our formula of these subformulas, then we are left with a formula in CNF. If we change Step 3 to distributivity for $\wedge$, then we would get a formula in DNF.

Example 1.58 We demonstrate the CNF algorithm with

$$F = (A \vee B) \to (\neg B \wedge A).$$

In Step 1, we get rid of $\to$, rewriting the formula as

$$\neg(A \vee B) \vee (\neg B \wedge A).$$

In Step 2, we apply DeMorgan's rule to obtain

$$(\neg A \wedge \neg B) \vee (\neg B \wedge A)$$

Proceeding to Step 3, we see that the formula in Step 2 is in DNF. In particular it has the form $(G \vee (H \wedge K))$ (taking $G = (\neg A \wedge \neg B)$). By distributivity, we get

$$((\neg A \wedge \neg B) \vee \neg B) \wedge ((\neg A \wedge \neg B) \vee A).$$

We still have two $\vee$'s that need to be distributed:

$$(\neg A \vee \neg B) \wedge (\neg B \vee \neg B) \wedge (\neg A \vee A) \wedge (\neg B \vee A).$$

Now there are no subformulas of the form $(G \vee (H \wedge K))$ or $((H \wedge K) \vee G)$ and so we are done with Step 3. We see that we have a formula in CNF as was promised. This formula is not written in the best form. Since $(\neg A \vee A)$ is a tautology, the above formula is equivalent to $(\neg A \vee \neg B) \wedge (\neg B) \wedge (\neg B \vee A)$ which is equivalent to $(A \vee \neg B) \wedge (\neg A \vee \neg B)$. Note that this is the same formula we obtained from the truth table following the proof of Theorem 1.57.

Inspecting the CNF algorithm, we see that Theorem 1.57 can be strengthened. This theorem states that for any formula F there exist formulas F_1 in CNF and F_2 in DNF that are equivalent to F. We now claim that F_1

and F_2 are provably equivalent to F. To see this, consider the algorithm step-by-step. In each step we replace certain subformulas with equivalent formulas. In each case we can formally prove the equivalence. For convenience, we use the notation $F \dashv\vdash G$ to abbreviate "F and G are provably equivalent."

Step 1:

$F \rightarrow G \dashv\vdash (\neg F \vee G)$ by $\rightarrow$-Definition

$F \leftrightarrow G \dashv\vdash (\neg F \vee G) \wedge (\neg G \vee F)$ by $\leftrightarrow$-Definition and $\rightarrow$-Definition.

Step 2:

$\neg\neg G \dashv\vdash G$ by Double negation and Example 1.43.

$\neg(G \wedge H) \dashv\vdash (\neg G \vee \neg H)$ by Proposition 1.44 (DeMorgan's rules).

$\neg(G \vee H) \dashv\vdash (\neg G \wedge \neg H)$ by Proposition 1.44 (DeMorgan's rules).

Step 3:

$(G \vee (H \wedge K)) \dashv\vdash ((G \vee H) \wedge (G \vee K))$ by Proposition 1.46($\vee$-Distributivity).

$((H \wedge K) \vee G) \dashv\vdash ((G \vee H) \wedge (G \vee K))$ by $\vee$-Symmetry and Proposition 1.46.

By Theorem 1.50, the result F_1 of this algorithm is provably equivalent to F. Likewise, F_2 and F are provably equivalent. We record this strengthening of Theorem 1.57 as follows.

Proposition 1.59 For every formula F there exist formulas F_1 in CNF and F_2 in DNF such that F, F_1, and F_2 are provably equivalent.

1.7 Horn formulas

A Horn formula is a particularly nice type of formula in CNF. There is a quick method for determining whether or not a Horn formula is satisfiable. We discuss both this method and what is meant by "quick."

Definition 1.60 A formula F is a *Horn formula* if it is in CNF and every disjunction contains at most one positive literal.

Clearly, the conjunction of two Horn formulas is again a Horn formula. This is not true for disjunctions.

Example 1.61 The formula $A \wedge (\neg A \vee \neg B \vee C) \wedge (\neg B \vee D) \wedge (\neg C \vee \neg D)$ is a Horn formula. The formula $A \vee B$ is not a Horn formula.

A *basic Horn formula* is a Horn formula that does not use $\wedge$. For example, $(\neg A \vee \neg B \vee C)$, A, and $(\neg B \vee \neg D)$ are basic Horn formulas. Every Horn formula is a conjunction of basic Horn formulas.

There are three types of basic Horn formulas: those that contain no positive literal (such as $(\neg B \vee \neg D)$), those that contain no negative literals (such as A), and those that contain both a positive literal and negative literals (such as $(\neg A \vee \neg B \vee C)$). If a basic Horn formula contains both positive and negative literals, then it can be written as an implication involving only positive literals. For example, $(\neg A \vee \neg B \vee C)$ is equivalent to $(A \wedge B) \rightarrow C$. If a basic Horn formula contains no positive literal, then it can be written as an implication involving a contradiction. For example, if $\bot$ is a contradiction, then $(\neg B \vee \neg D)$ is equivalent to $(B \wedge D) \rightarrow \bot$. Otherwise, if a basic Horn formula contains no negative literals, then it is an atomic formula. We can again write this as an implication if we wish. The atomic formula A is equivalent to $T \rightarrow A$, where T is a tautology. In this way every basic Horn formula can be written as an implication and every Horn formula can be written as a conjunction of implications.

Example 1.62 The Horn formula in Example 1.61 can be written as follows:

$$(T \rightarrow A) \wedge ((A \wedge B) \rightarrow C) \wedge (B \rightarrow D) \wedge ((C \wedge D) \rightarrow \bot).$$

Suppose we are given a Horn formula H and want to decide whether or not it is satisfiable. We refer to this decision problem as the *Horn satisfiability problem*. Unlike the other decision problems we have seen, there is an efficient algorithm for resolving the Horn satisfiability problem. There are three steps in this algorithm corresponding to the three types of basic Horn formulas. We assume that the Horn formula has been given as a conjunction of implications.

The Horn algorithm

Given a Horn formula H written as a conjunction of implications, list the atomic formulas occuring in H.

Step 1: Mark each atomic formula A in the list that is in a subformula of the form $(T \rightarrow A)$.

Step 2: If there is a subformula of the form $(A_1 \wedge A_2 \wedge \cdots \wedge A_m) \rightarrow C$ where each A_i has been marked and C has not been marked, then mark C. Repeat this step until there are no subformulas of this form and then proceed to step 3.

Step 3: Consider the subformulas of the form $(A_1 \wedge A_2 \wedge \ldots \wedge A_m) \rightarrow \bot$. If there exists such a subformula where each A_i has been marked, then conclude "No, H is not satisfiable." Otherwise, conclude "Yes, H is satisfiable."

Example 1.63 We demonstrate the Horn algorithm. Let H be the formula

$$(T \to A) \land (C \to D) \land ((A \land B) \to C) \land ((C \land D) \to \bot) \land (T \to B).$$

The atomic subformulas of H are A, B, C, and D.

In Step 1 of the algorithm, since H has subformulas $(T \to A)$ and $(T \to B)$ we mark both A and B.

In Step 2, since H has subformula $(A \land B) \to C$, we mark C. Now that C has been marked, we must also mark D because of the subformula $(C \to D)$.

In Step 3, since H has subformula $(C \land D) \to \bot$, the algorithm concludes "No, H is not satisfiable."

Note that for the Horn formula in Example 1.62, the Horn algorithm yields a different conclusion.

We want to show that, for any given Horn formula, the Horn algorithm works quickly. First we show that it works.

Proposition 1.64 The Horn algorithm concludes "Yes, H is satisfiable" if and only if H is satisfiable.

Proof Let $\mathcal{S} = \{C_1, C_2, \ldots, C_n\}$ be the set of atomic formulas occuring in H. After concluding the algorithm, some of these atomic formulas have been marked.

Suppose H is satisfiable. Then there exists an assignment $\mathcal{A}$ of $\mathcal{S}$ such that $\mathcal{A} \models H$. For each basic Horn subformula B of H, $\mathcal{A}(B) = 1$. If B has the form $(T \to C_i)$, then $\mathcal{A}(C_i) = 1$. If B has the form $(C_1 \land C_2 \land \cdots \land C_m) \to D$ where each $\mathcal{A}(C_i) = 1$, then $\mathcal{A}(D)$ also equals 1. It follows that $\mathcal{A}(C_i) = 1$ for each C_i that has been marked.

Suppose for a contradiction that the algorithm concludes "No, H is not satisfiable." This only happens if there exists a subformula B of the form $(A_1 \land A_2 \land \cdots \land A_m) \to \bot$ where each A_i has been marked. Since each A_i has been marked, $\mathcal{A}(A_i) = 1$ for each A_i. By the semantics of $\to$ (Table 1.4), we have $\mathcal{A}(B) = 0$ which is a contradiction. So if H is satisfiable, then the algorithm concludes "Yes, H is satisfiable."

Conversely, suppose that the algorithm concludes "Yes, H is satisfiable." Let $\mathcal{A}_0$ be the assignment of $\mathcal{S}$ defined by $\mathcal{A}_0(C_i) = 1$ if and only if C_i is marked. We claim that $\mathcal{A}_0 \models H$. It suffices to show that $\mathcal{A}_0$ models each basic Horn subformula of H.

Let B be a basic Horn formula that is a subformula of H. If B has the form $(T \to A)$, then A is marked in Step 1 of the algorithm and so $\mathcal{A}_0(B) = 1$. Otherwise B has the form $(A_1 \land A_2 \land \cdots \land A_n) \to G$ where G is either an atomic formula or a contradiction $\bot$. If $\mathcal{A}_0(A_i) = 0$ for some i, then $\mathcal{A}_0(B) = 1$. So assume that $\mathcal{A}_0$ models each A_i. Then each A_i has been marked. Since the algorithm concluded "Yes," G is not $\bot$. So G is an atomic formula. Since each

A_i is marked, G is also marked (Step 2 of the algorithm). Since $\mathcal{A}_0(G) = 1$, we have $\mathcal{A}_0(B) = 1$. □

So the Horn algorithm works. Given any Horn formula H, the algorithm correctly determines whether or not H is satisfiable. We now consider the following question. How many steps does it take the Horn algorithm to reach a conclusion? The answer depends on the length of the input H. Suppose that the formula H is a string of n symbols, where n is some large natural number. We claim that the Horn algorithm concludes in fewer than n^2 steps.

To verify this claim, we count the number of steps in the Horn algorithm. But what exactly is meant by a "step?" Looking at the algorithm, we see that there are three steps named Step 1, Step 2, and Step 3. This is not what is meant. We may have to repeat Step 2 more than once in which case it will take more than three steps to reach a "yes" or "no" answer. We precisely define what constitutes a "step of an algorithm" in Chapter 7. For the time being, let us count the number of times we must read the input H.

First we read the formula H symbol-by-symbol from left to right and list all of its atomic subformulas. Since H contains n symbols, there are at most n atomic formulas in our list. Then, in Step 1, we read through H again, this time looking for any occurences of the tautology T. We mark the appropriate atomic formulas. In Step 2, we are in search of subformulas of the form $(A_1 \wedge A_2 \wedge \cdots \wedge A_m) \rightarrow C$ where each A_i has been marked. If we find such a subformula where C has not been marked, then we mark C. Having marked a new atomic formula, we may have created new subformulas of the form $(A_1 \wedge A_2 \wedge \cdots \wedge A_m) \rightarrow C$ where each A_i has been marked. Each time we mark a formula in Step 2, we must go back and read H again. Since we can mark at most n atomic formulas, we must repeat Step 2 no more than n times. Finally, in Step 3, we must read H one more time (looking for $\perp$), to reach the conclusion. In all, we must read H at most $1 + 1 + n + 1 = n + 3$ times to arrive at a conclusion. Since $n^2 > n + 3$ for $n > 2$, this verifies our claim.

Definition 1.65 An algorithm is *polynomial-time* if there exists a polynomial $p(x)$ such that given input of size n, the algorithm halts in fewer than $p(n)$ steps.

The class of all decision problems that can be resolved by some polynomial-time algorithm is denoted by **P**.

If an algorithm is not polynomial-time, then by any measure, it is not quick. The previous discussion shows that the Horn algorithm is polynomial-time and so the Horn satisfiability problem is in **P**. In contrast, consider the following decision problems.

Validity problem: Given formula F, is F valid?
Satisfiability problem: Given formula F, is F satisfiable?

Consequence problem: Given formulas F and G, is G a consequence of F?
Equivalence problem: Given formulas F and G, are F and G equivalent?

In some sense, these four problems are really the same. Any algorithm that works for one of these problems also works for all of these problems. If we had an algorithm for the Validity problem, for example, then we could use it to resolve the Satisfiability problem since F is satisfiable if and only if $\neg F$ is not valid. Similarly, any algorithm for the Satisfiability problem can be used for the Consequence problem since G is a consequence of F if and only if $\neg(F \rightarrow G)$ is not satisfiable. Clearly, any algorithm for the Consequence problem can be used (twice) to resolve the Equivalence problem. Finally, given an algorithm that decides the Equivalence problem, we can check whether F is equivalent to a known tautology T to resolve the Validity problem. In particular, if one of these four problems is in **P** then all four are.

Truth tables provide an algorithm for solving each of these problems. For the Satisfiability problem, we first compute a truth table for F and then check to see if its truth value is ever one. This algorithm certainly works, but how many steps does it take? Computing the truth table is not just one step. Again, we count how many times we are required to read the input F. If F has n atomic formulas, then the truth table for F has 2^n rows. We must refer to F to compute each of these rows. So we must read the input at least 2^n times. This is exponential and not a polynomial. Given any polynomial $p(x)$, 2^n is larger than $p(n)$ for sufficiently big values of n. So this algorithm is not polynomial-time.

It is not known whether the Satisfiability problem (and the other three decision problems) is in **P**. We do not know of a polynomial-time algorithm for satisfiability, but this does not mean one does not exist. If someone could find such an algorithm, or prove that no such algorithm exists, then it would answer one of the most famous unsolved questions of mathematics: the **P** = **NP** question. We will define **NP** and discuss this problem in Chapter 7. For now, we merely point out that we do not present an efficient algorithm for the Satisfiability problem and such an algorithm probably does not exist.

We do, however, present an algorithm that is an alternative to truth tables for the Satisfiability problem. Formal proofs avoid truth tables, but do not always resolve this decision problem. Given a formula F, we can use formal proofs to show that F is unsatisfiable (by demonstrating that $\emptyset \vdash \neg F$), but we cannot show that F is satisfiable. Likewise, formal proofs can establish that a formula is valid or that one formula is a consequence of another, but they cannot show a formula to be not valid or not a consequence of another. If we find a formal proof for $\{F\} \vdash G$ then we can rightly conclude "yes, G is a consequence of F." But if G is not a consequence of F, then we will forever search in vain for a proof and never reach a conclusion. In the next section we present resolution, a refinement of formal proofs that does provide an algorithm (although not polynomial-time) for these decision problems.

1.8 Resolution

Resolution is a system of formal proof that involves a minimal number of rules. One of the rules is a variation of the *cut rule*. This rule states that from the formulas $(F \rightarrow G)$ and $(G \rightarrow H)$, we can deduce the formula $(F \rightarrow H)$. Another rule is a variation of the *Substitution rule* stated as follows.

> Let H be a formula that contains F as a subformula. If $G \equiv F$, then we can deduce H' form H where H' is the formula obtained by replacing some occurrence of F in H with G.

That is, we consider Theorem 1.49 as a rule for deduction. This is really many rules in one, so we are kind of cheating to get few rules. In particular, for any pair of equivalent formulas F and G, we can deduce G from F. It may seem that this defeats one of our purposes: the Equivalence problem. However, the Substitution rule can be relaxed somewhat. The main purpose of this rule is to put the formulas into CNF. The crux of resolution is that, once the formulas are in CNF, we need only two rules to deduce everything. This will provide an algorithm for the Equivalence problem and the other decision problems from the previous section. It also brings us one step closer to proving the Completeness theorem for propositional logic.

1.8.1 Clauses. Suppose F is a formula in CNF. Then F is a conjunction of disjunctions of literals. We refer to a disjunction of literals as a *clause*. For convenience, we write each clause as a set. We regard

$$L_1 \vee L_2 \vee \cdots \vee L_n \text{ as the set } \{L_1, L_2, \ldots, L_n\}.$$

Any formula that is a disjunction of literals uniquely determines such a set. However, the set does not uniquely determine the formula. Recall that two sets are equal if and only if they contain the same elements. Order and repetition do not matter. For example, the formulas $(L_1 \vee L_2)$, $(L_2 \vee L_1)$, and $(L_1 \vee L_2 \vee L_2)$ each give rise to the same set $\{L_1, L_2\}$. Although these formulas are not identical, they are equivalent.

Proposition 1.66 Let C and D be clauses. If C and D are the same when viewed as sets, then $C \equiv D$.

Proof Let $\mathcal{S}$ be the set of literals occuring in C. Both C and D are equivalent to the disjunction of the literals in $\mathcal{S}$. □

If F is in CNF, then F is a conjunction of clauses and we can write F as a set of sets. We regard F as the set $\{C_1, \ldots, C_n\}$ where the C_is are the clauses occuring in F (written as sets). For example, we regard the formula

$(A \vee B \vee \neg C) \wedge (C \vee D) \wedge \neg A \wedge (\neg B \vee \neg D)$, as the following set of four clauses $\{\{A, B, \neg C\}, \{C, D\}, \{\neg A\}, \{\neg B, \neg D\}\}$.

Proposition 1.67 Let F and G be two formulas in CNF. If F and G are the same when viewed as sets, then $F \equiv G$.

Proof Let $\mathcal{C}$ be the set of clauses occuring in F. Both F and G are equivalent to the conjunction of the clauses in $\mathcal{C}$. This proposition then follows from Proposition 1.66. □

Throughout this section, we regard any formula in CNF as both a formula and as a set of clauses. If F and G are formulas in CNF, then their conjunction may be written either as the formula $F \wedge G$ or as the set $F \cup G$. By the previous proposition, there is no ambiguity in regarding F as both set and formula. However, we stress that viewing formulas as sets only makes sense for formulas in CNF. In particular, there is no nice set theoretic counterpart for disjunction or negation. The formulas $F \vee G$ and $\neg F$ are not in CNF and cannot be viewed as sets of clauses.

1.8.2 Resolvents Given a formula in CNF, resolution repeatedly uses two rules to determine whether or not the formula is satisfiable. One of these rules states that any clause of F can be deduced from F. The other rule involves the *resolvent* of two clauses. We now define this notion.

Definition 1.68 Let C_1 and C_2 be two clauses. Suppose that $A \in C_1$ and $\neg A \in C_2$ for some atomic formula A. Then the clause $R = (C_1 - \{A\}) \cup (C_2 - \{\neg A\})$ is a *resolvent* of C_1 and C_2.

We represent this situation graphically by the following diagram:

$$
\begin{array}{ccc}
C_1 & & C_2 \\
& \searrow \quad \swarrow & \\
& R &
\end{array}
$$

Example 1.69 Let $C_1 = \{A_1, \neg A_2, A_3\}$ and $C_2 = \{A_2, \neg A_3, A_4\}$. Since $A_3 \in C_1$ and $\neg A_3 \in C_2$ we can find a resolvent.

$$
\begin{array}{ccc}
\{A_1, \neg A_2, A_3\} & & \{A_2, \neg A_3, A_4\} \\
& \searrow \quad \swarrow & \\
& \{A_1, A_2, \neg A_2, A_4\}. &
\end{array}
$$

Propositional logic 39

Example 1.70 The resolvent of two clauses is not necessarily unique. In the previous example, since $\neg A_2 \in C_1$ and $A_2 \in C_2$, we also have

$$\{A_1, \neg A_2, A_3\} \qquad\qquad \{A_2, \neg A_3, A_4\}$$
$$\{A_1, A_2, \neg A_3, A_4\}.$$

We now list the three rules for deduction used in resolution.

- Let G be any formula. Let F be the CNF formula resulting from the CNF algorithm when applied to G. Then F can be deduced from G.
- Let F be a formula in CNF. Any clause of F can be deduced from F.
- Let F be a formula in CNF. Any resolvent of two clauses of F can be deduced from F.

Remarkably, these three rules suffice for propositional logic. Resolution is complete. Prior to proving this fact, we must verify that these rules are sound. We show something stronger. We show that each of these rules can be derived using formal proofs. In the first rule, F can be derived from G by Proposition 1.59. If C is a clause of F, then we can derive C from F using $\wedge$-Symmetry and $\wedge$-Elimination.

It remains to be shown that R can be derived from F where R is a resolvent of two clauses of F. Note the similarity between this and the Cut rule. Let C_1 and C_2 be as in Example 1.69. Then C_1 is equivalent to $(\neg A_1 \wedge A_2) \to A_3$ and C_2 is equivalent to $A_3 \to (A_2 \vee A_4)$. The Cut rule states that from these formulas we can derive the formula $(\neg A_1 \wedge A_2) \to (A_2 \vee A_4)$. This formula is equivalent to the resolvent obtained in Example 1.69.

Proposition 1.71 Let C_1 and C_2 be clauses and let R be a resolvent of C_1 and C_2. Then $\{C_1, C_2\} \vdash R$.

Proof Since C_1 and C_2 have a resolvent, there must exist an atomic formula A such that A is in one of these clauses and $\neg A$ is in the other. With no loss of generality, we may assume that A is in C_1 and $\neg A$ is in C_2. So C_1 is equivalent to $(A \vee F)$ for some clause F and C_2 is equivalent to $(\neg A \vee G)$ for some clause G. The formula $(F \vee G)$ is a resolvent of C_1 and C_2. We may assume that R is this resolvent. We provide a formal proof for $\{C_1, C_2\} \vdash R$.

Premise: $\mathcal{F} \vdash (A \vee F)$ and $\mathcal{F} \vdash (\neg A \vee G)$
Conclusion: $\mathcal{F} \vdash (F \vee G)$.

Statement	Justification
1. $\mathcal{F} \vdash (A \vee F)$	Premise
2. $\mathcal{F} \cup \{\neg A\} \vdash (A \vee F)$	Monotonicity applied to 1
3. $\mathcal{F} \cup \{\neg A\} \vdash \neg A$	Assumption
4. $\mathcal{F} \cup \{\neg A\} \vdash F$	∨-Elimination applied to 2 and 3
5. $\mathcal{F} \cup \{\neg A\} \vdash (F \vee G)$	∨-Introduction applied to 4
6. $\mathcal{F} \vdash (\neg A \vee G)$	Premise
7. $\mathcal{F} \cup \{\neg\neg A\} \vdash (\neg A \vee G)$	Monotonicity applied to 1
8. $\mathcal{F} \cup \{\neg\neg A\} \vdash \neg\neg A$	Assumption
9. $\mathcal{F} \cup \{\neg\neg A\} \vdash G$	∨-Elimination applied to 7 and 8
10. $\mathcal{F} \cup \{\neg\neg A\} \vdash (G \vee F)$	∨-Introduction applied to 9
11. $\mathcal{F} \cup \{\neg\neg A\} \vdash (F \vee G)$	∨-Symmetry applied to 10
12. $\mathcal{F} \vdash (F \vee G)$	Proof by cases applied to 5 and 11

So anything that can be proved using resolution can be given a formal proof. It then follows from Theorem 1.37 that resolution is sound. In particular, if R is the resolvent of two clauses of a formula F in CNF, then R is a consequence of F. Ostensibly, resolution is a fragment of our formal proof system. As we now show, resolution is just as powerful as formal proofs.

1.8.3 Completeness of resolution. We show that resolution can be used to determine whether or not any given formula is satisfiable. We may assume that the formula is in CNF. Given any formula F in CNF, let $Res^0(F) = \{C | C \text{ is a clause of } F\}$. For each $n > 0$, let $Res^n(F) = Res^{n-1}(F) \cup \{R | R \text{ is a resolvent of two clauses of } Res^{n-1}(F)\}$. Since $Res^0(F) = F$ is a finite set, there are only finitely many clauses that can be derived from F using resolvents. In fact, there are only finitely many clauses that use the same atomic formulas as F. So, eventually, we will find some m so that $Res^m(F) = Res^{m+1}(F)$. Let $Res^*(F)$ denote such $Res^m(F)$. This is the set of all clauses that can be derived from F using resolvents. Viewing it as a formula, $Res^*(F)$ is the conjunction of all consequences of F that can be derived by resolvents.

Proposition 1.72 Let F be a formula in CNF. If $\emptyset \in Res^*(F)$, then F is unsatisfiable.

Propositional logic

Proof If $\emptyset \in Res^*(F)$, then $\emptyset \in Res^n(F)$ for some n. Since $\emptyset \notin Res^0(F)$ ($\emptyset$ is not a clause) there must be some m such $\emptyset \notin Res^m(F)$ and $\emptyset \in Res^{m+1}(F)$ in which case $\emptyset$ is the resolvent of two clauses of $Res^m(F)$. But $\emptyset$ can only be obtained as the resolvent of $\{A\}$ and $\{\neg A\}$ for atomic A. Both $\{A\}$ and $\{\neg A\}$ must be in $Res^m(F)$. By the previous proposition, both A and $\neg A$ are consequences of F. It follows that $A \wedge \neg A$ is a consequence of F and F is unsatisfiable. $\square$

Example 1.73 Let F be the formula

$$\{\{A, B, \neg C\}, \{\neg A\}, \{A, B, C\}, \{A, \neg B\}\}$$

We show that F is unsatisfiable using resolution.

Let C_1, C_2, C_3, and C_4 denote the four clauses of F in the order given above.

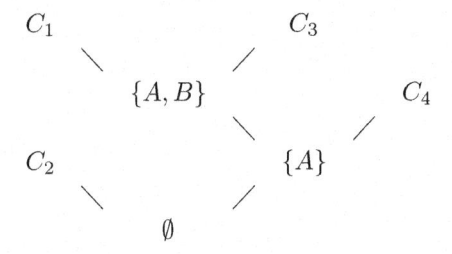

```
       C₁           C₃
         ╲        ╱
          {A, B}         C₄
               ╲      ╱
      C₂        {A}
         ╲    ╱
           ∅
```

We see that $\{A, B\} \in Res(F)$, $\{A\} \in Res^2(F)$, and $\emptyset \in Res^3(F)$. By Proposition 1.72, F is unsatisfiable. We can arrange this as a two-column proof as follows.

Consequence of F	Justification
C_1	Clause of F
C_3	Clause of F
$\{A, B\}$	Resolvent of C_1 and C_2
C_4	Clause in F
$\{A\}$	Resolvent of $\{A, B\}$ and C_4
C_2	Clause in F
$\emptyset$	Resolvent of $\{A\}$ and C_2

We now consider the converse of Proposition 1.72. Let F be a formula in CNF. If F is unsatisfiable, then must $\emptyset$ be in $Res^*(F)$? We show that the answer is "yes." Resolution is all we need to show unsatisfiability. This is not immediately apparent. After all, for the "Justification" column of these proofs, we have only two options. Either a clause is given, or it is a resolvent of two previously derived clauses. It may seem that this method of proof is too restrictive. We prove that it is not.

Proposition 1.74 Let F be a formula in CNF. If F is unsatisfiable, then $\emptyset \in Res^*(F)$.

Proof Let $F = \{C_1, \ldots, C_k\}$. We assume that none of the C_is is a tautology (otherwise we just throw away these clauses and show that $\emptyset$ can be derived from what remains). We will prove this proposition by induction on the number n of atomic formulas that occur in F.

Let $n = 1$. Let A be the only atomic formula occurring in F. Then there are only three possible clauses in F. Each C_i is either $\{A\}$, $\{\neg A\}$, or $\{A, \neg A\}$. The last clause is a tautology, and so, by our previous assumption, it is not a clause of F. So the only clauses in F are $\{A\}$ and $\{\neg A\}$. There are three possibilities, $F = \{\{A\}\}$, $F = \{\{\neg A\}\}$, or $F = \{\{A\}, \{\neg A\}\}$. The first two of these are satisfiable. So F must be $\{\{A\}, \{\neg A\}\}$. Clearly, $\emptyset \in Res^*(F)$.

Now suppose F has atomic subformulas $A_1, \ldots, A_{n+1}$. Suppose further that $\emptyset \in Res^*(G)$ for any unsatisfiable formula G that uses only the atomic formulas $A_1, \ldots, A_n$.

We define some new formulas.

Let $\tilde{F}_0$ be the conjunction of all C_i in F that do not contain $\neg A_{n+1}$.
Let $\tilde{F}_1$ be the conjunction of all C_i in F that do not contain A_{n+1}.
These are CNF formulas. We claim that, viewing these as sets,

$$\tilde{F}_0 \cup \tilde{F}_1 = F.$$

For suppose that there is some clause C_i of F that is not in $\tilde{F}_0 \cup \tilde{F}_1$. Then C_i must contain both A_{n+1} and $\neg A_{n+1}$. But then C_i is a tautology, contrary to our previous assumption. So $\tilde{F}_0 \cup \tilde{F}_1$ and F contain the same clauses.

Let $F_0 = \{C_i - \{A_{n+1}\} | C_i \in \tilde{F}_0\}$.
Let $F_1 = \{C_i - \{\neg A_{n+1}\} | C_i \in \tilde{F}_1\}$.

That is, F_0 is formed by throwing A_{n+1} out of each clause of $\tilde{F}_0$ in which it occurs. Likewise, F_1 is obtained by throwing $\neg A_{n+1}$ out of each clause of $\tilde{F}_1$.

We claim that if we replace A_{n+1} in F with a contradiction, then the resulting formula is equivalent to F_0. And if we replace A_{n+1} in F with a tautology, then the resulting formula is equivalent to F_1. We give an example to illustrate this, but leave the verification of this fact to the reader.

Example 1.75 Suppose $n = 2$ so that A_{n+1} is A_3.
Let $F = \{\{A_1, A_3\}, \{A_2\}, \{\neg A_1, \neg A_2, A_3\}, \{\neg A_2, \neg A_3\}\}$.
Then $\tilde{F}_0 = \{\{A_1, A_3\}, \{A_2\}, \{\neg A_1, \neg A_2, A_3\}\}$
and $\tilde{F}_1 = \{\{A_2\}, \{\neg A_2, \neg A_3\}\}$.
So $F_0 = \{\{A_1\}, \{A_2\}, \{\neg A_1, \neg A_2\}\}$

Propositional logic 43

and $F_1 = \{\{A_2\}, \{\neg A_2\}\}$.

Now F is the formula $(A_1 \vee A_3) \wedge (A_2) \wedge (\neg A_1 \vee \neg A_2 \vee A_3) \wedge (\neg A_2 \vee \neg A_3)$.
If we know A_3 has truth value 0, then this becomes
$(A_1 \vee 0) \wedge (A_2) \wedge (\neg A_1 \vee \neg A_2 \vee 0) \wedge (1)$ which is equivalent to F_0.
If we know that A_3 has truth value 1, then F reduces to
$(1) \wedge (A_2) \wedge (1) \wedge (\neg A_2 \vee 0)$ which is equivalent to F_1.

Since A_{n+1} must either have truth value 0 or 1, it follows that $F \equiv F_0 \vee F_1$. Since F is unsatisfiable, F_0 and F_1 are each unsatisfiable. The formulas F_0 and F_1 only use the atomic formulas $A_1, \ldots, A_n$. By our induction hypothesis, $\emptyset \in Res^*(F_0)$ and $\emptyset \in Res^*(F_1)$. (Note that $\emptyset$ can easily be derived from both F_0 and F_1 in our example.)

Now F_0 was formed from $\tilde{F}_0$ by throwing A_{n+1} out of each clause. Since we can derive $\emptyset$ from F_0, we can derive either $\emptyset$ or $\{A_{n+1}\}$ from $\tilde{F}_0$ (by reinstating $\{A_{n+1}\}$ in each clause of F_0). Likewise we can derive either $\emptyset$ or $\{\neg A_{n+1}\}$ from $\tilde{F}_1$. If we can derive $\{A_{n+1}\}$ form F_0 and $\{\neg A_{n+1}\}$ from F_1, then we can derive $\emptyset$ from $\tilde{F}_0 \cup \tilde{F}_1$. Since $F = \tilde{F}_0 \cup \tilde{F}_1$, we conclude that $\emptyset \in Res^*(F)$. □

This yields an algorithm for the Satisfiability problem. Given any formula G, we first find a formula F in CNF that is equivalent to G (using the CNF algorithm). We then compute the finite set $Res^*(F)$. If $\emptyset \in Res^*(F)$, then the algorithm concludes "No, G is not satisfiable." Otherwise, it concludes "Yes, G is satisfiable." By Propositions 1.72 and 1.74, this algorithm works. This algorithm is not necessarily quick. As we previously mentioned, there is no known polynomial-time algorithm for this decision problem. However, in certain instances, this algorithm can reach a quick conclusion. If F is unsatisfiable, then we do not necessarily have to compute all of $Res^*(F)$. As soon as $\emptyset$ makes an appearance, we know that it is not satisfiable. If F is satisfiable, on the other hand, then truth tables can reach a quick conclusion. We only need to compute the truth table until we find a truth value of 1.

We summarize the main results of this section in the following theorem. This theorem is a finite version of the Completeness theorem for propositional logic.

Theorem 1.76 Let F and G be formulas of propositional logic. Let H be the CNF formula obtained by applying the CNF algorithm to the formula $F \wedge \neg G$. The following are equivalent:

1. $F \models G$
2. $\{F\} \vdash G$
3. $\emptyset \in Res^*(H)$

Proof (2) implies (1) by Theorem 1.37.

(1) implies (3) by Proposition 1.74.
We must show that (3) implies (2). By Proposition 1.59, we have $\{F \wedge \neg G\} \vdash H$.
By $\wedge$-Introduction, $\{F, \neg G\} \vdash F \wedge \neg G$.
It follows that $\{F, \neg G\} \vdash H$.
Since $\emptyset \in Res^*(H)$, there must exist an atomic formula A such that both $\{A\}$ and $\{\neg A\}$ are in $Res^*(H)$. It follows from Proposition 1.71 that both $\{H\} \vdash A$ and $\{H\} \vdash \neg A$. Therefore, both

$$\{F, \neg G\} \vdash A \text{ and } \{F, \neg G\} \vdash \neg A.$$

By proof by contradiction, we have $\{F\} \vdash \neg\neg G$. Finally, $\{F\} \vdash G$ by Double negation. □

1.9 Completeness and compactness

Completeness and compactness are two properties that a logic may or may not possess. We conclude our study of propositional logic by showing that this logic does, in fact, have each of these properties.

A logic is a formal language that has rules for deducing the truth of one statement from that of another. If a sentence G can be deduced from a set of sentences $\mathcal{F}$ using these rules, then we write $\mathcal{F} \vdash G$. The notation $\mathcal{F} \models G$, on the other hand, means that whenever each sentence in $\mathcal{F}$ is true, G is also true. If $\mathcal{F} \vdash G$, then $\mathcal{F} \models G$. The opposite, however, is not necessarily true. Put another way, $\mathcal{F} \models G$ means that $\mathcal{F}$ implies G and $\mathcal{F} \vdash G$ means that we can prove that $\mathcal{F}$ implies G using the rules of the logic. But just because something is true does not mean we can prove it. Perhaps the rules of the logic are too weak to prove everything (or the expressive power of the logic is too strong). If we can prove everything that is true (that is, if $\mathcal{F} \models G$ does imply $\mathcal{F} \vdash G$), then we say that the logic is *complete*.

(**Completeness:**) $\mathcal{F} \models G$ if and only if $\mathcal{F} \vdash G$.

In Section 1.4, we defined the notation $\mathcal{F} \vdash G$ for propositional logic by listing a bunch of rules. However, completeness should be understood not as a statement about these specific rules, but as a statement about the logic itself. Completeness asserts the existence of a list of rules that allows us to deduce every consequence from any set of formulas of the logic. To prove this we need to demonstrate such a list of rules. We show that the rules in Tables 1.5 and 1.6, as well as the rules for resolution, suffice for propositional logic. As we will see in Chapter 9, second-order logic does not have completeness. We cannot give a nice list of rules that allow us to deduce every consequence from any set of second-order sentences.

To prove that propositional logic has completeness, we must pass from finite to infinite sets of formulas. If $\mathcal{F}$ is finite, then $\mathcal{F} \models G$ if and only if $\mathcal{F} \vdash G$ by Theorem 1.76. Suppose now that $\mathcal{F}$ is infinite. If $\mathcal{F}$ is a set of formulas in CNF, then it can be viewed as a set of clauses. The set $Res^n(\mathcal{F})$ is defined as it was for finite sets of clauses. Let $Res^*(\mathcal{F})$ denote the union of all of the sets $Res^n(\mathcal{F})$ (for $n \in \mathbb{N}$). Again, $Res^*(\mathcal{F})$ is the set of all clauses that can be derived from $\mathcal{F}$ using resolution. If $\mathcal{F}$ is infinite, then $Res^*(\mathcal{F})$ is infinite and cannot be viewed as a formula. Such an infinite set of clauses is *satisfiable* if and only if there exists an assignment that models each clause of the set. To prove that propositional logic has completeness, it suffices to prove the following.

Proposition 1.77 Let $\mathcal{F}$ be a set of formulas in CNF. Then $\emptyset \in Res^*(\mathcal{F})$ if and only if $\mathcal{F}$ is unsatisfiable.

For finite $\mathcal{F}$, this is a restatement of Propositions 1.72 and 1.74. Recall the proofs of these two statements. For Proposition 1.74, we assumed that F was unsatisfiable, and we proved that $\emptyset \in Res^*(F)$ by induction on the number of atomic formulas occurring in F. But mathematical induction proves only that something is true for all finite n. So the method we used to prove Proposition 1.74 does not work if $\mathcal{F}$ involves infinitely many atomic formulas.

Consider the other direction of Proposition 1.77. Suppose $\emptyset \in Res^*(\mathcal{F})$. Then $\emptyset \in Res^n(\mathcal{F})$ for some n. That is, we can derive $\emptyset$ from $\mathcal{F}$ in a finite number of steps. Therefore, we can derive $\emptyset$ from some finite subset F of $\mathcal{F}$. By Proposition 1.72, F is unsatisfiable. Since F is a subset of $\mathcal{F}$, $\mathcal{F}$ must be unsatisfiable also.

So one direction of Proposition 1.77 follows from the results of the previous section. We can deduce the infinite case from the finite case by observing that if $\emptyset$ can be derived from $\mathcal{F}$, then it can be derived from some finite subset of $\mathcal{F}$. To prove the other direction of Proposition 1.77 we need an analogous idea. We need to show that if $\mathcal{F}$ is unsatisfiable, then some finite subset of $\mathcal{F}$ is unsatisfiable. This is known as compactness.

Compactness: $\mathcal{F}$ is unsatisfiable if and only if some finite subset of $\mathcal{F}$ is unsatisfiable.

Put another way, compactness says that $\mathcal{F}$ is satisfiable if and only if every finite subset of $\mathcal{F}$ is satisfiable. As with completeness, one direction of compactness always holds. If $\mathcal{F}$ is satisfiable, then every finite subset of $\mathcal{F}$ must be satisfiable also. But just because every finite subset of a set is satisfiable does not necessarily mean that the set itself is satisfiable. Consider, for example, the following set of English sentences.

$F_0 =$ "There are finitely many objects in the universe."
$F_1 =$ "There is at least one object in the universe."
$F_2 =$ "There are at least two objects in the universe."
$F_3 =$ "There are at least three objects in the universe."
...

$F_n =$ "There are at least n objects in the universe."
...

Taken together, these sentences are contradictory. If there are more than n objects for each n, then there cannot possibly be finitely many objects as F_0 asserts. However, if we take only finitely many of the above statements, then there is no problem. Any finite set of these sentences is satisfiable, but the collection as a whole is not. Any logic that can express these sentences does not have compactness.

We prove that propositional logic does have compactness in Theorem 1.79. First, we prove the following lemma. This lemma may not seem relevant at the moment, but it is the key to proving Theorem 1.79.

Lemma 1.78 Let X be an infinite set of finite binary strings. There exists an infinite binary string $\bar{w}$ so that any prefix of $\bar{w}$ is also prefix of infinitely many $\bar{x}$ in X.

Proof A binary string is a sequence on 0s and 1s such as 1011. The strings 1, 10, 101, and 1011 are the prefixes of 1011. We have an infinite set X of such strings of finite length. We want to construct an infinite string $\bar{w}$ of 0s and 1s so that each prefix of $\bar{w}$ is also a prefix of infinitely many strings in X.

We construct $\bar{w}$ step-by-step from left to right. In each step we will do two things. In the nth step, we not only decide what the nth digit of $\bar{w}$ should be, we also delete strings from X that we do not like.

To determine what the first digit of $\bar{w}$ should be, look at the first digits of all the strings in X. Of course, there are infinitely many strings and you cannot look at all these digits at once, but suppose that you are somehow omniscient. There are two possibilities. Either you see infinitely many 1s or you do not. If infinitely many strings in X start with 1, then we let the first digit of $\bar{w}$ be a 1 and we delete all strings in X that begin with a 0 (we are still left with infinitely many). Otherwise, if only finitely many strings in X start 1, we delete these and let the first digit of $\bar{w}$ be a 0.

Now suppose we have determined the first n digits of $\bar{w}$. Suppose too that we have deleted all sequences from X that do not start with these same n digits and are left with an infinite subset X' of X. To determine the $(n+1)$th entry in $\bar{w}$ we look at the $(n+1)$th digits of all the strings in X'. Since X' is infinite, X' must have infinitely many strings of length $n+1$ or greater. So again, there are two possibilities. If infinitely many strings in X' have 1s in the $(n+1)$th place, then we let the $(n+1)$th digit of $\bar{w}$ be 1. Otherwise, we let the $(n+1)$th digit be 0. Either way, we delete all strings from X' that do not share the same first $n+1$ entries as $\bar{w}$. We are still left with an infinite subset of X.

Continuing this procedure, we obtain an infinite sequence $\bar{w}$ so that the first n digits of $\bar{w}$ agrees with the first n digits of infinitely many sequences in X. We have not really given a practical way of constructing $\bar{w}$, but we have proven that such a string exists. □

We are ready now to prove propositional logic has compactness.

Theorem 1.79 (Compactness of propositional logic) A set of sentences of propositional logic is satisfiable if and only if every finite subset is satisfiable.

Proof As we remarked earlier, only one direction of this requires proof. Suppose $\mathcal{F} = \{F_1, F_2, \ldots\}$ is a set of formulas and every finite subset of $\mathcal{F}$ is satisfiable. Let $A_1, A_2, A_3, \ldots$ be a list without repetition of the atomic formulas occurring in F_1 followed by the atomic formulas occurring in F_2 (but not F_1), and so on.

Since every finite subset of $\mathcal{F}$ is satisfiable, for each n there exists an assignment $\mathcal{A}_n$ such that $\mathcal{A}_n \models \bigwedge_{i=1}^{n} F_n$. So each F_i in $\mathcal{F}$ holds under all but finitely many of these assignments. We may assume that $\mathcal{A}_n$ is defined only on the atomic formulas occurring in $F_1, \ldots, F_n$. For each n, the truth values $\mathcal{A}_n$ assigns to $A_1, A_2, \ldots$ forms a finite sequence of 0s and 1s. So $X = \{\mathcal{A}_n | n = 1, 2, \ldots\}$ is an infinite set of finite binary sequences. By the previous lemma, there exists an infinite binary sequence $\bar{w}$ so that every prefix of $\bar{w}$ is a prefix of infinitely many sequences in X.

Define an assignment $\mathcal{A}$ on all the A_ns as follows: let $\mathcal{A}(A_n)$ be the nth digit of $\bar{w}$. We must show that every formula F in $\mathcal{F}$ holds under $\mathcal{A}$. This follows from the fact that F holds under all but finitely many of the assignments in X. Let m be such that F contains no atomic formula past A_m in our list. Then there is an $\mathcal{A}_n$ in X so that $\mathcal{A}_n \models F$ and the first m entries of $\mathcal{A}_n$ are the same as $\mathcal{A}$. It follows that $\mathcal{A}$ also models F. □

Proposition 1.77 follows from compactness. We can now prove that propositional logic has completeness. We could give a proof similar to that of Theorem 1.76 using Proposition 1.77 in place of Propositions 1.72 and 1.74. However, compactness yields a more direct proof.

Theorem 1.80 (Completeness of propositional logic) For any sentence G and set of sentences $\mathcal{F}$, $\mathcal{F} \models G$ if and only if $\mathcal{F} \vdash G$.

Proof By Theorem 1.37, if $\mathcal{F} \vdash G$, then $\mathcal{F} \models G$.

Conversely, suppose that $\mathcal{F} \models G$. Then $\mathcal{F} \cup \{\neg G\}$ is unsatisfiable. By compactness, some finite subset of $\mathcal{F} \cup \{\neg G\}$ is unsatisfiable. So there exists finite $\mathcal{F}_0 \subset \mathcal{F}$ such that $\mathcal{F}_0 \cup \{\neg G\}$ is unsatisfiable and, equivalently, $\mathcal{F}_0 \models G$. Since $\mathcal{F}_0$ is finite, we can apply Theorem 1.76 to get $\mathcal{F}_0 \vdash G$. Finally, $\mathcal{F} \vdash G$ by Monotonicity. □

Exercises

1.1. Show that $\neg$ and $\vee$ can be taken as primitive symbols in propositional logic. That is, show that each of the symbols $\wedge$, $\rightarrow$, and $\leftrightarrow$ can be defined in terms of $\neg$ and $\vee$.

1.2. Show that $\neg$ and $\rightarrow$ can be taken as primitive symbols in propositional logic. That is, show that each of the symbols $\wedge$, $\vee$, and $\leftrightarrow$ can be defined in terms of $\neg$ and $\rightarrow$.

1.3. Find the truth tables for each of the following formulas. State whether each is a tautology, a contradiction, or neither.
   (a) $(\neg A \rightarrow B) \vee ((A \wedge \neg C) \leftrightarrow B)$
   (b) $(A \rightarrow B) \wedge (A \rightarrow \neg B)$
   (c) $(A \rightarrow (B \vee C)) \vee (C \rightarrow \neg A)$
   (d) $((A \rightarrow B) \wedge C) \vee (A \wedge D)$.

1.4. In each of the following, determine whether the two formulas are equivalent.
   (a) $(A \wedge B) \vee C$ and $(A \rightarrow \neg B) \rightarrow C$
   (b) $(((A \rightarrow B) \rightarrow B) \rightarrow B)$ and $(A \rightarrow B)$
   (c) $(((A \rightarrow B) \rightarrow A) \rightarrow A)$ and $(C \rightarrow D) \vee C$
   (d) $A \leftrightarrow ((\neg A \wedge B) \vee (A \wedge \neg B))$ and $\neg B$.

1.5. Show that the following statements are equivalent.
   1. $F \models G$,
   2. $\models F \rightarrow G$,
   3. $F \wedge \neg G$ is unsatisfiable, and
   4. $F \equiv F \wedge G$.

1.6. Show that the following statements are equivalent.
   1. $F \equiv G$,
   2. $\models F \leftrightarrow G$, and
   3. $(F \wedge \neg G) \vee (\neg F \wedge G)$ is unsatisfiable.

Propositional logic

1.7. (a) Find a formula F in CNF which has the following truth table.

A	B	C	F
0	0	0	0
1	0	0	1
0	1	0	1
0	0	1	1
1	1	0	0
1	0	1	0
0	1	1	0
1	1	1	1

(b) Find a formula in DNF having the above truth table.

1.8. Find formulas in CNF equivalent to each of the following.
  (a) $(A \leftrightarrow B) \leftrightarrow C$
  (b) $(A \to (B \lor C)) \lor (C \to \neg A)$
  (c) $(\neg A \land \neg B \land C) \lor (\neg A \land \neg C) \lor (B \land C) \lor A$.

1.9. The Cut rule states that from the formulas $(F \to G)$ and $(G \to H)$ we can derive the formula $(F \to H)$. Verify this rule by giving a formal proof.

1.10. (a) Let $\leftrightarrow$-Symmetry be the following rule:

Premise: $\mathcal{F} \vdash (F \leftrightarrow G)$
Conclusion: $\mathcal{F} \vdash (G \leftrightarrow F)$
Verify this rule by giving a formal proof.

  (b) Give a formal proof demonstrating that $\{(F \leftrightarrow G)\} \vdash (\neg F \leftrightarrow \neg G)$.

1.11. Give formal proofs demonstrating that the formulas $(F \land (F \lor G))$ and $(F \lor (F \land G))$ are provably equivalent.

1.12. If $F \to G$ is a consequence of $\mathcal{F}$, then so is $\neg G \to \neg F$. We refer to this rule as $\to$-Contrapositive. Verify this rule by giving a formal proof.

1.13. Show that $\lor$-Symmetry follows from the other rules of Tables 1.5 and 1.6.

1.14. Show that $\to$-Elimination follows from the other rules of Tables 1.5 and 1.6.

1.15. Show that Double negation follows from Assumption, Monotonicity, and Proof by cases.

1.16. Suppose that we remove from Table 1.5 the following four rules:
∨-Elimination, ∨-Symmetry, →-Introduction, and →-Elimination and replace these with

DeMorgan's rules, ∨-Distributivity, the Cut rule (from Exercise 1.9), and the converse of Double negation (if $\mathcal{F} \vdash \neg\neg F$ then $\mathcal{F} \vdash F$).

Show that the resulting set of rules is complete.

1.17. Use resolution to verify each of the following statements:
(a) $\neg A$ is a consequence of $(A \to B) \land (A \to \neg B)$
(b) $(\neg A \land \neg B \land C) \lor (\neg A \land \neg C) \lor (B \land C) \lor A$ is a tautology
(c) $((A \to B) \land (A \to \neg B)) \to \neg A$ is a tautology.

1.18. For each formula in Exercise 1.3 find an equivalent formula in CNF.

1.19. For each formula in Exercise 1.3, verify your answer to that problem by using resolution.

1.20. Determine whether or not the following Horn formulas are satisfiable. If it is satisfiable, find an assignment that models the formula.
(a) $(T \to A_1) \land (T \to A_2) \land (A_1 \land A_2 \land A_3 \to A_4) \land (A_1 \land A_2 \land A_4 \to A_5)$
$\land (A_1 \land A_2 \land A_3 \land A_4 \to A_6) \land (A_5 \land A_6 \to A_7) \land (A_2 \to A_3) \land (A_7 \to \bot)$
(b) $(T \to A_1) \land (T \to A_2) \land (A_1 \land A_2 \land A_4 \to A_3) \land (A_1 \land A_5 \land A_6) \land (A_2 \land A_7 \to A_5)$
$\land (A_1 \land A_3 \land A_5 \to A_7) \land (A_2 \to A_4) \land (A_4 \to A_8) \land (A_2 \land A_3 \land A_4 \to A_9)$
$\land (A_3 \land A_9 \to A_6) \land (A_6 \land A_7 \to A_8) \land (A_7 \land A_8 \land A_9 \to \bot)$

1.21. Consider the following formula in DNF.

$$(A_1 \land B_1) \lor (A_2 \land B_2) \lor \cdots \lor (A_n \land B_n)$$

Given this formula as input, how many steps will it take the CNF algorithm to halt and output a formula in CNF? Is this algorithm polynomial-time?

1.22. Complete the proof of Theorem 1.37.

1.23. Complete the proof of Proposition 1.44.

1.24. Prove Proposition 1.45 by providing two formal proofs.

1.25. Prove Proposition 1.46.

1.26. What is wrong with the following claim? Why is the given "fake proof" not a proof?

Claim: $(A \to B) \lor C$ is not a subformula of any formula.

Proof [Fake proof] Let F be any formula. We show that $(A \to B) \lor C$ is not a subformula of F by induction on the complexity of F.

If F is atomic, then clearly $(A \to B) \lor C$ is not a subformula.

Let F_1 and F_2 be two formulas. Our induction hypothesis is that neither F_1 nor F_2 has $(A \to B) \lor C$ as a subformula.

Propositional logic 51

Suppose F is $\neg F_1$. If $(A \to B) \vee C$ is a subformula of F, then either $(A \to B) \vee C$ is a subformula of F_1 or is F itself. It is not a subformula of F_1 by our induction hypothesis. Moreover, since $(A \to B) \vee C$ does not contain the symbol $\neg$, it cannot be F.

Suppose F is $F_1 \wedge F_2$. If $(A \to B) \vee C$ is a subformula of F, then since $(A \to B) \vee C$ does not contain the symbol $\wedge$, it must be a subformula of either F_1 or of F_2. But by our induction hypothesis, this is not the case.

It follows that $(A \to B) \vee C$ is not a subformula of any formula. □

1.27. Prove Proposition 1.48 by mathematical induction. That is, given formulas $\{F_1, \ldots, F_n\}$ and $\{G_1, \ldots, G_m\}$, prove each of the following by induction on n.
   (a) $(\bigwedge_{i=1}^n F_i) \vee (\bigwedge_{j=1}^m G_j) \equiv \bigwedge_{i=1}^n (\bigwedge_{j=1}^m (F_i \vee G_j))$
   (b) $(\bigvee_{i=1}^n F_i) \wedge (\bigvee_{j=1}^m G_j) \equiv \bigvee_{i=1}^n (\bigvee_{j=1}^m (F_i \wedge G_j))$.

1.28. Prove Theorem 1.50 by induction on the complexity of H.

1.29. Let $\mathcal{F}$ and $\mathcal{G}$ be sets of formulas. We say that $\mathcal{F}$ is *equivalent* to $\mathcal{G}$, denoted $\mathcal{F} \equiv \mathcal{G}$, if for every assignment $\mathcal{A}$, $\mathcal{A} \models \mathcal{F}$ if and only if $\mathcal{A} \models \mathcal{G}$.
   (a) Show that the following is true:

   For any $\mathcal{F}$ and $\mathcal{G}$, $\mathcal{F} \equiv \mathcal{G}$ if and only if both:

   $\mathcal{F} \vdash G$ for each $G \in \mathcal{G}$ and

   $\mathcal{G} \vdash F$ for each $F \in \mathcal{F}$.

   (b) Demonstrate that the following is *not* true:

   For any $\mathcal{F}$ and $\mathcal{G}$, $\mathcal{F} \equiv \mathcal{G}$ if and only if both:

   for each $G \in \mathcal{G}$ there exists $F \in \mathcal{F}$ such that $G \models F$, and

   for each $F \in \mathcal{F}$ there exists $G \in \mathcal{G}$ such that $F \models G$.

1.30. If a contradiction can be derived from a set of sentences, then the set of sentences is said to be *inconsistent*. Otherwise, the set of sentences is *consistent*. Let $\mathcal{F}$ be a set of sentences. Show that $\mathcal{F}$ is consistent if and only if it is satisfiable.

1.31. Suppose that $\mathcal{F}$ is an inconsistent set of sentences (as defined in Exercise 1.30). For each $G \in \mathcal{F}$, let $\mathcal{F}_G$ be the set obtained by removing G from $\mathcal{F}$.
   (a) Prove that for any $G \in \mathcal{F}$, $\mathcal{F}_G \vdash \neg G$ by using the result of Exercise 1.30.
   (b) Prove that for any $G \in \mathcal{F}$, $\mathcal{F}_G \vdash \neg G$ by sketching a formal proof.

1.32. A set of sentences $\mathcal{F}$ is said to be *closed under conjunction* if for any F and G in $\mathcal{F}$, $F \wedge G$ is also in $\mathcal{F}$. Suppose that $\mathcal{F}$ is closed under conjunction and is inconsistent (as defined in Exercise 1.30). Prove that for any $G \in \mathcal{F}$ there exists $F \in \mathcal{F}$ such that $\{F\} \vdash \neg G$.

1.33. Call a set of sentences *minimal unsatisfiable* if it is unsatisfiable, but every proper subset is satisfiable.
  (a) Show that there exist minimal unsatisfiable sets of sentences of size n for any n.
  (b) Show that any unsatisfiable set of sentences has a minimal unsatisfiable subset.

1.34. (Craig's interpolation theorem) Suppose $\models (F \to G)$ and F is not a contradiction and G is not a tautology. Show that there exists a formula H such that every atomic in H is in both F and G and $\models (F \to H)$ and $\models (H \to G)$.

1.35. (Beth's definability theorem) Let H be a subformula of F. Let $A_1, \ldots, A_m$ be the atomic subformulas of F that do not occur in H. Suppose that, for any formula H', the formula $H \leftrightarrow H'$ is a consequence of the formula $F \wedge F'$ where F' is the formula obtained by replacing each occurrence of H in F with H'. Suppose also that $m \geq 1$. Show that there exists a formula G having no atomic subformulas other than $A_1, \ldots, A_m$ such that $\models F \to (H \leftrightarrow G)$.

2 Structures and first-order logic

2.1 The language of first-order logic

First-order logic is a richer language than propositional logic. Its lexicon contains not only the symbols $\land$, $\lor$, $\neg$, $\rightarrow$, and $\leftrightarrow$ (and parentheses) from propositional logic, but also the symbols $\exists$ and $\forall$ for "there exists" and "for all," along with various symbols to represent variables, constants, functions, and relations. These symbols are grouped into five categories.

- **Variables.** Lower case letters from the end of the alphabet $(\ldots x, y, z)$ are used to denote variables. Variables represent arbitrary elements of an underlying set. This, in fact, is what "first-order" refers to. Variables that represent sets of elements are called *second-order*. Second-order logic, discussed in Chapter 9, is distinguished by the inclusion of such variables.
- **Constants.** Lower case letters from the beginning of the alphabet $(a, b, c, \ldots)$ are usually used to denote constants. A constant represents a specific element of an underlying set.
- **Functions.** The lower case letters f, g, and h are commonly used to denote functions. The arguments may be parenthetically listed following the function symbol as $f(x_1, x_2, \ldots, x_n)$. First-order logic has symbols for functions of any number of variables. If f is a function of one, two, or three variables, then it is called *unary*, *binary*, or *ternary*, respectively. In general, a function of n variables is called *n-ary* and n is referred to as the *arity* of the function.
- **Relations.** Capital letters, especially P, Q, R, and S, are used to denote relations. As with functions, each relation has an associated arity.

We have an infinite number of each of these four types of symbols at our disposal. Since there are only finitely many letters, subscripts are used to accomplish this infinitude. For example, x_1, x_2, x_3, $\ldots$ are often used to denote variables. Of course, we can use any symbol we want in first-order logic. Ascribing the letters of the alphabet in the above manner is a convenient convention. If you turn to a random page in this book and see "$R(a, x, y)$," you can safely assume that R is a ternary relation, x and y are variables, and a is a constant. However, we may at times use symbols that we have not yet mentioned. We may use the symbol $\heartsuit$ if we please. However, if we do so, we must say what this symbol represents, whether it is a constant, a variable, a function of 23 variables, a ternary relation, or what.

- **Fixed symbols.** The fixed symbols are $\land$, $\lor$, $\neg$, $\rightarrow$, $\leftrightarrow$, (,), $\exists$, and $\forall$.

By "fixed" we mean that these symbols are always interpreted in the same way. If you look on page 189 of this book and see the symbol $\land$, it means the same thing as it did on page 8. It means "and." The same is true for each of the fixed symbols. In contrast, the interpretation of the function symbol f depends on the context. We may use this symbol to represent any function we choose.

The fixed symbols $\exists$ and $\forall$, called *quantifiers*, make the language of first-order logic far more expressive than propositional logic. They are called the *existential* and *universal* quantifiers, respectively. In any first-order formula, each quantifier is immediately followed by a variable. We read $\exists x$ as "there exists x such that" and $\forall x$ as "for all x."

The following is an example of a sentence of first-order logic: $\forall y \exists x R(f(x), y)$. This sentence says that for all y there exists x such that the relation R holds for the ordered pair $(f(x), y)$. Here f is a unary function and R is a binary relation. Whether this sentence is true or not depends on the context. If the relation R is equality, then this sentence is true if and only if the function f is onto.

Because of the ubiquity of equality in mathematics, we add to our list of fixed symbols the symbol $=$ for "equals." We still refer to this as "first-order logic" although it is often called "first-order logic with equality." The inclusion of equality allows the quantifiers to actually quantify. For example, the sentence

$$\exists x_1 \exists x_2 \exists x_3 (\neg(x_1 = x_2) \land \neg(x_1 = x_3) \land \neg(x_2 = x_3))$$

says that there exist at least three distinct elements. Likewise, we can write sentences that say there exist at least seven elements, or fewer than 23 elements, or exactly 45 elements.

We have now completely listed the symbols of first-order logic. Our next priority is to define the syntax and semantics. That is, we need to say which strings of these symbols are permissable as formulas and also how to interpret the formulas.

2.2 The syntax of first-order logic

The definition of a *formula* in first-order logic is analogous to the definition of formula in propositional logic. We first define *atomic formulas* and then give rules for constructing more complex formulas. We used upper case Roman letters such as F, G, and H to denote formulas in propositional logic. In first-order logic, we reserve these letters for other uses and instead use lower case Greek letters such as φ, ψ, and θ to denote formulas.

Prior to defining formulas, we must define the term *term*. Terms are defined inductively by the following two rules.

(T1) Every variable and constant is a term.
(T2) If f is an m-ary function and $t_1, \ldots, t_m$ are terms, then $f(t_1, \ldots, t_m)$ is also a term.

Definition 2.1 An *atomic formula* is a formula that has the form $t_1 = t_2$ or $R(t_1, \ldots, t_n)$ where R is an n-ary relation and $t_1, \ldots, t_n$ are terms.

As with propositional logic, we regard some of the fixed symbols as *primitive*. The other symbols are defined in terms of the primitive symbols. We view $\neg$, $\wedge$, and $\exists$ as primitive. Every formula of first-order logic is built from atomic formulas by repeated application of three rules. Each rule corresponds to a primitive symbol.

(R1) If φ is a formula then so is $\neg \varphi$.
(R2) If φ and ψ are formulas then so is $\varphi \wedge \psi$.
(R3) If φ is a formula, then so is $\exists x \varphi$ for any variable x.

Note that (R1) and (R2) were also rules for propositional logic and only the rule (R3) is new.

Definition 2.2 A string of symbols is a *formula* of first-order logic if and only if it is constructed from atomic formulas by repeated application of rules (R1), (R2), and (R3).

The definitions of $\vee$, $\rightarrow$, and $\leftrightarrow$ are the same as in propositional logic. We define $\forall x \varphi$ as $\neg \exists x \neg \varphi$. For any formula φ, the two formulas $\forall x \varphi$ and $\neg \exists x \neg \varphi$ are interchangeable. So from (R1) and (R3) we have the following: if φ is a formula, then so is $\forall x \varphi$ for any variable x.

Example 2.3 $\forall y P(x, y) \vee \exists y Q(x, y)$ is a formula of first-order logic and $x(Q \forall P)y \exists)(\vee$ is not.

In the next section, we discuss the semantics of first-order logic. For this we need to know the order of operations of the symbols. Parentheses dictate the order of operations in any formula. In absence of parentheses, we use the following rule: $\neg, \exists,$ and $\forall$ have priority over $\wedge, \vee, \rightarrow,$ and $\leftrightarrow$.

Example 2.4 $\exists x P(x,y) \vee Q(x,y)$ means $(\exists x P(x,y)) \vee (Q(x,y))$ and $\forall y P(x,y) \to Q(x,y)$ means $(\forall y P(x,y)) \to (Q(x,y))$.

We also use the following convention: the order in which to consider $\neg$, $\exists$, and $\forall$ is determined by the order in which they are listed. We again employ conventions (C1) and (C2) from Section 1.1. These allow us to drop parentheses that are not needed.

Example 2.5 We write $\neg \exists x \forall y \exists z R(x,y,z)$ instead of $\neg(\exists x(\forall y(\exists z(R(x,y,z)))))$.

Having defined formulas, we next define the notion of a subformula.

Definition 2.6 Let φ be a formula of first-order logic. We inductively define what it means for θ to be a *subformula* of φ as follows:

If φ is atomic, then θ is a subformula of φ if and only if $\theta = \varphi$.
If φ has the form $\neg \psi$, then θ is a subformula of φ if and only if $\theta = \varphi$ or θ is a subformula of ψ.
If φ has the form $\psi_1 \wedge \psi_2$, then θ is a subformula of φ if and only if $\theta = \varphi$ or θ is a subformula of ψ_1, or θ is a subformula of ψ_2.
If φ has the form $\exists x \psi$, then θ is a subformula of φ if and only if $\theta = \varphi$ or θ is a subformula of ψ.

Example 2.7 Let φ be the formula $\exists x \forall y P(x,y) \vee \forall x \exists y Q(x,y)$ where P and Q are binary relations. The subformulas of φ are $\exists x \forall y P(x,y)$, $\forall y P(x,y), P(x,y), \forall x \exists y Q(x,y), \exists y Q(x,y), Q(x,y)$ and φ itself.
Note that the formula $P(x,y) \vee \forall x \exists y Q(x,y)$, occurring as part of φ, is not a subformula of φ.

The *free variables* of a formula φ are those variables occurring in φ that are not quantified. For example, in the formula $\forall y R(x,y)$, x is a free variable, but y is not since it is quantified by $\forall$. For any first-order formula φ, let $free(\varphi)$ denote the set of free variables of φ. We can define $free(\varphi)$ inductively as follows:

If φ is atomic, then $free(\varphi)$ is the set of all variables occurring in φ,
if $\varphi = \neg \psi$, then $free(\varphi) = free(\psi)$,
if $\varphi = \psi \wedge \theta$, then $free(\varphi) = free(\psi) \cup free(\theta)$, and
if $\varphi = \exists x \psi$, then $free(\varphi) = free(\psi) - \{x\}$.

Definition 2.8 A *sentence* of first-order logic is a formula having no free variables.

Structures and first-order logic

Example 2.9 $\exists x \forall y P(x,y) \vee \forall x \exists y Q(x,y)$ is a sentence of first-order logic, whereas $\exists x P(x,y) \vee \forall x Q(x,y)$ is a formula but not a sentence (y is a free variable).

Example 2.10 Let φ be the formula $\forall y \exists x f(x) = y$. Then φ is a sentence since both of the variables occurring in φ are quantified. The formulas $f(x) = y$ and $\exists x f(x) = y$ are both subformulas of φ. Neither of these subformulas is a sentence.

In contrast to the free variables of a formula φ, the *bound* variables of φ are those variables that do have quantifiers. For any first-order formula φ, $bnd(\varphi)$ denotes the set of bound variables occurring in φ. Again, this notion can be precisely defined by induction.

If φ is atomic, then $bnd(\varphi) = \emptyset$,
if $\varphi = \neg \psi$, then $bnd(\varphi) = bnd(\psi)$,
if $\varphi = \psi \wedge \theta$, then $bnd(\varphi) = bnd(\psi) \cup bnd(\theta)$, and
if $\varphi = \exists x \psi$, then $bnd(\varphi) = bnd(\psi) \cup \{x\}$.

Every variable occurring in φ is in $free(\varphi)$ or $bnd(\varphi)$. As the next example shows, these two sets are not necessarily disjoint. A variable can have both free and bound occurrences within the same formula.

Example 2.11 Consider the formula $\exists x (R(x,y) \wedge \exists y R(y,x))$. The variable y occurs free in $R(x,y)$ and bound in $\exists y R(y,x)$. The variable x occurs only as a bound variable. So, if ψ denotes this formula, then $free(\psi) = \{y\}$ and $bnd(\psi) = \{x,y\}$.

Free variables are more important to us than bound variables. We often write formulas with its free variables. For example, $\varphi(x_1, x_2, x_3)$ denotes a formula having free variables x_1, x_2, and x_3. This notation is suggestive. We write $\varphi(t_1, t_2, t_3)$ to denote the formula obtained by replacing each free occurrence of x_i in φ with the term t_i. The presence of free variables distinguishes formulas from sentences. This distinction did not exist in propositional logic. The notion of truth is defined only for sentences. It does not make sense to ask whether the formula $y = x + 1$ is true or not. But we can ask whether $\forall y \exists x (y = x + 1)$ or $c_1 = c_2 + 1$ is true or not. The answer, as we have already indicated, depends on the context.

2.3 Semantics and structures

As with propositional logic, the semantics for $\wedge$ and $\neg$ can be described by saying $\wedge$ behaves like "and" and $\neg$ behaves like "negation." Likewise, the semantics for the quantifiers $\exists$ and $\forall$ can be inferred from the phrases "there exists" and "for all." However, we must be more precise when defining the semantics of a logic.

The goal of this section is to formally define the semantics of first-order logic. First, we intuitively describe the semantics with some examples.

Consider the first-order sentence

$$\forall y \exists x f(x) = y.$$

This sentence says that for all y there exists x so that $f(x) = y$. To determine whether this sentence is true or not, we need a context. It depends on what the variables represent and what the function f is. For example, suppose the variables are real numbers and f is defined by the rule $f(x) = x^2$. Then the above sentence is false since there is no x such that $f(x) = -1$. If the function f is defined by $f(x) = x^3$ (or, if the variables represent complex numbers) then the sentence is true.

Now consider $\forall x \forall y (R(x,y) \to \exists z (z \neq x \land z \neq y \land (R(x,z) \land R(z,y))))$. This sentence says that for any x and y, if $R(x,y)$ holds, then there exists some z other than x and y so that $R(x,z)$ and $R(z,y)$ both hold. Suppose again that the variables represent real numbers. If the relation $R(x,y)$ means $x < y$, then the above sentence is true since between any two real numbers there exists other real numbers. That is, the real numbers are *dense*. However, if the variables represent integers (or if R means $\leq$) then this sentence is false.

So whether a sentence is true or not depends on two things: our underlying set and our interpretation of the function, constant, and relation symbols. This observation leads us to the central concept of this chapter. A *structure* consists of an underlying set together with an interpretation of various functions, constants, and relations. The role of structures in first-order logic is analogous to the role played by assignments in propositional logic. Given any sentence φ and any structure M, we define what it means for M to *model* φ. Intuitively, this means that the sentence φ is true with respect to M. As in propositional logic, we write $M \models \varphi$ to denote this concept. The formal definition of this concept will be given later in this section.

Structures naturally arise in many branches of mathematics. For example, a vector space is a structure. The groups, rings, and fields of abstract algebra also provide examples of structures. In graph theory, the graphs can be viewed as first-order structures (we shall discuss this in detail in Section 2.4). The real numbers provide examples of structures that should be familiar to all readers. The real numbers form not one structure, but many. Recall that a structure has two components: and underlying set and an interpretation of certain functions, constants, and relations. When we refer to the "real numbers" we are only specifying the underlying set and not the symbols to be interpreted. We may want to consider the reals with the functions of addition and multiplication. That is one structure. Another structure is the reals with the relation $\leq$ and the constant 0. Depending on what aspect of the real numbers we wish to investigate, we may choose various functions, constants, and relations on the reals. The functions, constants, and relations that we choose to consider is called the *vocabulary* of the structure. Each choice of a vocabulary determines a different structure having the real numbers as an underlying set.

Definition 2.12 A *vocabulary* is a set of function, relation, and constant symbols.

Definition 2.13 Let $\mathcal{V}$ be a vocabulary. A $\mathcal{V}$-*structure* consists of a nonempty underlying set U along with an interpretation of $\mathcal{V}$. An *interpretation* of $\mathcal{V}$ assigns:

- an element of U to each constant in $\mathcal{V}$,
- a function from U^n to U to each n-ary function in $\mathcal{V}$, and
- a subset of U^n to each n-ary relation in $\mathcal{V}$.

We say M is a *structure* if it is a $\mathcal{V}$-structure for some vocabulary $\mathcal{V}$.

We present structures by listing the underlying set, or *universe*, followed by the function, relation, and constant symbols that it interprets.

Example 2.14 Let $\mathcal{V} = \{f, R, c\}$ where f is a unary function, R is a binary relation, and c is a constant. Then $M = (\mathbb{Z}|f, R, c)$ denotes a $\mathcal{V}$-structure. The universe of M is the set of integers $\mathbb{Z}$. To complete the description of M, we must say how the symbols of $\mathcal{V}$ are to be interpreted. We may say, for example, that M interprets $f(x)$ as x^2, $R(x, y)$ as $x < y$, and the constant c as 3. This completely describes the structure M.

Example 2.15 Let $\mathcal{V} = \{P, R\}$ where P is a unary relation and R is a binary relation. Then $M = (\mathbb{N}|P, R)$ denotes a $\mathcal{V}$-structure. The universe of M is the set of natural numbers $\mathbb{N}$. To complete the description of M, we must say how the symbols of $\mathcal{V}$ are to be interpreted. We may say, for example, that M interprets

$P(x)$ as "x is an even number," and
$R(x, y)$ as "$x + 1 = y$."

This information completely describes structure M.

Example 2.16 $\mathbf{R} = (\mathbb{R}|+, \cdot, 0, 1)$ denotes a structure in the vocabulary $\{+, \cdot, 0, 1\}$ where $+$ and $\cdot$ are binary functions and 0 and 1 are constants. The universe of $\mathbf{R}$ is the set of real numbers $\mathbb{R}$. To complete the description of $\mathbf{R}$, we must say how the symbols are to be interpreted. We may simply say that $\mathbf{R}$ interprets the symbols in the "usual way." This means that $\mathbf{R}$ interprets $+$ as plus, $\cdot$ as times, 0 as 0, and 1 as 1. This completely describes the structure $\mathbf{R}$.

Definition 2.17 Let $\mathcal{V}$ be a vocabulary. A $\mathcal{V}$-*formula* is a formula in which every function, relation, and constant is in $\mathcal{V}$. A $\mathcal{V}$-*sentence* is a $\mathcal{V}$-formula that is a sentence.

If M is a $\mathcal{V}$-structure, then each $\mathcal{V}$-sentence φ is either true or false in M. If φ is true in M, then we say M models φ and write $M \models \varphi$. Structures in first-order logic play an analogous role to assignments in propositional logic. But whereas, in propositional logic, there were only finitely many possible assignments for a sentence, there is no end to the number of structures that may or may not model a given sentence of first-order logic.

Intuitively, $M \models \varphi$ means that the sentence φ is true of M. We must precisely define this concept. Before doing so, we consider one more example.

Example 2.18 Consider again the structure **R** from Example 2.16. The vocabulary for this structure is $\{+, \cdot, 0, 1\}$ which we denote by $\mathcal{V}_{ar}$ (the vocabulary of arithmetic).

Consider the $\mathcal{V}$-sentence $\forall x \exists y (1 + x \cdot x = y)$. This sentence says that for any x there exists y that is equal to $x^2 + 1$. This is true in **R**. If we take any real number, square it, and add one, then the result is another real number. So $\mathbf{R} \models \forall x \exists y (1 + x \cdot x = y)$.

Consider next the $\mathcal{V}$-sentence $\forall y \exists x (1 + x \cdot x = y)$. This sentence asserts that for every y there is an x so that $1 + x^2 = y$. This sentence is not true in **R**. If we take $y = -2$, for example, then there is no such x. So the structure **R** does not model the sentence $\forall y \exists x (1 + x \cdot x = y)$.

Let M be a $\mathcal{V}$-structure and let φ be a $\mathcal{V}$-sentence. We now formally define what it means for M to model φ. First we define this concept for sentences φ that do not contain the abbreviations $\vee$, $\rightarrow$, $\leftrightarrow$, or $\forall$. We define $M \models \varphi$ by induction on the total number of occurrences of the symbols $\wedge$, $\neg$, and $\exists$. If φ has zero occurences of these symbols, then φ is atomic.

- If φ is atomic, then φ either has the form $t_1 = t_2$ or $R(t_1, \ldots, t_m)$ where $t_1, \ldots, t_m$ are terms and R is a relation in $\mathcal{V}$. Since φ is a sentence, φ contains no variables, and so each t_i is interpreted as some element a_i in the universe U of M. In this case,

  $M \models t_1 = t_2$ if and only if a_1 and a_2 are the same element of U, and

  $M \models R(t_1, \ldots, t_m)$ if and only if the tuple $(a_1, \ldots, a_m)$ is in the subset of U^m assigned to the m-ary relation R.

Now suppose that φ contains $m+1$ occurrences of $\wedge$, $\neg$, and $\exists$. Suppose that $M \models \psi$ has been defined for any sentence ψ containing at most m occurrences of these symbols. Since first-order formulas are constructed from atomic formulas using rules (R1), (R2), and (R3), there are three possibilities for φ.

- If φ has the form $\neg\psi$, then $M \models \varphi$ if and only if M does not model ψ.
- If φ has the form $\psi \wedge \theta$, then $M \models \varphi$ if and only if both $M \models \psi$ and $M \models \theta$.

The third possibility is that φ has the form $\exists x \psi$. If x is not a free variable of ψ then $M \models \varphi$ if and only if $M \models \psi$. Otherwise, let $\psi(x)$ be a formula having x as a free variable. Before defining $M \models \varphi$ in this case, we introduce the notion of *expansion*.

Definition 2.19 Let $\mathcal{V}$ be a vocabulary. An expansion of $\mathcal{V}$ is a vocabulary containing $\mathcal{V}$ as a subset.

Definition 2.20 Let M be a $\mathcal{V}$-structure. A structure M' is an *expansion* of M if M' has the same universe as M and interprets the symbols of $\mathcal{V}$ in the same way as M.

If M' is an expansion of M, then, reversing our point of view, we say that M is a *reduct* of M'.

If M' is an expansion of M, then the vocabulary of M' is necessarily an expansion of the vocabulary of M.

Example 2.21 The structure $M' = (\mathbb{R}|+, -, \cdot, <, 0, 1)$ is an expansion of $M = (\mathbb{R}|+, \cdot, <, 0)$ where each of these structures interpret the symbols in the usual way (see Example 2.16).

Example 2.22 Any structure is (trivially) an expansion of itself.

Our immediate interest is the expansion of a $\mathcal{V}$-structure M obtained by adding a new constant to the vocabulary for each element of the universe U_M of M. Let $\mathcal{V}(M)$ denote the vocabulary $\mathcal{V} \cup \{c_m | m \in U_m\}$ where each c_m is a constant. Let M_C denote the expansion of M to a $\mathcal{V}(M)$-structure that interprets each c_m as the element m.

- If φ has the form $\exists x \psi(x)$, then $M \models \varphi$ if and only if $M_C \models \psi(c)$ for some constant c of $\mathcal{V}(M)$.

We have now defined $M \models \varphi$ for any sentence φ that does not use $\vee$, $\rightarrow$, $\leftrightarrow$, or $\forall$. Since each of these symbols is defined in terms of $\neg$, $\wedge$, and $\exists$, the definition of $M \models \varphi$ can be extended to all sentences in a natural way. Suppose that, for some sentence φ', $M \models \varphi'$ has been defined. Suppose further that φ' has a subformula of the form $\neg(\neg\psi \wedge \neg\theta)$. Let φ be the sentence obtained by replacing an occurrence of the subformula $\neg(\neg\psi \wedge \neg\theta)$ in φ' with $(\psi \vee \theta)$. We define $M \models \varphi$ to mean the same as $M \models \varphi'$. Likewise, if φ is obtained from φ' by replacing a subformula of the form $(\psi \to \theta)$ with $(\neg\psi \vee \theta)$, $(\psi \leftrightarrow \theta)$ with $(\psi \leftarrow \theta) \wedge (\theta \leftarrow \psi)$, or $\neg\exists x \neg\psi(x)$ with $\forall x \psi(x)$ then, as definition, $M \models \varphi$ if and only if $M \models \varphi'$.

We have now defined what it means for a $\mathcal{V}$-structure M to be a model of a $\mathcal{V}$-sentence φ. We further extend the definition to apply to all $\mathcal{V}(M)$-sentences. Recall that $\mathcal{V}(M)$ is the expansion of $\mathcal{V}$ obtained by adding a new constant for each element in the universe of M. There is a natural expansion of M to a $\mathcal{V}(M)$-structure denoted by M_C. For any $\mathcal{V}(M)$-sentence φ, we define $M \models \varphi$ to mean $M_C \models \varphi$. For any $\mathcal{V}$-structure M, we refer to the constants of $\mathcal{V}(M)$ that are not in $\mathcal{V}$ as *parameters*.

Example 2.23 Let $\mathcal{V}_<$ be the vocabulary consisting of a single binary relation $<$. Let $\mathbf{R}_<$ be the $\mathcal{V}_<$-structure having underlying set $\mathbb{R}$ which interprets $<$ in the usual way. Then $\mathbf{R}_<$ models

$$\forall x \exists y \exists z ((y < x) \wedge (x < z)),$$
$$\forall x \forall y ((x < y) \to \exists z ((x < z) \wedge (z < y)))$$
$$(3 < 5) \wedge (-2 < 0), \quad and$$
$$\neg \exists x ((x < -2) \wedge (5 < x)).$$

The first two are $\mathcal{V}_<$-sentences. The other two are $\mathcal{V}_<(\mathbf{R}_<)$-sentences that are not $\mathcal{V}_<$-sentences. We regard $-2, 0, 3,$ and 5 as parameters.

Note that we have not defined the concept of "models" for formulas that are not sentences. Conventionally, when one says that a structure M models a *formula* $\varphi(x_1, \ldots, x_n)$, what is meant is that M models the sentence $\forall x_1 \ldots \forall x_n \varphi(x_1, \ldots, x_n)$. Of course, the formula $\varphi(x_1, \ldots, x_n)$ may be true for some values of $x_1, \ldots, x_n$ and not for others. The set of n-tuples for which the formula holds is called the *set defined by* φ.

Definition 2.24 Let $\varphi(x_1, \ldots, x_n)$ be a $\mathcal{V}$-formula. Let M be a $\mathcal{V}$-structure having underlying set U_M. The set of all n-tuples $(b_1, \ldots, b_n) \in (U_M)^n$ for which $M \models \varphi(b_1, \ldots, b_n)$ is denoted by $\varphi(M)$. The set $\varphi(M)$ is called a $\mathcal{V}$-*definable subset of* M (although it is actually a subset of $(U_M)^n$).

Typically, most subsets of a structure's universe are not definable (as we will see in Section 2.5). The definable subsets are special subsets and play a central role in model theory (Chapters 4–6). The $\mathcal{V}$-definable subsets are the subsets that the vocabulary $\mathcal{V}$ is capable of describing. For the sake of model theory, the notion of a *first-order structure* can be defined without reference to the syntax of first-order logic. In general, a "structure" can be defined as a set with together with special subsets having names. A *first-order structure* is a structure having names for those sets that are definable by a first-order formula (see Exercises 2.11 and 2.12).

Example 2.25 Let $\mathcal{V}_<$ and $\mathbf{R}_<$ be as in the previous example. Consider the $\mathcal{V}_<(\mathbf{R}_<)$-formulas

$$(x < y) \lor (x > y) \lor (x = y), \quad \text{and}$$
$$(\neg(x < 3) \land \neg(x = 3) \land (5 < x)) \lor (x = 5) \lor (x < -2).$$

Let $\varphi(x, y)$ denote the first formula and let $\psi(x)$ denote the second formula. Then $\mathbf{R}_< \models \varphi(x, y)$. By this we mean that $\mathbf{R}_<$ models the sentence $\forall x \forall y \varphi(x, y)$. It follows that the set defined by $\varphi(x, y)$ is all of $\mathbb{R}^2$. In contrast, the formula $\psi(x)$ does not hold for all x in $\mathbb{R}$. So $\mathbf{R}_<$ does not model this formula. The set $\psi(\mathbf{R}_<)$ defined by $\psi(x)$ is $(-\infty, -2) \cup (3, 5]$. Note that $\mathbf{R}_<$ also does not model the formula $\neg \psi(x)$. The set $\neg \psi(\mathbf{R}_<)$ is $[2, 3] \cup (5, \infty)$, the complement of $\psi(\mathbf{R}_<)$ in $\mathbb{R}$.

If M models φ, then we say φ *holds* in M, or simply, that φ is true in M. A sentence may be true in one structure and not in another. If a $\mathcal{V}$-sentence φ holds in every $\mathcal{V}$-structure, then it is *valid*, (or a *tautology*). If the sentence φ holds in some structure, then it is *satisfiable*. Otherwise, if there is no structure in which φ is true, then φ is *unsatisfiable* (or a *contradiction*).

We use the same terminology as in propositional logic. We give the analogous definitions for consequence and equivalence. For $\mathcal{V}$-sentences θ and φ, "θ is a consequence of φ" means that, for every $\mathcal{V}$-structure M, if $M \models \varphi$ then $M \models \theta$. And "θ is equivalent to φ" means that θ and φ are consequences of each other. Again, we use the following notation:

$\models \varphi$ means that φ is a tautology,
$\varphi \models \psi$ means ψ is a consequence of φ, and
$\varphi \equiv \theta$ means φ and ψ are equivalent.

The definition of satisfiability can be extended to apply to all formulas of first-order logic (not just sentences). The formula $\varphi(x_1, \ldots, x_m)$ is *satisfiable* if and only if the sentence $\forall x_1 \ldots \forall x_m \varphi(x_1, \ldots, x_m)$ is satisfiable. Therefore, the notions of unsatisfiability, tautology, and consequence also apply to formulas as well as sentences (see Exercise 2.6).

A primary aim of ours is to resolve the following decision problems.

Validity problem: Given formula φ, is φ valid?
Satisfiability problem: Given formula φ, is φ satisfiable?

Consequence problem: Given formulas φ and ψ, is ψ a consequence of φ?
Equivalence problem: Given formulas φ and ψ, are φ and ψ equivalent?

These are, in some sense, variations of the same problem. For this reason we focus on just one of these: the Satisfiability problem. If we could resolve this problem, then we could also resolve the Validity problem (by asking if $\neg\varphi$ is unsatisfiable), the Consequence problem (by asking if $\varphi \wedge \neg\psi$ is unsatisfiable), and the Equivalence problem (by asking if φ and ψ are consequences of each other).

The question of whether or not a given formula is satisfiable regards the syntax of the formula rather than the semantics. For example, consider the formula $(y + 1) < y$. If we interpret the vocabulary $\{+, <, 1\}$ in the usual manner, then this formula cannot be satisfied. The result of adding one to a number cannot be less than the number. Under a different interpretation, however, this formula is satisfiable (suppose that we interpret $<$ as "not equal"). For the same reason, $2 + 2 = 4$ is not a tautology. For an example of a formula that is not satisfiable, consider $\forall x R(x, y) \wedge \exists x \neg R(x, y)$. This formula is unsatisfiable by virtue of its structure. It has the form "p and not p." Regardless of how the binary relation R is interpreted, the formula is contradictory.

The Satisfiability problem for first-order logic is decidedly more difficult than the corresponding problem for propositional logic. In propositional logic we could, in theory, compute a truth table to determine whether or not a formula is satisfiable. In first-order logic, we would have to check *every* structure to do this. We have no systematic way for doing this. So, for now, we have no way of proving that a first-order formula is unsatisfiable. To show that a formula is satisfiable, however, can be easy. We need only to find one structure in which it is true.

Example 2.26 Let φ be the sentence $\forall x \exists y R(x, y) \wedge \exists y \forall x \neg R(x, y)$. To show that this is satisfiable, we must find a structure M that models φ.

Let $M = (\mathbb{N}|R)$ where $\mathbb{N}$ denotes the natural numbers and the binary relation R is interpreted as the successor relation. That is, $R(x, y)$ holds if and only if $y = x + 1$ (y is the successor of x).

Under this interpretation, φ says that every element has a successor and there exists an element that has no predecessor. This is true in M. Every natural number has a successor, but 0 has no predecessor.

So $M \models \varphi$ and φ is satisfiable.

It is also easy to see that φ is not a tautology. We need only to find one structure that models $\neg\varphi$. Consider, for example, the structure $N = (\mathbb{Z}|R)$

where $\mathbb{Z}$ denotes the set of integers and the binary relation R is interpreted as the successor relation. This structure does not model φ since every integer has a predecessor.

Example 2.27 Let $\mathcal{V}_E$ be the vocabulary $\{E\}$ consisting of one binary relation. Let M be a $\mathcal{V}_R$-structure. The relation E is an equivalence relation on M if and only if M models the three sentences

$$\forall x E(x,x)$$

$$\forall x \forall y (E(x,y) \to E(y,x))$$

$$\forall x \forall y \forall z ((E(x,y) \land E(y,z)) \to E(x,z))).$$

The first sentence, call it φ_1, says that E is reflexive, the second sentence, φ_2, says that E is symmetric, and the third sentence φ_3 says E is transitive.

We have seen equivalence relations before. "Equivalence" was the name we gave to the relation $\equiv$ between formulas of propositional logic. It is easy to see that this relation warrants the name we bestowed it. It clearly satisfies the three conditions of an equivalence relation. That is, the $\mathcal{V}_E$-structure $(U|E)$ models each φ_i, where U is the set of all formulas of propositional logic and E is interpreted as $\equiv$.

We can show that these three sentences are not redundant, that all three are needed to define the notion of equivalence relation. To do this, we show that none of these sentences is a consequence of the other two. For example, to show that φ_2 is not a consequence of φ_1 and φ_3, we must find a structure that is a model of $\varphi_1 \land \varphi_3 \land \neg \varphi_2$. That is, we must demonstrate a $\mathcal{V}_E$-structure where E is reflexive and transitive, but not symmetric. The $\mathcal{V}_E$-structure $(\mathbb{R}|E)$ where E is interpreted as $\leq$ on the real numbers is such a structure. Likewise, we can show that φ_1 is not a consequence of φ_2 and φ_3, and φ_3 is not a consequence of φ_1 and φ_2. We leave this as Exercise 2.4.

In these examples we are able to show that certain formulas are satisfiable by exhibiting structures in which they hold. Using this same idea, we can show that a given formula is *not* a tautology, that one formula is *not* a consequence of another, and that two given formulas are *not* equivalent. However, we have no way at present to show that a formula is unsatisfiable, or a tautology, or that one formula is a consequence of another. This is the topic of Chapter 3 where we define both formal proofs and resolution for first-order logic.

2.4 Examples of structures

Let us now examine some specific structures. We consider four types of structures that one encounters in mathematics and computer science: number systems, linear orders, databases, and graphs.

2.4.1 Graphs. Graph theory provides examples of mathematical structures that are both accessible and versatile.

Definition 2.28 A *graph* is a set of points, called *vertices*, and lines, called *edges* so that every edge starts at a vertex and ends at a vertex. Two vertices are said to be *adjacent* if they are connected by an edge

The following are examples of graphs:

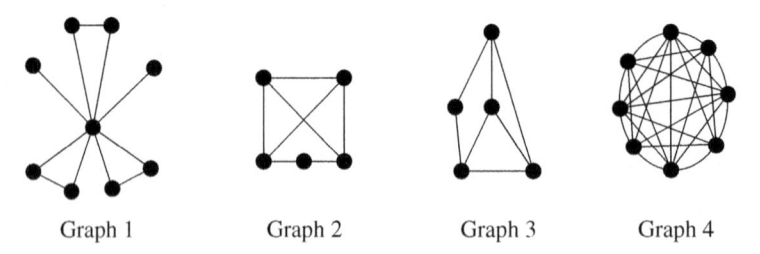

Graph 1   Graph 2   Graph 3   Graph 4

Instead of giving a picture, we can describe a graph by listing its vertices and edges. The following data completely describes a graph.

Vertices: a, b, c, d, e
Edges: ab, ad, ae, bc, cd, ce, de

This graph has five vertices (a, b, c, d, and e) and seven edges (between vertices a and b, a and d, and so forth). Note that both Graphs 2 and 3 fit this description. We regard Graphs 2 and 3 as two depictions of the same graph.

We can view any graph as a structure G as follows. The underlying set U of G is the set of vertices. The vocabulary V_G of G consists of a single binary relation R. The structure G interprets R as the edge relation. That is, for elements a and b of U, $G \models R(a,b)$ if and only if the graph has an edge between vertices a and b.

Each of the above graphs model each the following two V_G-sentences.

$$\forall x \neg R(x,x)$$

$$\forall x \forall y (R(x,y) \leftrightarrow R(y,x))$$

The first of these sentences says that the binary relation R is not reflexive (no vertex is adjacent to itself). The second sentence says that R is symmetric. Henceforth, when we speak of a *graph*, we mean a V_G-structure that models the above two sentences. Our notion of a "graph" is more accurately described in graph theoretic terms as an "undirected graph with neither multiple edges nor loops."

Graphs 1–4 also model the sentence $\forall x \exists y R(x,y)$ which asserts that each vertex is adjacent to some other vertex. However, this is not true of all graphs. For example, consider the following graph:

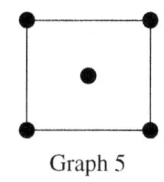

Graph 5

The vertex in the middle of the square is not adjacent to any vertex. Therefore, this graph models $\exists x \forall y \neg R(x,y)$ which is equivalent to the negation of $\forall x \exists y R(x,y)$. Any graph containing more than one vertex that models this negation must not be *connected*. We now define this terminology.

Definition 2.29 For any vertices a and b of a graph, a *path from a to b* is a sequence of vertices beginning with a and ending with b such that each vertex other than a is adjacent to the previous vertex in the sequence.

Definition 2.30 A graph is *connected* if for any two vertices a and b in G, there exists a path from a to b.

Each of the Graphs 1–4 is connected. Since each has more than one vertex, each models $\forall x \exists y R(x,y)$. On the other hand, none of these graphs models $\exists x \forall y R(x,y)$. This sentence asserts that there exists a vertex that is adjacent to every vertex. Since no vertex is adjacent to itself, no graph models this sentence (i.e. the negation of $\exists x \forall y R(x,y)$ is a consequence of $\forall x \neg R(x,x)$). However, Graph 1 contains a vertex that is adjacent to every vertex other than itself. This can be expressed in first-order logic as follows:

$$\exists x \forall y (\neg(x = y) \to R(x,y)).$$

Graph 4 also models this sentence.

To distinguish Graph 1 from Graph 4, we can say that Graph 1 contains a *unique* vertex that is adjacent to every vertex other than itself. This can be expressed as a sentence of first-order logic. To simplify this sentence, let $\varphi(x)$ denote the formula $\forall y(\neg(x = y) \to R(x,y))$. For any graph G and any vertex a of G, $G \models \varphi(a)$ if and only if a is adjacent to every vertex of G other than a itself. The following sentence says there is a unique such element:

$$\exists y \varphi(y) \land \forall z (\varphi(z) \to (z = y)).$$

This sentence distinguishes Graph 1 from Graphs 2–4.

Graph 4, on the other hand, is characterized by the following sentence that says that $\varphi(x)$ holds for every vertex x.

$$\forall x \forall y (\neg(x = y) \to R(x,y)).$$

Any graph that models this sentence is called a *clique* (or a *complete graph*). The clique having n vertices is called *the n-clique* and is denoted by K_n. So Graph 4 is the 8-clique K_8. Note that, when n is specified, we use the definite article when referring to *the n-clique*. This is because any two n-cliques are essentially the same. More precisely, they are *isomorphic*.

Definition 2.31 Graphs G_1 and G_2 are said to be *isomorphic* if there exists a one-to-one correspondence f from the set of vertices of G_1 onto the set of vertices of G_2 such that for any vertices a and b of G_1, a and b are adjacent in G_1 if and only $f(a)$ and $f(b)$ are adjacent in G_2. Such a function f is called an *isomorphism*.

Isomorphic graphs are essentially the same.

Example 2.32 Consider the following two graphs.

Graph G:
  Vertices: a, b, c, d
  Edges: ab, bc, cd, ad.

Graph H:
  Vertices: w, x, y, z
  Edges: wx, wy, xz, yz.

The function f defined by

$$f(a) = w, \ f(b) = x, \ f(c) = z, \ \text{and} \ \ f(d) = y$$

is an isomorphism from G onto H. Both of these graphs can be depicted as squares. The only difference between G and H are the letters used to represent the vertices.

We have demonstrated a $\mathcal{V}_G$-sentence distinguishing Graph 1 from Graph 4. We can do much better than this. There exists a $\mathcal{V}_G$-sentence distinguishing Graph 1 from all graphs that are not isomorphic to Graph 1. That is, there exists a $\mathcal{V}_G$-sentence φ_G such that for any graph H, $H \models \varphi_G$ if and only if H is isomorphic to G. We prove this in Section 2.6 as Proposition 2.81. In this sense, first-order logic is a powerful language for describing finite graphs.

In another sense, however, first-order logic is not a powerful language. Basic graph theoretic properties cannot be expressed using first-order logic. For example, there is no first-order sentence that says a graph has an even number of vertices. Also, first order logic cannot say that a graph is connected. Recall that the sentence $\forall x \exists y R(x,y)$ holds in any connected graph having more than one vertex. However, just because this sentence holds in a structure does not mean that it is connected. There is no $\mathcal{V}_G$-sentence φ such that $G \models \varphi$ if and only if G is a connected graph. These and other limitations of first-order logic are discussed in Section 4.7.

2.4.2 Relational databases. Relational databases provide concrete examples of structures. Any collection of data can be viewed as a database, whether it be a phone book, a CD catalog, or a family tree. A relational database is presented as a set of tables. For example, the three tables below form a relational database (Tables 2.1–2.3).

We now describe a structure D representing this relational database. The underlying set of D consists of all items occuring as an entry in some column of a table. So this set contains 13 names and four dates.

Table 2.1 Parent table

Parent	Child
Ray	Ken
Ray	Sue
Sue	Tim
Dot	Jim
Bob	Jim
Bob	Liz
Jim	Tim
Sue	Sam
Jim	Sam
Zelda	Max
Sam	Max

Table 2.2 Female table

Women
Dot
Zelda
Liz
Sue

Table 2.3 Birthday table

Person	Birthday
Ann	August 5
Leo	August 8
Max	July 28
Sam	August 1
Sue	July 24

The vocabulary $\mathcal{V}$ of D consists of a n-ary relation for each table where n is the number of columns in the table. That is, the vocabulary contains a unary relation F and binary relations P and B corresponding to the Female, Parent, and Birthday tables.

The $\mathcal{V}$-structure D interprets these relations as rows of the tables. For example, $D \models B(a,b)$ if and only if "ab" is a row of the Birthday table. This completely describes the $\mathcal{V}$-structure D. For example, we see that

$$D \models F(Dot), D \models P(Zelda, Max), \quad \text{and}$$
$$D \models \neg B(Zelda, July\ 28).$$

In addition to B, F, and P, we can define first-order formulas expressing various other relations in D. For example, the formula $\neg F(x)$ says that x is male. The formula $\exists z(P(x,z) \wedge P(z,y))$ says that x is a grandparent of y. The conjunction of this formula with $F(x)$ says that x is a grandmother of y. The formula $\exists y B(x,y)$ says that x is a date and the negation of this formula says that x is a person. The formula $\exists z(B(x,z) \wedge B(y,z))$ asserts that x and y share the same birthday. There is no end to the relations that can be defined (see Exercise 2.2).

We return to this example at the end of Chapter 3 where we discuss Prolog. Prolog is a programming language based on first-order Horn logic that can be used to present and search any relational database.

2.4.3 Linear orders. Next, we look at some structures in the vocabulary $\mathcal{V}_<$ consisting solely of the binary relation $<$. Rather than use the notation "$<(x,y)$" we use the more familiar "$x < y$" to express that the binary relation $<$ holds for the ordered pair (x,y). As our choice of symbols indicates, each of the structures we consider interprets $<$ as "less than."

We consider four $\mathcal{V}_<$-structures denoted by $\mathbf{N}_<$, $\mathbf{Z}_<$, $\mathbf{Q}_<$, and $\mathbf{R}_<$. To define each structure, we must state what the underlying set is and how the symbols are to be interpreted. The underlying sets of the above structures are, in order, the natural numbers, the integers, the rational numbers, and the real numbers.

Each of these structures interprets $<$ in the usual way. We can present these structures more concisely as follows:

- $\mathbf{N}_< = (\mathbb{N}|<)$
- $\mathbf{Z}_< = (\mathbb{Z}|<)$
- $\mathbf{Q}_< = (\mathbb{Q}|<)$
- $\mathbf{R}_< = (\mathbb{R}|<)$.

These four structures have a lot in common. They are all $\mathcal{V}_<$-structures and each of them models the following $\mathcal{V}_<$-sentences:

$$\forall x \forall y ((x < y) \to \neg(y < x))$$
$$\forall x (\neg(x < x))$$
$$\forall x \forall y ((x < y) \lor (y < x) \lor (x = y))$$
$$\forall x \forall y \forall z (((x < y) \land (y < z)) \to (x < z)).$$

Taken together, these sentences say that $<$ *linearly orders* the underlying set. Each of the four structures models each of these four sentences. However, this is not true for all $\mathcal{V}_<$-sentences. Let φ be the sentence $\forall x \exists y (y < x)$, saying that there is no smallest element. Clearly, $\mathbf{R}_<$, $\mathbf{Q}_<$, and $\mathbf{Z}_<$ are models of φ. However, $\mathbf{N}_<$ does have a smallest element, namely 1. So $\mathbf{N}_<$ does not model φ, rather $\mathbf{N}_<$ models the sentence $\exists x \forall y \neg(y < x)$, asserting that there is a smallest element. Call this sentence θ. Note that θ is equivalent to $\neg \varphi$. The sentence θ distinguishes $\mathbf{N}_<$ from the other three models.

Next let us find a first-order sentence distinguishing $\mathbf{Z}_<$ from the other three structures. Observe that $\mathbf{Z}_<$ has no smallest element and it is not dense. A linearly ordered set is *dense* if between any two elements, there is another element. This property can be expressed in first-order logic by the following $\mathcal{V}_<$-sentence

$$\forall x \forall y ((x < y) \to \exists z ((x < z) \land (z < y))).$$

Call this sentence δ. Both $\mathbf{Q}_<$ and $\mathbf{R}_<$ model δ. Between any two rational numbers a and b there exist infinitely many rational numbers $[(a+b)/2$ for one]. The same is true for the real numbers. However, the integers are not dense. Between 1 and 2 there are no other integers. So $\mathbf{Z}_< \models \neg \delta$. The $\mathcal{V}_<$-sentence $\varphi \land \neg \delta$ distinguishes $\mathbf{Z}_<$ from the other three structures.

Now suppose that we want to distinguish between $\mathbf{Q}_<$ and $\mathbf{R}_<$. We may use the fact that $\mathbf{R}_<$ is bigger than $\mathbf{Q}_<$. (In the next section, we discuss the size of a structure in detail and show that, is some precise sense, there are more real numbers than rational numbers.) Another distinguishing characteristic is order-completeness. A linear order is order-complete if it cannot be split into two open intervals. The set of rational numbers, for example, is the union of the intervals $(-\infty, \sqrt{2})$ and $(\sqrt{2}, \infty)$. The parentheses "(" and ")" indicate that the intervals do not contain the end points (this is what we mean by "open"). Since $\sqrt{2}$ is not a rational number, every rational number is in one of these two intervals. So $\mathbf{Q}_<$ does not have order-completeness. The structure $\mathbf{R}_<$, on the other hand, does have order-completeness. This is a distinguishing characteristic of the real numbers.

However, if we attempt to find a $\mathcal{V}_<$-sentence that distinguishes $\mathbf{R}_<$ from $\mathbf{Q}_<$, we will fail. For every $\mathcal{V}_<$-sentence φ, $\mathbf{R}_< \models \varphi$ if and only if $\mathbf{Q}_< \models \varphi$. We give an elementary proof of this in Section 5.2. Our first-order language is too weak to express any difference in these structures. We noted that $\mathbf{R}_<$ is order-complete whereas $\mathbf{Q}_<$ is not, but we cannot express this with a first-order sentence (try it). Rather, order-completeness is a second-order concept. In second-order logic we can express things like "there do not exist two subsets such that..." We also noted that $\mathbb{R}$ is bigger than $\mathbb{Q}$, but, as we will see in Chapter 4, first-order logic can not distinguish between one infinite number and another. Both $\mathbf{Q}_<$ and $\mathbf{R}_<$ are infinite, and that is all first-order logic can say. From the point of view of $\mathcal{V}_<$-sentences, the structures $\mathbf{R}_<$ and $\mathbf{Q}_<$ are identical.

2.4.4 Number systems. Although first-order logic cannot tell the difference between the $\mathcal{V}_<$-structures $\mathbf{Q}_<$ and $\mathbf{R}_<$, it can tell the difference between the real numbers and the rational numbers in vocabularies other than $\mathcal{V}_<$. Consider the vocabulary of arithmetic $\{+, \cdot, 0, 1\}$ having binary functions $+$ and $\cdot$, and constants 0 and 1. Let $\mathcal{V}_{ar}$ denote this vocabulary and consider the following $\mathcal{V}_{ar}$-structures:

- $\mathbf{A} = (\mathbb{Z}|+, \cdot, 0, 1)$,
- $\mathbf{Q} = (\mathbb{Q}|+, \cdot, 0, 1)$,
- $\mathbf{R} = (\mathbb{R}|+, \cdot, 0, 1)$,
- $\mathbf{C} = (\mathbb{C}|+, \cdot, 0, 1)$.

The underlying sets of these structures are, in order, the integers, the rational numbers, real numbers, and the complex numbers. Each of these structures interprets the symbols of $\mathcal{V}_{ar}$ in the usual way.

Rather than use the formal notation "$+(x, y) = z$" we use the more conventional "$x + y = z$." Likewise, we write $x \cdot y$ instead of $\cdot(x, y)$. We let 2 abbreviate $(1 + 1)$, x^2 abbreviate $x \cdot x$, and so on. Any polynomial having natural numbers as coefficients is a $\mathcal{V}_{ar}$-term. Equations such as (for example)

$$x^5 - 9x + 3 = 0$$

are $\mathcal{V}_{ar}$-formulas. Again, 3 and x^5 are not symbols in $\mathcal{V}_{ar}$, they are abbreviations for the $\mathcal{V}_{ar}$ terms $(1 + (1 + 1))$ and $x \cdot (x \cdot (x \cdot (x \cdot x)))$, respectively.

We still cannot express order-completeness in this vocabulary, but we can distinguish between the structures **R** and **Q**. The $\mathcal{V}_{ar}$-sentence $\exists x(x^2 = 2)$ asserts the existence of $\sqrt{2}$. It follows that **R** models this sentence and **Q** does not. Likewise, the equation $2x + 3 = 0$ has a solution in **Q** but not in **A**. So $\mathbf{Q} \models \exists \mathbf{x}(2\mathbf{x} + 3 = \mathbf{0})$, whereas $\mathbf{A} \models \neg \exists \mathbf{x}(2\mathbf{x} + 3 = \mathbf{0})$.

To progress from $\mathbb{N}$ to $\mathbb{Z}$ to $\mathbb{Q}$, we add solutions for more and more polynomials. We reach the end of the line with the complex numbers $\mathbb{C}$. The complex numbers are obtained by adding to the reals, the solution $i = \sqrt{-1}$ of the equation $x^2 + 1 = 0$. The $\mathcal{V}_{ar}$-sentence $\exists x(x^2 + 1 = 0)$ distinguishes **C** from the other structures in our list. The set $\mathbb{C}$ consists of all numbers of the form $a + bi$ where a and b are both real numbers. The Fundamental Theorem of Algebra states that for any nonconstant polynomial $P(x)$ having coefficients in $\mathbb{C}$, the equation $P(x) = 0$ has a solution in $\mathbb{C}$ (this is true even for polynomials of more than one variable). So there is no need to extend to a bigger number system. By virtue of adding a solution of $x^2 + 1 = 0$ to $\mathbb{R}$, we have added a solution for every polynomial.

The names of these number systems reflect historical biases. The counting numbers $1, 2, 3, \ldots$ are the "natural" numbers to consider in mathematics. Negative numbers are not natural, the square root of 2 is irrational, and the square root of -1 is imaginary. The names suggest that things get more complicated as we progress from "natural" numbers to "complex" numbers. From the point of view of first-order logic, however, this is backwards. The structure **C** is the most simple. The structure **R** is not simple like **C**, but it does have many desirable properties. We will discuss the properties of these two structures in Chapter 5. The structure **A** is not so nice. The "A" stands for arithmetic, which sounds quite elementary. However, from the point of view of first-order logic, **A** is most complex. We investigate the structure **A** in Chapter 8.

2.5 The size of a structure

For any set U, $|U|$ denotes the number of elements in U. For a $\mathcal{V}$-structure M, $|M|$ means $|U_M|$, the number of elements in the underlying set U_M of M. We refer to $|M|$ as the *size* of M. For example, if M_2 is Graph 2 from Section 2.1, then $|M_2| = 5$. If the underlying set of M is infinite, then we could just write

$|M| = \infty$ and say no more, but this oversimplifies the situation. It implies that any two infinite sets have the same size. This is not the case. To explain this, we need to say precisely what we mean by "same size."

Let A and B be two finite sets. Picture each set as a box of ping pong balls. Imagine reaching into box A with your left hand and box B with your right hand and removing one ball from each. Repeat this process. Reach in to the boxes and simultaneously remove a ball from each, and again, and again. Eventually, one of the boxes is emptied. If box B is emptied first, then we conclude that box A must have contained at least as many balls as box B at the outset. That is, $|B| \leq |A|$. Since A and B are finite, this is elementary. For infinite sets we take this idea as definition of "$|B|$ is less than or equal to $|A|$."

Definition 2.33 Let A and B be sets. We define "$|B| \leq |A|$" as follows: $|B| \leq |A|$ if there exists a one-to-one function f from B into A.

The function in this definition plays the same role as our right and left hands in the preceding discussion. The definition requires that f is one-to-one and has domain B. Given any element b in B, the function "picks out" an element $f(b)$ from A. If such a function exists, we conclude that $|B| \leq |A|$.

Example 2.34 Let P be the set of all prime natural numbers and let E be the set of all even natural numbers. Let $f: P \to E$ be defined by $f(p) = 2p$. This function is one-to-one. We conclude that $|P| \leq |E|$. That is, there are at least as many even numbers as there are prime numbers.

Example 2.35 Recall that $\mathbb{N} \times \mathbb{N}$ denotes the set of all ordered pairs (m, n) of natural numbers. Let $f: \mathbb{N} \to \mathbb{N} \times \mathbb{N}$ be defined by $f(n) = (n, 1)$ for all $n \in \mathbb{N}$. This function is one-to-one. We conclude that $|\mathbb{N}| \leq |\mathbb{N} \times \mathbb{N}|$. The reader should not be surprised by this fact. Less obvious is the fact that the opposite is true. Consider the function $g: \mathbb{N} \times \mathbb{N} \to \mathbb{N}$ defined by $g(m, n) = 2^m 3^n$. This too is a one-to-one function. So not only is $|\mathbb{N}|$ less than or equal to $|\mathbb{N} \times \mathbb{N}|$, but also $|\mathbb{N} \times \mathbb{N}|$ is less than or equal to $|\mathbb{N}|$. Naturally, we conclude that these two sets have the same size.

Definition 2.36 Let A and B be sets. We say A and B *have the same size* and write $|A| = |B|$ if both $|B| \leq |A|$ and $|A| \leq |B|$. We write $|A| < |B|$ if both $|A| \leq |B|$ and it is not the case that $|A| = |B|$.

So to show that two sets A and B have the same size we must demonstrate a one-to-one function from A to B and a one-to-one function from B to A. It suffices to show there exists a function f from A to B (or from B to A) that is both one-to-one and onto (since f^{-1} is also one-to-one and onto). Such a function is called a *one-to-one correspondence* or a *bijection*.

Example 2.37 Let $\mathbb{N}$ be the natural numbers and again let E denote the even natural numbers. In some sense there are "more" natural numbers than even numbers (since $E \subset \mathbb{N}$). However, these two sets have the same size. This is witnessed by the function $f(x) = 2x$ defining a bijection from $\mathbb{N}$ onto E.

Example 2.38 Let $\mathbb{R}$ be the real numbers and let I be $(0,1)$, the set of all reals between 0 and 1. The function $f \colon \mathbb{R} \to I$ defined by $f(x) = (2/\pi) \arctan x$ is a bijection from $\mathbb{R}$ onto I. So $|\mathbb{R}| = |I|$.

If sets A and B can be put into one-to-one correspondence with each other, then they must have the same size. The following theorem states that the converse is also true. If $|A| \leq |B|$ and $|B| \leq |A|$, then there must exist a bijection between A and B. This provides an alternative definition for "same size."

Theorem 2.39 Sets A and B have the same size if and only if there exists a bijection from A onto B.

Proof Only one direction requires proof. As we previously remarked, if there exists a bijection between A and B, then A and B must have the same size. We now prove the opposite: if $|A| = |B|$, then such a bijection necessarily exists.

Suppose A and B have the same size. By the definition of "same size" there exist one-to-one functions $f \colon A \to B$ and $g \colon B \to A$. Our goal is to demonstrate a bijection $h \colon A \to B$. Before defining h, we define some sequences.

Given any $a \in A$, we define a (possibly finite) sequence s_a as follows. Let $a_1 = a$. Now suppose $a_m \in A$ has been defined for some $m \in \mathbb{N}$. Take $b_m \in B$ such that $g(b_m) = a_m$. If no such b_m exists, then the sequence ends. Otherwise, if b_m does exist, the sequence continues. Take $a_{m+1} \in A$ such that $f(a_{m+1}) = b_m$. Again, if no such a_{m+1} exists, the sequence terminates. Note that the sequence alternates between elements of A and elements of B. The sequence s_a can be depicted as follows:

$$a_1 \xleftarrow{g} b_1 \xleftarrow{f} a_2 \xleftarrow{g} b_2 \xleftarrow{f} a_3 \xleftarrow{g} b_3 \cdots$$

There are three possibilities for the sequence s_a. Either it terminates with some element $a_i \in A$, or it terminates with some element $b_i \in B$, or it never terminates. These three possibilities partition the set A into three subsets.

- Let A_A be the set of all $a \in A$ such that s_a terminates in A.
- Let A_B be the set of all $a \in A$ such that s_a terminates in B.
- Let A_N be the set of all $a \in A$ such that s_a never terminates.

Similarly, we can define sequences s_b that begin with $b \in B$ and partition B as follows:

- Let B_A be the set of all $b \in B$ such that s_b terminates in A.

- Let B_B be the set of all $b \in B$ such that s_b terminates in B.
- Let B_N be the set of all $b \in B$ such that s_b never terminates.

The function f, when restricted to A_A, is a bijection $f\colon A_A \to B_A$. We know that f is one-to-one. To see that it is onto, take any $b \in B_A$. Since the sequence s_b terminates in A, there must exist $a \in A_A$ such that $f(a) = b$. (Otherwise, s_b would be the one-element sequence b). Likewise g, when restricted to B_B, forms a bijection $g\colon B_B \to A_B$. Finally, A_N and B_N are in one-to-one correspondence by either g or f. A bijection $h\colon A \to B$ can now be defined by putting these three parts together.

$$h(a) = \begin{cases} f(a), & a \in A_A \\ g^{-1}(a), & a \in A_B \\ f(a), & a \in A_N \end{cases}$$

□

For finite sets, Theorem 2.39 is elementary. To determine how many ping pong balls are in a given box, we put the ping pong balls into one-to-one correspondence with the set $\{1, 2, 3, \ldots, k\}$ for some $k \in \mathbb{N}$ (that is, we *count* them). We say that two boxes contain the same number of ping pong balls if each can be put into one-to-one correspondence with the same set $\{1, 2, 3, \ldots, k\}$ and, hence, with each other. If A and B are infinite, we may have difficulty visualizing them as boxes of ping pong balls. We extrapolate our definitions for infinite sets from the corresponding definitions for finite sets. Furthermore, we employ the following assumption.

Assumption: If A and B are sets, then $|A| \leq |B|$ or $|B| \leq |A|$.

For finite A and B, this assumption is a fact that can be proved. If we remove ping pong balls one at a time from each of two given boxes, eventually one (or both) of the boxes will be emptied. We must be careful, however, when handling boxes containing infinitely many ping pong balls (see Exercise 2.43). For infinite A and B, we accept this assumption without proof. It is equivalent to an axiom of mathematics known as the Axiom of Choice.

It follows from this assumption that, for any infinite set A, $|\mathbb{N}| \leq |A|$. This leads to a crucial dichotomy of infinite sets: either $|\mathbb{N}| = |A|$ or $|\mathbb{N}| < |A|$.

Definition 2.40 A set A is *denumerable* if there exists a bijection between A and $\mathbb{N}$.

Definition 2.41 A set A is *countable* if it is either finite or denumerable. Otherwise, A is *uncountable*.

Proposition 2.42 The set of rational numbers $\mathbb{Q}$ is countable.

Proof Clearly, $|\mathbb{N}| \leq |\mathbb{Q}|$ (since $\mathbb{N} \subset \mathbb{Q}$). Conversely, each nonzero element in $\mathbb{Q}$ can be written in a unique way as a reduced fraction of natural numbers times $(-1)^m$ for $m = 1$ or 2. Let $f: \mathbb{Q} \to \mathbb{N}$ be defined by $f(\frac{a}{b}(-1)^m) = 2^a 3^b 5^m$ where $\frac{a}{b}$ is reduced. Further, let $f(0) = 0$. Now f is a one-to-one function from $\mathbb{Q}$ into $\mathbb{N}$. By definition, $|\mathbb{Q}| \leq |\mathbb{N}|$. Hence $\mathbb{Q}$ and $\mathbb{N}$ have the same size. $\square$

In a similar manner, we showed in Example 2.35 that $\mathbb{N} \times \mathbb{N}$ has the same size as $\mathbb{N}$. So $\mathbb{N} \times \mathbb{N}$ is a countable set. We use this to prove the following useful fact.

Proposition 2.43 The union of countably many countable sets is countable.

Proof For each $n \in \mathbb{N}$, let A_n be a countable set. Let U denote the union of these sets. If the A_ns are each denumerable and are disjoint from one another, then U is as big as possible. Suppose this is the case. So each A_n can be enumerated as $\{a_1, a_2, a_3, \ldots\}$. Let $f(m, n)$ denote the mth element in the enumeration of A_n. This defines a bijection $f: \mathbb{N} \times \mathbb{N} \to U$. We conclude that U has the same size as $\mathbb{N} \times \mathbb{N}$. Since $\mathbb{N} \times \mathbb{N}$ is countable, so is U. $\square$

An example of an uncountable set is provided by the set of all subsets of $\mathbb{N}$. For any set A, the set of all subsets of A is called the *power set* of A, denoted by $\mathcal{P}(A)$. We show that $|\mathcal{P}(A)|$ is always strictly bigger than $|A|$.

Proposition 2.44 For any set A, $|A| < |\mathcal{P}(A)|$.

Proof To show that $|A| < |\mathcal{P}(A)|$ we must show that both $|A| \leq |\mathcal{P}(A)|$ and $|A| \neq |\mathcal{P}(A)|$.

The one-to-one function $f: A \to \mathcal{P}(A)$ defined by $f(a) = \{a\}$ (for each $a \in A$) shows that $|A| \leq |\mathcal{P}(A)|$.

To show that $|A| \neq |\mathcal{P}(A)|$, we must show that there does not exist a bijection between A and $\mathcal{P}(A)$. Let g be an arbitrary one-to-one function from A to $\mathcal{P}(A)$. We show that g is necessarily not onto. (Note that the above one-to-one function f is not onto.) For each element a in A, either a is in the set $g(a)$ or a is not in $g(a)$. Let X be the set of those elements a in A for which a is not in $g(a)$. Then $a \in X$ if and only if $a \notin g(a)$. For each $a \in A$, it cannot be the case that $g(a) = X$ (otherwise we would have $a \in X$ if and only if $a \notin X$ which is absurd). Since X is not in the range of g, g is not onto. Since g was arbitrary, we conclude that no one-to-one function from A to $\mathcal{P}(A)$ is onto. $\square$

Corollary 2.45 Any denumerable set has uncountably many subsets.

In particular, there are uncountably many subsets of $\mathbb{N}$. We use this fact to show that there are uncountably many real numbers.

Proposition 2.46 The set of real numbers $\mathbb{R}$ is uncountable.

Proof We define a one-to-one function f from $\mathcal{P}(\mathbb{N})$ into $\mathbb{R}$.

Let X be an element of $\mathcal{P}(\mathbb{N})$. Then, as a subset of the natural numbers, X contains at most 10 single-digit numbers, at most 90 two-digit numbers, at most 900 three-digit numbers, and so forth. Let r_X be the real number between 0 and 1 described as follows. The first two digits following the decimal point represent the number of single-digit numbers in X. These are succeeded by each of the single-digit numbers in X listed in ascending order. The next two digits in the decimal expansion of r_X represent the number of two-digit numbers in X. These are followed by the list of the two-digit numbers in X. The next three digits state how many three-digit numbers are in X, and so forth.

For example, let $X = \{2, 4, 5, 6, 7, 8, 9, 10, 24, 213, 3246\}$. There are 07 single-digit numbers in X (namely 2, 4, 5, 6, 7, 8, and 9), there are 02 two-digit numbers (namely 10 and 24), there is 001 three-digit number (213), and 0001 four-digit number (3246). So we have

$$r_X = 0.0724567890210240012130001324600000000\ldots$$

The number r_X contains a complete description of the set X. It follows that the function $f\colon \mathcal{P}(\mathbb{N}) \to \mathbb{R}$ defined by $f(X) = r_X$ is a one-to-one function. Hence $|\mathcal{P}(\mathbb{N})| \leq |\mathbb{R}|$. Since $\mathcal{P}(\mathbb{N})$ is uncountable, so is $\mathbb{R}$. □

We next show that there are only countably many $\mathcal{V}$-formulas for any countable vocabulary $\mathcal{V}$.

Proposition 2.47 If the vocabulary $\mathcal{V}$ is countable, then so is the set of all $\mathcal{V}$-formulas.

Proof We define a one-to-one function f from the set of all $\mathcal{V}$-formulas into $\mathbb{N}$.

Since $\mathcal{V}$ is countable, we can assign a different natural number to each symbol occurring in a $\mathcal{V}$-formula. Then to each $\mathcal{V}$-formula, there is an associated finite sequence of natural numbers. Suppose that a given $\mathcal{V}$-formula φ has $a_1, a_2, \ldots, a_m$ as its associated sequence of natural numbers. Define $f(\varphi)$ as the product

$$2^{a_1} \cdot 3^{a_2} \cdot 5^{a_3} \cdot \ldots \cdot p_n^{a_n}$$

where p_n denotes the nth prime number. We recall two basic facts about the natural numbers: there are infinitely many primes and there is a unique way to factor any given natural number into primes. So we can factor the natural number $f(\varphi)$ to recover the sequence $a_1, \ldots, a_n$ and the formula φ. It follows that f is a one-to-one function as was required. □

By Proposition 2.47, most subsets of $\mathbb{N}$ are not definable in any countable vocabulary. The same idea used to prove Proposition 2.47 can be used to show that there are countably many sentences in English or any other natural language. So there exist uncountably many real numbers that elude description in any natural language. Likewise, there exist uncountably many subsets of the natural numbers that cannot be defined. The following proposition shows that this is also true of functions on the natural numbers.

Proposition 2.48 The set of all functions from $\mathbb{N}$ to $\mathbb{N}$ is uncountable.

Proof Let F denote the set of all functions from $\mathbb{N}$ to $\mathbb{N}$. We show that $|I| \leq |F|$. Recall that I is the interval $(0,1)$ consisting of real numbers between 0 and 1. By Example 2.38, I and $\mathbb{R}$ have the same size. By the previous proposition, I is uncountable. Let r be an arbitrary element of I. Let $f_r \colon \mathbb{N} \to \mathbb{N}$ be defined by letting $f_r(n)$ be the nth digit in the decimal expansion of r. Clearly, if r_1 and r_2 are distinct numbers in I, then f_{r_1} and f_{r_2} are distinct functions. Therefore, the function assigning f_r to input r is a one-to-one function from I to F. It follows that $|I| \leq |F|$ and $|F|$ is uncountable. (In fact, we have shown that there exist uncountably many functions from $\mathbb{N}$ to the set $\{0, 1, 2, 3, 4, 5, 6, 7, 8, 9\}$). $\square$

A function $f(x)$ is said to be *computable* if there exists a computer program that outputs $f(x)$ when given input x. Applying Proposition 2.47 to computer languages, we see that there are only countably many possible computer programs. It follows that there are uncountably many functions from $\mathbb{N}$ to $\mathbb{N}$ that cannot be computed. This is also true for functions on the reals. Most functions are not computable. This fact defies empirical evidence. Most of the functions with which we are familiar (most functions one encounters in calculus, say) are computable. The notion of *computability* is discussed in detail in Chapter 7. In Section 7.6.1, we shall give examples of functions that are precisely defined but not computable.

At the outset of this section, we said that having a single notion of "infinity" is misleading. We have replaced this with two notions. An infinite set is either countable or uncountable. Many of the infinite sets we encounter either have the same size as $\mathbb{N}$ or the same size as $\mathbb{R}$. (Both $\mathcal{P}(\mathbb{N})$ and F have the same size as $\mathbb{R}$. See Exercises 2.41 and 2.42.) This dichotomy is still crude. Proposition 2.44 guarantees the existence of arbitrarily large uncountable sets, so having a single notion of "uncountable" is now misleading. In Section 4.2, we introduce cardinal numbers to represent the size of a set and study the plethora of uncountable numbers in more depth. For now, we end our digression into the infinite and return to our discussion of structures.

2.6 Relations between structures

We consider certain relations that may or may not hold between two structures in the same vocabulary.

2.6.1 Embeddings.
Let M and N be structures. The notation $f\colon M \to N$ is used to denote "f is a *function* from M to N." When using this notation, it is understood that f is not a symbol in the vocabularies of M or N. Each unary function in the vocabulary of M is interpreted as a function from the universe of M to itself. When we speak of a function from M to N, we actually mean a function from the underlying set of M to the underlying set of N. That is, to each element a from the universe U_M of M, f assigns an element $f(a)$ in the universe U_N of N. We are most interested in the case where, for some vocabulary $\mathcal{V}$, M and N are both $\mathcal{V}$-structures and f *preserves* certain $\mathcal{V}$-formulas.

Definition 2.49 Let $\mathcal{V}$ be a vocabulary and let M and N be $\mathcal{V}$-structures. A function $f\colon M \to N$ *preserves* the $\mathcal{V}$-formula $\varphi(\bar{x})$ if, for each tuple $\bar{a}$ of elements in M, $M \models \varphi(\bar{a})$ implies $N \models \varphi(f(\bar{a}))$.

Definition 2.50 Let M and N be $\mathcal{V}$-structures and let $f\colon M \to N$ be a function. If f preserves all $\mathcal{V}$-formulas that are literals, then f is a *literal embedding* (or just an *embedding*). If f preserves all $\mathcal{V}$-formulas, then f is an *elementary embedding*.

Example 2.51 Consider the following two graphs:

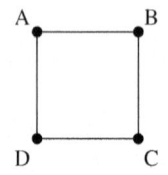

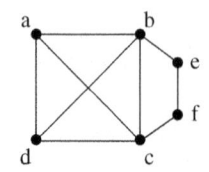

Let $f\colon M \to N$ be defined by
$$f(A) = a,\ f(B) = b,\ f(C) = c,\quad \text{and}\quad f(D) = d.$$

Let $g\colon M \to N$ be defined by
$$f(A) = b,\ f(B) = e,\ f(C) = d,\quad \text{and}\quad f(D) = f.$$

Then g is a literal embedding and f is not.

Example 2.52 Recall the structures $\mathbf{N}_<$, $\mathbf{Z}_<$, $\mathbf{Q}_<$, and $\mathbf{R}_<$ from Section 2.4.3.

Let $id\colon \mathbf{N}_< \to \mathbf{Z}_<$ be the identity function defined by $id(x) = x$. This is a literal embedding. Since $\mathbf{N}_< \models \neg\exists x(x < 0)$ and $\mathbf{Z}_< \models \exists x(x < 0)$ this embedding does not preserve the formula $\neg\exists x(x < y)$, and so it is not an elementary embedding.

The identity function $id\colon \mathbf{Z}_< \to \mathbf{Q}_<$ is also a literal embedding that is not elementary (it does not preserve the formula $\neg\exists x(y < x \wedge x < z)$).

The identity function from $\mathbf{Q}_<$ to $\mathbf{R}_<$, on the other hand, is an elementary embedding. This will be proved in Chapter 5.

We next show that literal embeddings necessarily preserve formulas other than literals.

Definition 2.53 A *quantifier-free formula* is a formula in which the quantifiers $\exists$ and $\forall$ do not occur.

Definition 2.54 An *existential formula* is a formula of the form $\exists y_1 \exists y_2 \ldots \exists y_m$ $\varphi(\bar{x}, y_1, y_2, \ldots, y_m)$, where $\varphi(\bar{x}, \bar{y})$ is a quantifier-free formula and $m \geq 0$.

We show that embeddings preserve existential formulas. First we prove the following proposition regarding quantifier-free formulas.

Proposition 2.55 Let $f : M \to N$ be an embedding. Then for any quantifier-free formula $\varphi(\bar{x})$ and any tuple $\bar{a}$ of elements from the universe of M,

$$M \models \varphi(\bar{a}) \text{ if and only if } N \models \varphi(f(\bar{a})).$$

Proof We proceed by induction on the complexity of φ.

Suppose $\varphi(\bar{x})$ is atomic. Then, since f preserves literals, if $M \models \varphi(\bar{a})$, then $N \models \varphi(f(\bar{a}))$. Conversely, if $N \models \varphi(f(\bar{a}))$ then, since $\neg\varphi(\bar{x})$ is a literal preserved by f, it must be the case that $M \models \varphi(\bar{a})$.

Now suppose that, for formulas ψ and θ,

$$M \models \psi(\bar{a}) \text{ if and only if } N \models \psi(f(\bar{a})), \quad \text{and}$$
$$M \models \theta(\bar{a}) \text{ if and only if } N \models \theta(f(\bar{a}))$$

for any tuple $\bar{a}$ of elements from the universe of M. This is our induction hypothesis. Since we want to prove the proposition only for quantifier-free formulas, the induction step, as in propositional logic, comprises three parts corresponding to $\neg$, $\wedge$, and $\equiv$. We must show that $M \models \varphi(\bar{a})$ if and only if $N \models \varphi(f(\bar{a}))$ when φ is $\neg\psi$, when φ is $\psi \wedge \theta$, and when $\varphi \equiv \psi$. The first two of these follow immediately from the semantics of first-order logic and the latter follows from the definition of $\equiv$. □

Proposition 2.56 Embeddings preserve existential formulas.

Proof Let $f : M \to N$ be an embedding and let $\varphi(\bar{x})$ be an existential formula. We must show that, for any tuple $\bar{a}$ of elements from the universe U_M of M, if $M \models \varphi(\bar{a})$ then $N \models \varphi(f(\bar{a}))$. Since $\varphi(\bar{x})$ is existential, it has the form $\exists y_1 \exists y_2 \ldots \exists y_m \varphi_0(\bar{x}, y_1, y_2, \ldots, y_m)$, where $\varphi_0(\bar{x}, \bar{y})$ is a quantifier-free formula and $m \geq 0$.

By the semantics of $\exists$, $M \models \varphi(\bar{a})$ means that $M \models \varphi_0(\bar{a}, \bar{b})$ for some tuple $\bar{b}$ of elements from U_M. Since φ_0 is quantifier-free, we have $N \models \varphi_0(f(\bar{a}), f(\bar{b}))$ by the previous proposition. Again by the semantics of $\exists$, $N \models \varphi(f(\bar{a}))$. □

Note that if $f : M \to N$ is a literal embedding then, by Proposition 2.55, $M \models a \neq b$ if and only if $N \models f(a) \neq f(b)$. It follows that any literal embedding is necessarily a one-to-one function. Note too that any elementary embedding is a literal embedding. In general, "elementary" is a much stronger adjective that "literal." However, if f happens to be onto, then these two notions coincide.

Proposition 2.57 Let M and N be $\mathcal{V}$-structures. If the function $f : M \to N$ is onto, then f is a literal embedding if and only if f is an elementary embedding.

Proof Let $f : M \to N$ be an literal embedding that is onto. Then f^{-1} is a one-to-one function from N onto M. We show that both f and f^{-1} preserve each $\mathcal{V}$-formula. That is, for each $\mathcal{V}$-formula $\varphi(\bar{x})$ and each tuple $\bar{a}$ of elements from M, $M \models \varphi(\bar{a})$ if and only if $N \models \varphi(f(\bar{a}))$. We prove this by induction on the complexity of $\varphi(\bar{x})$. If $\varphi(\bar{x})$ is atomic, then this is precisely Proposition 2.55.

Our induction hypothesis is that both f and f^{-1} preserve $\mathcal{V}$-formulas ψ and θ. If φ is equivalent to ψ then it is also preserved by f and f^{-1}. Moreover, if φ is either $\neg\psi$ or $\psi \wedge \theta$, then, by the semantics of $\neg$ and $\wedge$, φ is preserved by f and f^{-1}. It remains to be shown that φ is preserved in the case where φ is the formula $\exists y \psi$.

Let $\varphi(\bar{x})$ be the formula $\exists y \psi(\bar{x}, y)$. First we show that f preserves φ. Suppose that $M \models \varphi(\bar{a})$ for some tuple $\bar{a}$ of elements in M. Then, by the semantics of $\exists$, $M \models \psi(\bar{a}, b)$ for some element b of M. Since ψ is preserved by f, $N \models \psi(f(\bar{a}), f(b))$. Again by the semantics of $\exists$, $N \models \varphi(f(\bar{a}))$.

Now we show that f^{-1} preserves φ. Suppose that $N \models \varphi(f(\bar{a}))$. Then, by the semantics of $\exists$, $N \models \psi(f(\bar{a}), c)$ for some element c of N. Since f is onto, $c = f(b)$ for some element b of M. Since f^{-1} preserves ψ, $M \models \psi(\bar{a}, b)$. Finally, again by the semantics of $\exists$, $M \models \varphi(\bar{a})$. □

Definition 2.58 Let M and N be $\mathcal{V}$-structures. A function from M to N is an *isomorphism* if it is a one-to-one correspondence that preserves every $\mathcal{V}$-formula. If such an isomorphism exists, then M and N are *isomorphic*, denoted by $M \cong N$.

Definition 2.59 Let M and N be $\mathcal{V}$-structures. If M and N models the same $\mathcal{V}$-sentences, then M and N are said to be *elementarily equivalent*, denoted $M \equiv N$.

Example 2.60 The $\mathcal{V}_<$-structures $\mathbf{Q}_<$ and $\mathbf{R}_<$ from Section 2.4.3 are elementarily equivalent.

Proposition 2.61 Let M and N be $\mathcal{V}$-structures. If $M \cong N$, then $M \equiv N$.

Structures and first-order logic 83

Proof Let $f : M \to N$ be an isomorphism. Then both f and f^{-1} preserve every formula. In particular, for any sentence φ, $M \models \varphi$ if and only if $N \models \varphi$. □

If $\mathcal{V}$-structures M and N are elementarily equivalent, then we cannot distinguish them using first-order logic. Moreover, if M and N are isomorphic, then they are essentially the same. The only difference between isomorphic structures is the names given to the elements of the underlying sets (recall Example 2.32).

2.6.2 Substructures. If B is a set, then $A \subset B$ means that A is a subset of B. If N is a structure, then $M \subset N$ means that M is a *substructure* of N. We now define this concept.

Definition 2.62 For any structure N, M is a *substructure* of N, denoted $M \subset N$, if

1. M is a structure having the same vocabulary as N,
2. the underlying set U_M of M is a subset of the underlying set U_N of N, and
3. M interprets the vocabulary in the same manner as N on U_M.

Example 2.63 Recall the structures $\mathbf{N}_<$, $\mathbf{Z}_<$, $\mathbf{Q}_<$, and $\mathbf{R}$ from Section 2.4.3. We have $\mathbf{N}_< \subset \mathbf{Z}_< \subset \mathbf{Q}_< \subset \mathbf{R}$. Likewise, for the structures discussed in Section 2.4.4, $\mathbf{A} \subset \mathbf{Q} \subset \mathbf{R} \subset \mathbf{C}$.

Example 2.64 Let G be the following graph:

Vertices: A, B, C, D, E
Edges: AB, AC, AD, AE, BC, CD, DE

If we choose any subset of these vertices and any subset of edges involving the chosen vertices, then we obtain what is known in graph theory as a *subgraph*.
Let H be the following subgraph of G.

Vertices: A, B, C, D
Edges: AB, AD, BC, CD

Although H is a subgraph of G, H is not a substructure of G (viewing G and H as $\mathcal{V}_G$-structures). Since $G \models R(A, C)$ and $H \models \neg R(A, C)$, H does not interpret the binary relation R the same way as G does on the set $\{A, B, C, D\}$. The notion of substructure corresponds to the graph theoretic notion of *induced subgraph*.

Let N be a $\mathcal{V}$-structure and let U_N be the underlying set for N. Not every subset of U_N may serve as the universe for a substructure of N. Since a substructure is itself a $\mathcal{V}$-structure, it must interpret each constant and function in $\mathcal{V}$.

Since N is a $\mathcal{V}$-structure, it interprets each constant c in $\mathcal{V}$ as an element a_c of U_N. Let C be the subset of U_N defined by $C = \{a_c | c$ a constant in $\mathcal{V}\}$. Let f be an n-ary function in $\mathcal{V}$. A subset D of U_N is *closed under* f if and only if, for each n-tuple $\bar{a}$ of elements of D, $f(\bar{a})$ is also an element of D. For D to be the universe of a substructure of N, it is necessary and sufficient that D contains each element in C and is closed under each function in $\mathcal{V}$.

Example 2.65 Let N be the structure $(\mathbb{N}|S)$ that interprets the binary relation S as the successor relation. That is, for any a and b in $\mathbb{N}$, $N \models S(a, b)$ if and only if $b = a + 1$. Since the vocabulary contains neither constants nor functions, every subset of $\mathbb{N}$ is the universe for a substructure of N. It follows that there are uncountably many substructures of N. Moreover, there exist uncountably many substructures, no two of which are isomorphic. We leave the verification of this fact as Exercise 2.35.

Example 2.66 Let N be the structure $(\mathbb{N}|s)$ that interprets the unary function s as the successor function. That is, for any a and b in $\mathbb{N}$, $N \models s(a) = b$ if and only if $b = a + 1$. Only those subsets of $\mathbb{N}$ that are closed under s may serve as the universe of a substructure. The closed subsets of $\mathbb{N}$ are the sets of the form $\{n | n \geq d\}$ for some $d \in \mathbb{N}$. It follows that there are countably many substructures of N. Moreover, all of these substructures are isomorphic. So there is only one substructure up to isomorphism.

Example 2.67 Let N be the structure $(\mathbb{N}|s, 1)$ that interprets the unary function s as the successor function and the constant 1 as the element 1 in $\mathbb{N}$. If $D \subset \mathbb{N}$ is the universe of a substructure of N, then D must contain 1 and be closed under the function s. It follows that D must be all of $\mathbb{N}$. Therefore, the only substructure of N is N itself.

An alternative definition of substructure is provided by the notion of embedding.

Proposition 2.68 Let N and M be structures in the same vocabulary. Then M is a substructure of N if and only if the identity function $id : M \to N$ defined by $id(x) = x$ is an embedding.

Proof Exercise 2.26. □

If $M \subset N$, then, reversing our point of view, N is said to be an *extension* of M. Note the distinction between an "extension" and an "expansion" of a structure. A structure has both an underlying set and a vocabulary. An *expansion* of a structure has the same underlying set, but the vocabulary may be increased. An *extension* of a structure has the same vocabulary, but the underlying set may be enlarged.

Structures and first-order logic

Definition 2.69 The formula $\varphi(\bar{x})$ is said to be *preserved under extensions* if, whenever $M \subset N$ and $\bar{a}$ is a tuple of elements from the universe of M, if $M \models \varphi(\bar{a})$ then $N \models \varphi(\bar{a})$.

Definition 2.70 The formula $\varphi(\bar{x})$ is said to be *preserved under substructures* if, whenever $M \subset N$ and $\bar{a}$ is a tuple of elements from the universe of M, if $N \models \varphi(\bar{a})$ then $M \models \varphi(\bar{a})$.

Proposition 2.71 Quantifier-free formulas are preserved under substructures and extensions.

Proof This follows immediately from Proposition 2.55. □

Proposition 2.72 Existential formulas are preserved under extensions.

Proof This follows immediately from Proposition 2.56. □

In particular, existential sentences are preserved under extensions. Intuitively, an existential sentence asserts that a quantifier-free formula $\varphi_0(\bar{y})$ holds for some tuple $\bar{y}$ of elements in the universe. If this is true in M and $M \subset N$, then it must also be true in N since every tuple of elements from the universe of M is also a tuple of elements from the universe of N. Likewise, if $\varphi_0(\bar{y})$ holds for all tuples $\bar{y}$ of elements in the universe of N, then, in particular, it holds for all elements in any substructure of N. So sentences of the form $\forall \bar{y} \varphi_0(\bar{y})$ are preserved under substructures.

Definition 2.73 A *universal formula* is a formula of the form

$$\forall y_1 \forall y_2 \ldots \forall y_m \varphi(\bar{x}, y_1, y_2, \ldots, y_m),$$

where $\varphi(\bar{x}, \bar{y})$ is a quantifier-free formula and $m \geq 0$.

Proposition 2.74 Universal formulas are preserved under substructures.

Proof Exercise 2.32. □

In Chapter 4, we prove converses of these propositions. We show in Section 4.5.1 that if a formula φ is preserved under substructures, then φ is equivalent to an universal formula. Likewise, if φ is preserved under extensions, then φ is equivalent to an existential formula.

The notion of elementary embedding yields the following strengthening of the notion of substructure.

Definition 2.75 Let N and M be structures in the same vocabulary. Then M is an *elementary substructure* of N (or, equivalently, N is an *elementary extension* of M), denoted $M \prec N$, if and only if the identity function $id: M \to N$ defined by $id(x) = x$ is an elementary embedding.

If N is an elementary extension of M, then for any formula $\varphi(\bar x)$ and any tuple $\bar a$ of elements from the universe of M, $M \models \varphi(\bar a)$ if and only if $N \models \varphi(\bar a)$. It follows that if $M \prec N$, then $M \equiv N$. The converse of this does not hold. In the following example, M is a substructure of N and $M \equiv N$, but M is not an elementary substructure of N.

Example 2.76 Let N be the natural numbers with the successor function. That is, $N = (\mathbb{N}|s)$ from Example 2.66. Let M be the substructure of N having universe $\{2,3,4,\ldots\}$. Let $f: N \to M$ be defined by $f(n) = n+1$ for each n in $\mathbb{N}$. Then f is an isomorphism from N onto M. We have both $M \subset N$ and $M \cong N$. However, M is not an elementary substructure of N. There exists an elementary embedding of M into N, but it is not the identity function. In particular, let $\varphi(x)$ be the formula $\neg \exists y(s(y) = x)$ saying that x has no predecessor. Then $M \models \varphi(2)$, but $N \models \neg\varphi(2)$.

2.6.3 Diagrams. The concept of a *diagram* (and, more specifically, an *elementary diagram*) of a $\mathcal{V}$-structure M is a fundamental concept that we shall use repeatedly in this book (primarily in Chapter 4). Intuitively, a *diagram* of M is a set of first-order sentences that together say "M can be embedded into me." That is, M can be embedded into any model of the diagram of M. Likewise, the *elementary diagram* of M is a set of sentences such that M can be elementarily embedded into any model. We now explicitly define these sets of sentences. Recall that $\mathcal{V}(M)$ denotes the expansion of $\mathcal{V}$ obtained by adding a constant for each element of the underlying set of M and M_C denotes the expansion of M to a $\mathcal{V}(M)$-structure that interprets these constants in the natural way.

Definition 2.77 Let M be a $\mathcal{V}$-structure.

The *elementary diagram* of M, denoted $\mathcal{ED}(M)$ is the set of all $\mathcal{V}(M)$-sentences that hold in M_C.

The *literal diagram* of M, denoted $\mathcal{D}(M)$, is the set of all literals in $\mathcal{ED}(M)$. We often refer to the literal diagram of M as simply the *diagram* of M.

Example 2.78 Consider the graph defined by the following information:

Vertices: a, b, c, d
Edges: ab, bc, cd, bd.

Let G denote the $\mathcal{V}_G$-structure represented by this graph. The diagram $\mathcal{D}(G)$ contains the atomic formulas

$$R(a,b), R(b,c), R(c,d), \quad \text{and} \quad R(b,d)$$

stating the edges of G. It also contains the negated atomic formulas
$$\neg R(a,c) \quad \text{and} \quad \neg R(a,d)$$
stating the edges that are not in G. There are also negated atomic formulas indicating that a, b, c, and d are distinct:
$$\neg(a=b), \neg(a=c), \neg(a=d), \neg(b=c), \neg(b=d), \quad \text{and} \quad \neg(c=d).$$
Note that G can be embedded into any graph which models these 12 literals in $\mathcal{D}(G)$. Moreover, $\mathcal{D}(G)$ contains the literals
$$R(b,a), R(c,b), R(d,c), \quad \text{and} \quad R(d,b)$$
along with
$$\neg R(a,a), \neg R(b,b), \neg R(c,c), \neg R(d,d),$$
and
$$a=a, b=b, c=c, d=d, \neg(b=a), \neg(c=a), \dots$$
and so forth. In all, there are 32 different (although redundant) literals in $\mathcal{D}(G)$. Note that G can be embedded into any $\mathcal{V}_G$-structure that models all of these sentences.

Proposition 2.79 Let M and N be $\mathcal{V}$-structures. The following are equivalent:

(i) M can be embedded into N.

(ii) $\tilde{N} \models \mathcal{D}(M)$ for some expansion $\tilde{N}$ of N.

(iii) $N' \cong N$ for some extension N' of M.

Proof Let U_M and U_N denote the underlying sets of M and N, respectively.

First, we show (iii) implies (i). Suppose that $M \subset N$ and $N' \cong N$. Let $f : N' \to N$ be an isomorphism. Then f restricted to M is an embedding of M into N.

To see that (i) implies (ii), suppose that $f : M \to N$ is an embedding. Let $C = \{c_m : m \in U_M\}$ be constants not in $\mathcal{V}$. Let $\mathcal{V}(C)$ be the expansion $\mathcal{V} \cup C$ of $\mathcal{V}$. Let $\tilde{N}$ be the expansion of N to a $\mathcal{V}(C)$-structure that interprets each $c_m \in C$ as the element $f(m) \in U_N$. Then $\tilde{N} \models \mathcal{D}(M)$.

Finally, let $\tilde{N}$ be as in (ii). We want to show that (iii) holds. The set U_M might not be a subset of U_N. However, for each $m \in U_M$, there must exist $M' \in U_N$ that $\tilde{N}$ interprets as the constant c_m. Let $U_{N'}$ be the set obtained by replacing each $m' \in U_N$ with m. Now $U_M \subset U_{N'}$. Let N' be the $\mathcal{V}$-structure having underlying set $U_{N'}$ that interprets $\mathcal{V}$ in the same manner as N. Then the function f defined by $f(m) = m'$ for $m \in U_M$ and $f(x) = x$ for $x \in U_N - U_M$ is and isomorphism from N onto N'. □

Likewise we have the following.

Proposition 2.80 Let M and N be $\mathcal{V}$-structures. The following are equivalent:

(i) M can be elementarily embedded into N.
(ii) $\tilde{N} \models \mathcal{ED}(M)$ for some expansion $\tilde{N}$ of N.
(iii) $N' \cong N$ for some elementary extension N' of M.

Proof Exercise 2.27. □

If M is a finite structure in a finite vocabulary, then $\mathcal{D}(M)$ is finite. It follows that any finite structure is completely described by a single sentence of first-order logic.

Proposition 2.81 Let $\mathcal{V}$ be a finite vocabulary. For any finite $\mathcal{V}$-structure M, there exists a $\mathcal{V}$-sentence φ_M such that, for any $\mathcal{V}$-structure N, $N \models \varphi_M$ if and only if $N \cong M$.

Proof Let $\{a_1, a_2, \ldots, a_n\}$ be the underlying set of M. Let $\varphi(\bar{a})$ be the conjunction of the finitely many sentences in $\mathcal{D}(M)$ where $\bar{a}$ denotes the n-tuple $(a_1, a_2, \ldots, a_n)$. Let $\varphi(\bar{x})$ denote the $\mathcal{V}$-formula obtained by replacing each a_i in $\varphi(\bar{a})$ with the variable x_i (which we assume does not occur in $\varphi(\bar{a})$). We abbreviate the sentence $\exists x_1 \exists x_2 \ldots \exists x_n \varphi(\bar{x})$ by simply writing $\exists \bar{x} \varphi(\bar{x})$.

Let ψ_n be the sentence

$$\forall x_1 \forall x_2 \ldots \forall x_{n+1} \left(\bigvee_{i \neq j} (x_i = x_j) \right)$$

saying that, given any $n+1$ elements, there must exist two that are equal.

Now let φ_M be the sentence $\psi_n \wedge \exists \bar{x} \varphi(\bar{x})$. We must verify that this sentence works. Suppose $N \models \varphi_M$. Then, since $N \models \exists \bar{x} \varphi(\bar{x})$, N contains n elements $b_1, \ldots, b_n$ so that $N \models \varphi(b_1, \ldots, b_n)$. By Proposition 2.79, M can be embedded into N. Let $f : M \to N$ be an embedding. Since $N \models \psi_n$, $|N| \leq n$. It follows that f must be onto. By Proposition 2.57, f is elementary and, hence, an isomorphism. □

Corollary 2.82 If M is finite, then, for any structure N, $M \cong N$ if and only if $M \equiv N$.

As we previously mentioned, this corollary is not true for infinite structures. If M is infinite, then there exist many non-isomorphic structures N for which $M \equiv N$. This is proved in Chapter 4. Phrased another way, first-order logic is not capable of fully describing infinite structures. First-order logic is, in this sense, a weak language. Ironically, as a consequence of this weakness, first-order logic has many desirable properties (discussed in Chapter 4) that make it a prominent logic. The weakness of first-order logic gives rise to the subject of model theory.

2.7 Theories and models

Model theory is the branch of logic concerned with the interplay between mathematical structures and sentences of a formal language. First-order logic serves as a primary language for this subject. Any structure M determines a set of first-order sentences $Th(M)$ called the *theory of M*.

Definition 2.83 For any $\mathcal{V}$-structure M, the *theory of* M, denoted $Th(M)$, is the set of all $\mathcal{V}$-sentences φ such that $M \models \varphi$.

Conversely, any set of first-order sentences Γ determines a class of structures $Mod(\Gamma)$.

Definition 2.84 For any set of $\mathcal{V}$-sentences, a *model* of Γ is a $\mathcal{V}$-structure that models each sentence in Γ. The class of all models of Γ is denoted by $Mod(\Gamma)$.

Note: The word *class* is used instead of *set* for $Mod(\Gamma)$ because of the following technicality: $Mod(\Gamma)$ is sometimes unbounded. It is unbounded precisely when Γ has an infinite model. By *unbounded* we mean that for any set X, $Mod(\Gamma)$ is strictly bigger than X. If this is the case, then $Mod(\Gamma)$ must not be a set (it cannot be strictly bigger than itself).

Under certain conditions on Γ, the theory of any model of Γ is Γ itself. If this is the case, then $Th(M) = \Gamma$ if and only if $M \in Mod(\Gamma)$. This happens only if Γ is a *complete theory*, a notion that we presently define.

Definition 2.85 Let Γ be a set of $\mathcal{V}$-sentences. Then Γ is a *complete $\mathcal{V}$-theory* if, for any $\mathcal{V}$-sentence φ either φ or $\neg\varphi$ is in Γ and it is not the case that both φ and $\neg\varphi$ are in Γ.

Proposition 2.86 For any $\mathcal{V}$-structure M, $Th(M)$ is a complete $\mathcal{V}$-theory.

Proof We show that for any vocabulary $\mathcal{V}$, any $\mathcal{V}$-structure M, and any $\mathcal{V}$-sentence φ:

† either φ or $\neg\varphi$ is in $Th(M)$ and it is not the case that both φ and $\neg\varphi$ are in $Th(M)$.

With no loss of generality, we may assume that φ contains no occurrences of $\lor$, $\rightarrow$, $\leftrightarrow$, or $\forall$. This is because these symbols are defined in terms of the primitive symbols $\neg$, $\land$, and $\exists$. We proceed by induction on the number of total occurences of $\neg$, $\land$, and $\exists$ in φ.

If φ contains no occurrence of the primitive symbols, then φ has the form $R(t_1,\ldots,t_n)$ or $t_1 = t_2$ where $t_1,\ldots,t_n$ are $\mathcal{V}$-terms. That is, φ is atomic. Since φ is a sentence, each t_i is variable-free. Since M is a $\mathcal{V}$-structure and each t_i is a variable-free $\mathcal{V}$-term, M interprets each t_i as an element a_i of the universe U of M. By the definition of $\models$, $M \models t_1 = t_2$ if and only if a_1 and a_2 are the same element of U, and $M \models R(t_1,\ldots,t_n)$ if and only if the tuple $(a_1,\ldots,a_n)$ is in the subset of U^n that the interpretation of M assigns to R.

In either case, we see that $M \models \varphi$ or $M \models \neg\varphi$ and not both.

We have verified (†) for any vocabulary $\mathcal{V}$, any $\mathcal{V}$-structure M, and any atomic $\mathcal{V}$-sentence φ. Now suppose that we have shown this for any $\mathcal{V}$-sentence containing at most m total occurences of $\neg$, $\wedge$, and $\exists$. This is our induction hypothesis.

Suppose φ has the form $\neg\psi$ or $\psi \wedge \theta$. By our induction hypothesis, (†) holds for both ψ and θ. By the semantics of $\neg$ and $\wedge$, the above statement also holds for φ. Finally, suppose that φ has the form $\exists\psi(x)$. By the semantics of $\exists$, $M \models \varphi$ if and only if $M_C \models \psi(c)$ for some constant c in the vocabulary of M_C. Again by our induction hypothesis, the above statement holds for $\psi(c)$, and therefore it holds for φ as well.

It follows from induction that (†) holds for all sentences φ. □

This proposition, although quite elementary, is of fundamental importance. This proposition verifies that first-order logic avoids the ambiguities and paradoxes that arise in natural languages. In any set of first-order sentences describing a given structure, there is nothing contradictory.

Definition 2.87 A set of sentences Γ is said to be *consistent* if no contradiction can be derived from Γ.

The word "derived" is formally defined for first-order logic in the next chapter, but the idea is analogous to the notion of "derived" for propositional logic.

Definition 2.88 A *theory* is a consistent set of sentences. If T is a theory, then $Mod(T)$ is called an *elementary class*.

Let $\mathcal{V}$ be a vocabulary. Then a $\mathcal{V}$-theory is a consistent set of $\mathcal{V}$-sentences. A $\mathcal{V}$-theory T is a complete theory if it is maximal in the following sense: any set of $\mathcal{V}$-sentences that contains T as a proper subset is not consistent. This agrees with our previous definition of "complete theory."

Model theory studies theories and models and the interaction between them. Understanding the theory of a structure lends insight into the structure. The theory describes the structure. On the other hand, understanding the models of a theory lends insight into the theory. A theory T can be classified based on various properties of $Mod(T)$.

We continue our study of model theory in Chapters 4–6. Chapter 4 considers the properties of first-order logic that make it an appropriate language for model theory. In Chapter 5 we focus on theories and consider some properties that a theory may or may not posses. In Chapter 6, we consider individual models of a theory that have special properties. Prior to this, in Chapter 3, we consider the basic problem of determining whether a given sentence of first-order logic is satisfiable. Toward this aim we develop formal proofs and resolution for first-order logic.

Exercises

2.1. Let $\mathcal{V}$ be the vocabulary $\{+, <, 1, 2, 3\}$ where $+$ is a binary function, $<$ is a binary relation, and 1, 2, and 3 are constants. We write $(x+y)$ for $+(x, y)$ and $x < y$ for $<(x, y)$. Consider the following $\mathcal{V}$-formulas:

1. $\forall x \exists y ((x + y) = 1)$
2. $\forall x \neg (x < 1)$
3. $((1 + 1) = 2)$
4. $2 < 1$
5. $\forall x (2 < 1) \rightarrow (x + 2 < x + 1)$
6. $\forall x \forall y \exists z (x + y = z)$
7. $\forall x \forall y \forall z (((x + 3 = y) \wedge (x + 3 = z)) \rightarrow (y = z))$
8. $\forall x \forall y \forall z (((x + y = 3) \wedge (x + z = 3)) \rightarrow (y = z))$
9. $\forall x \forall y (((x + 3) < (y + 3)) \rightarrow (x < y))$
10. $\forall x \forall y ((x < 2) \rightarrow ((x + 3) = 4))$

(a) Which of these 10 formulas are sentences?
(b) Which of these 10 formulas are satisfiable?
(c) Which of these 10 formulas are tautologies?
(d) Let $\mathbf{N}^+$ be the $\mathcal{V}$-structure having universe $\mathbb{N}$ that interprets the symbols of $\mathcal{V}$ in the usual way. Which of the above sentences does $\mathbf{N}^+$ model?
(e) Let $\mathbf{R}^+$ be the $\mathcal{V}$-structure having universe $\mathbb{R}$ that interprets the symbols of $\mathcal{V}$ in the usual way. Which of the above sentences does $\mathbf{R}^+$ model?
(f) List the terms occurring in the above formula.
(g) For each of the ten formulas, state the number of subformulas. How many atomic subformulas does each formula have?

2.2. Let $\mathcal{V}$ be the vocabulary consisting of a binary relation P and a unary relation F. Interpret $P(x, y)$ as "x is a parent of y" and $F(x)$ as "x is female."
   (a) Define a $\mathcal{V}$-formula $\varphi_B(x, y)$ that says that x is a brother of y.
   (b) Define a $\mathcal{V}$-formula $\varphi_A(x, y)$ that says that x is an aunt of y.
   (c) Define a $\mathcal{V}$-formula $\varphi_C(x, y)$ that says that x and y are cousins.
   (d) Define a $\mathcal{V}$-formula $\varphi_O(x)$ that says that x is an only child.
   (e) Define a $\mathcal{V}$-formula $\varphi_T(x)$ that says that x has exactly two brothers.
   (f) Give an example of a family relationship that cannot be defined by a $\mathcal{V}$-formula.

2.3. The finite spectrum of a first-order sentence φ is the set of natural numbers n such that φ has a model of size M. Find a first-order sentence φ having S as a finite spectrum for each of the following sets S:
   (a) S is the set of even natural numbers.
   (b) S is the set of odd natural numbers.
   (c) S is the set of prime numbers.
   (d) S is the set of perfect squares.

2.4. Refer to Example 2.27.
   (a) Show that φ_1 is not a consequence of φ_2 and φ_3.
   (b) Show that φ_3 is not a consequence of φ_1 and φ_2.

2.5. Let $\mathcal{V}_{gp}$ be the vocabulary $\{+, 0\}$ where $+$ is a binary function and 0 is a constant. We use the notation $x + y$ to denote the term $+(x, y)$. Consider the following $\mathcal{V}$-sentences.

$$\forall x \forall y \forall z (x + (y + z) = (x + y) + z)$$

$$\forall x ((x + 0 = x) \land (0 + x = x))$$

$$\forall x (\exists y (x + y = 0) \land \exists z (z + x = 0)),$$

Let γ be the conjunction of these three sentences.
   (a) Show that γ is satisfiable by exhibiting a model.
   (b) Show that γ is not a tautology.
   (c) Let α be the sentence $\forall x \forall y ((x + y) = (y + x))$.
       Show that α is not a consequence of γ.
   (d) Show that γ is not equivalent to the conjunction of any two of the above three sentences.

Structures and first-order logic

2.6. A first-order formula $\varphi(x)$ is said to be satisfiable if and only if the sentence $\forall x \varphi(x)$ is satisfiable. Prove that a formula $\varphi(x)$ is a tautology if and only if the sentence $\exists x \varphi(x)$ is a tautology.

2.7. Let $\mathcal{V}_N = \{+, \cdot, 1\}$. Let **N** be the $\mathcal{V}_N$-structure having underlying set $\mathbb{N}$ that interprets this vocabulary in the usual manner.
   (a) Define a $\mathcal{V}_N$-formula $\varepsilon(x)$ such that, for any $a \in \mathbb{N}$, $\mathbf{N} \models \varepsilon(a)$ if and only if a is even.
   (b) Define a $\mathcal{V}_N$-formula $\pi(x)$ such that, for any $a \in \mathbb{N}$, $\mathbf{N} \models \pi(a)$ if and only if a is prime.
   (c) Define a $\mathcal{V}_N$-formula $\mu(x, y)$ such that, for any a and b in $\mathbb{N}$, $\mathbf{N} \models \mu(a, b)$ if and only if a and b are relatively prime (that is, the greatest common divisor of a and b is 1).
   (d) Define a $\mathcal{V}_N$-formula $\nu(x, y, z)$ such that, for any a, b, and c in $\mathbb{N}$, $\mathbf{N} \models \nu(a, b, c)$ if and only if c is the least number divisible by both a and b.

2.8. Goldbach's conjecture states that every even integer greater than 2 is the sum of two primes. Whether or not this is true is an open question of number theory. State Golbach's conjecture as a $\mathcal{V}_{ar}$-sentence where $\mathcal{V}_{ar} = \{+, \cdot, 0, 1\}$.

2.9. Let $\mathcal{V}_{ar} = \{+, \cdot, 0, 1\}$ be the vocabulary of arithmetic. Let **R** be the $\mathcal{V}_{ar}$-structure that has universe $\mathbb{R}$ and interprets the vocabulary in the usual manner.
   (a) Define a $\mathcal{V}_{ar}$-formula $\alpha(x)$ such that, for any $a \in \mathbb{R}$, $\mathbf{R} \models \alpha(a)$ if and only if a is positive.
   (b) Define a $\mathcal{V}_{ar}$-formula $\beta(x, y)$ such that, for any a and b in $\mathbb{R}$, $\mathbf{A} \models \beta(a, b)$ if and only if $a \leq b$.
   (c) Define a $\mathcal{V}_{ar}$-formula $\gamma(x)$ such that, for any a in $\mathbb{R}$, $\mathbf{R} \models \gamma(a)$ if and only if the absolute value of a is less than 1.

2.10. Let $\mathcal{V}_{ar}$ and **R** be as in the previous exercise. Let $\mathcal{V}^+ = \mathcal{V}_{ar} \cup \{f\}$ be the expansion of $\mathcal{V}_{ar}$ obtained by adding a unary function f. Define a $\mathcal{V}^+$-sentence ζ such that, for any expansion $\mathbf{R}^+$ of **R** to a $\mathcal{V}^+$-structure, $\mathbf{R}^+ \models \zeta$ if and only if $\mathbf{R}^+$ interprets f as a continuous function.

2.11. Let A and B be definable subsets of structure M. Suppose that A and B are both sets of n-tuples of elements from the underlying set of M.
   (a) Show that $A \cup B$ is definable.
   (b) Show that $A \cap B$ is definable.
   (c) Show that $A - B = \{a | a \in A \text{ and } a \notin B\}$ is definable.

2.12. Let U_M be the underlying set for structure M. Suppose that $A \subset (U_M)^3$ and $B \subset (U_M)^3$ are definable subsets of M.
   (a) Show that $A \times B \subset (U_M)^6$ is definable.
   (b) Suppose we rearrange the order of the n-tuples. Consider the set of all (z, x, y) such that (x, y, z) is in A. Show that this set is definable.
   (c) Show that $C \subset (U_M)^2$ is definable where C is the set of ordered pairs (x, y) such that (x, y, z) is in A for some z.
   (d) Show that $D \subset (U_M)^2$ is definable where D is the set of ordered pairs (x, y) such that both $(x, y, z) \in A$ for some z and $(x, y, z) \in B$ for some z.
   (e) Show that $E \subset (U_M)^2$ is definable where E is the set of ordered pairs (x, y) such that, for some z, (x, y, z) is in both A and B.

2.13. We define the distance $d(a, b)$ between two vertices a and b of a graph as the least number of edges in a path from a to b. If no such path exists, then $d(a, b) = \infty$. Recall that $\mathcal{V}_G$ is the vocabulary of graphs.
   (a) Show that, for any $n \in \mathbb{N}$, there exists a $\mathcal{V}_G$-formula $\delta_n(x, y)$ so that, for any graph G, $G \models \delta_n(a, b)$ if and only if $d(a, b) = n$. (Define the formulas $\delta_n(x, y)$ by induction on n.)
   (b) Does there exist a $\mathcal{V}_G$-formula $\delta_\infty(x, y)$ so that, for any graph G, $G \models \delta_\infty(a, b)$ if and only if $d(a, b) = \infty$? Explain your answer.

2.14. (a) Define a $\mathcal{V}_G$-sentence φ such that φ has arbitrarily large finite models and, for any model G, G is a connected graph.
   (b) Find a connected graph that does not model the sentence φ you found in part (a).

2.15. (a) Define a $\mathcal{V}_G$-sentence φ such that $\neg\varphi$ has arbitrarily large finite models and, $G \models \varphi$ for any connected graph G.
   (b) Find a graph that is not connected and models the sentence φ from part (a).

2.16. (a) Define a $\mathcal{V}_G$-sentence φ such that φ has arbitrarily large finite models and, for any finite model G of φ, $|G|$ is even.
   (b) Find a finite graph G such that $|G|$ is even and G does not model the sentence φ from part (a).

2.17. (a) Define a $\mathcal{V}_G$-sentence φ such that $\neg\varphi$ has arbitrarily large finite models and, for any finite graph G, if $|G|$ is even, then $G \models \varphi$.
   (b) Find a finite model G for the sentence φ from in part (a) such that $|G|$ is odd.

Structures and first-order logic

2.18. (a) Explain the difference between the first-order prefixes $\exists x \forall y$ and $\forall x \exists y$.

(b) Explain the difference between the first-order prefixes $\exists x \forall y \exists z$ and $\forall x \exists y \forall z$.

(c) Explain the difference between the first-order prefixes $\forall x \exists y \forall z \exists w$ and $\exists x \forall y \exists z \forall w$.

2.19. Show that the sentences $\forall x \exists y \forall z (R(x,y) \land R(x,z) \land R(y,z))$ and $\exists x \forall y \exists z (R(x,y) \land R(x,z) \land R(y,z))$
are not equivalent by exhibiting a graph that models one but not both of these sentences.

2.20. For each $n \in \mathbb{N}$, $\exists^{\geq n}$ denotes a *counting quantifier*. Intuitively, $\exists^{\geq n}$ means "there exists at least n such that." First-order logic with counting quantifiers is the logic obtained by adding these quantifiers (for each $n \in \mathbb{N}$) to the fixed symbols of first-order logic. The syntax and semantics of this logic are defined as follows.

> Syntax: for any formula φ of first-order logic with counting quantifiers, $\exists^{\geq n} x \varphi$ is also a formula.
> Semantics: $M \models \exists^{\geq n} \varphi(x)$ if and only if $M \models \varphi(a_i)$ for each of n distinct elements $a_1, a_2, \ldots, a_n$ in the universe of M.

(a) Using counting quantifiers, define a sentence φ_7 such that $M \models \varphi_7$ if and only if $|M| > 7$.

(b) Using counting quantifiers, define a sentence φ_{23} such that $M \models \varphi_{23}$ if and only if $|M| \leq 23$.

(c) Using counting quantifiers, define a sentence φ_{45} such that $M \models \varphi_{45}$ if and only if $|M| = 45$.

(d) Define a first-order sentence φ (not using counting quantifiers) that is equivalent to the sentence $\exists^{\geq n} x (x = x)$.

(e) Show that every formula using counting quantifiers is equivalent to a formula that does not use counting quantifiers. Conclude that first-order logic with counting quantifiers has the same expressive power as first-order logic.

2.21. Suppose we are presented with a graph G that has multiple edges. This means that there may be more than one edge between two vertices of G (so, by our strict definition of "graph," a graph with multiple edges is not a graph). Describe G as a first-order $\mathcal{V}$-structure for a suitable vocabulary $\mathcal{V}$.

2.22. Let K_n be the n-clique for some $n \in \mathbb{N}$. Then any graph having at most n vertices is a subgraph of K_n.

(a) How many substructures does K_n have?

(b) How many substructures does K_n have up to isomorphism?

(c) How many elementary substructures does K_n have?

2.23. Define an infinite structure having exactly n substructures where n is a natural number greater than 1.

2.24. Let G be Graph 1 from Section 2.4.1.
  (a) How many sentences are in the diagram of G?
  (b) Find a sentence φ_G such that $H \models \varphi_G$ if and only if $H \cong G$.

2.25. Repeat Exercise 2.24 with Graph 4 from Section 2.4.1.

2.26. Prove Proposition 2.68.

2.27. Prove Proposition 2.80.

2.28. (a) Let $N = (\mathbb{N}|S, 1)$. Show that any proper substructure of N is not elementarily equivalent to N.

  (b) Let $\mathbf{N}_<$ be the structure $(\mathbb{N}| <)$ from Section 2.4.3. Show that any infinite substructure of $\mathbf{N}_<$ is elementarily equivalent to $\mathbf{N}_<$ but no proper substructure is an elementary substructure of $\mathbf{N}_<$.

2.29. Let A, B, and C be $\mathcal{V}$-structures with $A \subset B \subset C$. For each of the following, either prove the statement or provide a counter-example.
  (a) If $A \prec B$ and $B \prec C$, then $A \prec C$.
  (b) If $A \prec C$ and $B \prec C$, then $A \prec B$.
  (c) If $A \prec B$ and $A \prec C$, then $B \prec C$.

2.30. Let $\mathcal{V}$ be the vocabulary $\{s, P\}$ consisting of a unary function s and a unary relation P. Let M be the $\mathcal{V}$-structure with universe $\mathbb{N}$ that interprets s as the successor function and P as the predicate "even." That is, for natural numbers a and b, $M \models s(a) = b$ if and only if $a + 1 = b$, and $M \models P(a)$ if and only if a is even.

Let N be the $\mathcal{V}$-structure with universe $\mathbb{N}$ that interprets s as the successor function and P as the predicate "odd." That is, N interprets s the same way as M, but $N \models P(a)$ if and only if a is odd.
  (a) Show that there exist embeddings $f_1 : M \to N$ and $f_2 : N \to M$.
  (b) Show that M and N are not isomorphic.

2.31. Define structures M and N in the same vocabulary so that there exist elementary embeddings $f : M \to N$ and $g : N \to M$, but $M \not\equiv N$.

2.32. Using the fact that existential formulas are preserved under extensions, prove that universal formulas are preserved under substructures.

Structures and first-order logic

2.33. Let M and N be $\mathcal{V}$-structures. A function $f : M \to N$ is said to be a *homomorphism* if it preserves atomic $\mathcal{V}$-formulas. Suppose that f is onto (i.e each element in the universe of N is in the range of f). Let φ be a $\mathcal{V}$-formula that does not contain the symbols $\neg$, $\to$, nor $\leftrightarrow$. Show that f preserves φ.

2.34. Let M be a $\mathcal{V}$-structure having underlying set U. For any n-tuple $\bar{a} = (a_1, \ldots, a_n)$ of elements from U, let $\langle \bar{a} \rangle$ be the substructure of M generated by $\bar{a}$. That is, the underlying set of $\langle \bar{a} \rangle$ is the smallest subset of U that contains each a_i and also contains all of the constants of $\mathcal{V}$ and is closed under each function of $\mathcal{V}$. Let $\bar{a}$ and $\bar{b}$ be two n-tuples of elements from U. Show that the following are equivalent:
  (i) For every quantifier-free $\mathcal{V}$-formula $\varphi(\bar{x})$, $M \models \varphi(\bar{a})$ if and only if $M \models \varphi(\bar{b})$.
  (ii) $\langle \bar{a} \rangle \cong \langle \bar{b} \rangle$.

2.35. Let N be the structure $(\mathbb{N}|S)$ that interprets the binary relation S as the successor relation. Show that N has uncountably many non-isomorphic substructures.

2.36. Let A be a set. Prove that the following are equivalent.
  (i) A is infinite.
  (ii) $|\mathbb{N}| \leq |A|$.
  (iii) $|A \cup B| = |A|$ for any finite set B.
  (iv) $|\mathcal{P}_F(A)| = |A|$ where $\mathcal{P}_F(A)$ is the set of all finite subsets of A.
  (v) There exists a function $f : A \to A$ that is one-to-one but not onto.
  (vi) For any B with $|B| < |A|$ and any function $f : A \to B$, there exists $b \in B$ such that $f(a) = b$ for infinitely many $a \in A$.

2.37. Find a $\mathcal{V}_<$-sentence φ so that the only models of φ interpret $<$ as a dense linear order. Show that φ has only infinite models.

2.38. Let $\mathcal{V}_f$ be the vocabulary consisting of a single unary function f. Find a $\mathcal{V}_f$-sentence that has only infinite models.

2.39. Find a set of sentences that has only uncountable models.

2.40. (a) Let F be the set of all finite strings of letters of the alphabet. Show that F is countable.
  (b) Let I be the set of all infinite strings of letters of the alphabet. Show that I is uncountable.

2.41. (a) Let $U = \{1, 2, 3\}$. List the elements of $\mathcal{P}(U)$.
  (b) Show that for any finite set U, if $|U| = n$ then $|\mathcal{P}(U)| = 2^n$.

(c) Show that the power set of the natural numbers $\mathcal{P}(\mathbb{N})$ and the real numbers $\mathbb{R}$ have the same size.

2.42. Let F be the set of all functions from $\mathbb{N}$ to $\mathbb{N}$. Show that F and $\mathbb{R}$ have the same size.

2.43. Box A contains infinitely many ping pong balls that are numbered $1, 2, 3, \ldots$

(a) Reach into box A and take out 100 balls and put them in your lap. Then put one back. Repeat this. Take out another 100 balls, put them in your lap, and then put one back. Suppose we do this countably many times. How many balls will you have in your lap?

(b) Suppose you began, in part (a), by taking out balls numbered 1–100 and then put ball 1 back. Suppose you then removed balls 101–200 and put ball 2 back. Then you took balls 201–300 into your lap, found ball 3, and put it back. And so forth. After doing this countably many times, which balls are left in your lap?

(c) Now suppose that we repeatedly remove 99 balls from box A and never return any of these balls to the box. First we take balls 1–99 into our lap and, instead of putting ball 1 back, we take a marker, add two zeros, and turn it into 100. We then take balls 101–199 out of A, take ball 2 from our lap, turn it into 200, and keep them all in our lap. After repeating this countably many times, how many balls are in your lap and what numbers do they have on them?

(d) Do the processes in (b) and (c) have different results? If so, explain why this is the case (if not, look at (b) and (c) again). Note that after each stage, we have the same numbered balls in our lap. Suppose someone else put the ping pong balls in our lap and we do not know if a marker was used or not. What then? Why should the use of a marker affect the outcome?

3 Proof theory

As with any logic, the semantics of first-order logic yield rules for deducing the truth of one sentence from that of another. In this chapter, we develop both formal proofs and resolution for first-order logic. As in propositional logic, each of these provides a systematic method for proving that one sentence is a consequence of another.

Recall the Consequence problem for propositional logic. Given formulas F and G, the problem is to decide whether or not G is a consequence of F. From Chapter 1, we have three approaches to this problem:

- We could compute the truth table for the formula $F \to G$. If the truth values are all 1s then we conclude that $F \to G$ is a tautology and G is a consequence of F. Otherwise, G is not a consequence of F.
- Using Tables 1.5 and 1.6, we could try to formally derive G from $\{F\}$. By the Completeness Theorem for propositional logic, G is a consequence of F if and only if $\{F\} \vdash G$.
- We could use resolution. By Theorem 1.76, G is a consequence of F if and only if $\emptyset \in Res(H)$ where H is a formula in CNF equivalent to $(F \land \neg G)$.

Using these methods not only can we determine whether one formula is a consequence of another, but also we can determine whether a given formula is a tautology or a contradiction. A formula F is a tautology if and only if F is a consequence of $(A \lor \neg A)$ if and only if $\neg F$ is a contradiction.

In this chapter, we consider the analogous problems for first-order logic. Given formulas φ and ψ, how can we determine whether ψ is a consequence of φ? Equivalently, how can we determine whether a given formula is a tautology or a contradiction? We present three methods for answering these questions.

- In Section 3.1, we define a notion of formal proof for first-order logic by extending Table 1.5.
- In Section 3.3, we "reduce" formulas of first-order logic to sets of formulas of propositional logic where we use resolution as defined in Chapter 1.
- Finally, in Section 3.4, we modify the notion of resolvents and develop resolution for first-order logic.

One aim of resolution is to provide an automated proof system. Toward this aim, we consider variations of resolution such as SLD-resolution. We close this chapter with a section on Prolog, a programming language that implements SLD-resolution.

3.1 Formal proofs

Let φ be a first-order formula and let Γ be a set of first-order formulas. We use the notation $\Gamma \vdash \varphi$ to express that φ can be formally derived from Γ. As with propositional logic, the definition of this notion consists of a list of several rules. For propositional logic, formal proofs were defined as sequences of statements each of which is justified by one of the rules in Tables 1.5 or 1.6. Changing the Roman letters to Greek letters yields Tables 3.1 and 3.2 below.

For first-order logic, this list of rules is incomplete. In contrast, if F and G are formulas of propositional logic and G is a consequence of F, then we can formally prove that G is a consequence of F using the rules of Table 1.5 or Table 1.6. This is the Completeness theorem for propositional logic. To obtain an

Table 3.1   Rules for derivations

Premise	Conclusion	Name
φ is in Γ	$\Gamma \vdash \varphi$	Assumption
$\Gamma \vdash \varphi$ and $\Gamma \subset \Gamma'$	$\Gamma' \vdash \varphi$	Monotonicity
$\Gamma \vdash \varphi$	$\Gamma \vdash \neg\neg\varphi$	Double negation
$\Gamma \vdash \psi, \Gamma \vdash \varphi$	$\Gamma \vdash (\psi \wedge \varphi)$	$\wedge$-Introduction
$\Gamma \vdash (\psi \wedge \varphi)$	$\Gamma \vdash \psi$	$\wedge$-Elimination
$\Gamma \vdash (\psi \wedge \varphi)$	$\Gamma \vdash (\varphi \wedge \psi)$	$\wedge$-Symmetry
$\Gamma \vdash \varphi$	$\Gamma \vdash (\varphi \vee \psi)$	$\vee$-Introduction
$\Gamma \vdash (\psi \vee \varphi),$ $\Gamma \cup \{\psi\} \vdash \theta, \Gamma \cup \{\varphi\} \vdash \theta$	$\Gamma \vdash \theta$	$\vee$-Elimination
$\Gamma \vdash (\psi \vee \varphi)$	$\Gamma \vdash (\varphi \vee \psi)$	$\vee$-Symmetry
$\Gamma \cup \{\varphi\} \vdash \psi$	$\Gamma \vdash (\varphi \to \psi)$	$\to$-Introduction
$\Gamma \vdash (\varphi \to \psi), \Gamma \vdash \varphi$	$\mathcal{F} \vdash \psi$	$\to$-Elimination
$\Gamma \vdash \psi$	$\Gamma \vdash (\psi)$	$(,)$-Introduction
$\Gamma \vdash (\psi)$	$\Gamma \vdash \psi$	$(,)$-Elimination
$\Gamma \vdash ((\psi \wedge \varphi) \wedge \theta)$	$\Gamma \vdash (\psi \wedge \varphi \wedge \theta)$	$\wedge$-Parentheses rule
$\Gamma \vdash ((\psi \vee \varphi) \vee \theta)$	$\Gamma \vdash (\psi \vee \varphi \vee \theta)$	$\vee$-Parentheses rule

Table 3.2 More rules for derivations

Rules	Name
$\Gamma \vdash (\varphi \vee \psi)$ if and only if $\Gamma \vdash \neg(\neg\varphi \wedge \neg\psi)$	$\vee$-Definition
$\Gamma \vdash (\varphi \rightarrow \psi)$ if and only if $\Gamma \vdash (\neg\varphi \vee \psi)$	$\rightarrow$-Definition
$\Gamma \vdash (\varphi \leftrightarrow \psi)$ if and only if both $\Gamma \vdash (\varphi \rightarrow \psi)$ and $\Gamma \vdash (\psi \rightarrow \varphi)$	$\leftrightarrow$-Definition

Table 3.3 Yet more rules for derivations

Premise	Conclusion	Name
$\Gamma \vdash \varphi(t)$	$\Gamma \vdash \exists y \varphi(y)$	$\exists$-Introduction
$\Gamma \vdash \varphi(c)$	$\Gamma \vdash \forall y \varphi(y)$	$\forall$-Introduction
$\Gamma \vdash \theta \rightarrow \psi$	$\Gamma \vdash \exists x \theta \rightarrow \exists x \psi$	$\exists$-Distribution
$\Gamma \vdash \theta \rightarrow \psi$	$\Gamma \vdash \forall x \theta \rightarrow \forall x \psi$	$\forall$-Distribution
$\Gamma \vdash Q_1 x (Q_2 y \theta)$	$\Gamma \vdash Q_1 x Q_2 y \theta$	Q-Parentheses rule
None	$\Gamma \vdash t = t$	Reflexivity
$\Gamma \vdash \varphi(t), \Gamma \vdash t = t'$	$\Gamma \vdash \varphi(t')$	Equality substitution

analogous result for first-order logic, we must add rules to this list pertaining to quantifiers and equality. For example, we certainly should include the definition of $\forall$:

$$\Gamma \vdash \forall x \varphi(x) \text{ if and only if } \Gamma \vdash \neg \exists x \neg \varphi(x).$$

In addition, we have the rules in Table 3.3. In this table, t is a term, c is a constant that does not occur in Γ, and Q_1 and Q_2 are quantifiers (each is either $\exists$ or $\forall$).

Recall that $\varphi(t)$ is the formula obtained by replacing each free occurrence of x in $\varphi(x)$ with the term t. In the above rules, $\varphi(x)$ may have free variables other than x. Also, we may use any letters in place of x and y. We demonstrate the rules in Table 3.3 with a couple of examples.

Example 3.1 We demonstrate the rules $\exists$-Introduction and $\forall$-Introduction. Suppose that $\Gamma \vdash R(a, b)$ where R is a binary relation and a and b are constants that do not occur in Γ. Then we can derive each of the following sentences (along with many others) from Γ:

$\exists z R(a, z)$ by $\exists$-Introduction

$\forall w R(w, b)$ by $\forall$-Introduction

$\forall w \forall z R(w, z)$ by ∀-Introduction (twice), and

$\exists z \forall w R(w, z)$ by ∀-Introduction followed by ∃-Introduction.

Suppose now that $\Gamma \vdash R(f(b), b)$ where f is a unary function. Since $f(b)$ is a term that is not a constant, we can derive from Γ the sentence $\exists z R(z, b)$ but not the sentence $\forall z R(z, b)$. Likewise, we cannot derive the sentence $\exists z \forall w R(w, z)$ from Γ. However, we can derive each of the following sentences:

$\forall w R(f(w), w)$ by ∀-Introduction

$\exists w \exists z R(w, z)$ by ∃-Introduction (twice), and

$\forall z \exists w R(w, z)$ by ∃-Introduction followed by ∀-Introduction.

Example 3.2 We illustrate the usefulness of ∃-Distribution. Suppose we want to formally prove that $\neg \exists x \psi(x)$ is a consequence of $\forall x \neg \psi(x)$. By ∀-Definition, we know that

$$\{\forall x \neg \psi(x)\} \vdash \neg \exists x \neg \neg \psi(x).$$

It remains to be shown that

$$\{\neg \exists x \neg \neg \psi(x)\} \vdash \neg \exists x \psi(x).$$

Using ∃-Distribution, we can formally prove this in three steps. First, show that $\psi(x) \to \neg \neg \psi(x)$ is a tautology. By the completeness of propositional logic, there exists a formal proof for this fact. Second, use ∃-Distribution to obtain the valid implication $\exists x \psi(x) \to \exists x \neg \neg \psi(x)$. Third, by →-Contrapositive (Exercise 1.12), $\neg \exists x \neg \neg \psi(x) \to \neg \exists x \psi(x)$ is also valid. We conclude that, if $\Gamma \vdash \forall x \neg \psi(x)$ then $\Gamma \vdash \neg \exists x \psi(x)$. This argument can be formalized into a two-column proof. We leave this to the reader.

Definition 3.3 A *formal proof* in first-order logic is a finite sequence of statements of the form "$\mathcal{X} \vdash Y$" each of which follows from the previous statements by one of the rules we have listed (including the definition of ∀ and the rules in Tables 3.1–3.3). We say that φ can be *derived* from Γ if there is a formal proof concluding with the statement $\Gamma \vdash \varphi$.

Our first priority is to show that this notion of formal proof is sound. We must show that if φ can be derived from Γ, then φ is in fact a consequence of Γ. We restate this as the following theorem.

Theorem 3.4 (Soundness) If $\Gamma \vdash \varphi$ then $\Gamma \models \varphi$.

Note: This theorem follows from the semantics of first-order logic (that is, the definition of "$\models$") given in Section 2.3. When we say something is true "by the semantics" the reader is referred to this section.

Proof We check that each rule for deduction is sound. In Theorem 1.37 we verified each of the rules in Table 1.5. It follows that each of the rules in Table 3.1 are also sound. Moreover, ∀-Definition and each of the rules in Table 3.2 are sound by the definition of the symbols. Reflexivity and Equality substitution are sound by the definition of $=$. The Q-Parentheses rule is one of our conventions regarding the use of parentheses. It remains to be shown that the first four rules of Table 3.3 are sound.

First, consider ∃-Introduction. This rule states that if $\Gamma \vdash \varphi(t)$, then $\Gamma \vdash \exists x \varphi(x)$. To show that this rule is sound, we must verify that if $\Gamma \models \varphi(t)$ then $\Gamma \models \exists x \varphi(x)$. It suffices to show that, for any structure M, $M \models \varphi(t)$ implies $M \models \exists x \varphi(x)$. This follows immediately from the semantics of ∃.

For ∀-Introduction, suppose that $\Gamma \models \varphi(c)$ where c is a constant that does not occur in Γ. Suppose that M is a $\mathcal{V}$-structure that models Γ. For any element a of the underlying set U_M of M, let $M_{c=a}$ be the structure having underlying set U_M that interprets c as a and interprets the other symbols of $\mathcal{V}$ in the same manner as M (if $c \notin \mathcal{V}$, then $M_{c=a}$ is an expansion of M). Since c does not occur in Γ, $M_{c=a}$ models Γ (since M does). Since $\Gamma \models \varphi(c)$, $M_{c=a} \models \varphi(c)$. It follows that $M \models \varphi(a)$. Since a is an arbitrary element from U_M, $M \models \forall x \varphi(x)$ by the semantics of ∀. This shows that $\Gamma \models \forall x \varphi(x)$ and verifies ∀-Introduction.

Now consider ∃-Distribution. Suppose that $M \models \theta \to \psi$ and $M \models \exists x \theta$. Let U_M denote the universe of M. We want to show that $M \models \exists x \psi$.

Case 1: x is not a free variable of θ. By the semantics of ∃, θ is equivalent to $\exists x \theta$. So if $M \models \exists x \theta$, then $M \models \theta$ and, by the semantics of $\to$, $M \models \psi$. Now if x is not a free variable of ψ, then $\psi \equiv \exists x \psi$. Otherwise, $M \models \psi(x)$ means $M \models \forall x \psi(x)$ which means $M \models \psi(a)$ for any a in U_M. Either way, we see that $M \models \exists x \psi$ as we wanted to show.

Case 2: x is a free variable of θ but not of ψ. In this case, $M \models \theta \to \psi$ means $M \models \forall x(\theta(x) \to \psi)$. By the semantics of ∀, $M \models \theta(a) \to \psi$ for any a in U_M. Since $M \models \exists x \theta$, $M \models \theta(a)$ for some $a \in U_M$. By the semantics of $\to$, $M \models \psi$. Finally, $M \models \exists x \psi$ since $\psi \equiv \exists x \psi$.

Case 3: x is a free variable of both θ and ψ. Here $M \models \theta \to \psi$ means $M \models \forall x(\theta(x) \to \psi(x))$. This means that, for all a in U_M, $M \models \theta(a) \to \psi(a)$. Since $M \models \exists x \theta$ it follows that $M \models \theta(a)$ for some a in U_M. Hence $M \models \psi(a)$. Again by the semantics of ∃, $M \models \exists x \psi$.

The verification of ∀-Distribution is similar and is left as Exercise 3.4. □

Corollary 3.5 If both $\{\varphi\} \vdash \psi$ and $\{\psi\} \vdash \varphi$, then $\varphi \equiv \psi$.

The Completeness theorem for first-order logic states that the converse of Theorem 3.4 is true. If φ is a consequence of Γ, then we can formally prove that it is a consequence. The rules for derivations we have given form a complete set of rules for first-order logic. It follows that the converse of Corollary 3.5 holds as well. However, the Completeness theorem will not be proved until the next chapter. For this reason, we presently do not assume that the converses of Theorem 3.4 and Corollary 3.5 hold. In the present chapter, just because two formulas are equivalent does not mean that we can formally prove that they are equivalent. For this, we again use the terminology *"provably equivalent"* previously defined in Section 1.5.

For the remainder of this section, we verify various instances of the converses of Theorem 3.4 and Corollary 3.5. For example, by the semantics of $\forall$, $\varphi(t)$ is a consequence of $\forall x \varphi(x)$ for any term t. We now show that $\varphi(t)$ can be formally derived from $\forall x \varphi(x)$.

Proposition 3.6 For any formula $\varphi(x)$ and any term t, $\{\forall x \varphi(x)\} \vdash \varphi(t)$.

Proof We use proof by Contradiction as defined in Example 1.36.

Premise: $\Gamma \vdash \forall x \varphi(x)$

Conclusion: $\Gamma \vdash \varphi(t)$

Statement	Justification
1. $\Gamma \vdash \forall x \varphi(x)$	Premise
2. $\Gamma \cup \{\neg \varphi(t)\} \vdash \forall x \varphi(x)$	Monotonicity applied to 1
3. $\Gamma \cup \{\neg \varphi(t)\} \vdash \neg \exists x \neg \varphi(x)$	$\forall$-Definition applied to 2
4. $\Gamma \cup \{\neg \varphi(t)\} \vdash \neg \varphi(t)$	Assumption
5. $\Gamma \cup \{\neg \varphi(t)\} \vdash \exists x \neg \varphi(x)$	$\exists$-Introduction applied to 4
6. $\Gamma \vdash \neg \neg \varphi(t)$	Proof by Contradiction applied to 3 and 5
7. $\Gamma \vdash \varphi(t)$	Double negation (from Example 1.43) applied to 6

□

Recall that $M \models \varphi(x_1, \ldots, x_n)$ means the same as $M \models \forall x_1 \cdots \forall x_n \varphi(x_1, \ldots, x_n)$. This is how the symbol $\models$ was defined in Section 2.3 for formulas having free variables. It follows that the formula $\varphi(x_1, \ldots, x_n)$ is equivalent to the sentence $\forall x_1 \cdots \forall x_n \varphi(x_1, \ldots, x_n)$. We now show that they are provably equivalent.

Proposition 3.7 The formulas $\varphi(x)$ and $\forall x \varphi(x)$ are provably equivalent.

Proof By Proposition 3.6, $\{\forall x \varphi(x)\} \vdash \varphi(t)$ for any term t. In particular, $\{\forall x \varphi(x)\} \vdash \varphi(x)$. We must prove the converse. We again utilize proof by Contradiction from Example 1.36.

Premise: $\Gamma \vdash \varphi(x)$ and c is a constant not occuring in Γ.

Conclusion: $\Gamma \vdash \forall x \varphi(x)$

Statement	Justification
1. $\Gamma \vdash \varphi(x)$	Premise
2. $\Gamma \cup \{\neg\varphi(c)\} \vdash \varphi(x)$	Monotonicity applied to 1
3. $\Gamma \cup \{\neg\varphi(c)\} \vdash \exists x \varphi(x)$	$\exists$-Introduction applied to 2
4. $\Gamma \cup \{\neg\varphi(c)\} \vdash \neg\varphi(c)$	Assumption
5. $\Gamma \cup \{\neg\varphi(c)\} \vdash \forall x \neg\varphi(x)$	$\forall$-Introduction applied to 4
6. $\Gamma \cup \{\neg\varphi(c)\} \vdash \neg\exists x \varphi(x)$	Example 3.2 applied to 5
7. $\Gamma \vdash \neg\neg\varphi(c)$	Proof by Contradiction applied to 3 and 6
8. $\Gamma \vdash \varphi(c)$	Double Negation
9. $\Gamma \vdash \forall x \varphi(x)$	$\forall$-Introduction

□

By the semantics $\forall$, $\forall x \varphi(x) \equiv \forall y \varphi(y)$ ($\varphi(x)$ holds for each element x of some model if and only if $\varphi(y)$ holds for each element y of that same model). We show that $\forall x \varphi(x)$ and $\forall y \varphi(y)$ are provably equivalent.

Corollary 3.8 Let x and y be variables that do not occur in the formula $\varphi(z)$. Then $\forall x \varphi(x)$ and $\forall y \varphi(y)$ are provably equivalent.

Proof By Proposition 3.6, $\{\forall x \varphi(x)\} \vdash \varphi(t)$ for any term t. In particular, $\{\forall x \varphi(x)\} \vdash \varphi(y)$. By Proposition 3.7, $\{\varphi(y)\} \vdash \forall y \varphi(y)$. Putting these two facts together, we see that $\{\forall x \varphi(x)\} \vdash \forall y \varphi(y)$. Likewise (switching the roles of x and y), we see that $\{\forall y \varphi(y)\} \vdash \forall x \varphi(x)$. □

Likewise, we have the following.

Corollary 3.9 Let x and y be variables that do not occur in formula $\varphi(z)$. Then $\exists x \varphi(x)$ and $\exists y \varphi(y)$ are provably equivalent.

We leave the proof of Corollary 3.9 to the reader (see Exercise 3.7).

Corollary 3.10 For any formula $\varphi(x)$, $\{\forall x \varphi(x)\} \vdash \exists x \varphi(x)$.

Proof $\{\forall x \varphi(x)\} \vdash \varphi(x)$ by Proposition 3.7.

$$\{\varphi(x)\} \vdash \exists x \varphi(x) \text{ by } \exists\text{-Introduction.}$$

Putting these two facts together, we see that $\{\forall x \varphi(x)\} \vdash \exists x \varphi(x)$. □

By the semantics of first-order logic, we know that $\exists x \varphi(x)$ is a consequence of $\forall x \varphi(x)$ (if $\varphi(x)$ holds for all elements of in a certain structure, then it holds for some elements in that structure). Corollary 3.10 states that we can formally prove this. Note that $\forall x \varphi(x)$ is not a consequence of $\exists x \varphi(x)$. So these formulas are not equivalent. However, if (and only if) the variable x has no free occurences in ψ, then $\exists x\psi$ and $\forall x\psi$ are equivalent formulas. Moreover, they are provably equivalent.

Proposition 3.11 Let x be a variable that does not occur as a free variable in the formula ψ. Then ψ, $\exists x\psi$, and $\forall x\psi$ are provably equivalent.

Proof We demonstrate that $\{\psi\} \vdash \forall x\psi$ and $\{\exists x\psi\} \vdash \psi$. The proposition then follows from Corollary 3.10 which implies $\{\forall x\psi\} \vdash \exists x\psi$.

First we show that $\{\psi\} \vdash \forall x\psi$

Premise: $\Gamma \vdash \psi$ and c is a constant that does not occur in Γ

Conclusion: $\Gamma \vdash \forall x\psi$

Statement	Justification
1. $\Gamma \vdash \psi$	Premise
2. $\Gamma \vdash (\psi \vee \neg(x = x))$	$\vee$-Introduction applied to 1
3. $\Gamma \vdash (\neg(x = x) \vee \psi)$	$\vee$-Symmetry applied to 2
4. $\Gamma \vdash (x = x) \rightarrow \psi$	$\rightarrow$-Definition applied to 3
5. $\Gamma \vdash \forall x(x = x) \rightarrow \forall x\psi$	$\forall$-Distribution applied to 4
6. $\Gamma \vdash (c = c)$	Reflexivity
7. $\Gamma \vdash \forall x(x = x)$	$\forall$-Introduction applied to 6
8. $\Gamma \vdash \forall x\psi$	$\rightarrow$-Elimination applied to 5 and 7

Next, we show that $\{\exists x\psi\} \vdash \psi$

Premise: $\Gamma \vdash \exists x\psi$

Conclusion: $\Gamma \vdash \psi$

Statement	Justification
1. $\Gamma \vdash \exists x\psi$	Premise
2. $\Gamma \cup \{\neg\psi\} \vdash \exists x\psi$	Monotonicity applied to 1
3. $\Gamma \cup \{\neg\psi\} \vdash \neg\psi$	Assumption
4. $\Gamma \cup \{\neg\psi\} \vdash \forall x\neg\psi$	The previous proof applied to 3
5. $\Gamma \cup \{\neg\psi\} \vdash \neg\exists x\psi$	Example 3.2 applied to 4
6. $\Gamma \vdash \neg\neg\psi$	Proof by Contradiction applied to 2 and 5
7. $\Gamma \vdash \psi$	Double negation (from Example 1.43) applied to 6

□

Proof theory

Proposition 3.12 The formulas $\forall x(\varphi(x) \wedge \psi(x))$ and $\forall x \varphi(x) \wedge \forall x \psi(x)$ are provably equivalent.

Proof We leave the verification of this as Exercise 3.8. □

It is not true that $\exists x(\varphi(x) \wedge \psi(x))$ and $\exists x \varphi(x) \wedge \exists x \psi(x)$ are provably equivalent. We can show that $\{\exists x(\varphi(x) \wedge \psi(x))\} \vdash \exists x \varphi(x) \wedge \exists x \psi(x)$, but not the converse. However, if (and only if) x does not occur as a free variable of ψ, the converse is true.

Proposition 3.13 If x does not occur as a free variable of ψ, then $\exists x \varphi(x) \wedge \exists x \psi$ and $\exists x(\varphi(x) \wedge \psi)$ are provably equivalent.

Proof We only prove this equivalence in one direction. The other direction is straight forward and is left as Exercise 3.13.

Premise: $\Gamma \vdash \exists x \varphi(x) \wedge \exists x \psi$

Conclusion: $\Gamma \vdash \exists x(\varphi(x) \wedge \psi)$

Statement	Justification
1. $\Gamma \vdash \exists x \varphi(x) \wedge \exists x \psi$	Premise
2. $\Gamma \vdash \exists x \psi$	$\wedge$-Elimination applied to 1
3. $\Gamma \vdash \psi$	Proposition 3.11 applied to 2
4. $\Gamma \vdash \neg \varphi(x) \vee \psi$	$\vee$-Introduction and $\vee$-symmetry applied to 3
5. $\Gamma \vdash \neg \varphi(x) \vee \varphi(x)$	Tautology rule (Example 1.32)
6. $\Gamma \vdash (\neg \varphi(x) \vee \varphi(x)) \wedge (\neg \varphi(x) \vee \psi)$	$\wedge$-Introduction applied to 4 and 5
7. $\Gamma \vdash \neg \varphi \vee (\varphi(x) \wedge \psi)$	$\vee$-Distributivity (Proposition 1.46) applied to 6
8. $\Gamma \vdash \varphi(x) \to (\varphi(x) \wedge \psi)$	$\to$-Definition applied to 7
9. $\Gamma \vdash \exists x \varphi(x) \to \exists x(\varphi(x) \wedge \psi)$	$\exists$-Distribution applied to 8
10. $\Gamma \vdash \exists x \varphi(x)$	$\wedge$-Symmetry and $\wedge$-Elimination applied to 1
11. $\Gamma \vdash \exists x(\varphi(x) \wedge \psi)$	$\to$-Elimination applied to 9 and 10

□

The previous propositions can be generalized as follows.

Proposition 3.14 Let $x_1, x_2, \ldots, x_n$ be variables that occur free in the formula φ but not in the formula ψ. Let $Q_1, \ldots, Q_n$ be quantifiers (that is, for each i, Q_i is either $\exists$ or $\forall$). Then the following two formulas are provably equivalent:

$$Q_1 x_1 Q_2 x_2 \cdots Q_n x_n \varphi(x_1, x_2, \ldots, x_n) \wedge \psi, \text{ and}$$

$$Q_1 x_1 Q_2 x_2 \cdots Q_n x_n (\varphi(x_1, x_2, \ldots, x_n) \wedge \psi).$$

Proof We prove this by induction on n. We use the following claim.

Claim If $\theta(x)$ and $\psi(x)$ are provably equivalent, then so are $Q_1 x \theta(x)$ and $Q_1 x \psi(x)$.

Proof of Claim If $\theta(x)$ and $\psi(x)$ are provably equivalent, then $\emptyset \vdash \theta(x) \to \psi(x)$. By $\exists$-Distribution or $\forall$-Distribution (depending on which quantifier is Q_1), we have $\emptyset \vdash Q_1 x \theta(x) \to Q_1 x \psi(x)$. Likewise, $\emptyset \vdash Q_1 x \psi(x) \to Q_1 x \theta(x)$. The claim follows.

We now prove the proposition. If $n = 1$ then this follows from Proposition 3.12 or 3.13 (depending on which quantifier is Q_1). Suppose now that $n = m+1$. Our induction hypothesis implies that the following two formulas are provably equivalent:

$$Q_2 x_2 \cdots Q_{m+1} x_{m+1} \varphi(x_1, x_2, \ldots, x_{m+1}) \wedge \psi, \text{ and}$$

$$Q_2 x_2 \cdots Q_{m+1} x_{m+1} (\varphi(x_1, x_2, \ldots, x_{m+1}) \wedge \psi).$$

It follows from the claim that the following two formulas are provably equivalent:

$$Q_1 x_1 (Q_2 x_2 \cdots Q_{m+1} x_{m+1} \varphi(x_1, x_2, \ldots, x_{m+1}) \wedge \psi), \text{ and}$$

$$Q_1 x_1 (Q_2 x_2 \cdots Q_n x_n (\varphi(x_1, x_2, \ldots, x_n) \wedge \psi)).$$

The former of these, again by Proposition 3.12 or 3.13, is provably equivalent with

$$Q_1 x_1 Q_2 x_2 \cdots Q_{m+1} x_{m+1} \varphi(x_1, x_2, \ldots, x_{m+1}) \wedge \psi.$$

The latter of the above two formulas, by the Q-Parentheses rule, is provably equivalent with

$$Q_1 x_1 Q_2 x_2 \cdots Q_{m+1} x_{m+1} (\varphi(x_1, x_2, \ldots, x_{m+1}) \wedge \psi).$$

This completes the induction step and the proposition follows. □

Similarly, we have the following.

Proposition 3.15 Let $Q_1, \ldots, Q_n$ denote quantifiers. For each i, let $\overline{Q}_i$ denote the quantifier that is not Q_i. That is, for each i, $\{Q_i, \overline{Q}_i\} = \{\exists, \forall\}$. For any formula $\varphi(x_1, \ldots, x_n)$,

$$\neg Q_1 x_1 \cdots Q_n x_n \varphi(x_1, \ldots, x_n) \text{ is provably equivalent to}$$

$$\overline{Q}_1 x_1 \cdots \overline{Q}_n x_n \neg \varphi(x_1, \ldots, x_n).$$

Proof It suffices to show that both

$\neg \forall x_1 \varphi(x_1)$ is provably equivalent to $\exists x_1 \neg \varphi(x_1)$, and

$\neg \exists x_1 \varphi(x_1)$ is provably equivalent to $\forall x_1 \neg \varphi(x_1)$ (see Example 3.2).

The proposition can then be proved by induction on n in a similar manner to Proposition 3.14. We leave the details as Exercise 3.15. □

It follows from the previous propositions that any formula is provably equivalent to a formula in which the quantifiers preceed all other fixed symbols. Informally, the quantifiers can be "pulled out in front" of any formula. We make this idea precise and prove it in the following section.

3.2 Normal forms

One of our goals in this chapter is to develop resolution for first-order logic. Recall that, in propositional logic, we needed to have the formulas in CNF before we could proceed with resolution. Likewise, in first-order logic the formulas will need to be in a nice form. In this section, we define what we mean by "nice."

3.2.1 Conjunctive prenex normal form.

Definition 3.16 A formula φ is in *prenex normal form* (PNF) if it has the form $Q_1x_1 \cdots Q_nx_n\psi$ where each Q_i is a quantifier (either $\exists$ or $\forall$) and ψ is a quantifier-free first-order formula. Moreover, if ψ is a conjunction of disjunctions of literals (atomic or negated atomic formulas), then φ is in *conjunctive prenex normal form*.

So a formula is in prenex normal form if all of its quantifiers are in front.

Example 3.17 $\forall y \exists x (f(x) = y)$ is in PNF, and $\neg \forall x \exists y P(x,y,z)$ and $\exists x \forall y \neg P(x,y,z) \land \forall x \exists y Q(x,y,z)$ are not.

Theorem 3.18 For any formula of first-order logic, there exists an equivalent formula in conjunctive prenex normal form.

Proof Let φ be an arbitrary formula. First we show that there exists an equivalent formula φ' in prenex normal form. We prove this by induction on the complexity of φ.

If φ is atomic, then φ is already in PNF, so we can just let φ' be φ.

Suppose ψ and θ are formulas and there exist ψ' and θ' in PNF such that $\psi \equiv \psi'$ and $\theta \equiv \theta'$. Clearly, if $\varphi \equiv \psi$ then we can let φ' be ψ'. To complete the induction step, we must consider three cases corresponding to $\neg$, $\land$, and $\exists$.

First, suppose φ is the formula $\neg\psi$. Then $\varphi \equiv \neg\psi'$. Since ψ' is in PNF, ψ' has the form $Q_1x_1 \cdots Q_mx_m\psi_0$ for some quantifier-free formula ψ_0 and quantifiers $Q_1,\ldots,Q_m$. So $\varphi \equiv \neg Q_1x_1 \cdots Q_mx_m\psi_0$. By Proposition 3.15, this is equivalent to $\overline{Q}_1x_1 \cdots \overline{Q}_mx_m \neg\psi_0$ where $\{Q_i, \overline{Q}_i\} = \{\exists, \forall\}$. This formula is in PNF, and so it may serve as φ'.

Next, suppose φ is the formula $\psi \wedge \theta$. Then $\varphi \equiv \psi' \wedge \theta'$. Since ψ' and θ' are in PNF,

$$\psi' \text{ is } Q_1x_1 \cdots Q_mx_m\psi_0(x_1,\ldots,x_m), \text{ and}$$

$$\theta' \text{ is } q_1x_1 \cdots q_nx_n\theta_0(x_1,\ldots,x_n)$$

for some quantifiers Q_i and q_i and some quantifier-free formulas ψ_0 and θ_0. Let $y_1,\ldots,y_m$ and $z_1,\ldots,z_n$ be new variables (that is, variables not occurring in ψ' or θ'). Then by Corollaries 3.8 and 3.9,

$$\psi' \equiv Q_1y_1 \cdots Q_my_m\psi_0(y_1,\ldots,y_m),$$

$$\theta' \equiv q_1z_1 \cdots q_nz_n\theta_0(z_1,\ldots,z_n), \text{ and so}$$

$$\varphi \equiv Q_1y_1 \cdots Q_my_m\psi_0(y_1,\ldots,y_m) \wedge q_1z_1 \cdots q_nz_n\theta_0(z_1,\ldots,z_n).$$

Applying Proposition 3.14 twice,

$$\varphi \equiv Q_1y_1 \cdots Q_my_mq_1z_1 \cdots q_nz_n(\psi_0(y_1,\ldots,y_m) \wedge \theta_0(z_1,\ldots,z_n))$$

which is in PNF. Let φ' be this formula.

Finally, suppose φ is the formula $\exists x\psi$. Then $\varphi \equiv \exists x_0\psi'$ for some variable x_0. Since ψ' is in PNF, $\exists x_0\psi'$ is in PNF. So in this case, we can let φ' be $\exists x_0\psi'$.

Given an arbitrary formula φ we have shown that there exists an equivalent formula φ' in prenex normal form. Let $Q_1x_1 \cdots Q_nx_n\varphi_0$ be the formula φ'. Each Q_i denotes a quantifier and φ_0 is a quantifier-free formula. We want to show that φ is equivalent to a formula in conjunctive prenex normal form. It remains to be shown that φ_0 is equivalent to a formula that is a conjunction of disjunctions. This can be done by induction on the complexity of φ_0. Since it is quantifier-free, we do not have to consider the part of the induction step corresponding to $\exists$. Therefore, the proof is identical to the proof of Theorem 1.57 where it was shown that every formula of propositional logic is equivalent to a formula in CNF. □

Example 3.19 Let φ be the formula $\neg(\forall x \exists y P(x,y,z) \vee \exists x \forall y \neg Q(x,y,z))$ having free variable z. By the previous theorem, there exists a formula φ' in PNF that is equivalent to φ. Moreover, the proof of the theorem indicates a method for finding such φ'. First, noting that φ has the form $\neg\psi$, we distribute the negation to obtain

$$\varphi \equiv \exists x \forall y \neg P(x,y,z) \wedge \forall x \exists y Q(x,y,z).$$

So φ is equivalent to a formula of the form $\psi \wedge \theta$. By renaming variables, we get

$$\varphi \equiv \exists x \forall y \neg P(x,y,z) \wedge \forall u \exists v Q(u,v,z).$$

By applying Proposition 3.14 twice,

$$\varphi \equiv \exists x \forall y \forall u \exists v (\neg P(x,y,z) \wedge Q(u,v,z))$$

which is in PNF. Moreover, this formula is in conjunctive PNF.

Our goal is to find a method for determining whether a given formula is satisfiable or not. By Theorem 3.18, it suffices to have a method that works for formulas in conjunctive prenex normal form (although, as we shall see in later chapters, no method "works" entirely). Next we show that we can simplify our formulas further. We show that we need only consider formulas that are *universal*: formulas in PNF in which the existential quantifier ∃ does not occur.

3.2.2 Skolem normal form.

Definition 3.20 A sentence is in *Skolem normal form* (SNF), if it is universal and in conjunctive prenex normal form.

Given any sentence φ of first-order logic we define a sentence φ^S that is in SNF. We prove in Theorem 3.22 that φ is satisfiable if and only if φ^S is satisfiable. The sentence φ^S is called a *Skolemization* of φ. The following is a step-by-step procedure for finding φ^S.

- First we find a sentence φ' in conjunctive prenex normal form such that $\varphi' \equiv \varphi$. So
$$\varphi' \text{ is } Q_1 x_1 \cdots Q_m x_m \varphi_0(x_1, \ldots, x_m)$$
for some quantifier-free formula φ_0 and quantifiers $Q_1, Q_2, \ldots, Q_m$.
- If each Q_i is ∀, then φ' is a universal sentence. In this case let φ^S be φ'.
- Otherwise, φ' has existential quantifiers. In this case we define a sentence $s(\varphi')$ that has fewer existential quantifiers than φ'. (So if φ' has just one existential quantifier, then $s(\varphi')$ is universal.) Let i be least such that Q_i is ∃.

  If $i = 1$, then φ' is $\exists x_1 Q_2 x_2 \cdots Q_m x_m \varphi_0(x_1, \ldots, x_m)$.
  Let $s(\varphi')$ be $Q_2 x_2 \cdots Q_m x_m \varphi_0(c, x_2, \ldots, x_m)$ where c is a constant symbol that does not occur in φ'.

  If $i > 1$, then φ' is $\forall x_1 \cdots \forall x_{i-1} \exists x_i Q_{i+1} x_{i+1} \cdots Q_m x_m \varphi_0(x_1, \ldots, x_m)$. Let $s(\varphi')$ be the sentence
$$\forall x_1 \cdots \forall x_{i-1} Q_{i+1} x_{i+1} \cdots Q_m x_m$$
$$\varphi_0(x_1, \ldots, x_{i-1}, f(x_1, \ldots, x_{i-1}), x_{i+1}, \ldots, x_m),$$

  where f is an $(i-1)$-ary function symbol that does not occur in φ'.
  So if the first quantifier in φ' is ∃, we replace x_1 with a new constant. And if the ith quantifier in φ' is ∃ and all previous quantifiers are ∀, replace x_i with $f(x_1, \ldots, x_{i-1})$ where f is a new function symbol.

- Since $s(\varphi')$ has fewer existential quantifiers than φ', by repeating this process, we will eventually obtain the required universal sentence φ^S. That is, φ^S is $s^n(\varphi') = s(s(s \cdots s(\varphi')))$ for some n.

Example 3.21 Suppose φ is the sentence $\exists z (\neg(\forall x \exists y P(x,y,z) \lor \exists x \forall y \neg Q(x,y,z)))$. First, we find a sentence φ' in conjunctive prenex normal form that is equivalent to φ. From Example 3.19 we see that φ is equivalent to

$$\exists z \exists x \forall y \forall u \exists v (\neg P(x,y,z) \land Q(u,v,z)).$$

Let φ' be this sentence.

Next we find $s(\varphi')$ as defined above. Then we find $s(s(\varphi'))$ and $s(s(s(\varphi')))$ and so forth, until we get a sentence in SNF. In this example, since φ' has three existential quantifiers, we will stop at $s(s(s(\varphi')))$.

We have $s(\varphi')$ is $\exists x \forall y \forall u \exists v (\neg P(x,y,c) \land Q(u,v,z))$, and $s(s(\varphi'))$ is $\forall y \forall u (\neg P(d,y,c) \land Q(u,f(y,u),c))$.

Finally, $s(s(s(\varphi')))$ is $\forall y \forall u (\neg P(d,y,c) \land Q(u,f(y,u),c))$ which is in SNF. So we have successfully *Skolemized* the given sentence φ and obtained the sentence $\forall y \forall u (\neg P(d,y,c) \land Q(u,f(y,u),c))$. This is the sentence denoted by φ^S.

Theorem 3.22 Let φ be a sentence of first-order logic and let φ^S be the Skolemization of φ. Then φ is satisfiable if and only if φ^S is satisfiable.

Proof By Theorem 3.18, we may assume that φ is in conjunctive prenex normal form. By induction, it suffices to show that φ is satisfiable if and only if $s(\varphi)$ is satisfiable. There are two possibilities for $s(\varphi)$.

Case 1: If φ' has the form $\exists x_1 Q_2 x_2 \cdots Q_m x_m \varphi_0(x_1, \ldots, x_m)$, then $s(\varphi')$ is $Q_2 x_2 \cdots Q_m x_m \varphi_0(c, x_2, \ldots, x_m)$ for some constant c. Let $\psi(x_1)$ be the formula

$$Q_2 x_2 \cdots Q_m x_m \varphi_0(x_1, \ldots, x_m)$$

so that φ' is $\exists x_1 \psi(x_1)$ and $s(\varphi')$ is $\psi(c)$. By the semantics for $\exists$,

$$M \models \exists x_1 \psi(x_1) \text{ if and only if } M_C \models \psi(c),$$

where M_C is an expansion of M by constants one of which is c. It follows that $\exists x_1 \psi(x_1)$ is satisfiable if and only if $\psi(c)$ is satisfiable.

Case 2: If φ' is $\forall x_1 \cdots \forall x_{i-1} \exists x_i Q_{i+1} x_{i+1} \cdots Q_m x_m \varphi_0(x_1, \ldots, x_m)$ then $s(\varphi')$ is the sentence

$$\forall x_1 \cdots \forall x_{i-1} Q_{i+1} x_{i+1} \cdots Q_m x_m \varphi_0(x_1, \ldots, x_{i-1}, f(x_1, \ldots, x_{i-1}), x_{i+1}, \ldots, x_m)$$

where f is an $(i-1)$-ary function symbol that does not occur in φ'.

Now let $\psi(x_1, \ldots, x_i)$ be the formula $Q_{i+1} x_{i+1} \cdots Q_m x_m \varphi_0(x_1, \ldots, x_m)$. Suppose that $\forall x_1 \cdots \forall x_{i-1} \exists x_i \psi(x_1, \ldots, x_i)$ is satisfiable. Let M be a model.

Let M_f be an expansion of M_C that interprets f in such a way that for all constants $c_1, \ldots, c_{i-1}$, $M_f \models \psi(c_1, \ldots, c_{i-1}, f(c_1, \ldots, c_{i-1}))$. Then

$$M_f \models \forall x_1 \cdots \forall x_{i-1} \psi(x_1, \ldots, x_{i-1}, f(x_1, \ldots, x_{i-1})).$$

So if $\forall x_1 \cdots \forall x_{i-1} \exists x_i \psi(x_1, \ldots, x_i)$ is satisfiable, then so is

$$\forall x_1 \cdots \forall x_{i-1} \psi(x_1, \ldots, x_{i-1}, f(x_1, \ldots, x_{i-1})).$$

Conversely, if $M \models \forall x_1 \cdots \forall x_{i-1} \psi(x_1, \ldots, x_{i-1}, f(x_1, \ldots, x_{i-1}))$, then, by the meaning of $\exists$, $M \models \forall x_1 \cdots \forall x_{i-1} \exists x_i \psi(x_1, \ldots, x_i)$.

It follows that φ is satisfiable if and only if $s(\varphi)$ is satisfiable. □

Note that φ and φ^S are not necessarily equivalent. Theorem 3.22 merely states that one is satisfiable if and only if the other is. For example, if φ is the sentence $\exists x \psi(x)$ for atomic $\psi(x)$, then φ^S is $\psi(c)$ which is equivalent to $\forall x \psi(x)$. Of course, $\exists x \psi(x)$ and $\forall x \psi(x)$ are not equivalent sentences, but if one of these sentences is satisfiable, then so is the other. For our purposes, this is all we need. To determine whether φ is satisfiable, it suffices to determine whether φ^S is satisfiable.

3.3 Herbrand theory

In this section we "reduce" sentences of first-order logic to sets of sentences in propositional logic. More precisely, given φ in SNF we find a (possibly infinite) set $E(\varphi)$ of sentences of propositional logic such that φ is satisfiable if and only if $E(\varphi)$ is satisfiable. We know $E(\varphi)$ is unsatisfiable if and only if $\emptyset \in Res^*(E(\varphi))$. So we can use the method of resolution from propositional logic to show that a first-order sentence φ in SNF is unsatisfiable. By Theorem 3.22, we can use this method to determine whether *any* sentence of first-order logic is unsatisfiable.

The method we describe in this section will not necessarily tell us if a sentence φ is satisfiable. Since $E(\varphi)$ may be infinite, there may be no way to tell whether $\emptyset$ is *not* in $Res^*(E(\varphi))$. But if $\emptyset$ is in $Res^*(E(\varphi))$, then, by the compactness of propositional logic, we can derive it in a finite number of steps. Recall that to show that φ is satisfiable, we must exhibit a model for φ. We have done this in previous examples. But to show that φ is unsatisfiable, we must show that it does not hold in *any* structure. Previously, we had no way of doing this. Theorem 3.25 provides the key. We show that, in certain circumstances, it suffices to show that φ does not hold in a specific type of structure called a *Herbrand structure*.

3.3.1 Herbrand structures.

Definition 3.23 Let $\mathcal{V}$ be a vocabulary. The *Herbrand universe* for $\mathcal{V}$ is the set of all variable free $\mathcal{V}$-terms.

For example, if $\mathcal{V}$ contains constant a and unary function f, then the Herbrand universe for $\mathcal{V}$ contains a, $f(a)$, $f(f(a))$, and so forth. If, in addition, $\mathcal{V}$ contains a binary function g, then the Herbrand universe will also contain $g(a,a)$, $g(a, f(a))$, $g(g(f(a), f(a)), f(f(a)))$, $f(g(f(a), a))$, and so forth.

Recall that a $\mathcal{V}$-structure is a set together with an interpretation for each of the symbols in $\mathcal{V}$. Suppose that we take the Herbrand universe for $\mathcal{V}$ as our underlying set. Call this set H. If $\mathcal{V}$ has no constant symbols, then H is empty. Suppose this is not the case. Then we can turn the Herbrand universe H into a $\mathcal{V}$-structure by giving an interpretation for $\mathcal{V}$. There is a natural interpretation for each of the constants and functions on H. Any $\mathcal{V}$-structure that has H as its underlying set and interprets the constants and functions in this "natural" way is called a *Herbrand $\mathcal{V}$-structure*.

For example, suppose $\mathcal{V} = \{f, R, c\}$ where f is a unary function, R is a binary relation, and c is a constant. Then the Herbrand universe for $\mathcal{V}$ is

$$H = \{c, f(c), f(f(c)), f(f(f(c))), \ldots\}.$$

Let $M = (H \,|\, f, R, c)$ be a $\mathcal{V}$-structure having underlying set H. The set H has an element called c (the first element in the above listing of H). If M interprets the constant c as any element of H other than the one denoted by c, there would be serious ambiguity. If M is a Herbrand structure, then there is no ambiguity, the constant c in $\mathcal{V}$ is interpreted as the element c of H. This is the natural interpretation for the constant c. Likewise, there is a natural interpretation for the function f. The interpretation assigns to f a function from H to H. Given an element of H as input, f outputs an element of H. If M is a Herbrand structure, then the function f when applied to the element c outputs the element $f(c)$ (the second element in the above listing of H). Likewise, given input $f(c)$, f outputs the element of H denoted by $f(f(c))$. This is the natural interpretation of f on H.

So a $\mathcal{V}$-structure M is a Herbrand structure if it has universe H and interprets the constants and functions in the manner suggested by the names given to the elements of H. It is a Herbrand structure regardless of how the relations are interpreted. So, if $\mathcal{V}$ contains a relation and a constant, then there are many Herbrand $\mathcal{V}$-structures.

Let H be the Herbrand universe and let M be a Herbrand structure for the vocabulary $\mathcal{V}$. We list a few basic facts:

- H is empty if and only if $\mathcal{V}$ contains no constants.
- H is finite if and only if $\mathcal{V}$ contains no functions.
- M is the unique Herbrand $\mathcal{V}$-structure if and only if $\mathcal{V}$ contains no relations or H is empty.

Proof theory 115

Definition 3.24 Let Γ be a set of sentences. The *Herbrand vocabulary for* Γ, denoted $\mathcal{V}_\Gamma$, is defined as follows. Let $\mathcal{V}_0$ be the set of functions, relations, and constants occurring in Γ. If $\mathcal{V}_0$ contains no constants, then $\mathcal{V}_\Gamma = \mathcal{V}_0 \cup \{c\}$. Otherwise, $\mathcal{V}_\Gamma = \mathcal{V}_0$. The *Herbrand universe for* Γ, denoted $H(\Gamma)$, is the Herbrand universe for $\mathcal{V}_\Gamma$. M is a *Herbrand model of* Γ, if M is a Herbrand $\mathcal{V}_\Gamma$-structure and $M \models \varphi_i$ for each φ_i in Γ.

In the case where Γ contains a single sentence φ, we will replace Γ in the above notation with φ.

Consider, for example, the sentence $\forall x ((f(x) \neq x) \wedge (f(f(x)) = x))$. Call this sentence φ. The Herbrand vocabulary for φ is $\mathcal{V}_\varphi = \{f, c\}$, where f is a unary function and c is a constant. The Herbrand universe for φ is $H(\varphi) = \{c, f(c), f(f(c)), \ldots\}$. In any Herbrand $\mathcal{V}_\varphi$-structure, c and $f(f(c))$ are distinct elements of the universe $H(\varphi)$. Since φ asserts that for all x, $f(f(x)) = x$, the sentence φ has no Herbrand model. Yet φ is satisfiable (find a model for φ). The following theorem shows that this only happens when φ uses the symbol "=". So if φ is a satisfiable sentence that is equality-free, then φ has a Herbrand model.

Theorem 3.25 Let $\Gamma = \{\varphi_1, \varphi_2, \ldots\}$ be a set of equality-free sentences in SNF. Then Γ is satisfiable if and only if Γ has a Herbrand model.

Proof If Γ has a Herbrand model, then, of course, Γ is satisfiable.

Conversely, suppose Γ is satisfiable. Let $\mathcal{V}_\Gamma$ be the Herbrand vocabulary for Γ. Let N be a $\mathcal{V}_\Gamma$-structure that models each $\varphi_i \in \Gamma$. Let M' be a Herbrand $\mathcal{V}_\Gamma$-structure.

We define a $\mathcal{V}_\Gamma$-structure M that is a hybrid of N and M'. The universe of M is $H(\Gamma)$, the Herbrand universe for $\mathcal{V}_\Gamma$. Let M interpret functions and constants the same way as M' and relations the same way as N. Since, M and N may have different universes, this requires some explaining.

"M interprets functions and constants the same way as M'" means that M is a Herbrand $\mathcal{V}_\Gamma$-structure. To complete our description of M we must say how M interprets relations. For any n-ary relation R in $\mathcal{V}_\Gamma$ and $t_1, \ldots, t_n$ in the universe $H(\Gamma)$ of M, we must say whether $M \models R(t_1, \ldots, t_n)$ or $M \models \neg R(t_1, \ldots, t_n)$. Since each $t_i \in H(\Gamma)$ is a variable free $\mathcal{V}_\Gamma$-term and N is a $\mathcal{V}_\Gamma$-structure, either $N \models R(t_1, \ldots, t_n)$ or $N \models \neg R(t_1, \ldots, t_n)$. We define M so that $M \models R(t_1, \ldots, t_n)$ if and only if $N \models R(t_1, \ldots, t_n)$.

The theorem follows from two claims.

Claim 1 For any $\mathcal{V}_\Gamma$-sentence ψ that is both quantifier-free and equality-free, $M \models \psi$ if and only if $N \models \psi$.

Claim 2 For any SNF $\mathcal{V}_\Gamma$-sentence ψ that is equality-free, if $N \models \psi$ then $M \models \psi$.

If Claim 2 is true, then M must model Γ. This is because, for each $\varphi_i \in \Gamma$, $N \models \varphi_i$ and φ_i is in SNF and equality-free. Since M is a Herbrand $\mathcal{V}_\Gamma$-structure, M is a Herbrand model of Γ. So if we can prove Claim 2, then the theorem follows. We first prove Claim 1, and then show that Claim 2 follows from Claim 1.

Proof of Claim 1 Let ψ be quantifier-free. We show that $M \models \psi$ if and only if $N \models \psi$ by induction on the complexity of ψ.

If ψ is atomic, then, since ψ does not use "=", ψ must be $R(t_1, \ldots, t_n)$ for some n-ary R in $\mathcal{V}_\Gamma$ and $\mathcal{V}_\Gamma$-terms t_i. Since ψ is a sentence, each t_i must be variable free. That is, each t_i is in $H(\Gamma)$. By the definition of M, $M \models \psi$ if and only if $N \models \psi$.

Suppose $M \models \psi_1$ if and only if $N \models \psi_1$ and $M \models \psi_2$ if and only if $N \models \psi_2$. Then clearly, $M \models \neg\psi_1$ if and only if $N \models \neg\psi_1$ and $M \models \psi_1 \wedge \psi_2$ if and only if $N \models \psi_1 \wedge \psi_2$. It follows that $M \models \psi$ if and only if $N \models \psi$ for any quantifier-free sentence ψ, completing the proof of Claim 1.

Proof of Claim 2 We prove this claim by induction on the number of quantifiers in ψ. If ψ has no quantifiers, then by Claim 1, $M \models \psi$ if and only if $N \models \psi$.

Suppose ψ is $\forall x_1 \cdots \forall x_n \psi_0(x_1, \ldots, x_n)$ where ψ_0 is quantifier (and equality) free. Our induction hypothesis is that Claim 2 holds for any equality-free sentence in SNF having fewer than n quantifiers. Let t be a variable free $\mathcal{V}_\Gamma$-term. Let $\psi'(x_1)$ be the formula $\forall x_2 \cdots \forall x_n \psi_0(x_1, x_2, \ldots, x_n)$ obtained by removing the first quantifier from the sentence ψ. Let t be any variable free $\mathcal{V}_\Gamma$-term. That is, t is in $H(\Gamma)$. We have

$N \models \psi$ implies $N \models \psi'(t)$ (by the semantics of $\forall$)

which implies $M \models \psi'(t)$ (by our induction hypothesis).

So if $N \models \psi$ then $M \models \psi'(t)$. But t was an arbitrary element of $H(\Gamma)$. So, if $N \models \psi$, then $M \models \psi'(t)$ for all $t \in H(\Gamma)$. Since $H(\Gamma)$ is the universe of M, $M \models \forall x_1 \psi'(x_1)$ (by the semantics of $\forall$). Since $\forall x_1 \psi'(x_1)$ and ψ are the same, the proof of Claim 2 is complete. $\square$

In particular, if Γ from the previous theorem contains a single sentence φ, we get the following.

Corollary 3.26 Let φ be an equality-free sentence in SNF. Then φ is satisfiable if and only if φ has a Herbrand model.

3.3.2 Dealing with equality.
Now suppose φ is in SNF and does use "=". We define a formula φ_E that does not use equality. Whereas Corollary 3.26 does not

apply to φ, it does apply to φ_E. Moreover, we prove that φ is satisfiable if and only if φ_E is satisfiable.

Let $\mathcal{V}$ be the vocabulary of φ. That is, $\mathcal{V}$ is the finite set of constants, relations, and functions that occur in φ. Let E be a binary relation that is not in $\mathcal{V}$. Let $\varphi_{\neq}$ be the sentence obtained by replacing each occurrence of $t_1 = t_2$ in φ (for $\mathcal{V}$-terms t_1 and t_2) with $E(t_1, t_2)$. Let φ_{ER} be the following sentence.

$$\forall x \forall y \forall z (E(x,x) \land (E(x,y) \leftrightarrow E(y,x)) \land (E(x,y) \land E(y,z) \rightarrow E(x,z))).$$

This sentence says "E is an equivalence relation."

For each relation R in $\mathcal{V}$, let φ_R be the formula

$$\forall x_1 \cdots \forall x_n \forall y_1 \cdots \forall y_n \left(\left(\bigwedge_{i=1}^n E(x_i, y_i) \land R(x_1, \ldots, x_n) \right) \rightarrow R(y_1, \ldots, y_n) \right),$$

where n is the arity of R. Let φ_1 be the conjunction of all φ_R taken over all relations $R \in \mathcal{V}$.

Likewise, for each function f in $\mathcal{V}$, let φ_f be the formula

$$\forall x_1 \cdots \forall x_n \forall y_1 \cdots \forall y_n \left(\bigwedge_{i=1}^n E(x_i, y_i) \rightarrow E(f(x_1, \ldots, x_n), f(y_1, \ldots, y_n)) \right),$$

where n is the arity of f. Let φ_2 be the conjunction of all φ_f taken over all functions $f \in \mathcal{V}$.

Now let φ'_E be the sentence $\varphi_{\neq} \land \varphi_{ER} \land \varphi_1 \land \varphi_2$.

The formulas φ_{ER}, φ_1, and φ_2 together say that the binary relation E behaves like equality. Note that $\varphi_{\neq}$, φ_{ER}, φ_1, and φ_2 are each equality-free formulas in SNF. If we put φ'_E into prenex normal form (by pulling the quantifiers out front, renaming variables if need be) we obtain an equality-free formula φ_E that is in SNF.

Lemma 3.27 For any formula φ in SNF, φ is satisfiable if and only if φ_E is satisfiable.

Proof Let $\mathcal{V}$ be the vocabulary of φ and let $\mathcal{V}_E = \mathcal{V} \cup \{E\}$ where E is a binary relation that is not in $\mathcal{V}$.

If $M \models \varphi$, then we can obtain a model for φ_E by interpreting E as equality in M.

Conversely, suppose φ_E has a model N. Then E is an equivalence relation on N. Let U be the underlying set of N and let U/E be the set of all E-equivalence classes in U. We define a $\mathcal{V}$-structure N_E having U/E as an underlying set. We must say how N_E interprets the constants, relations, and functions of $\mathcal{V}$.

For each $a \in N$, let $[a]$ denote the E-equivalence class containing a.

For each constant c in $\mathcal{V}$, N_E interprets c as $[c]$, the E-equivalence class of the interpretation of c in N.

Let R be an n-ary relation in $\mathcal{V}$. For any n-tuple $([a_1], \ldots, [a_n])$ of elements of U/E,

$$N_E \models R([a_1], \ldots, [a_n]) \text{ if and only if } N \models R(a_1, \ldots, a_n).$$

Let f be an n-ary relation in $\mathcal{V}$. For any $[b] \in U/E$ and n-tuple $([a_1], \ldots, [a_n])$ of elements of U/E,

$$N_E \models f([a_1], \ldots, [a_n]) = [b] \text{ if and only if } N \models f(a_1, \ldots, a_n) = b.$$

Because N models both φ_1 and φ_2, the structure N_E is well defined.

Finally, it can be shown that $N_E \models \varphi_E$ by induction on the complexity of φ. □

Example 3.28 Consider the sentence $\forall x((f(x) \neq x) \land (f(f(x)) = x))$. If this sentence is φ, then $\varphi_{\neq}$ is the sentence

$$\forall x(\neg E(f(x), x) \land E(f(f(x)), x))$$

and φ_2 is the sentence

$$\forall x \forall y (E(x, y) \to E(f(x), f(y))).$$

Since φ contains no relations, we need not consider φ_1. The conjunction φ'_E of $\varphi_{\neq}$, φ_2, and φ_{ER}, is equivalent to the following sentence φ_E in SNF.

$$\forall x \forall y \forall z (\neg E(f(x), x) \land E(f(f(x)), x) \land (E(x, y) \to E(f(x), f(y)))$$
$$\land E(x, x) \land (E(x, y) \leftrightarrow E(y, x)) \land ((E(x, y) \land E(y, z)) \to E(x, z))).$$

Now, by Corollary 3.26, φ_E has a Herbrand model. That is, there is a model for φ_E having universe $H(\varphi) = \{c, f(c), f(f(c)), \ldots\}$. Indeed, we may interpret E on $H(\varphi)$ to be the equivalence relation having the following two classes:

$$C_{\text{odd}} = \{t \in H(\varphi) \,|\, t \text{ has an odd number of } fs\}, \text{ and}$$
$$C_{\text{even}} = \{t \in H(\varphi) \,|\, t \text{ has an even number of } fs\}.$$

It follows that φ has a model having only two elements. Let N be the structure having universe $\{c_{\text{odd}}, c_{\text{even}}\}$ that interprets the function f by the rule $f(c_{\text{odd}}) = c_{\text{even}}$ and $f(c_{\text{even}}) = c_{\text{odd}}$. Clearly, $N \models \varphi$.

3.3.3 The Herbrand method. We now describe a method for determining whether an arbitrary sentence φ of first-order logic is unsatisfiable. We have shown that we may assume φ is equality-free and is in SNF. Let φ be

$$\forall x_1 \cdots \forall x_n \varphi_0(x_1, \ldots, x_n),$$

where φ_0 is quantifier-free and equality-free. Let $H(\varphi)$ be the Herbrand universe of φ. Let $E(\varphi)$ be the set

$$\{\varphi_0(t_1, \ldots, t_n) \mid t_1, \ldots, t_n \in H(\varphi)\}.$$

So $E(\varphi)$ is the set obtained by substituting terms from $H(\varphi)$ for the variables of φ_0 in every possible way. Let $\{\varphi_1, \varphi_2, \ldots\}$ be an enumeration of $E(\varphi)$.

We claim that φ is satisfiable if and only if $E(\varphi)$ is satisfiable. If M is a model of φ, then $M \models \forall x_1 \cdots \forall x_n \varphi_0(x_1, \ldots, x_n)$. In particular, $M \models \varphi_0(t_1, \ldots, t_n)$ for all variable free $\mathcal{V}_\varphi$-terms t_i. That is, M models each φ_i in $E(\varphi)$ and so $E(\varphi)$ is satisfiable.

Conversely, suppose $E(\varphi)$ is satisfiable. Then, by Theorem 3.25, $E(\varphi)$ has a Herbrand model M. Note that the Herbrand vocabulary for $E(\varphi)$ is the same as the Herbrand vocabulary for φ. So the universe of M is $H(\varphi)$. For each $t_1, \ldots, t_n$ in $H(\varphi)$, M models $\varphi_0(t_1, \ldots, t_n)$ since this sentence is in $E(\varphi)$. It follows from the semantics of $\forall$ that $M \models \forall x_1 \cdots \forall x_n \varphi_0(x_1, \ldots, x_n)$. That is, $M \models \varphi$ and φ is satisfiable.

So φ is unsatisfiable if and only if $E(\varphi)$ is unsatisfiable. Since $E(\varphi)$ contains sentences with no quantifiers, we can view $E(\varphi)$ as a set of sentences of propositional logic. Since φ is in SNF, each φ_i in $E(\varphi)$ is in CNF. We know from propositional logic that the set $E(\varphi)$ is unsatisfiable if and only if $\emptyset \in Res^*(E(\varphi))$. By the compactness of propositional logic, $E(\varphi)$ is unsatisfiable if and only if some finite subset $\{\varphi_1, \ldots, \varphi_m\}$ is unsatisfiable. So if φ is unsatisfiable, then $\emptyset \in Res^*(\{\varphi_1, \ldots, \varphi_m\})$ for some m.

This gives us a method for showing that φ is unsatisfiable. Check if $\emptyset$ is in $Res^*(\{\varphi_1, \ldots, \varphi_m\})$ for some m. Recall that $Res^*(\{\varphi_1, \ldots, \varphi_m\})$ is a finite set. If $\emptyset$ is in $Res^*(\{\varphi_1, \ldots, \varphi_m\})$ we stop and conclude that φ must be unsatisfiable. Otherwise we continue and check $Res^*(\{\varphi_1, \ldots, \varphi_m, \varphi_{m+1}\})$. If φ is unsatisfiable, then this method will eventually find $\emptyset$ and conclude that φ is unsatisfiable in a finite number of steps. If φ is satisfiable, however, this procedure will continue forever.

So, in principle, we have a method to show that a given sentence of first-order logic is unsatisfiable. The first step is to find φ that is in SNF and does not use "=". This can be done relatively quickly (in polynomial time). But to show that $\emptyset \in Res^*(E(\varphi))$ can take an arbitrarily large amount of time. This method is far from efficient. Even if $\emptyset$ is in $Res^*(E(\varphi))$, it may take a very long time to find it.

In the next section, we define another way to show that a formula is unsatisfiable. We define resolution for first-order logic. This method is not polynomial time, but it is more systematic than the method described here. Herbrand theory will be useful in proving that the resolution we define works.

3.4 Resolution for first-order logic

We now define resolution for first-order logic. Let φ be any sentence in SNF. Then φ has the form $\forall x_1 \forall x_2 \cdots \forall x_m \varphi_0$ where φ_0 is a conjunction of disjunctions of literals. In particular, φ_0 is quantifier-free, and so it can be viewed as a formula of propositional logic that is in CNF. Let $\mathcal{C}(\varphi_0)$ denote the set of all clauses in the CNF formula φ_0. We define $\mathcal{C}(\varphi)$ to be $\mathcal{C}(\varphi_0)$. That is, $\mathcal{C}(\varphi) = \{C_1, \ldots, C_m\}$ where C_i is the set of all literals occurring in the ith disjunction.

For example, if φ is the sentence

$$\varphi = \forall x \forall y \forall z ((P(x,y) \vee \neg Q(x,z)) \wedge ((R(x,y,z) \vee \neg P(f(x,y),z)),$$

then $\mathcal{C}(\varphi)$ is the set

$$\{\{P(x,y), \neg Q(x,z)\}, \{R(x,y,z), \neg P(f(x,y),z)\}\}.$$

Note that a sentence φ in SNF uniquely determines $\mathcal{C}(\varphi)$. Conversely, by Proposition 1.67, $\mathcal{C}(\varphi_0)$ determines φ_0 up to equivalence. It follows that $\mathcal{C}(\varphi)$ determines φ up to equivalence. That is, if $\mathcal{C}(\varphi) = \mathcal{C}(\psi)$ for sentences φ and ψ in SNF, then $\varphi \equiv \psi$. For this reason, we need not distinguish between formulas in SNF and sets of clauses.

We want to say what it means for a clause R to be a *resolvent* of two clauses C_1 and C_2. As in propositional logic, a resolvent of C_1 and C_2 is a consequence of the conjunction of C_1 and C_2. Before giving a formal definition for resolvents, we consider a couple of examples.

Example 3.29 Let $C_1 = \{\neg Q(x,y), P(f(x),y)\}$ and $C_2 = \{\neg P(f(x),y), R(x,y,z)\}$. The clause $R = \{\neg Q(x,y), R(x,y,z)\}$ is a resolvent of C_1 and C_2. This works the same way as in propositional logic. Since the literal $P(f(x),y)$ occurs in one clause and the negation of this same literal occurs in the other, the resolvent can be formed by taking the union of C_1 and C_2 less $P(f(x),y)$ and $\neg P(f(x),y)$.

Example 3.30 Let $C_1 = \{\neg Q(x,y), P(f(x),y)\}$ and $C_2 = \{\neg P(z,y), R(x,y,z)\}$. Then we cannot directly find a resolvent of C_1 and C_2 as in the previous example. Let C_2' be the clause obtained by substituting $f(x)$ for z in the clause C_2. That is, $C_2' = \{\neg P(f(x),y), R(x,y,f(x))\}$. We make two observations. First, we can easily find a resolvent of C_1 and C_2', namely $R = \{\neg Q(x,y), R(x,y,f(x))\}$. Second, note that C_2' is a consequence of C_2. This is because the SNF sentence represented by C_2 asserts that the formula $\neg P(z,y) \vee R(x,y,z)$ holds for every

x, y, and z. In particular, this formula holds in the specific case where $z = f(x)$. That is, C_2 implies C_2'. Hence, R, which is a consequence of $\{C_1, C_2'\}$, is also a consequence of $\{C_1, C_2\}$. We define resolvents so that R is a resolvent of C_1 and C_2 (and of C_1 and C_2' as well). We diagram this situation as follows:

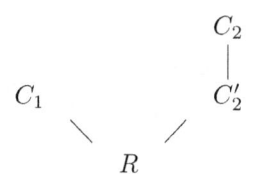

$$\begin{array}{ccc} & & C_2 \\ & & | \\ C_1 & & C_2' \\ & \searrow \quad \swarrow & \\ & R & \end{array}$$

So prior to finding a resolvent, we must first make substitutions for variables to make certain literals look the same. In the previous example, we did a substitution that made $P(f(x), y)$ and $P(z, y)$ identical. This process is called *unification* and we postpone the formal definition of "resolvent" until after we have discussed unification in detail.

3.4.1 Unification. Let $\mathbb{L} = \{L_1, \ldots, L_n\}$ be a set of literals. We say $\mathbb{L}$ is *unifiable* if there exist variables $x_1, \ldots, x_m$ and terms $t_1, \ldots, t_m$ such that substituting t_i for x_i (for each i) makes each literal in $\mathbb{L}$ look the same. We denote such a substitution by $sub = (x_1/t_1, x_2/t_2, \ldots, x_m/t_m)$. For any sentence φ in SNF, we denote the result of applying this substitution to φ by φsub.

For example, if $sub = (x/w, y/f(a), z/f(w))$ and $\varphi = \{\neg Q(x, y), R(a, w, z)\}$, then $\varphi sub = \{\neg Q(w, f(a)), R(a, w, f(w))\}$.

If $\mathbb{L}$ is a set of literals, then $\mathbb{L}sub$ denotes the set of all $L_i sub$ such that $L_i \in \mathbb{L}$. So $\mathbb{L}$ is unifiable if and only if there exists a substitution sub such that $\mathbb{L}sub$ contains only one literal. If this is the case, we call sub a *unifier* for $\mathbb{L}$ and say that sub unifies $\mathbb{L}$.

Example 3.31 Let $\mathbb{L} = \{P(f(x), y), P(f(a), w)\}$. Let $sub_1 = (x/a, y/w)$ and $sub_2 = (x/a, y/a, w/a)$. Then both sub_1 and sub_2 unify $\mathbb{L}$. We have $\mathbb{L}sub_1 = \{P(f(a), w)\}$ and $\mathbb{L}sub_2 = \{P(f(a), a)\}$. Note that, by making another substitution, we can get $\mathbb{L}sub_2$ from $\mathbb{L}sub_1$. Namely, if $sub_3 = (w/a)$, then $sub_1 sub_3$ (sub_1 followed by sub_3) has the same effect as sub_2. However, we cannot generate $\mathbb{L}sub_1$ from $\mathbb{L}sub_2$ since $\mathbb{L}sub_2$ has no variables. So, in some sense, the unifier sub_1 is better for our purposes. It is more versatile. "Our purposes" will be resolution, and if we choose sub_2 as our unifier instead of sub_1, we might needlessly limit our options.

Definition 3.32 Let $\mathbb{L}$ be a set of literals. The substitution sub is a *most general unifier* for $\mathbb{L}$ if it unifies $\mathbb{L}$ and for any other unifier sub' for $\mathbb{L}$, we have $sub\, sub' = sub'$.

In Example 3.31, sub_1 is the most general unifier. As we pointed out, this is the best unifier for our purposes.

Proposition 3.33 A finite set of literals is unifiable if and only if it has a most general unifer.

There are two possiblities for a finite set $\mathbb{L}$ of literals, either it is unifiable or it is not. Proposition 3.33 asserts that if $\mathbb{L}$ is unifiable, then it automatically has a most general unifier. We prove this by exhibiting an algorithm that, given $\mathbb{L}$ as input, outputs "not unifiable" if no unifier exists and otherwise ouputs a most gerneal unifier for $\mathbb{L}$. The algorithm runs as follows.

The unification algorithm

Given: a finite set of literals $\mathbb{L}$.

Let $\mathbb{L}_0 = \mathbb{L}$ and $sub_0 = \emptyset$.

Suppose we know $\mathbb{L}_k$ and sub_k. If $\mathbb{L}_k$ contains just one literal, output "$sub_0 sub_1 \cdots sub_k$ is a most general unifier for $\mathbb{L}$."

Otherwise, there exist L_i and L_j in $\mathbb{L}_k$ such that the nth symbol of L_i differs from the nth symbol of L_j (for some n). Suppose n is least in this regard. If the nth symbol of L_i is a variable v and the nth symbol of L_j is the first symbol of a term t that does not contain v or vice versa (with L_i and L_j reversed) then:

Let $sub_{k+1} = (v/t)$ and $\mathbb{L}_{k+1} = \mathbb{L}_k sub_{k+1}$.

If any of the hypotheses of the previous sentence do not hold, output "$\mathbb{L}$ is not unifiable."

We must verify that this algorithm works. First we give a demonstration.

Example 3.34 Let $\mathbb{L} = \{R(f(g(x)), a, x), R(f(g(a)), a, b), R(f(y), a, z)\}$. First set $\mathbb{L}_0 = \mathbb{L}$ and $sub_0 = \emptyset$.

As we read each of the three literals in $\mathbb{L}_0$ from left to right, we see that each begins with "$R(f(\ldots$", but then there is a discrepency. Whereas the second literal continues with "$g(a)$", the third literal has "y". We check that one of these two terms is a variable and the other is a term that does not contain that variable. This is the case and so we let

$sub_1 = (y/g(a))$, and
$\mathbb{L}_1 = \mathbb{L}_0 sub_1 = \{R(f(g(x)), a, x), R(f(g(a)), a, b), R(f(g(a)), a, z)\}$.

We note that $\mathbb{L}_1$ contains more than one literal and proceed. Now all literals begin with $R(f(g(\ldots$, but then the first literal has "x" and the second has "a".

One of these is a variable and the other is a term that does not contain that variable, and so we let

$sub_2 = (x/a)$, and
$\mathbb{L}_2 = \mathbb{L}_1 sub_2 = \{R(f(g(a)), a, a), R(f(g(a)), a, b), R(f(g(a)), a, z)\}$.

The set $\mathbb{L}_2$ still contains more than one literal, and so we continue. Each literal in $\mathbb{L}_2$ looks the same up to $R(f(g(a)), a, \ldots$, but then the first literal has "a" and the second has "b." Neither of these is a variable, and so the algorithm concludes with output "$\mathbb{L}$ is not unifiable."

If the algorithm outputs "not unifiable," it is for one of two reasons. One is illustrated by the previous example. Here we had a discrepency between two literals that did not involve a variable. Where one literal had the constant a, the other had b. Clearly, this cannot be reconciled by a substitution and the set is, in fact, not unifiable. The other possibility is that the dicrepency involves a variable and a term, but the variable occurs in the term. For example, the set $\{P(x, y), P(x, f(y))\}$ is not unifiable. No matter what we substitute for the variables x and y, the second literal will have one more occurrence of f than the first literal. The algorithm, noting a discrepency occurs with y and $f(y)$, will terminate with "not unifiable" because the variable y occurs in the term $f(y)$. Both reasons for concluding "not unifiable" are good reasons. If the algorithm yields this output, then the set must not be unifiable.

Note that, when applied to the set $\mathbb{L}$ from Example 3.31, this algorithm outputs sub_1 as the most general unifier. So, in these examples, the algorithm works. We want to show that it always works.

If the set $\mathbb{L}$ is a finite set, then only finitely many variables occur in $\mathbb{L}$. It follows that the algorithm when applied to $\mathbb{L}$ must terminate in a finite number of steps. If it terminates with "$\mathbb{L}$ is not unifiable," then, as we have already mentioned, $\mathbb{L}$ must not be unifiable. Otherwise, the algorithm outputs "$sub_0 sub_1 \cdots sub_k$ is a most genral unifier." We must show that, when this statement is the output, it is true.

The algorithm outputs "$sub_0 sub_1 \cdots sub_k$ is the most genral unifier" only if $\mathbb{L}_k = \mathbb{L} sub_0 sub_1 \cdots sub_k$ contains just one literal. If this is the output, then $sub_0 sub_1 \cdots sub_k$ is a unifier. We must show that it is a most general unifier.

Let sub' be any other unifier for $\mathbb{L}$. We know that $sub_0 sub' = sub'$ because sub_0 is empty. Now suppose that we know $sub_0 \cdots sub_m sub' = sub'$ for some m, $0 \leq m < k$. Then $\mathbb{L}_m sub' = \mathbb{L} sub_0 \cdots sub_m sub' = \mathbb{L} sub' = \{L\}$. That is, since sub' unifies $\mathbb{L}$, it also unifies $\mathbb{L}_m$.

Suppose sub_{m+1} is (x/t). By the definition of the algorithm, t must be a term in which the variable x does not occur. Moreover, for some literals L_i and L_j in $\mathbb{L}_m$, x occurs in the nth place of L_i and t begins in the nth place of L_j

(for some n). Since sub' unifies $\mathbb{L}_m$, sub' must do the same thing to both x and t. That is, $xsub' = tsub'$. It follows that $sub_{m+1}sub' = (x/t)sub' = sub'$. By induction, we have $sub_0 \cdots sub_{m+1}sub' = sub'$ for all $m < k$. In particular, $sub_0 \cdots sub_k sub' = sub'$ and $sub_0 \cdots sub_k$ is the most general unifier for $\mathbb{L}$.

3.4.2 Resolution. We now define resolution for first-order logic. Recall that for any literal L, $\overline{L}$ is the literal defined by $\overline{L} = \neg L$ or $\neg \overline{L} = L$.

Definition 3.35 Let C_1 and C_2 be two clauses. Let s_1 and s_2 be any substitutions such that $C_1 s_1$ and $C_2 s_2$ have no variables in common. Let $L_1, \ldots, L_m \in C_1 s_1$ and $L'_1, \ldots, L'_n \in C_2 s_2$ be such that $\mathbb{L} = \{\overline{L}_1, \ldots, \overline{L}_m, L'_1, \ldots, L'_n\}$ is unifiable. Let sub be a most general unifier for $\mathbb{L}$.

Then $R = [(C_1 s_1 - \{L_1, \ldots, L_m\}) \cup (C_2 s_2 - \{L'_1, \ldots, L'_n\})]sub$ is a *resolvent* of C_1 and C_2.

Let φ be a sentence in SNF. Then $\varphi = \{C_1, \ldots, C_n\}$ for some clauses $C_1, \ldots, C_n$.
Let $Res(\varphi) = \{R|\ R$ is a resolvent of some C_i and C_j in $\varphi\}$.
Let $Res^0(\varphi) = \varphi$, and $Res^{n+1} = Res(Res^n(\varphi))$.
Let $Res^*(\varphi) = \bigcup_n Res^n(\varphi)$.

The same notation was used in propositional logic. However, unlike the propositional case, $Res^*(\varphi)$ may be an infinite set. To justify this notation and the definition of "resolvent" we need to show that $\emptyset \in Res^*(\varphi)$ if and only if φ is unsatisfiable. First we look at an example.

Example 3.36 Let $C_1 = \{Q(x,y), P(f(x), y)\}$, and $C_2 = \{R(x,c), \neg P(f(c), x), \neg P(f(y), h(z))\}$.

Suppose we want to find a resolvent of C_1 and C_2. First, we need to rename some variables since x and y occur in both C_1 and C_2. Let $s_1 = (x/u, y/v)$. Then $C_1 s_1 = \{Q(u,v), P(f(u), v)\}$ which has no variables in common with C_2.

Second, note that $C_1 s_1$ contains a literal of the form $P(\_,\_)$ and C_2 contains literals of the form $\neg P(\_,\_)$. Namely, $P(f(u), v)$ is in $C_1 s_1$ and $\neg P(f(c), x)$ and $\neg P(f(y), h(z))$ are in C_2. Let

$$\mathbb{L} = \{P(f(u), v), P(f(c), x), P(f(y), h(z))\}.$$

By applying the unification algorithm, we see that $\mathbb{L}$ is unifiable and $sub = (u/c, y/c, v/h(z), x/h(z))$ is a most general unifier. We conclude that C_1 and C_2 have resolvent

$$R = [(C_1 s_1 - \{P(f(u), v)\}) \cup (C_2 - \{\neg P(f(c), x), \neg P(f(y), h(z))\})]sub$$
$$= \{Q(u,v), R(x,c)\}sub = \{Q(c, h(z)), R(h(z), c)\}.$$

We verify that the resolvent R from the previous example is in fact a consequence of C_1 and C_2. Recall that C_1 and C_2 represent sentences in SNF.

C_1 represents $\forall x \forall y (Q(x,y)) \vee P(f(x),y))$, and

C_2 represents $\forall x \forall y \forall z (R(x,c)) \vee \neg P(f(c),x) \vee \neg P(f(y), h(z))$.

Suppose C_1 and C_2 hold (in some structure). Then, since these sentences are universal, they hold no matter what we plug in for the variables. In particular,

$$C_1 s_1 sub \equiv \forall z(Q(c, h(z)) \vee P(f(c), h(z))), \text{ and}$$
$$C_2 sub \equiv \forall z(R(h(z),c)) \vee \neg P(f(c), h(z))$$

both hold. That is, $C_1 s_1 sub$ is a consequence of C_1 and $C_2 sub$ is a consequence of C_2. Put another way,

$$C_1 s_1 sub \equiv \forall z(\neg Q(c, h(z)) \to P(f(c), h(z))), \text{ and}$$
$$C_2 sub \equiv \forall z(P(f(c), h(z)) \to R(h(z), c)).$$

From these two sentences, we can deduce

$$\forall z(\neg Q(c, h(z)) \to R(h(z), c))$$

which is equivalent to

$$\forall z(Q(c, h(z)) \vee R(h(z), c))$$

which is the sentence represented by R. Hence, R is a consequence of the conjunction of C_1 and C_2.

In a similar manner, we can show that any resolvent of any two clauses is necessarily a consequence of the conjunction of the two clauses. It follows that if $\emptyset \in Res^*(\varphi)$, then φ must be unsatisfiable. Conversely, suppose φ is unsatisfiable. We need to show that $\emptyset \in Res^*(\varphi)$.

At the end of the previous section we showed that φ is unsatisfiable if and only if the set $E(\varphi)$ is unsatisfiable. Recall that $E(\varphi)$ is the set of all sentences obtained by replacing each variable of φ with a term from the Herbrand universe. These sentences can be viewed as sentences of propositional logic. Suppose that C_1' and C_2' are in $E(\varphi)$ and R' is a resolvent of C_1' and C_2' in the sense of propostional logic. Then there are some clauses C_1 and C_2 of φ such that $C_1' = C_1 sub_1$ and $C_2' = C_2 sub_2$. In the following lemma we show that there exists a resolvent R of C_1 and C_2 (in the sense od first-order logic) and a substitution sub such that $Rsub = R'$. So, essentially, this lemma says that any R' that can be derived from $E(\varphi)$ using propositional resolution can also be derived from φ using first-order resolution.

Lemma 3.37 (Lifting lemma) Let φ be a sentence in SNF. If $R' \in Res(E(\varphi))$, then there exists $R \in Res(\varphi)$ such that such that $Rsub' = R'$ for some substitution sub'.

This is called the "Lifting lemma" because we are "lifting" the resolvent R' from propositional logic to first-order logic.

Let φ be a sentence in SNF and let C_1 and C_2 be two clauses of φ. Let s_1 be a substitution such that $C_1 s_1$ and C_2 have no variables in common. Let C'_1 and C'_2 in $E(\varphi)$ be such that $C_1 s_1 sub_1 = C'_1$ and $C_2 sub_2 = C'_2$ for some substitutions sub_1 and sub_2. Let R' be a resolvent (in propositional logic sense) of C'_1 and C'_2. This setup can be diagramed as follows:

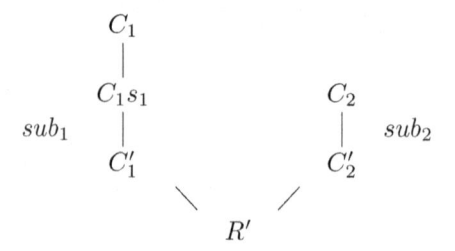

$$\begin{array}{ccc} & C_1 & \\ & | & \\ & C_1 s_1 & \quad C_2 \\ sub_1 \quad | & & | \quad sub_2 \\ & C'_1 \quad\quad C'_2 \\ & \diagdown \quad \diagup \\ & R' & \end{array}$$

The lemma says that if this setup holds, then there exists a resolvent R of $C_1 s_1$ and C_2 (in the sense of first-order logic) such that $Rsub' = R'$ for some substitution sub'. This conclusion can be diagramed as follows:

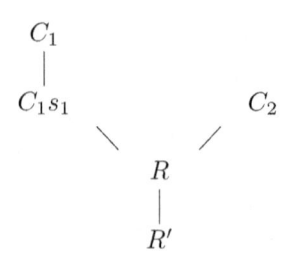

$$\begin{array}{c} C_1 \\ | \\ C_1 s_1 \quad\quad C_2 \\ \diagdown \quad \diagup \\ R \\ | \\ R' \end{array}$$

In the first diagram, the resolvent is taken as in propositional logic. In the second diagram, the resolvent R is as in Definition 3.35. The vertical lines in each diagram refers to a substitution. The lemma can be summarized as saying "if the first diagram holds, then so does the second diagram."

Proof of Lemma Supose the first diagram holds. Then there must exist some literal $L \in C'_1$ such that $\overline{L} \in C'_2$ and $R' = (C'_1 - \{L\}) \cup (C'_2 - \{\overline{L}\})$. This is the definition of resolvent for propositional logic.

Let $sub' = sub_1 sub_2$. Since $C_1 s_1$ and C_2 have no variables in common, $C_1 s_1 sub' = C_1 s_1 sub_1 = C_1'$ and $C_2 sub' = C_2 sub_2 = C_2'$.

Let $\mathbb{L}_1 = \{L_1, \ldots, L_n\}$ be the set of all L_i in $C_1 s_1$ such that $L_i sub' = L$. Likewise, let $\mathbb{L}_2 = \{L_1', \ldots, L_m'\}$ be the set of all L_i' in C_2 such that $L_i' sub' = \overline{L}$. We have the following diagram:

$$\begin{array}{ccccccc} & \mathbb{L}_1 & \subset & C_1 s_1 & & C_2 & \supset & \mathbb{L}_2 \\ sub' & \big| & & \big| & & \big| & & \big| & sub' \\ & L & \in & C_1' & & C_2' & \ni & \overline{L} \\ & & \searrow & & & & \swarrow & \\ & & & & R' & & & \end{array}$$

Let $\mathbb{L} = \{\overline{L}_1, \ldots, \overline{L}_n, L_1', \ldots, L_m'\}$ (that is $\mathbb{L} = \bar{\mathbb{L}}_1 \cup \mathbb{L}_2$). This set is unifiable since $\mathbb{L} sub' = \{\overline{L}\}$. Let sub be a most general unifier for $\mathbb{L}$. Then we can apply Definition 3.35 to find the following resolvent of C_1 and C_2:

$$R = [(C_1 s_1 - \mathbb{L}_1) \cup (C_2 - \mathbb{L}_2)] sub.$$

Referring to the second diagram of the lemma, we see that it remains to be shown that R' can be obtained from R by a substitution. We complete the proof of the lemma by showing that $R sub' = R'$. By applying sub' we get

$$R sub' = [(C_1 s_1 - \mathbb{L}_1) \cup (C_2 - \mathbb{L}_2)] sub\, sub'.$$

Since sub' is a unifier for $\mathbb{L}$ and sub is a most general unifier for $\mathbb{L}$, we know $sub\, sub' = sub'$. So we have

$$R sub' = [(C_1 s_1 - \mathbb{L}_1) \cup (C_2 - \mathbb{L}_2)] sub'$$
$$= (C_1 s_1 sub' - \mathbb{L}_1 sub') \cup (C_2 sub' - \mathbb{L}_2 sub')$$
$$= (C_1' - \{L\}) \cup (C_2' - \{\overline{L}\}) = R'. \qquad \square$$

Corollary 3.38 Let φ be a sentence in SNF. If $C' \in Res^*(E(\varphi))$, then there exists $C \in Res^*(\varphi)$ and a substitution sub' such that $C sub' = C'$.

Proof If $C' \in Res^*(E(\varphi))$, then $C' \in Res^n(E(\varphi))$ for some n. We prove the corollary by induction on n. If $n = 0$, then $C' \in E(\varphi)$. Then, by the definition of $E(\varphi)$, C' is obtained by substituting variable free terms in for the variables of some $C \in \varphi$.

For the induction step, we utilize the Lifting lemma. Suppose that for some m, each clause of $Res^m(E(\varphi))$ is obtained from some clause of $Res^*(\varphi)$ via substitution. Let $\tilde{\varphi} \subset Res^*(\varphi)$ be such that every clause of $Res^m(E(\varphi))$ comes from some clause in $\tilde{\varphi}$. Then $Res^m(E(\varphi)) \subset E(\tilde{\varphi})$. If $C' \in Res^{m+1}(E(\varphi))$, then $C' \in Res(E(\tilde{\varphi}))$. By the Lifting lemma, there is some $C \in Res(\tilde{\varphi})$ such that $C sub' = C'$ for some substitution sub'. Since $\tilde{\varphi} \subset Res^*(\varphi)$, $C \in Res^*(\varphi)$. $\square$

In particular, if $\emptyset \in Res^*(E(\varphi))$, then there exists some $C \in Res^*(\varphi)$ such that $Csub' = \emptyset$ for some substitution sub'. But this is only possible if $C = \emptyset$. So if $\emptyset \in Res^*(E(\varphi))$, then $\emptyset \in Res^*(\varphi)$. We conclude that if φ is unsatisfiable, then $\emptyset \in Res^*(\varphi)$. We have shown that the notion of resolution defined in this section works. We state this as a theorem.

Theorem 3.39 Let φ be a sentence in SNF. Then φ is unsatisfiable if and only if $\emptyset \in Res^*(\varphi)$.

3.5 SLD-resolution

One purpose of resolution is to provide a method of proof that can be done by a computer. Toward this aim, we refine resolution in this section. Our goal is to find a version of resolution that can be completely automated. The advantage of resolution over other formal proof systems is that it rests on a single rule. Resolution proofs may not be the most succinct. They will not lend insight as to why, say, a sentence φ is unsatisfiable. The benefit of resolution is precisely that it does not require any insight. To show that φ is unsatisfiable, we can blindly compute $Res^*(\varphi)$ until we find $\emptyset$. However, this method is not practical. If $\emptyset$ is in $Res^*(\varphi)$, then calculating the clauses in $Res^*(\varphi)$ one-by-one in no particular order is not an efficient way of finding it. The first two theorems of this section show that it is not necesssary to compute all of $Res^*(\varphi)$. We show that we only need to compute resolvents R of clauses C_1 and C_2 that have certain forms. We refer to C_1 and C_2 as the *parents* of R.

Definition 3.40 *N-resolution* requires that one parent contain only negative literals.

We look at example from propositional logic. Let

$$\varphi = \{\{A, B\}, \{\neg A, C\}, \{\neg B, D\}, \{\neg C\}, \{\neg D\}\}.$$

We show that φ is unsatisfiable using N-resolution.

```
    {¬A, C}        {¬C}
         \          |
      {A, B}       {¬A}        {¬B, D}      {¬D}
           \        |              |         /
                   {B}            {¬B}
                     \            /
                          ∅
```

Note that each resolvent has a parent that contains only negative literals.

Definition 3.41 *Linear resolution* requires that one parent be the resolvent from the previous step.

The word "linear" refers to the diagram. The previous diagram is not linear because it has two "branches." The following diagram illustrates a linear resolution for this same example.

$$
\begin{array}{cc}
\{\neg A, C\} & \{\neg C\} \\
\searrow & | \\
\{A, B\} & \{\neg A\} \\
\searrow & | \\
\{\neg B, D\} & \{B\} \\
\searrow & | \\
\{\neg D\} & \{D\} \\
\searrow & | \\
& \emptyset
\end{array}
$$

So we can derive the emptyset from φ either by N-resolution or linear resolution. The next two theorems show that this is true for any unsatisfiable φ. That is, to show that φ is unsatisfiable, we can restrict our computations to N-resolution or linear resolution.

Theorem 3.42 Let φ be a sentence in SNF. Then φ is unsatisfiable if and only if $\emptyset$ can be derived from φ by N-resolution.

Proof If $\emptyset$ can be derived by N-resolution, then $\emptyset \in Res^*(\varphi)$ and so φ is unsatisfiable. Conversely, suppose φ is unsatisfiable.

By the Lifting lemma, it suffices to prove this theorem for propositional logic. This requires some explanation. If $\emptyset$ can be derived from $E(\varphi)$ by N-resolution, then we can "lift" this to a derivation of $\emptyset$ from φ. By "lift" we mean that the former can be obtained from the latter via substitutions. Note that a clause C contains only negative literals if and only if $Csub$ does (for any substitution sub). So an N-resolution derivation in propositional logic lifts to an N-resolution derivation in first-order logic.

Having said this, we assume φ is an unsatisfiable set of sentences of propositional logic in CNF. By compactness, we may assume φ is finite. We showed that $\emptyset \in Res^*(\varphi)$ in Proposition 1.74. Following that same proof, we show that $\emptyset$ can be derived by N-resolution.

Let $\varphi = \{C_1, \ldots, C_k\}$. We assume that none of the C_is is a tautology (otherwise we just throw away these clauses and show that $\emptyset$ can be derived from what remains). We will prove this proposition by induction on the number n of atomic subformulas of φ.

First suppose $n = 1$. Let A be the only atomic formula that occurs in φ. Then there are only three possible clauses in φ. Each C_i is either $\{A\}$, $\{\neg A\}$, or $\{A, \neg A\}$. The last clause is a tautology, and so, by our previous assumption, it is not a clause of φ. So the only clauses in φ are $\{A\}$ and $\{\neg A\}$. There are three possibilities, $\varphi = \{\{A\}\}$, $\varphi = \{\{\neg A\}\}$, or $\varphi = \{\{A\}, \{\neg A\}\}$. The first two of these are satisfiable. So φ must be $\{\{A\}, \{\neg A\}\}$. Clearly, $\emptyset$ can be derived by N-resolution.

Now suppose φ has atomic subformulas $A_1, \ldots, A_{n+1}$. Suppose further that $\emptyset$ can be derived by N-resolution from any unsatisfiable formula ψ that uses only the atomic formulas $A_1, \ldots, A_n$.

Let $\tilde{\varphi}_0$ be the conjunction of all C_i in φ that do NOT contain $\neg A_{n+1}$.
Let $\tilde{\varphi}_1$ be the conjunction of all C_i in φ that do NOT contain A_{n+1}.

If A_{n+1} and $\neg A_{n+1}$ are both in a clause, then that clause is a tautology. By our assumption, there is no such clause and $\tilde{\varphi}_0 \cup \tilde{\varphi}_1 = \varphi$.

Let $\varphi_0 = \{C_i - \{A_{n+1}\}| C_i \in \tilde{\varphi}_0\}$.
Let $\varphi_1 = \{C_i - \{\neg A_{n+1}\}| C_i \in \tilde{\varphi}_1\}$.

That is, φ_0 is formed by throwing A_{n+1} out of each clause of $\tilde{\varphi}_0$ in which it occurs. Likewise, φ_1 is obtained by throwing $\neg A_{n+1}$ out of each clause of $\tilde{\varphi}_1$.

Note that φ_0 is the formula obtained by replacing A_{n+1} in φ with a contradiction, and φ_1 is obtained by replacing A_{n+1} in φ with a tautology. Since A_{n+1} must either have truth value 0 or 1, it follows that $\varphi \equiv \varphi_0 \vee \varphi_1$. Since φ is unsatisfiable, φ_0 and φ_1 are each unsatisfiable. The formulas φ_0 and φ_1 only use the atomic formulas $A_1, \ldots, A_n$. By our induction hypothesis, we can derive $\emptyset$ from both φ_0 and φ_1 using N-resolution.

Now φ_0 was formed from $\tilde{\varphi}_0$ by throwing A_{n+1} out of each clause. Since we can derive $\emptyset$ from φ_0 by N-resolution, we can derive either $\emptyset$ or $\{A_{n+1}\}$ from $\tilde{\varphi}_0$ by N-resolution (by reinstating $\{A_{n+1}\}$ in each clause of φ_0). Likewise, we can derive either $\emptyset$ or $\{\neg A_{n+1}\}$ from $\tilde{\varphi}_1$ by N-resolution. In any case, we can use N-resolution to derive $\emptyset$ from $\varphi = \tilde{\varphi}_0 \cup \tilde{\varphi}_1$. □

Theorem 3.43 Let φ be a sentence in SNF. Then φ is unsatisfiable if and only if $\emptyset$ can be derived from φ by linear resolution.

Proof As with Theorem 3.42, only one direction of this theorem requires proof. Suppose that φ is unsatisfiable. We want to show that $\emptyset$ can be derived from φ by linear resolution. By the Lifting lemma, it suffices to prove this for propositional logic. So suppose φ is a set of sentences in propositional logic that are in CNF. We say that a set of sentences is *minimal unsatisfiable* if it is unsatisfiable and every proper subset is satisfiable. By Exercise 1.33(b), φ contains a minimal unsatisfiable subset φ'. The following lemma states that for any $C \in \varphi'$, we can derive $\emptyset$ by linear resolution begining with C as one parent. The proof of this lemma completes the proof of Theorem 3.43. □

Lemma 3.44 Let $\mathcal{F}$ be a minimal unsatisfiable set of sentences of propositional logic that are in CNF. For any $C \in \mathcal{F}$, we can derive $\emptyset$ from $\mathcal{F}$ by linear resolution begining with C as one parent.

Proof By the Compactness of propositional logic, it suffices to prove this for finite $\mathcal{F}$. So we may view $\mathcal{F}$ as a formula in CNF. We proceed by induction on the number n of atomic subformulas of $\mathcal{F}$. If $n = 1$, then $\mathcal{F} = \{\{A\}, \{\neg A\}\}$ for some atomic formula A. In this case, the conclusion of the lemma is obvious. Now suppose that, for some $n \in \mathbb{N}$, $\mathcal{F}$ contains $n + 1$ atomic subformulas. Our induction hypothesis is that the lemma holds for any formula in CNF containing at most n atomic subformulas.

Let L be a literal in C. This literal partitions $\mathcal{F}$ into three subsets as follows.

Let $\mathcal{C} = \{C_1, \ldots, C_i\}$ be the set of clauses in $\mathcal{F}$ that contain L.

Let $\mathcal{D} = \{D_1, \ldots, D_j\}$ be the set of clauses in $\mathcal{F}$ that contain $\overline{L}$.

Let $\mathcal{E} = \{E_1, \ldots, E_k\}$ be the set of clauses in $\mathcal{F}$ that contain neither L nor $\overline{L}$.

Any clause that contains both L and $\overline{L}$ is a tautology. Since $\mathcal{F}$ is minimal unsatisfiable, $\mathcal{F}$ contains no such clauses. So every clause in $\mathcal{F}$ is in exactly one of the above three sets. Note that C is in $\mathcal{C}$. We may assume that $C_1 = C$.

Case 1: C = {L}.

For any clause $D \in \mathcal{F}$, we can find a resolvent of C and D if and only if D is in $\mathcal{D}$. If D is in $\mathcal{D}$, then let D' denote the resolvent of C and D. That is, D' is the formula obtained by removing $\overline{L}$ from the formula D. Since $\mathcal{F}$ is unsatisfiable, we can derive $\emptyset$ from $\mathcal{F}$ by resolution. Since it is minimal unsatisfiable the clause C, as well as each clause of $\mathcal{F}$, is needed in this derivation. It follows that the clause $D_1 \in \mathcal{D}$ must exist.

Let $\mathcal{F}_L$ be the set $\{D'_1, \ldots, D'_j, E_1, \ldots, E_k\}$.

Note that $\mathcal{F}_L$ is equivalent to the formula obtained from $\mathcal{F}$ by replacing L with a tautology and $\overline{L}$ with a contradiction. Either L or $\overline{L}$ is an atomic formula that occurs in $\mathcal{F}$ but not $\mathcal{F}_L$. If $\mathcal{F}$ contains $n + 1$ atomic subformulas, then $\mathcal{F}_L$

contains n atomic subformulas. Our induction hypothesis applies to any minimal unsatisfiable subset of $\mathcal{F}_L$.

Claim There is a minimal unsatisfiable subset of $\mathcal{F}_L$ containing D_1'.

First we show that $\mathcal{F}_L$ is unsatisfiable. Suppose to the contrary that $\mathcal{F}_L$ is satisfiable. Then there exists an assignment $\mathcal{A}$ defined on the n atomic formulas of $\mathcal{F}_L$ and not defined on L. Let $\mathcal{A}'$ be an extension of $\mathcal{A}$ such that $\mathcal{A}'(L) = 1$. Then $\mathcal{A}'$ models each clause of $\mathcal{F}$. This is a contradiction. Since $\mathcal{F}$ is unsatisfiable, so is $\mathcal{F}_L$. By Exercise 1.33(b), $\mathcal{F}_L$ contains a minimal unsatisfiable subset.

Suppose we remove D_1' from $\mathcal{F}_L$. We show that the resulting set $\mathcal{F}_L - \{D_1'\}$ is satisfiable. To see this, consider the set $\{C, D_2, \ldots, D_j, E_1, \ldots, E_k\}$. Since this is a proper subset of $\mathcal{F}$ (not containing $D_1 \in \mathcal{F}$), this set must be satisfiable. Let $\mathcal{A}$ be an assignment that models this set. Since $\mathcal{A}$ models both C and D_2, it also models their resolvent D_2'. Likewise, $\mathcal{A}$ models each formula of $\mathcal{F}_L - \{D_1'\}$ and this set is satisfiable. It follows that any minimal unsatisfiable subset of $\mathcal{F}_L$ must contain D_1' as claimed.

By the induction hypothesis, $\emptyset$ can be derived from $\mathcal{F}_L$ by linear resolution beginning with D_1' as one parent. The lemma states that $\emptyset$ can be derived from $\mathcal{F}$ by linear resolution beginning with C. We begin this derivation by taking C and D_1 as parents of the resolvent D_1'. Consider now the set $\{D_1', D_2, \ldots, D_j, E_1, \ldots, E_k\}$. Note that this set is obtained from $\mathcal{F}_L$ by reinstating $\overline{L}$ to the clauses $D_2', \ldots, D_j'$. Since $\emptyset$ can be derived from $\mathcal{F}_L$ by linear resolution beginning with D_1', either $\emptyset$ or $\overline{L}$ can be derived after reinstating L to these clauses. If $\emptyset$ is derived, then we are done. Otherwise, if $\overline{L}$ is derived, then $\emptyset$ is obtained from $\overline{L}$ and C to successfully conclude the linear resolution.

Case 2: C contains literals other than L.

For this case, consider the set $\mathcal{F}_{\overline{L}}$ defined as follows:

$$\mathcal{F}_{\overline{L}} = \{C_1', \ldots, C_i', E_1, \ldots, E_k\},$$

where each C_s' is obtained by removing L from $C_s \in \mathcal{C}$. Note that $\mathcal{F}_{\overline{L}}$ is equivalent to the formula obtained from $\mathcal{F}$ by replacing $\overline{L}$ with a tautology and L with a contradiction. As with $\mathcal{F}_L$, the set $\mathcal{F}_{\overline{L}}$ contains n atomic subformulas and we can apply our induction hypothesis to any minimal unsatisfiable subset of $\mathcal{F}_{\overline{L}}$.

Claim There is a minimal unsatisfiable subset of $\mathcal{F}_{\overline{L}}$ containing C_1'.

The set $\mathcal{F}_{\overline{L}}$ is unsatisfiable for the same reason that $\mathcal{F}_L$ is unsatisfiable. Any assignment that models $\mathcal{F}_{\overline{L}}$ can be extended to an assignment that models $\mathcal{F}$. Since $\mathcal{F}$ is unsatisfiable, no such assignment exists.

Now suppose that we remove C_1' from $\mathcal{F}_{\overline{L}}$. We show that the resulting set $\mathcal{F}_{\overline{L}} - \{C_1'\}$ is satisfiable. Since $\mathcal{F}$ is minimal unsatisfiable, there exists an assignment $\mathcal{A}$ that models every formula of $\mathcal{F}$ other than C. Since $\mathcal{F}$ is unsatisfiable, $\mathcal{A} \models \neg C$. Since $\mathcal{A} \models \neg C$ and $L \in C$, it must be the case that $\mathcal{A} \models \overline{L}$. Since $\mathcal{A}$ models C_2 and not L, $\mathcal{A}$ must model the formula C_2' obtained by removing L from C_2. Likewise, $\mathcal{A}$ models each formula in $\mathcal{F}_{\overline{L}} - \{C_1'\}$. It follows that any minimal unsatisfiable subset of $\mathcal{F}_{\overline{L}}$ must contain C_1' as claimed.

By the induction hypothesis, $\emptyset$ can be derived from $\mathcal{F}_{\overline{L}}$ by linear resolution begining with C_1' as one parent. The set

$$\{C_1, \ldots, C_i, E_1, \ldots, E_k\}$$

is obtained from $\mathcal{F}_{\overline{L}}$ by reinstating L to some of the clauses. Either $\emptyset$ or L can be derived from this subset of $\mathcal{F}$ by linear resolution begining with $C = C_1$. If $\emptyset$ is derived, we are done. Suppose L is derived. Consider the set

$$\{L, D_1, \ldots, D_j, E_1, \ldots, E_k\}.$$

If we remove L from this set, then we have a proper subset of $\mathcal{F}$ which must be satisfiable. Having L in this set, however, makes it unsatisfiable. Any assignment $\mathcal{A}$ that models L also models each clause of $\mathcal{C}$. Since $\mathcal{F}$ is unsatisfiable, so is the above set. So there exists a minimal unsatisfiable subset of this set containing L. By case 1, $\emptyset$ can be derived from this set by linear resolution begining with L as one parent. This completes the linear resolution and the proof. □

So if φ is unsatisfiable, then we can prove that it is unsatisfiable using either linear resolution or N-resolution. Anything that can be proved using resolution can also be proved using either of these restricted versions of resolution. Suppose we restrict further to resolution that is both linear resolution *and* N-resolution. Call this *LN-resolution*.

Question 1 Can we derive $\emptyset$ from any unsatisfiable φ using LN-resolution?

The answer is "no." Consider again the example

$$\varphi = \{\{A, B\}, \{\neg A, C\}, \{\neg B, D\}, \{\neg C\}, \{\neg D\}\}.$$

We showed that φ is unsatisfiable using both N-resolution and linear resolution. Note that neither of these derivations was by LN-resolution. In fact, $\emptyset$ cannot be derived from φ using LN-resolution (try it). So LN-resolution is too weak to prove everything. However, suppose we restrict our attention to Horn sentences.

Definition 3.45 Let φ be a sentence in SNF. If each clause in φ contains at most one positive literal, then φ is a *Horn sentence*.

This is the same definition we gave for sentences of propositional logic in CNF. Note that φ in our example is not Horn since it contains the clause $\{A, B\}$. The following theorem shows that if we require φ in Question 1 to be a Horn sentence, then the answer becomes "yes."

Theorem 3.46 Let φ be a sentence in SNF. If φ is a Horn sentence, then φ is unsatisfiable if and only if $\emptyset$ can be derived from φ by *LN-resolution*.

Proof Suppose φ is an unsatisfiable Horn sentence. Again by the Lifting Lemma, we may assume that φ is a sentence of propositional logic. Let φ' be a minimal unsatisfiable subset of φ. By Theorem 3.42, there is an N-resolution derivation of $\emptyset$ from φ'. In particular, φ' must contain a negative clause N. By Lemma 3.44, there exists a linear resolution derivation of $\emptyset$ begining with N. So we have

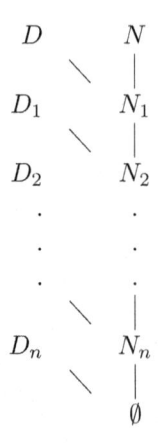

$$
\begin{array}{cc}
D & N \\
\searrow & | \\
D_1 & N_1 \\
\searrow & | \\
D_2 & N_2 \\
\cdot & \cdot \\
\cdot & \cdot \\
\cdot & \cdot \\
& \searrow \quad | \\
D_n & N_n \\
& \searrow \quad | \\
& \emptyset
\end{array}
$$

for some clauses $D, D_1, \ldots, D_n$ and $N_1, \ldots, N_n$. Since φ is Horn, D contains at most one positive literal. It follows that N_1 is contains only negative literals. Likewise, each N_i is negative and this is an *LN*-resolution. □

Clauses that contain exactly one positive literal are called *definite*. Note that each D_i in the previous proof is necessarily definite. The positive literal in D_i "cancels" with one of the negative literals in N_i yeilding the negative resolvent N_{i+1}. For each N_i there may be many possibilities for D_i depending on which literal in N_i we want to cancel. A *selector function* chooses which literal in N_i to cancel at each stage. For example, we may say cancel the leftmost literal of N_i at each stage. Or we may require that we cancel a literal that has

Proof theory

appeared in the most N_j for $j < i$. We demand only that the selector function is invariant under substitutions of variables. That is, suppose that, for a given set of negative literals N, the selector function chooses $L \in N$. Then for any substitution of variables sub, we require that the selector function chooses $Lsub$ from $Nsub$.

Definition 3.47 *SLD-resolution* is LN-resolution with a selector function.

The "S" stands for selector, the "L" for linear, and the "D" for definite.

Example 3.48 Let $\varphi = \{\{\neg A, C\}, \{A, \neg B\}, \{B\}, \{\neg D, \neg C\}, \{D, \neg E\}, \{E\}\}$. We will perform *SLD*-resolution on φ twice. First, we use the "leftmost" selector function. At each stage, we underline the literal that we seek to eliminate.

$$
\begin{array}{cc}
\{D, \neg E\} & \{\underline{\neg D}, \neg C\} \\
& \searrow \quad | \\
\{E\} & \{\underline{\neg E}, \neg C\} \\
& \searrow \quad | \\
\{\neg A, C\} & \{\underline{\neg C}\} \\
& \searrow \quad | \\
\{A, \neg B\} & \{\underline{\neg A}\} \\
& \searrow \quad | \\
\{B\} & \{\underline{\neg B}\} \\
& \searrow \quad | \\
& \emptyset
\end{array}
$$

Now we use the "rightmost" selector function.

$$
\begin{array}{cc}
\{\neg A, C\} & \{\neg D, \underline{\neg C}\} \\
& \searrow \quad | \\
\{D, \neg E\} & \{\neg A, \underline{\neg D}\} \\
& \searrow \quad | \\
\{A, \neg B\} & \{\neg E, \underline{\neg A}\} \\
& \searrow \quad | \\
\{E\} & \{\neg B, \underline{\neg E}\} \\
& \searrow \quad | \\
\{B\} & \{\underline{\neg B}\} \\
& \searrow \quad | \\
& \emptyset
\end{array}
$$

Theorem 3.49 Let φ be a sentence in SNF. If φ is a Horn sentence, then φ is unsatisfiable if and only if $\emptyset$ can be derived from φ by SLD-resolution for any choice of selector function that is invariant under substitutions.

Proof Since the selector function is invariant under substitutions, we may apply the Lifting Lemma and assume that φ is a sentence of propositional logic. By Theorem 3.46, we can derive $\emptyset$ from φ by LN-resolution as follows:

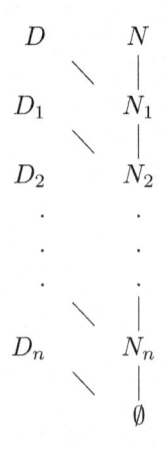

$$
\begin{array}{cc}
D & N \\
\searrow & | \\
D_1 & N_1 \\
\searrow & | \\
D_2 & N_2 \\
\vdots & \vdots \\
\searrow & | \\
D_n & N_n \\
& \searrow | \\
& \emptyset
\end{array}
$$

In this derivation, N is a negative clause of φ and $D, D_1, \ldots, D_n$ are definite clauses of φ. We proceed by induction on n. If $n = 0$, then N contains only one literal L. Any concievable selector function must choose the literal L from N (there is no other choice). So in this case, the above LN-resolution is also SLD-resolution regardless of the choice of selector function.

Now suppose that n in the above derivation equals $m+1$ for some integer $m \geq 0$. Then there are $m+2$ steps in this derivation. Suppose further that if $\emptyset$ can be derived by LN-resolution in $m+1$ steps, then it can also be derived by SLD-resolution. This is our induction hypothesis.

Let L be the literal of N chosen by the selector function. Then $\overline{L}$ must occur in D or some D_i. If it happens to occur in D, then SLD-resolution begins with the same resolvent N_1 as the above LN-resolution. From this point, $\emptyset$ is derived in $m+1$ steps from N_1 in the the above LN-resolution. By our induction hypothesis, it can be derived by SLD-resolution.

Now suppose that $\overline{L}$ is in D_i for some $i = 1, \ldots, n$. Then SLD-resolution begins by finding the resolvent N' of D_i and N. Consider the following two derivation by LN-resolution:

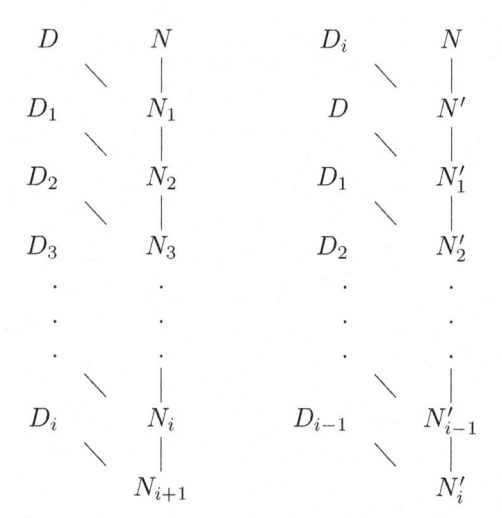

$$
\begin{array}{cccc}
D & N & D_i & N \\
\searrow & | & \searrow & | \\
D_1 & N_1 & D & N' \\
\searrow & | & \searrow & | \\
D_2 & N_2 & D_1 & N'_1 \\
\searrow & | & \searrow & | \\
D_3 & N_3 & D_2 & N'_2 \\
\vdots & \vdots & \vdots & \vdots \\
D_i & N_i & D_{i-1} & N'_{i-1} \\
\searrow & | & \searrow & | \\
& N_{i+1} & & N'_i \\
\end{array}
$$

The derivation on the left is the same as before. Since the derivation on the right also begins with N and involves precisely the same definite clauses (in a different order) the result N'_i is the same as the result N_{i+1} of the derivation on the left. The derivation on the left can be continued as above to obtain $\emptyset$. Since, $N'_i = N_{i+1}$, we can conclude the derivation on the right in exactly the same manner. So, in this new derivation, we derive $\emptyset$ in the same number of steps ($m+2$) as before. Beginning this derivation from the second step, we see that, using LN-resolution, $\emptyset$ can be derived from $\{N', D, D_1, \ldots, D_n\}$ in $m + 1$ steps. By our induction hyothesis, we can derive $\emptyset$ from this set using SLD-resolution. □

3.6 Prolog

There are many conceivable ways to implement resolution into a programming language. Prolog and Otter are two examples. The language of Prolog is based on first-order Horn logic. Otter allows far more expressions. Otter can take a set of first-order sentences, put them into CNF, and derive consequences using resolution and other methods. Whereas Otter is used as a theorem prover for elementary mathematics and has successfully obtained new results, Prolog is primarily a search engine for databases (Prolog is closely related to Datalog). Several versions of Prolog and Otter are freely available on the internet. (Otter may be downloaded from the pages of the Mathematics and Computer Science Division of Argonne National Laboratory where it was developed.)

In this section, we give some examples of Prolog programs and discuss how Prolog uses SLD-resolution. We refer the reader to Ref. [20] for details on Otter.

We begin by defining the basic syntax of Prolog. We consider only a fragment of Prolog called "pure Prolog." In pure Prolog, there are symbols for conjunction and implication, but not disjunction, negation, or equality. Lower case letters are used for relations and functions. Commas are used for conjunction and "q :- p" is used for $p \to q$. The Horn sentence

$$\forall x \forall y ((P((x,y)) \land E(f(x), y) \to Q(x)))$$

is written in Prolog as

```
q(X) :- p(X,Y), e(f(X),Y).
```

Note that if this were written as a disjuntion of literals, then the literal $q(x)$ on the left would occur as a positive literal and the literals on the right would occur as negative literals.

Recall that a Horn clause is a clause that contains at most one positive literal. So in Prolog, there is at most one literal to the left of :-. If it contains no positive literal, it is called a *goal clause*. Otherwise, it is called a *program clause*. There are two varieties of program clauses. If there are no negative literals, then it is called a *fact*. Otherwise it is called a *rule*. A *program* in Prolog is a set of program clauses. For example, the following is a program.

```
p(ray,ken)
p(ray,sue)
p(sue,tim)
p(dot,jim)
p(bob,jim)
p(bob,liz)
p(jim,tim)
p(sue,sam)
p(jim,sam)
p(zelda,max)
p(sam,max)
gp(X,Y) :- p(X,Z),p(Z,Y).
```

This program consists of 11 facts and one rule. The facts, since they contain no variables, are sentences of propositional logic. These form a database. If we interpret the predicate p(X,Y) as "X is a parent of Y," then the facts just list pairs of parents and children. The rule defines the relation gp(X,Y) as "X is a grandparent of Y." Using first-order logic, we can define gp(X,Y) in a single sentence without having to list all pairs of grandparents and grandchildren.

Proof theory

We can ask certain questions in Prolog. Questions are presented as a list of positive literals. We use "?-" for the Prolog prompt. The following is an example of a question we may ask:

$$\text{?- gp(ray,X),p(X,max)}$$

This can be interpreted as "does there exist an X such that both gp(ray,X) and p(X,max) hold?" That is, is Ray a great-grandparent of Max? Prolog will not only answer this question, it will output all values for X for which the statement is true. We describe how Prolog does this.

Let P denote the set of all sentences in the program. Let Q denote the sentence $\exists X(gp(ray, X) \wedge p(X, max))$. Our question asked if Q is a consequence of P. That is, we want to know if $P \wedge \neg Q$ is unsatisfiable. Note that $\neg Q$ is equivalent to the Horn sentence $\forall X(gp(ray, X) \wedge p(X, max) \to 0)$. In Prolog this sentence is written as

$$\text{:- gp(ray,X),p(X,max)}$$

which is a goal clause.

Prolog proceeds with SLD-resolution on $P \cup \{\neg Q\}$ using the "leftmost" selection rule. Since, $\neg Q$ is the only negative clause of $P \cup \{\neg Q\}$, SLD-resolution must begin with this clause. Prolog searches the program for something to unify with gp(ray,X), the leftmost literal of $\neg Q$. Prolog searches until it finds the rule gp(X,Y) :- p(X,Z),p(Z,Y) which has a positive occurrence of gp(X,Y). Prolog computes the following resolvent:

```
gp(X,Y)   :- p(X,Z),p(Z,Y)              :- gp(ray,X),p(X,max)
    |                                        |
gp(ray,Y) :- p(ray,Z),p(Z,Y)            :- gp(ray,Y),p(Y,max)
                       \         /
                :- p(ray,Z),p(Z,Y),p(Y,max).
```

Prolog then searches and finds p(ray,ken) which can be unified with the leftmost literal of the above resolvent. Prolog then calculates the resolvent p(ken,Y),p(Y,max). Searching again, Prolog finds nothing that can be unified with p(ken,Y). So this is a dead end and Prolog must backtrack. Prolog goes back to the previous step and, instead of taking p(ray,ken), proceeds down the list to p(ray,sue). After computing the resolvent, Prolog will next find p(sue,tim). This is another dead end, and so Prolog will backtrack and take p(sue,sam) instead. This choice yields the following SLD-resolution:

```
gp(X,Y) :- p(X,Z),p(Z,Y)                :- gp(ray,X),p(X,max)
           \                                |
        p(ray,sue)          :- p(ray,Z),p(Z,Y),p(Y,max)
              \                             |
           p(sue,sam)           :- p(sue,Y),p(Y,max)
                 \                          |
              p(sam,max)            :- p(sam,max)
                    \                       |
                                            ∅
```

It follows that $P \wedge \neg Q$ is unsatisfiable, and so Q is a consequence of P. So Prolog can give an affirmative answer to our question ?- gp(ray,X),p(X,max). Moreover, by keeping track of the substitutions that were made, Prolog can output the appropriate values for X. In the first step of the above resolution, the substitution (X/Y) was made (as indicated by the vertical lines in the first diagram). Later, the substitution (Y/sam) was made. So X=sam works. Prolog will backtrack again and continue to find all values for X that work. In this example, X = sam is the only solution.

Some other questions we might ask are as follows:

Input	Output
?- p(bob,liz)	yes
?- gp(X,sam)	X=ray,X=bob,X=dot
?- gp(tim,max)	no
?- p(tim,X)	no
?- p(X,Y),p(Y,max)	X=jim,X=sue,Y=sam

The output "no" means Prolog has computed all possible SLD-resolutions and did not come across ∅.

Suppose now we want to know who is the grandmother of whom. We will have to add some relations to be able to ask such a question. Let gm(X,Y) mean X is the grandmother of Y. We need to add sentences to the program that define this relation. One way is to list as facts all pairs for which gm holds. This is how the relation p was defined in the original program. But this defeats the point. If we could produce such a list, then we would not need to ask Prolog the question. Another way is to introduce a unary relation f(X) for "female." The following rule defines gm.

$$gm(X,Y) :- gp(X,Y),f(X)$$

But now we need to define `f(X)`. We do this by adding the following facts to the program:

> `f(dot)`
>
> `f(sue)`
>
> `f(liz)`
>
> `f(zelda).`

We can now ask the question

> `?- gm(X,Y)`

to which Prolog responds

> `X=dot,Y=tim`
>
> `X=dot,Y=sam`
>
> `X=sue,Y=max.`

There is one caveat that must be mentioned. Prolog does not use the Unification algorithm as we stated it. This algorithm is not polynomial time. Recall that, to unify a set of literals, we look at the symbols one-by-one until we find a discrepancy. If this discrepancy involves a variable and a term *that does not include the variable*, then we substitute the term for that variable and proceed. We must check that the variable does not occur in the term. This checking procedure can take exponentially long. Prolog avoids this problem simply by not checking. Because it excludes a step of the Unification algorithm, Prolog may generate incorrect answers to certain queries. This is a relatively minor problem that can be avoided in practice.

Example 3.50 Consider the program consisting of the fact `p(X,X)` and the rule `q(a) :- p(f(X),X)`. If we ask `?- q(a)`, then Prolog will output the incorrect answer of "yes." Since it fails to check whether X occurs in `f(X)`, Prolog behaves as though `p(X,X)` and `p(f(X),X)` are unifiable.

In "pure Prolog," we have been ignoring many features of Prolog. Prolog has many built in relation and function symbols. "Impure Prolog" includes `is(X,Y)` which behaves like $X = Y$. It also has binary function symbols for addition and multiplication. A complete list of all such function and relations available in Prolog would take several pages. We merely point out that we can express equations and do arithmetic with Prolog.

So we can answer questions regarding the natural numbers using Prolog. There are three limitations to this. One is the before mentioned caveat regarding the Unification algorithm. Another limitation is that Prolog uses Horn clauses. This is a restrictive language (try phrasing Goldbach's conjecture from

Exercise 2.8 as a goal clause for some Prolog program). Third, suppose that Prolog was capable of carrying out full resolution for first-order logic (as Otter does). Since resolution is complete, it may seem that we should be able to prove every first-order sentence that is true of the natural numbers. This is not the case. Even if we had a computer language capable of implementing all of the methods discussed in this chapter, including formal proofs, Herbrand theory, and resolution, there would still exist theorems of number theory that it would be incapable of proving. This is a consequence of Gödel's incompleteness theorems that are the topic of Chapter 8.

Exercises

3.1. Let φ be a $\mathcal{V}$-sentence and let Γ be a set of $\mathcal{V}$-sentences such that $\Gamma \vdash \varphi$. Show that there exists a derivation φ from Γ that uses only $\mathcal{V}$-formulas.

3.2. Let Γ be a set of $\mathcal{V}$-sentences. Show that the following are equivalent:
   (i) For every universal $\mathcal{V}$-formula φ, there exists an existential $\mathcal{V}$-formula ψ such that $\Gamma \vdash \varphi \leftrightarrow \psi$.
   (ii) For every $\mathcal{V}$-formula φ, there exists an existential $\mathcal{V}$-formula ψ such that $\Gamma \vdash \varphi \leftrightarrow \psi$.
   (iii) For every existential $\mathcal{V}$-formula φ, there exists an universal $\mathcal{V}$-formula ψ such that $\Gamma \vdash \varphi \leftrightarrow \psi$.
   (iv) For every $\mathcal{V}$-formula φ, there exists a universal $\mathcal{V}$-formula θ such that $\Gamma \vdash \varphi \leftrightarrow \theta$.

3.3. Let Γ be a set of $\mathcal{V}$-sentences. Show that the following are equivalent:
   (i) For every quantifier-free $\mathcal{V}$-formula $\varphi(x_1, \ldots, x_n, y)$ (for $n \in \mathbb{N}$), there exists a quantifier-free $\mathcal{V}$-formula $\psi(x_1, \ldots, x_n)$ such that $\Gamma \vdash \exists y \varphi(x_1, \ldots, x_n, y) \leftrightarrow \psi(x_1, \ldots, x_n)$.
   (ii) For every formula $\varphi(x_1, \ldots, x_n)$ (for $n \in \mathbb{N}$), there exists a quantifier-free formula $\theta(x_1, \ldots, x_n)$ such that $\Gamma \vdash \varphi(x_1, \ldots, x_n) \leftrightarrow \theta(x_1, \ldots, x_n)$.

3.4. Complete the proof of Theorem 3.4 by verifying the soundness of $\forall$-Distribution.

3.5. Verify each of the following by providing a formal proof:
   (a) $\{\forall y \exists x \varphi(x, y)\} \vdash \forall x \exists y \varphi(x, y)$
   (b) $\{\forall x \exists y \forall z \psi(x, y, z)\} \vdash \exists x \forall z \exists y \psi(x, y, z)$
   (c) $\{\exists x \forall y \exists z \forall w \theta(x, y, z, w)\} \vdash \forall y \exists x \forall w \exists z \theta(x, y, z, w)$.

3.6. Verify that the following pairs of formulas are provably equivalent by sketching formal proofs:
   (a) $\exists x \exists y \varphi(x,y)$ and $\exists y \exists x \varphi(x,y)$.
   (b) $\forall x \exists y \exists z \psi(x,y,z)$ and $\forall x \exists z \exists y \psi(x,y,z)$.
   (c) $\exists x_1 \forall x_2 \forall x_3 \exists x_4 \exists x_5 \exists x_6 \theta(x_1,x_2,x_3,x_4,x_5,x_6)$ and
   $\exists x_1 \forall x_3 \forall x_2 \exists x_6 \exists x_5 \exists x_4\; \theta(x_1,x_2,x_3,x_4,x_5,x_6)$.

3.7. Let x and y be variables that do not occur in the formula $\varphi(z)$. Show that $\exists x \varphi(x)$ and $\exists y \varphi(y)$ are provably equivalent by giving formal proofs.

3.8. Show that $\forall x(\varphi(x) \land \psi(x))$ and $\forall x \varphi(x) \land \forall x \psi(x)$ are provably equivalent by providing formal proofs.

3.9. (a) Show that $\{\exists x(\varphi(x) \land \psi(x))\} \vdash \exists x \varphi(x) \land \exists x \psi(x)$.
   (b) Show that the sentences $\exists x(\varphi(x) \land \psi(x))$ and $\exists x \varphi(x) \land \exists x \psi(x)$ are not provably equivalent.

3.10. Show that $\exists x(\varphi(x) \lor \psi(x))$ and $\exists x \varphi(x) \lor \forall x \psi(x)$ are provably equivalent by providing formal proofs.

3.11. (a) Show that $\{\forall x \varphi(x) \lor \forall x \psi(x)\} \vdash \forall x(\varphi(x) \lor \psi(x))$.
   (b) Show that the sentences $\forall x(\varphi(x) \lor \psi(x))$ and $\forall x \varphi(x) \lor \forall x \psi(x)$ are not provably equivalent.

3.12. Let $\mathcal{V}_{gp}$ be the vocabulary $\{+, 0\}$ where $+$ is a binary function and 0 is a constant. We use the notation $x + y$ to denote the term $+(x,y)$. Let Γ be the set consisting of the following three $\mathcal{V}_{gp}$-sentences from Exercise 2.5:

$$\forall x \forall y \forall z (x + (y + z) = (x + y) + z)$$
$$\forall x((x + 0 = x) \land (0 + x = x))$$
$$\forall x \exists y (x + y = 0) \land \exists z (z + x = 0)$$

   (a) Show that $\Gamma \vdash \forall x \forall y \forall z (x + y = x + z \to (y = z))$.
   (b) Show that $\Gamma \vdash \forall x (\forall y (x + y = y) \to (x = 0))$.
   (c) Show that $\Gamma \vdash \forall x \exists y \forall z ((x + y = 0) \land (z + x = 0)) \to (y = z)$.

3.13. Complete the proof of Proposition 3.13 by deriving $\exists x \varphi(x) \land \exists x \psi$ from $\exists x(\varphi(x) \land \psi)$.

3.14. Verify that $\forall x \exists y (f(x) = y)$ is a tautology by giving a formal proof.

3.15. Complete the proof of Proposition 3.15.

3.16. For each $n \in \mathbb{N}$, let Φ_n be the sentence

$$\exists x_1 \cdots \exists x_n \left(\bigwedge_{i \neq j} x_i \neq x_j \right)$$

asserting that there exist at least n elements.

Let φ_1 be the sentence
$$\forall x \forall y ((f(x) = f(y)) \to (x = y))$$
saying that f is one-to-one and let φ_2 be the sentence
$$\forall y \exists x (f(x) = y)$$
saying that f is onto.

Show that $\{\varphi_1, \varphi_2, \Phi_n\} \vdash \Phi_{n+1}$ by giving a sketch of a formal proof.

3.17. For each $n \in \mathbb{N}$, let Φ_n be as defined in the previous exercise. Consider the following three sentences:
$$\forall x \forall y ((x < y) \to \neg(x = y)))$$
$$\forall x \forall y \forall z (((x < y) \land (y < z)) \to (x < z))$$
$$\forall x \forall y ((x < y) \to \exists z ((x < z) \land (z < y))).$$

Each of these sentences hold in any structure that interprets $<$ as a dense linear order (such as $\mathbf{Q}_< = (\mathbb{Q}| <)$ or $\mathbf{R}_< = (\mathbb{R}| <)$). Let ψ_1, ψ_2 and ψ_3 denote these three sentences in the order they are given.

(a) Show that $\{\psi_1, \psi_2, \psi_3, \Phi_n\} \vdash \Phi_{n+1}$.

(b) Show that it is not the case that $\{\psi_1, \psi_3, \Phi_n\} \vdash \Phi_{n+1}$.

3.18. For each of the following formulas, find an equivalent formula in Conjunctive Prenex Normal Form. Note that each of these formulas have x and y as free variables.

(a) $\neg \exists z Q(x, y, z) \lor \forall z \exists w P(w, x, y, z)$

(b) $\forall z (R(x, z) \land R(x, y) \to \exists w (R(x, w) \land R(y, w) \land R(z, w)))$

(c) $\exists z (S(y, z) \land \exists y (S(z, y) \land \exists z (S(x, z) \land (S(z, y)))))$.

3.19. Find the Skolemization of $\exists x \forall y \psi(x, y)$ where $\psi(x, y)$ is any one of the formulas from the previous exercise.

3.20. Under what conditions on φ will the Skolemization φ^S be equivalent to φ?

3.21. Let $\mathcal{V}$ be the vocabulary $\{f, P\}$ consisting of a unary function f and a unary relation P. Let φ be the formula
$$\forall x (P(x) \to P(f(x)) \land \exists x P(x) \land \exists x \neg P(x)).$$

(a) Show that φ does not have a Herbrand model.

(b) Find a Herbrand model for the Skolemization φ^s.

3.22. Let φ_1 and φ_2 be the sentences in the vocabulary $\{f\}$ defined in Exercise 3.16. Show that φ_1 has a Herbrand model, but neither φ_2 nor the Skolemization of φ_2 has a Herbrand model.

Proof theory   145

3.23. Let $\varphi(x)$ be a formula that is both quantifier-free and equality-free. Show that $\exists x \varphi(x)$ is a tautology if and only if $\varphi(t_1) \vee \varphi(t_2) \vee \cdots \vee \varphi(t_n)$ is a tautology for some $n \in (N)$ and terms t_i in the Herbrand universe for φ.

3.24. Use the Herbrand method to show that the following sentence is not satisfiable: $\forall x \neg R(x,x) \wedge \forall x R(x, f(x)) \wedge \exists x \forall y (R(x,y) \rightarrow R(f(x), y))$.

3.25. A first-order sentence is a *Horn sentence* if it is in SNF and each clause contains at most one positive literal. Describe a polynomial-time algorithm that determines whether or not a given Horn sentence is satisfiable. (Use the Herbrand method and the Horn algorithm from Section 1.7.)

3.26. Is the following set of literals unifiable?

$$\{Q(f(g(w)), h(x)), y, f(z, a)), Q(f(y, h(x)), g(w), f(g(w), x)),$$
$$Q(f(g(z), h(a)), z, f(y, x))\}.$$

If so, give the most general unifier and another unifier that is not most general.

3.27. Is the following set of literals unifiable?

$$\{R(f(x), g(z)), R(y, g(x)), R(v, w), R(w, g(x))\}.$$

If so, give the most general unifier and another unifier that is not most general.

3.28. (a) Using the Unification algorithm, find a most general unifier for the set $\{R(x,y,z), R(f(w,w), f(x,x), f(y,y)\}$.

(b) Now consider the set $\{R(x_1, \ldots, x_{n+1}), R(f(x_0, x_0)), \ldots, f(x_n, x_n)\}$. Given this set as input, how many steps will it take the Unification algorithm to halt and output a most general unifier? Is this algorithm polynomial time?

3.29. Use resolution to prove that the following are tautologies:
  (a) $(\exists x \forall y Q(x,y) \wedge \forall x (Q(x,x) \rightarrow \exists y R(y,x))) \rightarrow \exists y \exists x R(x,y)$
  (b) $(\exists x \forall y R(x,y)) \leftrightarrow (\neg \forall x \exists y \neg R(x,y))$
  (c) $(\forall x((P(x) \rightarrow Q(x)) \rightarrow \forall y R(x,y)) \wedge \forall y(\neg R(a,y) \rightarrow \neg P(a))) \rightarrow R(a,b)$.

3.30. Let $\mathcal{V}_E$ be the vocabulary $\{E\}$ consisting of one binary relation. Let Γ be the set consisting of the following three $\mathcal{V}_\mathcal{E}$-sentences from Example 2.27.
$$\forall x E(x,x)$$
$$\forall x \forall y (E(x,y) \rightarrow E(y,x))$$
$$\forall x \forall y \forall z ((E(x,y) \wedge E(y,z)) \rightarrow E(x,z))).$$

Using resolution, derive $\forall x(E(x,y) \leftrightarrow E(x,z))$ from $\Gamma \cup \{\exists x(E(x,y) \wedge E(x,z))\}$. (First put these sentences in SNF.)

3.31. Refer to the proof of Lemma 3.44.
   (a) Show that if $C = \{L\}$, then $\mathcal{C} = \{C\}$.
   (b) Show that if $C = \{L\}$, then $\mathcal{F}_L$ is minimal unsatisfiable.

3.32. *P-resolution* is the refinement of resolution that requires that one parent contains only positive literals. Let φ be a sentence in SNF. Show that φ is unsatisfiable if and only if $\emptyset$ can be derived from φ by P-resolution.

3.33. *T-resolution* is the refinement of resolution that requires that neither parent is a tautology. Let φ be a sentence in SNF. Show that φ is unsatisfiable if and only if $\emptyset$ can be derived from φ by T-resolution.

3.34. Let F be a formula of propositional logic in CNF. Let $\mathcal{A}$ be an assignment defined on the atomic subformulas of F. Let $\mathcal{A}$-*resolution* be the refinement of resolution that requires that $\mathcal{A}(C) = 0$ for one parent C. Show that F is unsatisfiable if and only if $\emptyset$ can be derived from F by $\mathcal{A}$-resolution.

3.35. Let F be a formula of propositional logic in CNF. Suppose that the atomic subformulas of F are among $\{A, B, C, \ldots, X, Y, Z\}$. Let *alphabetical-resolution* be the refinement of resolution having the following requirement. We only allow the resolvent R of C_1 and C_2 if there exists an atomic subformula of both C_1 and C_2 that precedes every atomic subformula of R alphabetically. Show that F is unsatisfiable if and only if $\emptyset$ can be derived from F by alphabetical-resolution.

4 Properties of first-order logic

We show that first-order logic, like propositional logic, has both completeness and compactness. We prove a countable version of these theorems in Section 4.1. We further show that these two properties have many useful consequences for first-order logic. For example, compactness implies that if a set of first-order sentences has an infinite model, then it has arbitrarily large infinite models. To fully understand completeness, compactness, and their consequences we must understand the nature of infinite numbers. In Section 4.2, we return to our discussion of infinite numbers that we left in Section 2.5. This digression allows us to properly state and prove completeness and compactness along with the Upward and Downward Löwenhiem–Skolem theorems. These are the four central theorems of first-order logic referred to in the title of Section 4.3. We discuss consequences of these theorems in Sections 4.4–4.6. These consequences include amalgamation theorems, preservation theorems, and the Beth Definability theorem.

Each of the properties studied in this chapter restrict the language of first-order logic. First-order logic is, in some sense, weak. There are many concepts that cannot be expressed in this language. For example, whereas first-order logic can express "there exist n elements" for any finite n, it cannot express "there exist countably many elements." Any sentence having a countable model necessarily has uncountable models. As we previously mentioned, this follows from compactness. In the final section of this chapter, using graphs as an illustration, we discuss the limitations of first-order logic. Ironically, the weakness of first-order logic makes it the fruitful logic that it is. The properties discussed in this chapter, and the limitations that follow from them, make possible the subject of model theory.

All formulas in this chapter are first-order unless stated otherwise.

4.1 The countable case

Many of the properties of first-order logic, including completeness and compactness, are consequences of the following fact:

Every model has a theory and every theory has a model.

Recall that a set of sentences is a "theory" if it is consistent (i.e. if we cannot derive a contradiction). "Every theory has a model" means that if a set of sentences is consistent, then it is satisfiable. Recall too that, for any $\mathcal{V}$-structure M, the "theory of M," denoted $Th(M)$, is the set of all $\mathcal{V}$-sentences that hold in M. "Every model has a theory" asserts that $Th(M)$ is consistent. Put another way, the above fact states that any set of sentences Γ is consistent if and only if it is satisfiable. In this section, we prove this fact for countable Γ and derive countable versions of completeness and compactness.

Proposition 4.1 Let Γ be a set of sentences. If Γ is satisfiable then Γ is consistent.

Proof If Γ is satisfiable then $M \models \Gamma$ for some structure M. By Theorem 2.86, $Th(M)$ is a complete theory. In particular, $Th(M)$ is consistent. Since Γ is a subset of $Th(M)$, Γ is consistent. □

Now consider the converse. If Γ is consistent, then it is satisfiable. One way to prove this is to demonstrate a model for Γ. Since Γ is an arbitrary set of sentences, this may seem to be a daunting task. However, there exists a remarkably elementary way to construct such a model. We use a technique known as a *Henkin construction*. This versatile technique will be utilized again in Sections 4.3 and 6.2.

Theorem 4.2 Let Γ be a countable set of sentences. If Γ is consistent then Γ is satisfiable.

Proof Suppose Γ is consistent. We will demonstrate a structure that models Γ.

Let $\mathcal{V}$ be the vocabulary of Γ. Let $\mathcal{V}^+ = \mathcal{V} \cup \{c_1, c_2, c_3, \ldots\}$, where each c_i is a constant that does not occur in $\mathcal{V}$. We let C denote the set $\{c_1, c_2, c_3, \ldots\}$. Since both $\mathcal{V}$ and C are countable, so is $\mathcal{V}^+$ (by Proposition 2.43).

We shall define a complete $\mathcal{V}^+$-theory T^+ with the following properties.

Property 1 Every sentence of Γ is in T^+.

Property 2 For every $\mathcal{V}^+$-sentence in T^+ of the form $\exists x \theta(x)$, the sentence $\theta(c_i)$ is also in T^+ for some $c_i \in C$.

The second property allows us to find a model M^+ of T^+. By the first property, M^+ is also a model of Γ. If we can define such T^+ and M^+, then this will prove the theorem.

We define T^+ in stages. Let T_0 be Γ. Enumerate the set of all $\mathcal{V}^+$-sentences as $\{\varphi_1, \varphi_2, \ldots\}$. This is possible since the set of $\mathcal{V}^+$-sentences is countable (by Proposition 2.47). Suppose that, for some $m \geq 0$, T_m has been defined in such a way that T_m is consistent and only finitely many of the constants in C occur in T_m. To define T_{m+1}, consider the sentence φ_{m+1}. There are two cases:

(a) If $T_m \cup \{\neg \varphi_{m+1}\}$ is consistent, then define T_{m+1} to be $T_m \cup \{\neg \varphi_{m+1}\}$.

(b) If $T_m \cup \{\neg\varphi_{m+1}\}$ is not consistent, then $T_m \cup \{\varphi_{m+1}\}$ is consistent. We divide this case into two subcases:
  (i) If φ_{m+1} does not have the form $\exists x \theta(x)$ for some formula $\theta(x)$, then just let T_{m+1} be $T_m \cup \{\varphi_{m+1}\}$.
  (ii) Otherwise φ_{m+1} has the form $\exists x \theta(x)$. In this case let T_{m+1} be $T_m \cup \{\varphi_{m+1}\} \cup \{\theta(c_i)\}$, where i is such that c_i does not occur in $T_m \cup \{\varphi_{m+1}\}$.

So given T_m which is consistent and uses only finitely many constants from C, we have defined T_{m+1}. In any case, T_{m+1} is obtained by adding at most two sentences to T_m. Since T_m uses only finitely many constants from C, so does T_{m+1}. Moreover, we claim that T_{m+1} is consistent.

Claim T_{m+1} is consistent.

Proof of Claim If T_{m+1} is as in (a) or (b)(i), then T_{m+1} is consistent by its definition. So assume that T_{m+1} is as in part (ii) of (b). We know that $T_m \cup \{\varphi_{m+1}\}$ is consistent.

Suppose for a contradiction that $T_{m+1} = T_m \cup \{\varphi_{m+1}\} \cup \{\theta(c_i)\}$ is inconsistent. Then we have

$$T_m \cup \{\varphi_{m+1}\} \vdash \neg\theta(c_i).$$

Since c_i does not occur in $T_m \cup \{\varphi_{m+1}\}$ we have

$$T_m \cup \{\varphi_{m+1}\} \vdash \forall x \neg\theta(x) \quad \text{by } \forall\text{-Introduction.}$$

Since φ_{m+1} is the formula $\exists x \theta(x)$, we have

$$T_m \cup \{\varphi_{m+1}\} \vdash \exists x \theta(x) \text{by Assumption.}$$

By the definition of $\forall$ we have

$$T_m \cup \{\varphi_{m+1}\} \vdash \neg\forall x \neg\theta(x).$$

We see that we can derive both $\forall x \neg\theta(x)$ and its negation from $T_m \cup \{\varphi_{m+1}\}$. This contradicts our assumption that $T_m \cup \{\varphi_{m+1}\}$ is consistent. Our supposition that T_{m+1} is inconsistent must be incorrect. We conclude that T_{m+1}, like T_m, is consistent.

Recall that T_0 is Γ which is consistent and uses no variables from C. We can apply the above definition of T_{m+1} with $m = 0$ to get T_1. By the claim, T_1 is also consistent and it uses at most finitely many constants from C. And so we can again apply the definition of T_{m+1}, this time with $m = 1$. This process generates the sequence $T_0 \subset T_1 \subset T_2, \ldots$.

We now define the $\mathcal{V}^+$-theory T^+. Let T^+ be the set of all $\mathcal{V}^+$-sentences φ that occur in T_i for some i. Put another way, T^+ is the *union* of all of the T_is.

Put yet another way T^+ is the *limit* of the sequence $T_0, T_1, T_2, \ldots$ If we continue this process forever, then T^+ is the end result.

We must verify that T^+ has all of the desired properties. First of all, T^+ is consistent. To see this, let Δ be any finite subset of T^+. Then Δ is a subset of T_m for some m. Since T_m is consistent, so is Δ. So every finite subset of T^+ is consistent. If T^+ were inconsistent, then we could derive a contradiction from T^+. Since formal proofs are finite, we could derive a contradiction from a finite subset of T^+. Since every finite subset of T^+ is consistent, so is T^+.

So T^+ is a theory. We next show that T^+ is a complete theory. Let φ be an arbitrary $\mathcal{V}^+$-sentence. Then φ is φ_i for some i. Since either φ_i or $\neg \varphi_i$ is in $T_i \subset T^+$, T^+ is complete.

Finally, we must show that T^+ has Properties 1 and 2. Since $T_0 = \Gamma$, every sentence of Γ is in T^+. So T^+ has Property 1. To show that T^+ has Property 2, let $\exists x \theta(x)$ be a $\mathcal{V}^+$-sentence in T^+. This sentence is φ_{m+1} for some m. Since this sentence is in T^+, $T_m \cup \{\neg \varphi_{m+1}\}$ is inconsistent. In this case, T_{m+1} is defined as $T_m \cup \{\varphi_{m+1}\} \cup \{\theta(c_i)\}$ for some constant c_i. So $\theta(c_i)$ is in T^+ and T^+ has Property 2.

Having successfully defined T^+, we next define a $\mathcal{V}^+$-structure M^+ that models T^+. The underlying set U^+ of M^+ is a set of variable-free $\mathcal{V}^+$-terms. Let t_1 and t_2 be two $\mathcal{V}^+$-terms that do not contain variables. We say t_1 and t_2 are the "same" if T^+ says they are. That is, t_1 and t_2 are the same if and only if $T^+ \vdash t_1 = t_2$. (Note that, since T^+ is complete, either $T^+ \vdash t_1 = t_2$ or $T^+ \vdash \neg(t_1 = t_2)$.) Let U^+ be such that every variable-free $\mathcal{V}^+$-term is the same as some term in U^+ and no two terms of U^+ are the same. So if t_1 and t_2 are the same, then U^+ does not contain both of them, it contains exactly one term that is the same as these terms.

To complete our description of M^+, we must say how M^+ interprets $\mathcal{V}^+$. Since the elements of U^+ are $\mathcal{V}^+$-terms, there is a natural interpretation. For any constant $c \in \mathcal{V}^+$, there exists a unique term $t \in U^+$ such that $T^+ \vdash t = c$. The structure M^+ interprets the constant c as the element t in its underlying set. Moreover, M^+ interprets the relations and functions of $\mathcal{V}^+$ in the manner described by T^+. More precisely,

- for any n-ary relation $R \in \mathcal{V}^+$ and any n-tuple $\bar{t}$ of elements from U^+, $M^+ \models R(\bar{t})$ if and only if $T^+ \vdash R(\bar{t})$, and

- for any n-ary function $f \in \mathcal{V}^+$, any element $s \in U^+$, and any n-tuple $\bar{t}$ of elements in U^+, $M^+ \models f(\bar{t}) = s$ if and only if $T^+ \vdash f(\bar{t}) = s$.

This completes our description of M^+.

Properties of first-order logic 151

Claim For any $\mathcal{V}^+$-sentence φ, $M^+ \models \varphi$ if and only if $T^+ \vdash \varphi$.

Proof Since every first-order sentence is equivalent to a sentence that uses only the fixed symbols $\neg$, $\wedge$, $\exists$, and $=$ (and neither $\vee$, $\leftarrow$, $\leftrightarrow$, nor $\forall$), we may assume with no loss of generality that these are the only fixed symbols occurring in φ. We proceed by induction on the total number of occurrences of $\neg$, $\wedge$, and $\exists$ in φ.

If φ has no occurrences of these three fixed symbols, then φ must be atomic. In this case, $M^+ \models \varphi$ if and only if $T^+ \vdash \varphi$ by the definition of M^+.

Suppose now that φ has a total of $m+1$ occurrences of $\neg$, $\wedge$, and $\exists$. Our induction hypothesis is that the claim holds for any sentence having m or fewer occurrences of these symbols.

If φ has the form $\psi \wedge \theta$, then $T^+ \vdash \varphi$ if and only if both $T^+ \vdash \psi$ and $T^+ \vdash \theta$ (since T^+ is a complete theory). By our induction hypothesis, this happens if and only if M^+ models both ψ and θ and, therefore, φ as well.

If φ has the form $\neg \psi$, then $T^+ \vdash \varphi$ if and only if ψ is not in T^+ (since T^+ is a complete theory). By our induction hypothesis, this happens if and only if M^+ does not model ψ. By the semantics of $\neg$, M^+ does not model ψ if and only if $M^+ \models \varphi$.

Lastly, suppose that φ has the form $\exists x \theta(x)$. By Property 2 and our definition of U^+, $T^+ \vdash \varphi$ if and only if $T^+ \vdash \theta(s)$ for some term s in U^+ (since T^+ is a complete theory). By our induction hypothesis, $T^+ \vdash \theta(s)$ if and only if $M^+ \models \theta(s)$. Finally, by the semantics of $\exists$, $M^+ \models \theta(s)$ for some term $s \in U^+$ if and only if $M^+ \models \varphi$.

This completes the proof of the claim.

It follows from this claim that $M^+ \models T^+$. Hence, we have demonstrated a model for Γ as was required. □

Corollary 4.3 Let Γ be a countable set of formulas. If Γ is consistent, then Γ has a countable model.

Proof The structure M^+ from the proof of Theorem 4.2 is countable. □

The following corollary is a countable version of the Compactness theorem for first-order logic.

Corollary 4.4 A countable set of formulas is satisfiable if and only if every finite subset is satisfiable.

Proof Let Γ be a countable set of formulas. We prove that Γ is unsatisfiable if and only if there exists a finite subset of Γ that is unsatisfiable. Clearly, if there exists a finite subset of Γ that is not satisfiable, then Γ is not satisfiable either. So suppose Γ is not satisfiable. By Theorem 4.2, Γ is inconsistent. That is, $\Gamma \vdash \bot$

for some contradiction $\bot$. Since formal proofs are finite, $\Delta \vdash \bot$ for some finite subset Δ of Γ. By Theorem 3.4, $\Delta \models \bot$ and Δ is unsatisfiable. □

The following corollary is a countable version of the Completeness theorem for first-order logic.

Corollary 4.5 For any countable set of formulas Γ, $\Gamma \vdash \varphi$ if and only if $\Gamma \models \varphi$.

Proof If $\Gamma \vdash \varphi$, then $\Gamma \models \varphi$ by Theorem 3.4. Conversely, suppose that $\Gamma \models \varphi$. Then $\Gamma \cup \{\neg\varphi\}$ is unsatisfiable. By Theorem 4.2, $\Gamma \cup \{\neg\varphi\}$ is inconsistent. That is,

$$\Gamma \cup \{\neg\varphi\} \vdash \bot$$

for some contradiction $\bot$. By Contrapositive,

$$\Gamma \cup \{T\} \vdash \neg\neg\varphi,$$

where T is the tautology $\neg \bot$. Finally,

$$\Gamma \vdash \varphi$$

by the Tautology rules and Double negation. □

All of the results of this section can be extended to include uncountable sets of sentences. We state and prove both the Compactness theorem and the Completeness theorem in their full generality in Section 4.3. This requires familiarity with cardinal numbers.

4.2 Cardinal knowledge

We return to our discussion of infinite sets. In Section 2.5, we defined what it means for two sets to have the "same size." We now introduce numbers to represent the size of a set. These numbers are called *cardinals* and the size of a set is called the *cardinality* of the set. If a set is finite, then its size is some natural number (or zero if the set is empty). So each natural number is a cardinal. The Hebrew letter $\aleph$ (aleph) is used with subscripts to denote infinite cardinals. The smallest infinite cardinal is $\aleph_0$. This is the cardinality of the set $\mathbb{N}$ and, therefore, of every countably infinite set.

The cardinality of set A is denoted $|A|$. In Section 2.5 we made the assumption that for any sets A and B, either $|A| \leq |B|$ or $|B| \leq |A|$. This assumption allows us to list the cardinals in ascending order as follows:

$$0, 1, 2, 3, \ldots, \aleph_0, \aleph_1, \aleph_2, \ldots$$

The cardinals $\aleph_1$, $\aleph_2$, and beyond are uncountable cardinals. We showed that the set of real numbers is uncountable in Section 2.5. This raises a new question: where in the above list does $|\mathbb{R}|$ fall? Is it equal to $\aleph_1$ or some other uncountable cardinal? We address this question and state some surprising results at the end of the present section. As we shall see, it is possible that the cardinality of the reals is bigger than $\aleph_n$ for each natural number n. The above list of cardinals is only a partial list. To extend this list we must discuss ordinal numbers.

4.2.1 Ordinal numbers. There are two types of numbers: cardinals and ordinals. Whereas cardinals regard quantity, ordinals regard the length of an ordered list. The difference between cardinals and ordinals is the difference between 7 and 7th. This distinction is mere pedantry for finite numbers. For infinite numbers, however, the distinction between cardinals and ordinals is essential.

Example 4.6 Consider the following ordered lists of natural numbers:

$$A = \{1, 2, 3, \ldots\}$$
$$B = \{2, 3, 4, \ldots\} \cup \{1\}$$
$$C = \{3, 4, 5, \ldots\} \cup \{1, 2\}$$
$$D = \{1, 3, 5, 7, \ldots\} \cup \{2, 4, 6, 8, \ldots\}.$$

As sets, each of these is identical to $\mathbb{N}$. The cardinality of each of these sets is $\aleph_0$. However, the order in which these sets are listed differs. In B, the number 1 follows infinitely many numbers. In this sense, B is longer than A. Likewise C is longer than B and D is the longest of the four lists. Ordinal numbers recognize this distinction. The ordinal number ω describes the length of the natural numbers with the usual order. So ω describes the ordered set A. The length of B is denoted $\omega+1$. Likewise, the ordinal $\omega+2$ describes C. Finally, the ordinal representing the length of D is $\omega + \omega$.

Whereas every set has a cardinality, not every set has an ordinality. Ordinality is defined only for sets that are *well ordered*. A *linearly ordered set* is a set X with a binary relation $<$ so that

1. for all a and b in X, exactly one of the following hold: either $a < b$, $b < a$, or $a = b$, and

2. for all a, b and c in X, if $a < b$ and $b < c$ then $a < c$.

That is, a linearly ordered set is a set equipped with a notion of "less than" by which any two nonequal elements can be compared. A *well-ordered set* is a linearly ordered set that is ordered in such a way that every nonempty subset has a least element.

Example 4.7 The natural numbers $\mathbb{N}$ with the usual ordering is a well ordered set. Any given set of natural numbers must contain a smallest number. The rational numbers $\mathbb{Q}$ with the usual ordering is a linearly ordered set that is not well ordered. To see this, consider the set $\{1/n \,|\, n \in \mathbb{N}\}$. This subset of the rational numbers does not contain a smallest element.

Any finite linearly ordered set is well ordered. The ordinality of a finite set does not depend on the particular order of the set. If ten people are standing in a queue, then, regardless of their arrangement, one thing is certain: the tenth person is last. As Example 4.6 demonstrates, the same cannot be said for infinite sets.

Definition 4.8 Let A be a finite well ordered set. The *ordinality* of A is the same as the cardinality of the set.

So the ordinals, listed in ascending order, begin with the finite ordinals $0, 1, 2, 3, \ldots$. To continue the list we apply the following rule.

> Given any nonempty set of ordinals, there exists a least ordinal greater than each ordinal in that set.

All ordinals are generated by repeated application of this single rule. The least ordinal greater than each finite ordinal is denoted by the Greek letter ω (omega). So ω is the smallest infinite ordinal. This is the ordinality of $\mathbb{N}$ with the usual ordering. The least ordinal greater than ω is denoted $\omega + 1$. The least ordinal greater than $\omega + 1$ is $\omega + 2$. These ordinals were illustrated in Example 4.6.

For any ordinal α, the least ordinal greater than α is called the *successor* of α and is denoted $\alpha + 1$. Let A be a well ordered set having ordinality α. Then $\alpha + 1$ is the ordinality of the well ordered set $A \cup \{b\}$ where b is a new element (not in A) that is greater than each element of A. Every ordinal has a successor, but not every ordinal has an immediate predecessor. An ordinal that has an immediate predecessor is called a *successor ordinal*. A nonzero ordinal that is not the successor of any ordinal is called a *limit ordinal*. For example, ω is the smallest limit ordinal.

The ordinals have a natural order. For any ordinal α, the successor of α and all subsequent ordinals are *greater* than α. We let $<$ denote this order and refer to this as the *usual* order for the ordinals.

Proposition 4.9 Any set of ordinals with the usual order is a well ordered set.

Proof Let A be a set of ordinals. It is clear that A is a linearly ordered set. To show that it is a well ordered set, we must show that any nonempty subset X of A contains a least element. If X happens to contain 0, then X certainly has a least element. So suppose that $0 \notin X$. Let L be the set of all ordinals that are less than every ordinal in X. Since $0 \notin X$, L is nonempty. So there exists a least ordinal greater than each ordinal in L. This is the least ordinal in X. □

Since the set of all ordinals less than α is well ordered, it has an ordinality. We naturally define the ordinality of this set to be α. For example, the set $\{0, 1, 2, 3\}$ of ordinals less than 4 has ordinality (and cardinality) 4. More generally, we now define the ordinality of an arbitrary well ordered set.

Definition 4.10 The well ordered set A has *ordinality* α if there exists a one-to-one correspondence f from A onto the set $\{\beta \,|\, \beta < \alpha\}$ that preserves the order. By "preserves the order" we mean that, for any x and y in A, $x < y$ if and only if $f(x) < f(y)$.

This is an unambiguous definition of ordinality that agrees with all of the facts we have previously stated about ordinality. (In particular, the reader can verify the ordinalities stated in Example 4.6.)

The ordinal α is identified with the set $\{\beta \,|\, \beta < \alpha\}$. Clearly, any ordinal α uniquely determines the set $\{\beta \,|\, \beta < \alpha\}$. Conversely, given $\{\beta \,|\, \beta < \alpha\}$, we can define α as the least ordinal greater than each ordinal in this set. In light of this association, we consider α and $\{\beta \,|\, \beta < \alpha\}$ to be interchangeable entities. So the ordinal 4 *is* the set $\{0, 1, 2, 3\}$. The purpose of this is to facilitate our notation. In particular, we write $|\alpha|$ to denote the cardinality of the set $\{\beta \,|\, \beta < \alpha\}$. We refer to α as being *countable* or *uncountable* depending on whether $|\alpha|$ is countable or uncountable.

Whereas there is only one countably infinite cardinal, there are many countably infinite ordinals (see Exercise 4.21). We proceed now to list some countable ordinals. The first ordinal is 0. After 0, we have the successors 1,2,3... followed by the limit ordinal ω. This is then followed by $\omega + 1$, $\omega + 2$, $\omega + 3$, and so forth. The least ordinal greater than each ordinal in the set $\{\omega + n \,|\, n \in \mathbb{N}\}$ is the limit ordinal $\omega + \omega$ also known as $\omega \cdot 2$. This has successor $\omega \cdot 2 + 1$ which has successor $\omega \cdot 2 + 2$. Continuing in this manner we arrive at the limit ordinals $\omega \cdot 3$, $\omega \cdot 4$, and so forth. The least ordinal greater than each ordinal in the set $\{\omega \cdot n \,|\, n \in \mathbb{N}\}$ is the ordinal $\omega \cdot \omega$ also known as ω^2. Likewise, ω^3, ω^4, and the limit ω^ω are each ordinals as are ω^{ω^ω} and

$$\omega^{\omega^{\omega^{\omega^{\omega^\omega}}}}.$$

Each of these ordinals is countable. The least ordinal greater than each countable ordinal is denoted ω_1. The cardinal $\aleph_1$ is defined as $|\omega_1|$. Likewise, ω_2 denotes the least ordinal greater than each ordinal of cardinality $\aleph_1$. We define $\aleph_2$ as $|\omega_2|$. Whereas $\aleph_2$ is the cardinal immediately following $\aleph_1$, ω_2 does not immediately follow ω_1. Rather, ω_1 is followed by $\omega_1 + 1$, $\omega_1 + 2$, and so forth.

The list of ordinals cannot be exhausted. Given any set of ordinals, there exist ordinals greater than all of those in that set. So it is nonsense to speak of the totality of all ordinals. When we refer to the *list* of ordinals, it should be understood that this is not a complete list. There necessarily exist ordinals beyond those in any list, no matter how extensive. In particular, we forbid ourselves from referring to the set (or list) of all ordinals. Although it is alluring terminology, "the set of all ordinals" does not make sense.

We conclude our discussion of ordinal numbers by introducing the Well Ordering Principle. Consider the set $\mathbb{Q}$. With its usual order, this set does not have an ordinality. As demonstrated in Example 4.7, this is not a well ordered set. With another order, however, $\mathbb{Q}$ is a well ordered set. Since $\mathbb{Q}$ has the same size as $\mathbb{N}$, we can enumerate $\mathbb{Q}$ as $\{q_1, q_2, q_3, \ldots\}$. The rational numbers with this order has ordinality ω. As Example 4.6 shows, $\mathbb{Q}$ may have different ordinalities when arranged in a different order. Likewise, we can impose a well ordering on any set. This is the Well Ordering Principle

Proposition 4.11 (Well Ordering Principle) Any set X can be enumerated as $\{x_\beta \mid \beta < \alpha\}$ for some ordinal α. Moreover, we may require that α be the least ordinal such that $|\alpha| = |X|$.

Proof First we show that there exists an order $<$ that makes X a well ordered set. Since there exist arbitrarily large ordinals, there exists an ordinal γ with $|X| \leq |\gamma|$. By the definition of $|X| \leq |\gamma|$, there exists a one-to-one function f from X into $\{\beta \mid \beta < \gamma\}$. For any x and y in X, we define $x < y$ to mean $f(x) < f(y)$. Since $\{\beta \mid \beta < \gamma\}$ is well ordered, so is X with this order. Let α' be the ordinality of this well ordered set.

Now consider the set of all ordinals δ with $|\delta| = |X|$. Since it contains α', this is a nonempty set of ordinals. By Proposition 4.9, there exists a least ordinal α in this set. Since $|\alpha| = |X|$, there exists (by Theorem 2.39), a one-to-one correspondence g from $\{\beta \mid \beta < \alpha\}$ onto X. For each ordinal $\beta < \alpha$, let x_β denote $g(\beta)$. This provides the required enumeration $\{x_\beta \mid \beta < \alpha\}$ of X. □

The Well Ordering Principle is in fact equivalent to the statement that every set has a cardinality. It is also equivalent to our earlier assumption that, for any sets A and B, either $|A| \leq |B|$ or $|B| \leq |A|$. Each of these statements is equivalent to an axiom of mathematics known as the Axiom of Choice. This axiom can be stated as follows: the Cartesian product of nonempty sets is nonempty. We view this as a reasonable axiom and employ it without further comment.

4.2.2 Cardinal arithmetic. The list of cardinal numbers begins with

$$0, 1, 2, 3, \ldots, \aleph_0, \aleph_1, \aleph_2, \ldots$$

We extend this list indefinitely by using ordinal numbers as subscripts.

We define the infinite cardinals by induction on the ordinals. Prior to stating this definition, we discuss what we mean by "induction on the ordinals."

This version of induction, known as *transfinite induction*, can be used to show that some property P holds for each ordinal α. Like other forms of induction, transfinite induction consists of two steps: the base step and the induction step. First, we show that P holds for 0. This is the base step. Second, we show that if P holds for all $\beta < \alpha$, then it holds for α as well. This is the induction step. If we successfully complete these two steps, then we can rightly conclude that P does, in fact, hold for each ordinal α (since there is no least ordinal for which property P does not hold).

Definition 4.12 We define the infinite cardinals by transfinite induction. First we define (again) $\aleph_0$ to be $|\mathbb{N}|$. Let α be a nonzero ordinal. Suppose that $\aleph_\iota$ has been defined for each $\iota < \alpha$. Let γ be the least ordinal such that $|\gamma| > \aleph_\iota$ for all $\iota < \alpha$. We define $\aleph_\alpha$ to be $|\gamma|$.

Having defined the cardinal numbers, we now define arithmetic operations for these numbers. Cardinal arithmetic must not be confused with ordinal arithmetic. Previous reference was made to $\omega+\omega$, $\omega\cdot 2$, and ω^ω. Since ω is an ordinal, these are expressions of ordinal arithmetic (each represents a countable ordinal). We turn now to cardinal arithmetic.

Definition 4.13 Let κ and λ be cardinals. Let A and B be disjoint sets with $|A| = \kappa$ and $|B| = \lambda$.

- *Addition:* $\kappa + \lambda = |A \cup B|$.
- *Multiplication:* $\kappa \cdot \lambda = |A \times B|$.
- *Exponentiation:* $\kappa^\lambda = |F(B, A)|$ where $F(B, A)$ is the set of all functions $f : B \to A$ having B as a domain and a subset of A as a range.

Note that these definitions are independent of our choice of A and B. The requirement that A and B are disjoint is needed only for adding finite cardinals.

If κ and λ are finite cardinals, then these definitions correspond to the familiar notions of addition, multiplication, and exponentiation. We demonstrate (but do not prove) this fact with an example.

Example 4.14 Let $A = \{a_1, a_2, a_3\}$ and let $B = \{b_1, b_2, b_3, b_4\}$.

Addition. We have $A \cup B = \{a_1, a_2, a_3, b_1, b_2, b_3, b_4\}$. Clearly $|A| + |B| = 3 + 4 = 7 = |A \cup B|$.

Multiplication. Recall that $A \times B$ is the set $\{(a_i, b_j)| 1 \leq i \leq 3, 1 \leq j \leq 4\}$. We list the elements of this set as follows:

(a_1, b_1) (a_1, b_2) (a_1, b_3) (a_1, b_4)
(a_2, b_1) (a_2, b_2) (a_2, b_3) (a_2, b_4)
(a_3, b_1) (a_3, b_2) (a_3, b_3) (a_3, b_4).

Observing the above arrangement of the elements in $A \times B$, we see that the size of $A \times B$ is $3 \cdot 4 = 12$. So $|A| \cdot |B| = 3 \cdot 4 = 12 = |A \times B|$.

Exponentiation. The set $F(B, A)$ consists of all functions $f : B \to A$. Each function is determined by the values of $f(b)$ for $b \in B$. For each of the four elements in B, there three possible values for $f(b)$ in A. It follows that there are $3 \cdot 3 \cdot 3 \cdot 3 = 3^4$ functions in $F(B, A)$. We see that $|A|^{|B|} = 3^4 = 81 = |F(B, A)|$.

So for finite cardinals, addition, multiplication, and exponentiation are nothing new. We now consider these operations for infinite cardinals. It turns out that adding and multiplying two infinite cardinals are remarkably easy tasks (easier than adding and multiplying finite cardinals). In contrast, exponentiation for infinite cardinals is remarkably hard. We deal with the two easier operations first. All there is to know about the addition and multiplication of infinite cardinals stems from the following result.

Theorem 4.15 Let κ be an infinite cardinal. Then $\kappa \cdot \kappa = \kappa$.

Proof We prove that this holds for $\kappa = \aleph_\alpha$ by transfinite induction on α. If $\alpha = 0$, then this follows from Example 2.35 where it was shown that $|\mathbb{N} \times \mathbb{N}| = |\mathbb{N}|$.

Suppose now that $\kappa = \aleph_\alpha$ for $\alpha > 0$. Our induction hypothesis is that $\lambda \cdot \lambda = \lambda$ for all infinite cardinals λ smaller than κ.

Let δ be the least ordinal such that $|\delta| = \kappa$. We regard δ as the set of ordinals less than δ. By the definition of cardinal multiplication, $\kappa \cdot \kappa$ is the cardinality of the set $\delta \times \delta$ of ordered pairs of ordinals less than δ. We show that $|\delta \times \delta| = |\delta|$ by arranging the elements of $\delta \times \delta$ into a well ordered set having ordinality δ.

Now, $\delta \times \delta$ is well ordered by the lexicographical order defined as follows: (β_1, β_2) precedes (γ_1, γ_2) lexicographically if and only if either $\beta_1 < \gamma_1$ or both $\beta_1 = \gamma_1$ and $\beta_2 < \gamma_2$ where $<$ is the usual order for ordinals. This order is analogous to the alphabetical order of words in a dictionary. The ordinality of this well ordered set is δ^2 which is bigger than δ.

We now impose a new order on $\delta \times \delta$. We claim that the new order makes $\delta \times \delta$ a well ordered set having ordinality δ. We denote this order by $\triangleleft$ and define it as follows:

$$(\beta_1, \beta_2) \triangleleft (\gamma_1, \gamma_2)$$

if and only if

either (β_1, β_2) precedes (γ_1, γ_2) lexicographically

OR

γ_2 is larger than both β_1 and β_2.

The set $\delta \times \delta$ with the order $\triangleleft$ is a well ordered set. We leave the verification of this as Exercise 4.18. This is also true of the lexicographical order. The crucial feature of $\triangleleft$ is that, with this order, each element of $\delta \times \delta$ has fewer than κ predecessors. This is not true of the lexicographical order.

Let (β_1, β_2) be an arbitrary element of $\delta \times \delta$. To see that this element has fewer than κ predecessors, first note that $(\beta_1, \beta_2) \triangleleft (\beta, \beta)$ where β is the larger of β_1 and β_2. Further, (γ_1, γ_2) does not preceed (β, β) if either γ_1 or γ_2 is larger than β. Because of this, the predecessors of (β_1, β_2) are contained in $(\beta+1) \times (\beta+1)$. For example, suppose $(\beta_1, \beta_2) = (1, 3)$. Then $(\beta_1, \beta_2) \triangleleft (3, 3)$. The set of all elements of $\delta \times \delta$ that preceed $(3, 3)$ are contained in the following square:

$$(0,0) \quad (0,1) \quad (0,2) \quad (0,3)$$
$$(1,0) \quad (1,1) \quad (1,2) \quad (1,3)$$
$$(2,0) \quad (2,1) \quad (2,2) \quad (2,3)$$
$$(3,0) \quad (3,1) \quad (3,2) \quad (3,3).$$

Note that this is the set 4×4 (recalling that the ordinal 4 is identified with $\{0, 1, 2, 3\}$). So, with the order $\triangleleft$, there are fewer than $|4 \times 4| = 16$ predecessors of the ordered pair $(3, 3)$. Likewise, for any $\beta < \delta$, there are fewer than $|(\beta+1) \times (\beta+1)|$ elements of $\delta \times \delta$ that preceed (β, β) in the order $\triangleleft$.

Since δ is least such that $|\delta| = \kappa$ and $\beta < \delta$, we have $|\beta + 1| = |\beta| < \kappa$. By our induction hypothesis, $|(\beta+1) \times (\beta+1)| = |\beta+1|$. It follows that each element of $\delta \times \delta$ has fewer than κ predecessors in the order $\triangleleft$ as was claimed.

Let γ denote the ordinality of $\delta \times \delta$ with $\triangleleft$. If γ were larger than δ, then there would necessarily exist elements with $\kappa = |\delta|$ predecessors. Since we have shown that this is not the case, we conclude that $\gamma \leq \delta$. It follows that $\kappa \cdot \kappa = |\delta \times \delta| = |\gamma| \leq |\delta| = \kappa$. Since it is clear that $\kappa \leq \kappa \cdot \kappa$, we have $\kappa \cdot \kappa = \kappa$ as was desired. By induction, this holds for $\kappa = \aleph_\alpha$ for each ordinal α. □

Corollary 4.16 Let κ and λ be nonzero cardinals. If either κ or λ is infinite, then $\lambda \cdot \kappa$ is the larger of κ and λ.

Proof Suppose that κ is infinite and $\lambda \leq \kappa$. We have $\kappa \leq \lambda \cdot \kappa \leq \kappa \cdot \kappa$. Since, $\kappa \cdot \kappa = \kappa$ by Theorem 4.15, we conclude that $\lambda \cdot \kappa = \kappa$. Likewise, if λ is infinite and $\kappa \leq \lambda$, then $\lambda \cdot \kappa = \lambda$. □

Corollary 4.17 Let κ and λ be cardinals. If either κ or λ is infinite, then $\lambda + \kappa$ is the larger of κ and λ.

Proof If one of κ and λ is infinite and the other is finite, then this corollary follows from Exercise 2.36. So suppose that κ and λ are both infinite. If $\lambda \leq \kappa$, then $\lambda + \kappa \leq \lambda \cdot \kappa$ (this is true for any κ and λ with $2 \leq \lambda$). We have

$$\kappa \leq \lambda + \kappa \leq \lambda \cdot \kappa \leq \kappa \cdot \kappa = \kappa.$$

We conclude that each of these inequalities must in fact be equalities. In particular, $\lambda + \kappa = \lambda \cdot \kappa = \kappa$. The proof is identical for $\kappa \leq \lambda$. □

We already know how to add and multiply finite cardinals. The previous corollaries tell us how to add and multiply infinite cardinals: simply take the larger of the two numbers. So cardinal addition is easy:

$$5 + 2 = 7,\ 2 + \aleph_0 = \aleph_0,\ \aleph_1 + 7 = \aleph_1,\ \text{and}\ \aleph_7 + \aleph_{23} = \aleph_{23}.$$

Cardinal multiplication is equally easy:

$$5 \cdot 2 = 10,\ 2 \cdot \aleph_0 = \aleph_0,\ \aleph_1 \cdot 7 = \aleph_1,\ \text{and}\ \aleph_7 \cdot \aleph_{23} = \aleph_{23}.$$

Note that the same result is obtained either by adding or by multiplying two infinite cardinals. This is also true for any finite number of infinite cardinals. If we have n cardinals (for $n \in \mathbb{N}$) at least one of which is infinite, then whether we add them together or multiply them, we obtain the largest of the n cardinals. This is no longer true if we have infinitely many cardinals. We extend the definitions of addition and multiplication to infinite sums and products in an obvious way.

Definition 4.18 Let α be an infinite ordinal and let $\{\kappa_\iota \,|\, \iota < \alpha\}$ be a set of cardinals. For each $\iota < \alpha$, let A_ι be a set of cardinality κ_ι. We assume that the A_ιs are disjoint from each other.

- Infinite sums: $\sum_{\iota < \alpha} \kappa_\iota = |\bigcup_{\iota < \alpha} A_\iota|$.
- Infinite products: $\Pi_{\iota < \alpha} \kappa_\iota = |\Pi_{\iota < \alpha} A_\iota|$ where $\Pi_{\iota < \alpha} A_\iota$ denotes the Cartesian product $A_0 \times A_1 \times A_2 \times \cdots$

Just as multiplication can be viewed as repeated applications of addition, exponentiation can be viewed as repeated applications of multiplication. That is, $\underbrace{\kappa + \kappa + \kappa + \ldots}_{\lambda \text{ times}} = \kappa \cdot \lambda$ and $\underbrace{\kappa \cdot \kappa \cdot \kappa \cdots}_{\lambda \text{ times}} = \kappa^\lambda$.

We leave the verification of this as Exercise 4.20. Whereas an infinite sum is as easy as multiplication, an infinite product is as difficult as exponentiation.

We turn now to cardinal exponentiation.

Proposition 4.19 For any set A, $|\mathcal{P}(A)| = 2^{|A|}$.

Proof Recall that $\mathcal{P}(A)$ is the set of all subsets of A. Let $F(A, 2)$ denote the set of all functions from A to the set $\{0, 1\}$. We define a one-to-one correspondence between $\mathcal{P}(A)$ and $F(A, 2)$. Each B in $\mathcal{P}(A)$ corresponds to its *characteristic*

Properties of first-order logic 161

function $\chi_B(x)$ in $F(A, 2)$ defined as follows:

$$\chi_B(x) = \begin{cases} 0 & \text{if } x \notin B \\ 1 & \text{if } x \in B. \end{cases}$$

Each set in $\mathcal{P}(A)$ uniquely determines its characteristic function and, in the other direction, each function f in $F(A, 2)$ is the characteristic function of the set $\{a \in A | f(a) = 1\}$. It follows that this is a one-to-one correspondence and $\mathcal{P}(A)$ and $F(A, 2)$ have the same size. Since $|F(A, 2)| = 2^{|A|}$, so does $|\mathcal{P}(A)| = 2^{|A|}$. □

Proposition 4.20 For any infinite cardinal κ and any cardinal λ with $2 \leq \lambda < \kappa$, $\lambda^\kappa = \kappa^\kappa$.

Proof Let A be a set of cardinality κ. Then κ^κ is the cardinality of the set of functions from A to A. The graph of any such a function is a subset of $A \times A$. It follows that $\kappa^\kappa \leq \mathcal{P}(A \times A)$. Also, since $|A| = |A \times A|$, we have $|\mathcal{P}(A)| = |\mathcal{P}(A \times A)|$. Putting this together we have

$$2^\kappa \leq \lambda^\kappa \leq \kappa^\kappa \leq |\mathcal{P}(A \times A)| \leq |\mathcal{P}(A)| = 2^\kappa.$$

It follows that $2^\kappa = \lambda^\kappa = \kappa^\kappa$. □

These two propositions reveal some basic facts regarding cardinal exponentiation. Suppose we want to compute λ^κ for $2 \leq \lambda \leq \kappa$. By the latter proposition, $\lambda^\kappa = 2^\kappa$. By the former proposition, this is $|\mathcal{P}(A)|$ where $|A| = \kappa$. This tells us that $\kappa < 2^\kappa$ (by Proposition 2.44), but it does not tell us precisely what 2^κ is.

This brings us back to a fundamental question posed at the outset of this section: how many real numbers are there? By Proposition 2.46, we know that $|\mathbb{R}| > \aleph_0$. Moreover, the proof of Proposition 2.46 shows that $\mathbb{R}$ has the same size as $\mathcal{P}(\aleph_0)$. So $|\mathbb{R}| = 2^{\aleph_0}$. But the question remains, which cardinal is this? Does $2^{\aleph_0}$ equal $\aleph_1$ or $\aleph_2$ or $\aleph_{23}$ or what?

4.2.3 Continuum hypotheses. How many points lie on a continuous line segment? We have shown in Proposition 2.42 that the rational number line contains only countably many points. But this line is not continuous. It has gaps. For example, $\sqrt{2}$ is not a rational number. So the rational numbers can be split into the two intervals $(-\infty, \sqrt{2})$ and $(\sqrt{2}, \infty)$. A continuous line cannot be split in this manner. If a continuous line is split into two sets A and B so that each element of A is to the left of each element of B, then this split must occur at some point of the line. We follow Richard Dedekind and take this as the definition of *continuous*.

The continuum problem. Let L be a continuous line segment. We regard L as a set of points. Does there exist an uncountable subset P of L such that $|P| < |L|$?

The real number line is continuous. This is another way of saying that the reals are order-complete (mentioned in Section 2.4.3). So we may assume that the line segment L is an interval of real numbers. We showed in Example 2.38 that the interval $(0,1)$ of real numbers has the same size as $\mathbb{R}$. It follows that any interval of real numbers has the same size as $\mathbb{R}$. So the continuum Problem can be rephrased as follows.

The continuum problem. Is there a subset of $\mathbb{R}$ that is bigger than $\mathbb{Q}$ and smaller than $\mathbb{R}$?

We know that $|\mathbb{R}| = 2^{\aleph_0} > \aleph_0$. The previous question asks if there exist any cardinals between $\aleph_0$ and $2^{\aleph_0}$. That is, is the following true?

The continuum hypothesis. $2^{\aleph_0} = \aleph_1$.

More generally, is this how cardinal exponentiation behaves for all cardinals?

The general continuum hypothesis. For each ordinal α, $2^{\aleph_\alpha} = \aleph_{\alpha+1}$.

As the word "hypothesis" suggests, this statement has neither been proved nor disproved. Remarkably, it *cannot* be proved or disproved from the standard axioms of mathematics. It is *independent* from these axioms. This has been proved!

The standard axioms of mathematics are Zermeleo–Frankel set theory with the previously mentioned Axiom of Choice. These axioms are denoted ZFC. The study of ZFC is the subject of set theory. Set theory is one branch of logic that we do not treat in depth in this book. We have touched on the basics of set theory in Section 2.5 and the present section. We conclude our discussion of set theory by stating without proof some of the subject's striking results. References are provided at the end of the section.

In 1937, Kurt Gödel showed that the general continuum hypothesis is consistent with ZFC. So this hypothesis cannot be disproved. In 1963, Paul Cohen showed that there are models of ZFC in which the General Continuum Hypothesis is false. So this hypothesis cannot be proved from the axioms in ZFC. Cohen introduced a method known as *forcing* to obtain his result. Using this method one can find models of ZFC in which $2^{\aleph_0} = \aleph_\alpha$ for any finite ordinal α (this is true for most infinite ordinals α as well). The question of whether or not the general continuum hypothesis is true in specific standard models of ZFC remains unanswered. Indeed, by the results of Gödel and Cohen, such questions cannot be resolved from the axioms of ZFC alone.

So how many real numbers are there? Or equivalently, how many subsets of $\mathbb{N}$ are there? Although this appears to be a precise and fundamental question, we cannot provide a definite answer. In the wake of Cohen's forcing, the possibilities are endless. Gödel showed that the hypothesis $2^{\aleph_0} = \aleph_1$ is consistent.

Cohen's methods show that the hypothesis $2^{\aleph_0} = \aleph_{23}$ is also consistent. More generally, one can use forcing to prove the following result.

Theorem 4.21 (Easton 1970) Let $a_0 < a_1 < a_2 < \cdots$ be any increasing sequence of natural numbers. The assertion that $2^{\aleph_\alpha} = \aleph_{a_\alpha}$ for each finite ordinal α is consistent with ZFC.

The possibilities are endless, but not everything is possible. Proved in 1974, Silver's theorem restricts the possibilities.

Theorem 4.22 (Silver) If $2^{\aleph_\alpha} = \aleph_{\alpha+1}$ for each $\alpha < \omega_1$, then $2^{\aleph_{\omega_1}} = \aleph_{\omega_1+1}$.

Phrased another way, this theorem says it is impossible for the general continuum hypothesis to hold for all cardinals up to $\aleph_{\omega_1}$ and to fail for $\aleph_{\omega_1}$. Shelah later proved that this is also true for $\aleph_\omega$. If $2^{\aleph_n} = \aleph_{n+1}$ for all finite n, then $2^{\aleph_\omega} = \aleph_{\omega+1}$. Whereas Easton showed that we can choose the values of $2^{\aleph_n}$ to be almost anything we want, these choices restrict the possible values of $2^{\aleph_\omega}$. In fact, given any sequence $a_0 < a_1 < a_2 < \cdots$ as in Easton's theorem, the possible values of $2^{\aleph_\omega}$ are bounded. Moreover, the values are uniformly bounded (regardless of our choice of a_is) by the number $\aleph_{\omega_4}$. This remarkable fact is due to Shelah.

Theorem 4.23 (Shelah) If $2^{\aleph_n} < \aleph_\omega$ for each finite n, then $2^{\aleph_\omega} \leq \aleph_{\omega_4}$.

Not only is this statement consistent with ZFC, it can be proved from the axioms of ZFC. To say that the proof of this theorem is not within the scope of this book is an understatement. For proofs of these results, we refer the reader to books dedicated to set theory. Both [18] and [25] are recommended. Kunen's book [25] is an excellent introduction to forcing and contains a proof of Cohen's result. Jech's book [18] contains a proof of Silver's theorem. Readers who have a strong background in set theory are referred to Shelah's book [44] for a proof of theorem 4.23 (in particular, refer to the section titled "Why in the HELL is it four?").

4.3 Four theorems of first-order logic

In this section, we prove four fundamental results for first-order logic. We prove the Completeness theorem, the Compactness theorem, the Upward Löwenhiem–Skolem theorem, and the Downward Löwenhiem–Skolem theorem. The first three of these four are consequences of the fact that, in first-order logic, every model has a theory and every theory has a model. In particular, any consistent set of sentences (any theory) is satisfiable (has a model). This was proved for countable sets of sentences in Theorem 4.2. The first objective of the present section is to extend this result to arbitrary sets of sentences.

Suppose that Γ is an uncountable set of sentences. Since there are only countably many sentences in any countable vocabulary (by Proposition 2.47), the vocabulary of Γ must be uncountable. Although we have not previously encountered uncountable vocabularies, such vocabularies naturally arise in model theory.

Example 4.24 Given any structure M, we may wish to consider the expansion M_C of M to a vocabulary $\mathcal{V}(M)$ containing a constant for each element of the underlying set of M. If M is uncountable, then so is the vocabulary $\mathcal{V}(M)$.

In particular, consider the structure $\mathbf{R} = (\mathbb{R}|+,\cdot,\mathbf{0},\mathbf{1})$ (the real numbers in the vocabulary $\mathcal{V}_{ar}$ of arithmetic). We may want to consider a set of $\mathcal{V}_{ar}$-sentences having parameters from the underlying set $\mathbb{R}$ of $\mathbf{R}$. For example, we may want to study polynomials having real coefficients. Such a set of sentences has vocabulary $\mathcal{V}_{ar}(\mathbf{R})$ containing a constant for each real number. This is an uncountable vocabulary.

Example 4.25 We may wish to consider a vector space as a first-order structure. One basic way of doing this is to use a vocabulary containing the constant 0, the binary function $+$, and a unary function m_r for each scalar r. Let the set of vectors serve as the underlying set. Let V be the structure having this underlying set and interpreting 0 as the zero vector, $+$ as vector addition, and m_r as scalar multiplication by r.

In particular, consider the vector space $\mathbb{R}^2$ of ordered pairs of real numbers. For this vector space, the scalars are real numbers. Let $\mathcal{V}_{vs}$ be the vocabulary consisting of the constant 0, the binary function $+$, and a unary function m_r for each real number r. In this case, V is the $\mathcal{V}_{vs}$-structure having $\mathbb{R}^2$ as an underlying set and interpreting 0 as the vector $(0,0)$, $(a,b)+(c,d)$ as $(a+c,b+d)$, and $m_r(a,b)$ as (ra,rb) for all real numbers a,b,c,d, and r. This is an example of a basic mathematical structure that requires an uncountable vocabulary.

For any vocabulary $\mathcal{V}$, let $||\mathcal{V}||$ denote the cardinality of the set of $\mathcal{V}$-formulas. In Proposition 2.47, it was shown that $||\mathcal{V}|| = \aleph_0$ for any countable $\mathcal{V}$. The following proposition extends this result to uncountable vocabularies.

Proposition 4.26 For any vocabulary $\mathcal{V}$, $||\mathcal{V}|| = |\mathcal{V}| + \aleph_0$.

Proof By Corollary 4.17, $|\mathcal{V}| + \aleph_0$ is the larger of $|\mathcal{V}|$ and $\aleph_0$. If $\mathcal{V}$ is countable, then this sum is $\aleph_0$. This agrees with Proposition 2.47. Now suppose $|\mathcal{V}| = \kappa$ for some uncountable κ. We want to show that $||\mathcal{V}|| = \kappa$. For each $n \in \mathbb{N}$, let F_n denote the set of $\mathcal{V}$-formulas having length n. Then $|F_n| \leq \kappa^n$. By repeatedly applying Corollary 4.16 we see that $\kappa^n = \kappa$. So we have

$$||\mathcal{V}|| = \left|\bigcup_{n\in\mathbb{N}} F_n\right| = \sum_{n\in\mathbb{N}} |F_n| \leq \sum_{n\in\mathbb{N}} \kappa^n = \sum_{n\in\mathbb{N}} \kappa = \aleph_0 \cdot \kappa = \kappa. \qquad \square$$

We now prove that every theory has a model. We follow the same Henkin Construction used to prove Theorem 4.2.

Theorem 4.27 If a set of first-order sentences is consistent, then it is satisfiable.

Proof Let Γ be a consistent set of first-order sentences. If Γ is countable, then Γ is satisfiable by Theorem 4.2. We generalize the proof of Theorem 4.2 to include uncountable Γ.

Suppose that the cardinality κ of Γ is uncountable.
Let α be the least ordinal with $|\alpha| = \kappa$.
Let $\mathcal{V}^+ = \mathcal{V} \cup \{c_\iota | \iota < \alpha\}$, where $\mathcal{V}$ is the vocabulary of Γ and each c_ι is a constant that does not occur in $\mathcal{V}$.
Let C be the set $\{c_\iota | \iota < \alpha\}$.

By Corollary 4.17, $|\mathcal{V}^+| = |\mathcal{V}| + |C| = \kappa + \kappa = \kappa$. Moreover, by Proposition 4.26, the set of all $\mathcal{V}^+$-sentences also has cardinality κ. By the Well Ordering Principle, the set of all $\mathcal{V}^+$-sentences can be enumerated as $\{\varphi_\iota | \iota < \alpha\}$.

As in the proof of Theorem 4.2, our goal is to define a complete $\mathcal{V}^+$-theory T_α with the following two properties.

Property 1 Every sentence of Γ is in T_α.

Property 2 For every $\mathcal{V}^+$-sentence in T_α of the form $\exists x \theta(x)$, the sentence $\theta(c)$ is also in T^α for some $c \in C$.

Prior to defining T_α, we inductively define $\mathcal{V}^+$-theories T_ι for $\iota < \alpha$. Let T_0 be Γ. Now suppose that, for some nonzero $\beta < \alpha$, T_γ has been defined for each $\gamma < \beta$. We want to define T_β.

We assume that for each $\gamma < \beta$, T_γ uses at most $|\gamma| + \aleph_0$ of the constants in C. Since $|\gamma| + \aleph_0 < \kappa = |C|$, most of the constants in C are not used in T_γ (here we are using the fact that κ is uncountable). Note that T_0 is a $\mathcal{V}$-theory and so contains none of these constants. We must define T_β so that T_β uses at most $|\beta| + \aleph_0$ of the constants in C.

We now define T_β. There are two possibilities: either β is a successor ordinal or it is not.

If β is a successor ordinal, then $\beta = \gamma + 1$ for some γ. By assumption, T_γ has been defined. Consider the $\mathcal{V}^+$-sentence φ_γ. We define $T_\beta = T_{\gamma+1}$ in the same manner that T_{m+1} was defined in the proof of Theorem 4.2.

(a) If $T_\gamma \cup \{\neg\varphi_\gamma\}$ is consistent, then define $T_{\gamma+1}$ to be $T_\gamma \cup \{\neg\varphi_\gamma\}$.

(b) If $T_\gamma \cup \{\neg\varphi_\gamma\}$ is not consistent, then $T_\gamma \cup \{\varphi_\gamma\}$ is consistent. We divide this case into two subcases.

    (i) If φ_γ does not have the form $\exists x \theta(x)$ for some formula $\theta(x)$, then just let $T_{\gamma+1}$ be $T_\gamma \cup \{\varphi_\gamma\}$.

(ii) Otherwise φ_γ has the form $\exists x \theta(x)$. In this case let $T_{\gamma+1}$ be $T_\gamma \cup \{\varphi_\gamma\} \cup \{\theta(c)\}$ where c is a constant in C that does not occur in $T_\gamma \cup \{\varphi_\gamma\}$. Since T_γ contains fewer than κ constants of C, such a c exists.

So if $\beta = \gamma + 1$, then $T_\beta = T_{\gamma+1}$ is obtained by adding at most a sentence or two to T_γ. Since T_γ contains at most $|\gamma| + \aleph_0$ of the constants in C, so does T_β. Moreover, T_β can be shown to be consistent in the same manner that T_{m+1} was shown to be consistent in the first claim in the proof of Theorem 4.2.

Now suppose that β is not a successor ordinal. Then it is a limit ordinal. In this case, define T_β as the set of all $\mathcal{V}^+$-sentences that occur in T_γ for some $\gamma < \beta$. Again, we claim that T_β is consistent and contains at most $|\beta| + \aleph_0$ of the constants in C.

Claim 1 T_β is consistent.

Proof Suppose T_β is not consistent. Then $T_\beta \vdash \bot$ for some contradiction $\bot$. Since formal proofs are finite, $\Delta \vdash \bot$ for some finite subset Δ of T_β. Since it is finite, $\Delta \subset T_\gamma$ for some $\gamma < \beta$. But this contradicts our assumption that any such T_γ is consistent. We conclude that T_β must be consistent as was claimed.

Claim 2 T_β contains at most $|\beta|$ of the constants in C.

Proof For each $\gamma < \beta$, let C_γ be the set of constants in C that occur in T_γ. Then the constants occurring in T_β are $\bigcup_{\gamma < \beta} C_\gamma$. By assumption, $|C_\gamma| \leq |\gamma| + \aleph_0 \leq |\beta| + \aleph_0$. Since we are assuming that β is a limit ordinal, β is infinite. In particular, $|\beta| + \aleph_0 = |\beta|$. So each $|C_\gamma| \leq |\beta|$. It follows that the number of constants from C occurring in T_β is

$$\left| \bigcup_{\gamma < \beta} C_\gamma \right| \leq \sum_{\gamma < \beta} |C_\gamma| \leq \sum_{\gamma < \beta} |\beta| = |\beta| \cdot |\beta| = |\beta|.$$

This completes the proof of the claim.

So for each $\beta < \alpha$ we have successfully defined a $\mathcal{V}^+$-theory T_β. These have been defined in such a way that $T_{\beta_1} \subset T_{\beta_2}$ for $\beta_1 < \beta_2 < \alpha$.

We now define T_α as the set of all $\mathcal{V}^+$-sentences that occur in T_β for some $\beta < \alpha$. Like each T_β, T_α is a theory. This can be proved in the same manner as Claim 1 above. Unlike T_β for $\beta < \alpha$, T_α is a complete theory. This is because each $\mathcal{V}^+$-sentence is enumerated as φ_ι for some $\iota < \alpha$. Either φ_ι or $\neg \varphi_\iota$ is in $T_{\iota+1}$ and, hence, in T_α as well.

Since $\Gamma = T_0 \subset T_\alpha$, T_α has Property 1. Moreover, part (b)ii of the definition of $T_{\gamma+1} \subset T_\alpha$ guarantees that T_α has Property 2. It was shown in the proof of Theorem 4.2 that any complete theory with Property 2 has a model. Therefore T_α has a model and Γ is satisfiable. □

We have now established that a set of sentences is consistent if and only if it is satisfiable (Theorems 4.1 and 4.27). Every model has a theory and every theory has a model. With this fact at hand, we can prove the completeness and compactness of first-order logic.

Theorem 4.28 (Completeness) For any sentence φ and any set of sentences Γ, $\Gamma \models \varphi$ if and only if $\Gamma \vdash \varphi$.

Proof That $\Gamma \vdash \varphi$ implies $\Gamma \models \varphi$ is Theorem 3.4.

Conversely, suppose $\Gamma \models \varphi$. Then $\Gamma \cup \{\neg \varphi\}$ does not have a model. Since every theory has a model, $\Gamma \cup \{\neg \varphi\}$ must not be a theory. That is, $\Gamma \cup \{\neg \varphi\} \vdash \bot$ for some contradiction $\bot$. By the Contradiction rule (Example 1.33), we can derive φ from $\Gamma \cup \{\neg \varphi\}$. Since we can also derive φ from $\Gamma \cup \{\varphi\}$ (by Assumption), we have $\Gamma \vdash \varphi$ by Proof by cases (Example 1.35). □

Theorem 4.29 (Compactness) Let Γ be a set of sentences. Every finite subset of Γ is satisfiable if and only if Γ is satisfiable.

Proof Any model of Γ is a model of every finite subset of Γ. We must prove the opposite. Suppose that Γ has no model. Then Γ must not be a theory. This means that we can derive a contradiction $\bot$ from Γ. Since derivations are finite, we can derive $\bot$ from a finite subset of Γ. So if Γ is unsatisfiable, then some finite subset of Γ must be unsatisfiable. □

Theorem 4.30 (Upward Löwenhiem–Skolem) If a theory T has an infinite model, then T has arbitrarily large models.

Proof Let M be an infinite model of T. Let κ be any cardinal. We show that there exists a model N of T with $|N| \geq \kappa$. To do this, we expand the vocabulary $\mathcal{V}$ of T by constants. Let C be a set of constants such that $|C| = \kappa$ and each constant in C does not occur in $\mathcal{V}$. Let $\mathcal{V}^+$ denote $\mathcal{V} \cup C$.

Let Γ be the set of all $\mathcal{V}^+$-sentences having the form $\neg(c = d)$ where c and d are distinct constants from C. Any $\mathcal{V}^+$-structure that models Γ must have at least κ elements in its underlying set.

We claim that $T \cup \Gamma$ is satisfiable.

If $\kappa \leq |M|$, then $T \cup \Gamma$ is satisfiable by an expansion of M. If M^+ is any expansion of M that interprets the constants of C as distinct elements of the underlying set of M, then $M^+ \models T \cup \Gamma$. In particular, if κ is finite, then, since M is infinite, such an M^+ exists.

If κ is bigger than $|M|$, then no expansion of M can model Γ. It is still the case, however, that $T \cup \Gamma$ is satisfiable. Any finite subset of $T \cup \Gamma$ will contain only finitely many constants from C. It follows that any finite subset of $T \cup \Gamma$ is satisfiable by an expansion of M. By compactness, $T \cup \Gamma$ is satisfiable.

Let N model $T \cup \Gamma$. Since $N \models \Gamma$, $|N| \geq \kappa$ as was required. □

The Downward Löwenhiem–Skolem theorem, unlike the Upward Löwenhiem–Skolem Theorem, is not an immediate consequence of the Compactness theorem. Rather, this theorem follows from the Tarski–Vaught criterion for elementary substructures. This criterion along with the Downward Löwenhiem–Skolem theorem could have been stated and proved immediately following the definition of elementary substructure in Section 2.6.2. Recall that for $\mathcal{V}$-structures M and N with $M \subset N$, M is an *elementary substructure* of N means that $M \models \varphi(\bar{a})$ if and only if $N \models \varphi(\bar{a})$ for any $\mathcal{V}$-formula $\varphi(\bar{x})$ and tuple $\bar{a}$ of elements from the underlying set of M. We use $M \prec N$ to denote this important concept. The Tarski–Vaught criterion states that, to show $M \prec N$, it suffices to only consider formulas $\varphi(\bar{x})$ that begin with $\exists$.

Proposition 4.31 (The Tarski–Vaught criterion) Let M and N be $\mathcal{V}$-structures with $N \subset M$. Suppose that for any $\mathcal{V}$-formula $\psi(\bar{x}, y)$ and any tuple $\bar{a}$ of elements from the underlying set of N, the following is true:

$$M \models \exists y \psi(\bar{a}, y) \quad \text{implies} \quad N \models \exists y \psi(\bar{a}, y).$$

Then $N \prec M$.

Proof To show that $N \prec M$, we must show that, for every $\mathcal{V}$-formula $\varphi(\bar{x})$ and every tuple $\bar{a}$ of elements from the underlying set U_N of N:

$$N \models \varphi(\bar{a}) \quad \text{if and only if} \quad M \models \varphi(\bar{a}).$$

This can be done by induction on the complexity of φ. It is true for atomic φ since $N \subset M$. Clearly, if it is true for φ, then it is true for any formula equivalent to φ. Now suppose it is true for formulas ψ and θ. That is, suppose

$$N \models \psi(\bar{a}) \quad \text{if and only if} \quad M \models \psi(\bar{a}), \text{ and}$$
$$N \models \theta(\bar{a}) \quad \text{if and only if} \quad M \models \theta(\bar{a})$$

for any tuple $\bar{a}$ of elements of U_N. This is our induction hypothesis. We must show that this is also true for $\psi \wedge \theta$, $\neg \theta$, and $\exists y \theta$. The first two of these follow from the semantics of $\wedge$ and $\neg$. Now suppose that $\varphi(\bar{x})$ has the form $\exists y \theta(\bar{x}, y)$.

Suppose that $N \models \exists y \theta(\bar{c}, y)$ for some tuple $\bar{c}$ of elements from U_N.
By the semantics of $\exists$, $N \models \theta(\bar{c}, b)$ for some $b \in U_N$.
By the induction hypothesis, $M \models \theta(\bar{c}, b)$.
Again by the semantics of $\exists$, $M \models \exists y \theta(\bar{c}, y)$.

We must also show that the reverse is true: that if $M \models \exists y \theta(\bar{c}, y)$, then $N \models \exists y \theta(\bar{c}, y)$. But this is exactly the condition stipulated in the proposition. If this condition holds, then we can complete the induction step and conclude that $N \models \varphi(\bar{a})$ if and only if $M \models \varphi(\bar{a})$ for all $\mathcal{V}$-formulas $\varphi(\bar{x})$ as we wanted to show. □

The Tarski–Vaught criterion, as stated in the previous proposition, can be strengthened. We do not need $N \subset M$ in the hypothesis. That N is a substructure of M follows from the other hypotheses of Proposition 4.31.

Corollary 4.32 Let M be a $\mathcal{V}$-structure. Let U be a subset of the underlying set U_M of M. Suppose that for any $\mathcal{V}$-formula $\varphi(\bar{x}, y)$ and any tuple $\bar{a}$ of elements from U, if $M \models \exists y \varphi(\bar{a}, y)$, then $M \models \varphi(\bar{a}, b)$ for some $b \in U$. Then U is the underlying set of an elementary substructure of M.

Proof Recall that not every subset of U_M may serve as the universe for a substructure of M. We must show that U contains each constant and is closed under each function of $\mathcal{V}$. Given any constant c in $\mathcal{V}$, $M \models \exists x (x = c)$. It follows from our hypothesis on U that $M \models (b = c)$ for some $b \in U$. Now let f be an n-ary function in $\mathcal{V}$ and let $\bar{a}$ be any n-tuple of elements from U. Since $M \models \exists x (f(\bar{a}) = x)$, $M \models (f(\bar{a}) = b)$ for some $b \in U$. So it makes sense to define N as the structure having underlying set U that interprets the symbols of $\mathcal{V}$ in the same manner as M. Since it interprets the constants and functions as well as the relations of $\mathcal{V}$, N is a $\mathcal{V}$-structure. Moreover, since $N \subset M$, we have $N \prec M$ by the Tarski–Vaught criterion. □

Theorem 4.33 (Downward Löwenhiem–Skolem) Let M be a structure having vocabulary $\mathcal{V}$ and underlying set U_M. For any $X \subset U_M$, there exists an elementary substructure N of M such that

1. X is a subset of the universe of N, and
2. $|N| \leq |X| + ||\mathcal{V}||$.

Proof We define a sequence $X_1 \subset X_2 \subset X_3 \subset \cdots$ of subsets of U_M.
Let $X_1 = X$.
Now suppose X_m has been defined for some $m \in \mathbb{N}$. Suppose that $|X_m| \leq |X| + ||\mathcal{V}||$. Let $\mathcal{V}(X_m)$ be the expansion of $\mathcal{V}$ obtained by adding new constants for each element of X_m. Let M_m be the natural expansion of M to this vocabulary. Let E_m be the set of all $\mathcal{V}(X_m)$-sentences of the form $\exists x \varphi(x)$ such that $M_m \models \exists x \varphi(x)$.
Let α be the least ordinal such that $|\alpha| = |E_m|$. By the Well Ordering Principle, E_m can be enumerated as $\{\exists x \varphi_\iota(x) | \iota < \alpha\}$. For each $\iota < \alpha$, there exists an element a_ι in U_M such that $M \models \varphi(a_\iota)$. Let $A = \{a_\iota | \iota < \alpha\}$. Note that $|A| \leq |E| \leq ||\mathcal{V}||$.
Let X_{m+1} be $X_m \cup A$. We have
$$|X_{m+1}| = |X_m \cup A| \leq |X_m| + |A| \leq (|X| + ||\mathcal{V}||) + (||\mathcal{V}||) = |X| + ||\mathcal{V}||.$$
Now let $U = \bigcup_{m < \omega} X_m$.

Claim 1 $|U| \leq |X| + ||\mathcal{V}||$.

Proof We have

$$|U| = \left| \bigcup_{m<\omega} X_m \right| \leq \sum_{m<\omega} |X_m| \leq \sum_{m<\omega} (|X| + ||\mathcal{V}||) = (|X| + ||\mathcal{V}||) \cdot \aleph_0 = |X| + ||\mathcal{V}||.$$

Claim 2 For any $\mathcal{V}$-formula $\varphi(\bar{x}, y)$ and any tuple $\bar{a}$ of elements from U, if $M \models \exists y \varphi(\bar{a}, y)$, then $M \models \varphi(\bar{a}, b)$ for some $b \in U$.

Proof Let $\varphi(\bar{x}, y)$ be any $\mathcal{V}$-formula. Suppose that $M \models \exists y \varphi(\bar{a}, y)$ for some tuple $\bar{a}$ of elements from U. Then $\bar{a}$ must be a tuple of elements from X_m for some m. It follows that $\exists y \varphi(\bar{a}, y)$ is in E_m. By the definition of X_{m+1}, $M \models \varphi(\bar{a}, b)$ for some $b \in X_{m+1} \subset U$.

By Corollary 4.32, U is the underlying set of an elementary substructure N of M. We have both $X \subset U$ and $|N| \leq |X| \cup ||\mathcal{V}||$ as was required. $\square$

Corollary 4.34 Let T be a theory having an infinite model. Then T has a model of size κ for each infinite cardinal κ with $\kappa \geq |T|$.

Proof We want to show that there exists a model M of T with $|M| = \kappa$. By the Upward Löwenhiem–Skolem theorem, there exists a model N of T with $|N| \geq \kappa$. Let X be a subset of the universe of N with $|X| = \kappa$. By the Downward Löwenhiem–Skolem theorem, there exists an elementary substructure M of N such that X is contained in the universe of M and $|M| \leq |X| + ||\mathcal{V}||$ where $\mathcal{V}$ is the vocabulary of T. Moreover, $||\mathcal{V}|| \leq |T|$ by Exercise 4.17. Since $|T| \leq \kappa$, we have $||\mathcal{V}|| \leq \kappa = |X|$. It follows that $|X| + ||\mathcal{V}|| = |X|$ and $|M| \leq |X|$. Since X is a subset of the universe of M, $|M| = |X| = \kappa$. $\square$

4.4 Amalgamation of structures

In first-order logic, we can amalgamate many structures into one. By "amalgamate" we simply mean to combine in some manner. There are various ways to make this idea precise. An *amalgamation theorem* of first-order logic is a theorem that can be diagramed as follows:

$$\begin{array}{ccc} & D & \\ \nearrow & & \nwarrow \\ M_1 & & M_2 \\ \nwarrow & & \nearrow \\ & C & \end{array}$$

In this diagram, M_1 and M_2 are given first-order structures, C is a set, and $f_1 : C \to M_1$ and $f_2 : C \to M_2$ are one-to-one functions having C as a domain.

An amalgamation theorem states that given these structures, sets, and functions, there exists structure D and functions $g_1 : M_1 \to D$ and $g_2 : M_2 \to D$ so that $g_1(f_1(c)) = g_2(f_2(c))$ for each $c \in C$. That is, given the bottom half of this diagram, an amalgamation theorem asserts the existence of the top half.

We prove several amalgamation theorems in this section. The above diagram depicts each of these theorems. Different amalgamation theorems arise from the various restrictions we may place on the structures, sets, and functions in this diagram. For example, in Theorem 4.38 we require that C is a structure and f_1 and f_2 are elementary embeddings. We refer to this theorem as *Elementary Amalgamation over Structures*. The conclusion of this theorem states that the functions g_1 and g_2 in the diagram are in fact elementary embeddings. This theorem, as with all of the amalgamation theorems, is a consequence of compactness. We repeatedly use the following corollary of compactness.

Definition 4.35 A set of sentences Γ is said to be *closed under conjunction* if for any sentences φ and ψ in Γ, the sentence $\varphi \wedge \psi$ is also in Γ.

Corollary 4.36 Let Γ be a set of sentences that is closed under conjunction. Let T be any consistent set of sentences. The set $T \cup \Gamma$ is inconsistent if and only if $T \vdash \neg \varphi$ for some φ in Γ.

Proof Clearly, if T entails the negation of a sentence that is in Γ, then $T \cup \Gamma$ is not consistent. The converse is a direct consequence of compactness. If $T \cup \Gamma$ inconsistent, then, by compactness, there exists an inconsistent finite subset Δ of $T \cup \Gamma$. Since T is consistent, there must exists sentences from Γ in Δ. Let Φ be the conjunction of the sentences in both Δ and Γ. Then $T \cup \{\Phi\}$ is inconsistent. By Proof by Contradiction, we have $T \vdash \neg \Phi$. Since Γ is closed under conjunction, Φ is a sentence in Γ as was required. $\square$

This corollary provides an alternative version of compactness. From now on, when we say that something is true "by compactness" we mean that it follows either from the Compactness theorem 4.29 or, equivalently, from Corollary 4.36.

Our first amalgamation theorem is known as the Joint Embedding lemma. This lemma states that any two models of a complete theory can be elementarily embedded into some other model of the same theory. This is a basic way to amalgamate many structures into one. In the above diagram, $M_1 \equiv M_2$, g_1 and g_2 are elementary embeddings, and C is the empty set.

Lemma 4.37 (Joint Embedding) Let M and N be models of a complete theory T. There exists a model D of T such that both M and N can be elementarily embedded into D. Moreover, if M or N is infinite, then we can take D so that $|D|$ is the same as the larger of $|M|$ and $|N|$.

Proof Let $\mathcal{V}$ be the vocabulary of T. Consider the elementary diagrams $\mathcal{ED}(M)$ and $\mathcal{ED}(N)$. We may assume that the added constants in each of these sets are distinct. That is, we assume that the only constants occurring in both $\mathcal{ED}(M)$ and $\mathcal{ED}(N)$ are those constants occurring in $\mathcal{V}$. We show that $\mathcal{ED}(M) \cup \mathcal{ED}(N)$ is consistent.

Suppose not. Suppose that $\mathcal{ED}(M) \cup \mathcal{ED}(N)$ is contradictory. By compactness, $\mathcal{ED}(M) \vdash \neg\varphi$ for some sentence $\varphi \in \mathcal{ED}(N)$. As a sentence in $\mathcal{ED}(N)$, φ has the form $\psi(\bar{b})$ where $\psi(\bar{x})$ is a $\mathcal{V}$-formula and $\bar{b}$ is an n-tuple of constants not in $\mathcal{V}$. Since M and N are elementarily equivalent $\mathcal{V}$-structures, n must be at least 1.

Since the parameters $\bar{b}$ do not occur in $\mathcal{ED}(M)$, we have $\mathcal{ED}(M) \vdash \forall \bar{x} \neg \psi(\bar{x})$ by $\forall$-Introduction. We have

$M \models \forall \bar{x} \neg \psi(\bar{x})$ which implies
$M \models \neg \exists \bar{x} \psi(\bar{x})$ which implies
$N \models \neg \exists \bar{x} \psi(\bar{x})$ since $M \equiv N$.

But this contradicts the fact that $\psi(\bar{b}) \in \mathcal{ED}(N)$. We conclude that our supposition must be wrong and $\mathcal{ED}(M) \cup \mathcal{ED}(N)$ is consistent. By Theorem 4.27, there exists a model D of $\mathcal{ED}(M) \cup \mathcal{ED}(N)$. By Proposition 2.80(b), both M and N can be elementarily embedded into D as required.

The "moreover" clause in this lemma is a direct consequence of the Downward Löwenhiem–Skolem theorem. □

In fact, any number of models of a theory can be elementarily embedded into a single model of that theory. We leave this generalization of the Joint Embedding lemma as Exercise 4.23.

We now prove the previously mentioned Elementary Amalgamation over Structures theorem.

Theorem 4.38 (Elementary Amalgamation over Structures) Let M_1, M_2 and N be models of a complete theory T. Let $f_1 : N \to M_1$ and $f_2 : N \to M_2$ be elementary embeddings. There exists a model D of T such that both M_1 and M_2 can be elementarily embedded into D in a manner that agrees on N. That is, there exists $D \models T$ and elementary embeddings $g_1 : M_1 \to D$ and $g_2 : M_2 \to D$ such that $f_2(f_1(c)) = g_2(g_1(c))$ for each c in the universe of N.

Proof Let $\mathcal{V}$ be the vocabulary of T. Let $\mathcal{V}(N)$ be the expansion of $\mathcal{V}$ that includes a constant c_a for each element a of the underlying set of N.

Let M_1' be the expansion of M_1 to a $\mathcal{V}(N)$-structure that interprets each c_a as $f_1(a)$.

Let M_2' be the expansion of M_2 to a $\mathcal{V}(N)$-structure that interprets each c_a as $f_2(a)$.

Since $M_1 \equiv M_2$ and f_1 and f_2 are both elementary, $M_1' \equiv M_2'$. Let T' be the complete theory of these structures. By the Joint Embedding lemma, there exists a model D of T' such that both M_1' and M_2' can be elementarily embedded into D. □

From the proof of Theorem 4.38, we see that something stronger is true. Nowhere in this proof did we use the fact that N is a model of T. In fact, N does not even have to be a structure. We need only that $M_1' \equiv M_2'$ where the primes denote expansions by constants representing elements of N. This suffices to show that D and the two elementary embeddings exist.

Theorem 4.39 (Elementary Amalgamation over Sets) Let M_1 and M_2 be models of a complete $\mathcal{V}$-theory T. Let C be a set of constants not in the vocabulary $\mathcal{V}$ of M_1 and M_2. Let $\mathcal{V}(C)$ be $\mathcal{V} \cup C$. Let $M_1(C)$ be an expansion of M_1 to a $\mathcal{V}(C)$-structure and let $M_2(C)$ be an expansion of M_2 to a $\mathcal{V}(C)$-structure. If $M_1(C) \equiv M_2(C)$, then there exists a $\mathcal{V}(C)$-structure $D(C)$ into which both $M_1(C)$ and $M_2(C)$ can be elementarily embedded.

Proof The proof is the same as the proof of Theorem 4.38. □

If we do not require the two embeddings into D to be elementary, then we can relax the condition that the two structures are elementarily equivalent. The following lemma is a modified version of the Joint Embedding lemma. Instead of requiring that M_1 models every sentence that M_2 models, we require only that M_1 models every existential sentence that M_2 models. Under this hypothesis, we still obtain a structure D and embeddings of M_1 and M_2 into D, but now only one of these embeddings is elementary.

Lemma 4.40 Let M and N be $\mathcal{V}$-structures. Suppose that for any existential $\mathcal{V}$-sentence φ, if $N \models \varphi$ then $M \models \varphi$. Then there exists a $\mathcal{V}$-structure D such that N can be embedded into D and M can be elementarily embedded into D.

Proof Consider the literal diagram $\mathcal{D}(N)$ and the elementary diagram $\mathcal{ED}(M)$. We may assume that the added constants in each of these sets are distinct. That is, we assume that the only constants occurring in both $\mathcal{ED}(M)$ and $\mathcal{D}(N)$ are those constants occurring in $\mathcal{V}$. We show that $\mathcal{ED}(M) \cup \mathcal{D}(N)$ is consistent.

Suppose not. Suppose that $\mathcal{ED}(M) \cup \mathcal{D}(N)$ is contradictory. By compactness, $\mathcal{ED}(M) \vdash \neg \varphi$ for some sentence $\varphi \in \mathcal{D}(N)$. As a sentence in $\mathcal{D}(N)$, φ has the form $\psi(\bar{b})$ where $\psi(\bar{x})$ is a literal and $\bar{b}$ is an n-tuple of constants that do not occur in $\mathcal{ED}(M)$.

Since the parameters $\bar{b}$ do not occur in $\mathcal{ED}(M)$, we have $\mathcal{ED}(M) \vdash \forall \bar{x} \neg \psi(\bar{x})$ by ∀-Introduction. We have

$M \models \forall \bar{x} \neg \psi(\bar{x})$ which implies
$M \models \neg \exists \bar{x} \psi(\bar{x})$.

Since $\psi(\bar{x})$ is a literal, $\exists \bar{x} \psi(\bar{x})$ is existential. So, by the hypothesis of the theorem, if $N \models \exists \bar{x} \psi(\bar{x})$, then $M \models \exists \bar{x} \psi(\bar{x})$. Since this is not the case, $N \models \neg \exists \bar{x} \psi(\bar{x})$. But this cannot be the case either. It contradicts the fact that $\psi(\bar{b}) \in \mathcal{D}(N)$. We conclude that our supposition must be wrong and $\mathcal{ED}(M) \cup \mathcal{D}(N)$ is consistent.

By Theorem 4.27, there exists a model D of $\mathcal{ED}(M) \cup \mathcal{D}(N)$. By Propositions 2.79 and 2.80, N can be embedded into D and M can be elementarily embedded into D. □

The following theorem follows from Lemma 4.40 just as Theorem 4.39 follows from the Joint Embedding lemma.

Theorem 4.41 (Existential Amalgamation over Sets) Let M_1, M_2 be $\mathcal{V}$-structures and let C be a set of constants not in $\mathcal{V}$. Let $\mathcal{V}(C)$ be $\mathcal{V} \cup C$. Let $M_1(C)$ be an expansion of M_1 to a $\mathcal{V}(C)$-structure and let $M_2(C)$ be an expansion of M_2 to a $\mathcal{V}(C)$-structure. If $M_2(C)$ models every existential $\mathcal{V}$-sentence that $M_1(C)$ models, then there exists a $\mathcal{V}(C)$-structure D into which $M_1(C)$ can be embedded and $M_2(C)$ can be elementarily embedded.

Proof Apply Lemma 4.40 with $M_1(C)$ as N and $M_2(C)$ as M. □

4.5 Preservation of formulas

If a formula is equivalent to an existential formula, then it is preserved under extensions by Proposition 2.72 of Section 2.6.2. Using Theorem 4.41, we prove the converse.

Proposition 4.42 If a formula is preserved under extensions, then it is equivalent to an existential formula.

We in fact prove something stronger.

Definition 4.43 Let T be a theory (not necessarily complete). We say that a formula $\varphi(\bar{x})$ is *preserved under supermodels of T* if for any two models M and N of T with $M \subset N$ and any tuple $\bar{a}$ of elements from the universe of M,
$M \models \varphi(\bar{a})$ implies $N \models \varphi(\bar{a})$.
If instead
$N \models \varphi(\bar{a})$ implies $M \models \varphi(\bar{a})$,
then we say that $\varphi(\bar{x})$ is *preserved under submodels of T*.

Properties of first-order logic 175

Definition 4.44 Formulas $\varphi(x_1,\ldots,x_n)$ and $\psi(x_1,\ldots,x_n)$ are said to be T-equivalent if $T \models \forall x_1 \ldots \forall x_n (\varphi(x_1,\ldots,x_n) \leftrightarrow \psi(x_1,\ldots,x_n))$.

In this section, we prove that a formula φ is preserved under supermodels of T if and only if φ is T-equivalent to an existential formula. In particular, taking T to be the empty set of sentences, Proposition 4.42 holds. As a corollary to this, a formula is preserved under submodels of T if and only if it is T-equivalent to a universal formula. In the second part of this section, we define the notion of a chain of models and prove a preservation theorem regarding formulas of the form $\forall \bar{x} \exists \bar{y} \varphi$ for quantifier-free φ.

4.5.1 Supermodels and submodels. Let T be a theory. We show that a formula φ is preserved under supermodels of T if and only if φ is T-equivalent to an existential formula. First, we show this is true in the case where φ is a sentence. Note that this is only interesting if neither φ nor $\neg\varphi$ is in T. Otherwise, φ is T-equivalent to either the existential tautology $\exists x(x=x)$ or the contradiction $\exists x \neg(x=x)$.

Proposition 4.45 Let T be a theory. If a sentence is preserved under supermodels of T, then it is T-equivalent to an existential sentence.

Proof Suppose that φ is a sentence that is preserved under supermodels of T. Let $\mathcal{V}$ be the vocabulary of $T \cup \{\varphi\}$.

Let $\mathcal{C}$ be the set of all existential $\mathcal{V}$-sentences ψ such that $T \cup \{\varphi\} \vdash \psi$. We want to show that $T \cup \mathcal{C} \cup \{\neg\varphi\}$ is inconsistent.

Let $\mathcal{D}$ be the set of all existential $\mathcal{V}$-sentences that are not in $\mathcal{C}$. So $\mathcal{C} \cup \mathcal{D}$ equals the set of all existential $\mathcal{V}$-sentences. Let Γ be the set of all $\mathcal{V}$-sentences that are equivalent to the negation of some sentence in $\mathcal{D}$.

Our goal is to show that $T \cup \mathcal{C} \cup \{\neg\varphi\}$ is inconsistent. It suffices to show that $T \cup \Gamma \cup \{\varphi\}$ is consistent.

Claim If $T \cup \Gamma \cup \{\varphi\}$ is consistent, then $T \cup \mathcal{C} \cup \{\neg\varphi\}$ is inconsistent.

Proof Suppose that $T \cup \Gamma \cup \{\varphi\}$ is consistent. Let N be a model. Then N models each existential sentence in $\mathcal{C}$ (since these are consequences of φ) and N models none of the existential sentences in $\mathcal{D}$ (since these are equivalent to the negation of sentences in Γ).

Suppose for a contradiction that $T \cup \mathcal{C} \cup \{\neg\varphi\}$ is also consistent. Let M be a model. Since the only existential sentences that N models are in $\mathcal{C}$, M models every existential sentence that N models. By Theorem 4.41, there exists a structure D into which N can be embedded and M can be elementarily embedded.

Since M can be elementarily embedded into D and $M \models T$, D is a model of T. Since $M \models \neg\varphi$, $D \models \neg\varphi$.

Since N can be embedded into D, D has a substructure N' that is isomorphic to N. Since N models φ, so does N'.

We have $N' \subset D$, $N' \models \varphi$, and $D \models \neg\varphi$. But D and N' are both models of T. This contradicts the assumption that φ is preserved under extentions of models of T. This contradiction proves the claim.

Claim $T \cup \Gamma \cup \{\varphi\}$ is consistent.

Proof Note that Γ is closed under conjunction. If $T \cup \Gamma \cup \{\varphi\}$ is inconsistent, then $T \cup \{\varphi\} \vdash \neg\gamma$ for some $\gamma \in \Gamma$ (If φ is contradictory, then this is the Contradiction rule. Otherwise, this is Corollary 4.36.) By the definition of Γ, $\neg\gamma$ is T-equivalent to a sentence ψ in $\mathcal{D}$. Since $T \cup \{\varphi\} \vdash \psi$, $\psi \in \mathcal{C}$. This contradicts the fact that $\mathcal{C}$ and $\mathcal{D}$ are disjoint sets of sentences. We conclude that $T \cup \Gamma \cup \{\varphi\}$ is consistent as claimed.

By the two claims, $T \cup \mathcal{C} \cup \{\neg\varphi\}$ is inconsistent. So $T \cup \mathcal{C} \vdash \varphi$. By compactness, $T \cup \{\psi_1 \wedge \psi_2 \wedge \cdots \wedge \psi_n\} \vdash \varphi$ for some $\psi_1, \ldots, \psi_n$ in $\mathcal{C}$. Since each ψ_i is existential, their conjunction is equivalent to an existential sentence Ψ. Since $T \cup \{\varphi\} \vdash \psi_i$ for each ψ_i, $T \cup \{\varphi\} \vdash \Psi$. By $\rightarrow$-Introduction, we have both $T \vdash \varphi \rightarrow \Psi$ and $T \vdash \Psi \rightarrow \varphi$. So φ is T-equivalent to the existential sentence Ψ. □

Using Proposition 4.45, we now prove two preservation theorems for formulas. Note that the sentence φ in Proposition 4.45 is not necessarily in the same vocabulary as T.

Theorem 4.46 Let T be a $\mathcal{V}$-theory. A $\mathcal{V}$-formula $\varphi(x_1, \ldots, x_n)$ is preserved under supermodels of T if and only if $\varphi(x_1, \ldots, x_n)$ is T-equivalent to an existential formula.

Proof If $\varphi(x_1, \ldots, x_n)$ is T-equivalent to an existential formula, then it is preserved under extensions by Proposition 2.72. We must prove the other direction of the theorem.

Suppose that $\varphi(x_1, \ldots, x_n)$ is preserved under supermodels of T. If $n = 0$, then φ is a sentence and we may apply Proposition 4.45. Otherwise, for $n \in \mathbb{N}$, let $c_1, \ldots, c_n$ be constants not contained in $\mathcal{V}$. Let $\mathcal{V}(C)$ be the expansion $\mathcal{V} \cup \{c_1, \ldots, c_n\}$ of $\mathcal{V}$.

Consider the $\mathcal{V}(C)$-sentence $\varphi(c_1, \ldots, c_n)$. Since the formula $\varphi(x_1, \ldots, x_n)$ is preserved under supermodels of T, so is the sentence $\varphi(c_1, \ldots, c_n)$. By Proposition 4.45, $\varphi(c_1, \ldots, c_n)$ is T-equivalent to a universal sentence $\psi(c_1, \ldots, c_n)$ (This sentence may or may not contain each constant c_i). We have

$$T \vdash \varphi(c_1, \ldots, c_n) \leftrightarrow \psi(c_1, \ldots, c_n).$$

Since the constants c_i do not occur in T,

$$T \vdash \forall x_1 \ldots \forall x_n (\varphi(x_1, \ldots, x_n) \leftrightarrow \psi(x_1, \ldots, x_n)) \text{ by } \forall\text{-Introduction.}$$

□

Theorem 4.47 Let T be a $\mathcal{V}$-theory. A $\mathcal{V}$-formula $\varphi(x_1, \ldots, x_n)$ is preserved under submodels of T if and only if $\varphi(x_1, \ldots, x_n)$ is T-equivalent to a universal formula.

Proof A formula φ is preserved under submodels of T if and only if its negation $\neg\varphi$ is preserved under supermodels of T. If this is the case, then, by Theorem 4.46, $\neg\varphi$ is T-equivalent to an existential sentence. Finally, $\neg\varphi$ is T-equivalent to an existential sentence if and only if φ is T-equivalent to a universal sentence. □

We now turn our attention to quantifier-free formulas. These formulas are preserved under both supermodels and submodels (this follows from Proposition 2.71). Conversely, suppose that a given formula φ is preserved under both supermodels and submodels of T. Then, by the previous two theorems, φ is T-equivalent to both an existential formula and a universal formula. This does not necessarily mean that φ is T-equivalent to a quantifier-free formula as the following example shows.

Example 4.48 Let $\mathcal{V}_S$ be the vocabulary consisting of a single binary relation. Consider the $\mathcal{V}_S$-structure $Z_S = (\mathbb{Z}|S)$. This structure has the set of integers as its underlying set and interprets S as the successor relation. That is, for any integers a and b, $Z_S \models S(a, b)$ if and only if $b = a + 1$. Let T be $Th(Z_S)$.

Consider the formula $\exists z(S(x, z) \wedge S(z, y))$. This formula says that y is the successor of the successor of x. We claim that this existential formula is not only preserved under supermodels of T, but also under submodels of T. To see this, consider the universal formula $\forall z_1 \forall z_2 (S(x, z_1) \wedge S(z_2, y) \rightarrow z_1 = z_2)$. This formula says that there is at most one element between x and y. Since the theory T says that every element has a unique successor and no element is a successor of itself, this formula implies that there is exactly one element between x and y. So $\exists z(S(x, z) \wedge S(z, y))$ is T-equivalent to this universal formula.

We now argue that $\exists z(S(x, z) \wedge S(z, y))$ is not T-equivalent to a quantifier-free formula. Consider the ordered pairs $(0, 2)$ and $(4, 7)$ in $\mathbb{Z}^2$. Since the only atomic $\mathcal{V}_S$-formulas are $S(x, y)$ and $x = y$, each of these pairs satisfy the same atomic formulas in the structure Z_S. It follows that

$$Z_S \models \psi(0, 2) \quad \text{if and only if} \quad Z_S \models \psi(4, 7)$$

for any quantifier-free $\mathcal{V}_S$-formula ψ (by induction on the complexity of ψ). However,

$$Z_S \models \exists z(S(0,z) \wedge S(z,2)) \text{ and } Z_S \models \neg \exists z(S(4,z) \wedge S(z,7)).$$

This shows that the formula $\exists z(S(x,z) \wedge S(z,y))$, although it is preserved under both submodels and supermodels of T, is not T-equivalent to a quantifier-free formula.

The following theorem provides a sufficient criterion for a formula to be T-equivalent to a quantifier-free formula (provided T is complete). As the previous example shows, the property of being preserved under both submodels and supermodels of T is not sufficient.

Theorem 4.49 Let T be a complete $\mathcal{V}$-theory and let $\varphi(x_1,\ldots,x_n)$ be a $\mathcal{V}$-formula. The following are equivalent:

(i) The formula $\varphi(x_1,\ldots,x_n)$ is T-equivalent to a quantifier-free formula.

(ii) Let M be a model of T. Let $\bar{a}$ and $\bar{b}$ be n-tuples from the universe of M such that $\bar{a}$ and $\bar{b}$ satisfy the same atomic $\mathcal{V}$-formulas in M. Then, $M \models \varphi(\bar{a})$ if and only if $M \models \varphi(\bar{b})$.

Proof Clearly (i) implies (ii). We must prove the converse. Suppose (ii) holds. We want to show that $\varphi(\bar{x})$ is T-equivalent to a quantifier-free formula.

Let $\bar{c} = (c_1,\ldots,c_n)$ be a tuple of constants that are not in $\mathcal{V}$. Let $\mathcal{V}(C)$ be $\mathcal{V} \cup \{c_1,\ldots,c_n\}$. Let $\mathcal{Q}$ be the set of all quantifier-free $\mathcal{V}(C)$-sentences ψ such that $T \cup \{\varphi(\bar{c})\} \vdash \psi$.

Claim $T \cup \mathcal{Q} \vdash \varphi(\bar{c})$.

Proof Suppose not. Then $T \cup \mathcal{Q} \cup \{\neg \varphi(\bar{c})\}$ is consistent. By Theorem 4.27, there is a model M' of this set of $\mathcal{V}(C)$-sentences. Let $\mathcal{P}$ be the set of all quantifier-free $\mathcal{V}(C)$-sentences that hold in M'. Note that $\mathcal{P}$ is closed under conjunction and $\mathcal{Q} \subset \mathcal{P}$.

Subclaim $T \cup \mathcal{P} \cup \{\varphi(\bar{c})\}$ is consistent.

Proof Otherwise, by compactness, $T \cup \{\varphi(\bar{c})\} \vdash \neg \psi$ for some $\psi \in \mathcal{P}$. By the definition of $\mathcal{Q}$ we have $\neg \psi \in \mathcal{Q}$. Since $\mathcal{Q} \subset \mathcal{P}$, $\neg \psi$ is in $\mathcal{P}$. But ψ is also in $\mathcal{P}$. This contradicts the fact that $\mathcal{P}$ has a model M'. This contradiction proves the subclaim.

By Theorem 4.27, there is a model N' of $T \cup \mathcal{P} \cup \{\varphi(\bar{c})\}$. Both M' and N' are $\mathcal{V}(C)$-structures. Let $\bar{a} = (a_1,\ldots,a_n)$ be the n-tuple of elements from the underlying set of M' that M' interprets as the constants $\bar{c}$. Let $\bar{b} = (b_1, \ldots, b_n)$ be the n-tuple that N' interprets as $\bar{c}$. Let M and N be the reducts of M' and N' to the vocabulary $\mathcal{V}$.

Since both M and N model the complete theory T, we can apply the Joint Embedding lemma 4.37. There exists a model D of T and elementary embeddings $f: M \to D$ and $g: N \to D$. Consider the two n-tuples $f(\bar{a}) = (f(a_1), \ldots, f(a_n))$ and $g(\bar{b}) = (g(b_1), \ldots, g(b_n))$ of elements from the universe of D. Each of these tuples satisfy the same atomic formulas in D, namely those from $\mathcal{P}$. However, $D \models \neg\varphi(f(\bar{a}))$ and $D \models \varphi(g(\bar{b}))$. This contradicts (ii). This contradiction proves the claim.

By compactness $T \cup \{\psi\} \vdash \varphi(\bar{c})$ for some $\psi \in \mathcal{Q}$ (since $\mathcal{Q}$ is closed under conjunction). Moreover, ψ, like every $\mathcal{V}(C)$-sentence, has the form $\psi_0(\bar{c})$ for some $\mathcal{V}$-formula $\psi_0(\bar{x})$. We have

$$T \vdash \psi_0(\bar{c}) \to \varphi(\bar{c}) \text{ by } \to\text{-Introduction.}$$

Since $\psi \in \mathcal{Q}$, we also have

$$T \vdash \varphi(\bar{c}) \to \psi_0(\bar{c}).$$

And so $T \vdash \varphi(\bar{c}) \leftrightarrow \psi_0(\bar{c})$
and $T \vdash \forall \bar{x}(\varphi(\bar{x}) \leftrightarrow \psi_0(\bar{x}))$ by $\forall$-Introduction. $\square$

4.5.2 Unions of chains.

Definition 4.50 A sequence $M_0 \subset M_1 \subset M_2 \subset \cdots$ of $\mathcal{V}$-structures is called a *chain*. The *length* of the chain is the least ordinal α such that $\beta < \alpha$ for each M_β in the sequence.

Proposition 4.51 Let $M_0 \subset M_1 \subset M_2 \subset \cdots$ be a chain of $\mathcal{V}$-structures of length α (for some ordinal α). Suppose that, for some $\beta < \alpha$, $\bar{a}$ is an n-tuple of elements in the universe of M_β that are not in the universe of M_ι for $\iota < \beta$. For any quantifier-free $\mathcal{V}$-formula $\varphi(x_1, \ldots, x_n)$,

$M_\gamma \models \varphi(\bar{a})$ for some γ such that $\beta \leq \gamma < \alpha$ if and only if
$M_\gamma \models \varphi(\bar{a})$ for all γ such that $\beta \leq \gamma < \alpha$.

Proof This follows immediately from the fact that quantifier-free formulas are preserved under both extensions and substructures (Proposition 2.55). $\square$

Definition 4.52 We define the *union of a chain* of $\mathcal{V}$-structures $M_0 \subset M_1 \subset M_2 \subset \cdots$. Let α be the length of this chain. The union of the chain is the $\mathcal{V}$-structure M defined as follows. The underlying set of M is $\bigcup_{\beta < \alpha} U_\beta$ where U_β is the underlying set of M_β. Given any atomic $\mathcal{V}$-formula $\varphi(x_1, \ldots, x_n)$ and any n-tuple $(a_1, \ldots, a_n)$ of elements from the universe of M, $M \models \varphi(a_1, \ldots, a_n)$ if and only if $M_\beta \models \varphi(a_1, \ldots, a_n)$ for all M_β containing each a_i in its universe (by the previous proposition, we can replace "all" with "some"). This describes how M interprets the vocabulary $\mathcal{V}$ and completes our definition of this structure.

Example 4.53 Let $\mathcal{V}_<$ be the vocabulary consisting of a single binary relation $<$. We define a chain of $\mathcal{V}_<$-structures of length ω. For each finite ordinal i, let M_i be the $\mathcal{V}_<$-structure that has underlying set $\{-i, -i+1, -i+2, -i+3, \ldots, \}$ and interprets $<$ as the usual order.

So the underlying set of M_0 is $\{0, 1, 2, \ldots\}$,
the underlying set of M_1 is $\{-1, 0, 1, 2, \ldots\}$,
the underlying set of M_2 is $\{-2, -1, 0, 1, 2, \ldots\}$, and so forth.

This forms a chain of $\mathcal{V}_<$-structures. The length of this chain is ω. The union of this chain is the $\mathcal{V}_<$-structure M that interprets $<$ as the usual order on the integers (that is, M is the structure $\mathbf{Z}_<$ from Section 2.4.3).

Note that each M_i is necessarily a substructure of the union M. However, the structure M can be quite different from the M_is. In the previous example, each M_i is isomorphic to the structure $\mathbf{N}_< = (\mathbb{N}, <)$, but the union M is not even elementarily equivalent to this structure (this was shown in Section 2.3.4).

Definition 4.54 An *elementary chain* is a chain of the form $M_0 \prec M_1 \prec M_2 \prec \cdots$

Unlike the situation in Example 4.53, if a chain is an elementary chain, then the union of the chain is elementarily equivalent to each structure in the chain.

Proposition 4.55 The union of an elementary chain is an elementary extension of each structures in the chain.

Proof Let $M_0 \prec M_1 \prec M_2 \prec \cdots$ be an elementary chain of length α. Let M be the union of this chain. Given $\beta < \alpha$, we apply the Tarski–Vaught criterion (Corollary 4.31) to show that $M_\beta \prec M$.

Let $\psi(x_1, \ldots, x_n, y)$ be an $\mathcal{V}$-formula and let $\bar{a}$ be an n-tuple of elements from the underlying set of M_β. Suppose that $M \models \exists y \psi(\bar{a}, y)$. It suffices to show that $M_\beta \models \exists y \psi(\bar{a}, y)$. By the semantics of $\exists$, $M \models \psi(\bar{a}, b)$ for some b in the universe of M. By the definition of M, b must be in the universe of M_γ for some $\gamma < \alpha$. So, $M_\gamma \models \exists y \psi(\bar{a}, y)$. Since the chain is elementary, $M_\beta \models \exists y \psi(\bar{a}, y)$ as was required to show. □

Definition 4.56 A formula $\varphi(x_1, \ldots, x_n)$ is said to be *preserved under unions of chains* if for any chain $M_0 \subset M_1 \subset M_2 \subset \cdots$ and any n-tuple $\bar{a}$ of elements from the universe of M_0, if each M_i models $\varphi(\bar{a})$, then so does the union M of this chain.

If $M_0 \subset M_1 \subset M_2 \subset \cdots$ is an elementary chain, then each M_i models $\varphi(\bar{a})$ if and only if the union M models $\varphi(\bar{a})$ for any n-tuple of elements from the

Properties of first-order logic 181

universe of M_0. For the formula $\varphi(\bar{x})$ to be preserved under unions of chains, this must be true for arbitrary chains. We want to determine which formulas have this property. Clearly, by the definition of the union of a chain, every atomic formula is preserved under unions of chains. Moreover, every existential formula is preserved under unions of chains since they are preserved under extensions (Proposition 2.72). Example 4.53 demonstrates that not all formulas are preserved under unions of chains. In that example, each M_i in the chain models the sentence $\exists x \forall y \neg (y < x)$, but the union M does not.

Definition 4.57 A formula is said to be $\forall_2$ if it has the form $\forall x_1 \cdots \forall x_n \exists y_1 \cdots \exists y_m \varphi$ for some quantifier-free formula φ.

More generally, we can define a hierarchy for all formulas in prenex normal form. A formula is $\forall_1$ if it is universal and $\exists_1$ if it is an existential formula. For each $n \in \mathbb{N}$, we define the $\forall_{n+1}$ formulas inductively. A formula is $\forall_{n+1}$ if it has the form $\forall x_1 \cdots \forall x_m \varphi$ for some $\exists_n$ formula φ. Likewise, a formula is $\exists_{n+1}$ if it has the form $\exists x_1 \cdots \exists x_m \varphi$ for some $\forall_n$ formula φ.

Example 4.58 Let $\varphi(\bar{x})$ be a quantifier-free formula.

$\exists x_1 \forall x_2 \exists x_3 \forall x_4 \exists x_5 \forall x_6 \exists x_7 \forall x_8 \varphi(\bar{x})$ is a $\exists_8$ formula, and
$\forall x_1 \forall x_2 \exists x_3 \exists x_4 \exists x_5 \forall x_6 \forall x_7 \forall x_8 \varphi(\bar{x})$ is a $\forall_3$ formula.

Note that for $m < n$, a $\exists_m$ formula is both a $\forall_n$ formula and a $\exists_n$ formula. The $\forall_2$ formulas were singled out in the previous definition because these are the formulas of immediate interest. The following proposition shows that these formulas are preserved under unions of chains. As demonstrated by the sentence $\exists x \forall y \neg (y < x)$, the same cannot be said of $\exists_2$ sentences nor for $\forall_n$ formulas for $n > 2$.

Proposition 4.59 $\forall_2$ formulas are preserved under unions of chains.

Proof Let $\varphi(x_1, \ldots, x_n)$ be a $\forall_2$ formula. Let $M_0 \subset M_1 \subset M_2 \subset \cdots$ be a chain and let M be the union of this chain. Let $\bar{a}$ be an n-tuple of elements from the universe of M_0. Suppose that each $M_i \models \varphi(\bar{a})$. We must show that $M \models \varphi(\bar{a})$.

Since it is a $\forall_2$ formula, $\varphi(\bar{x})$ has the form $\forall z_1 \cdots \forall z_l \exists y_1 \cdots \exists y_m \varphi_0(\bar{x}, \bar{y}, \bar{z})$ where φ_0 is quantifier-free. Let $\bar{c} = (c_1, \ldots, c_l)$ be an arbitrary l-tuple of elements from the universe U_M of M. Each of these elements is contained in the universe of some structure M_β in the chain. Since $\varphi(\bar{a})$ holds in M_β, $M_\beta \models \varphi_0(\bar{a}, \bar{b}, \bar{c})$ for some m-tuple $\bar{b}$ of elements from its universe. Since quantifier-free formulas are preserved under extensions,

$$M \models \varphi_0(\bar{a}, \bar{b}, \bar{c}).$$

By the semantics of $\exists$,

$$M \models \exists y_1 \cdots \exists y_m \varphi_0(\bar{a}, \bar{y}, \bar{c}).$$

Since $\bar{c}$ is arbitrary,
$$M \models \forall z_1 \cdots \forall z_l \exists y_1 \cdots \exists y_m \varphi_0(\bar{a}, \bar{y}, \bar{z}).$$
by the semantics of $\forall$. Thus, we have shown that $M \models \varphi(\bar{a})$. □

Let T be a theory. Let $M_0 \subset M_1 \subset M_2 \subset \cdots$ be a chain of models of T. If a formula is T-equivalent to a $\forall_2$ formula, then, by Proposition 4.59, it is preserved under the union of this chain. We next prove that the converse of this also holds. If a formula is preserved under unions of chains of models of T, then that formula must be T-equivalent to a $\forall_2$ formula. The following proposition shows that this is true for sentences. As with Proposition 4.45, this is only interesting for sentences that are not in T.

Proposition 4.60 Let T be a theory. If a sentence is preserved under unions of chains of models of T, then it is T-equivalent to a $\forall_2$ sentence.

Proof Suppose that the sentence φ is preserved under unions of chains of models of T. Let $\mathcal{C}$ be the set of all $\forall_2$ sentences ψ such that $T \cup \{\varphi\} \vdash \psi$.

Claim $T \cup \mathcal{C} \vdash \varphi$.

Proof Suppose not. Then $T \cup \mathcal{C} \cup \{\neg\varphi\}$ has a model M_0.

We aim to construct a chain $M_0 \subset N_1 \subset M_1 \subset N_2 \subset M_2 \subset \cdots$ such that, for each $i \in \mathbb{N}$

- $M_{i-1} \prec M_i$, and
- each N_i models $T \cup \{\varphi\}$.

The existence of such a chain suffices to prove the claim. To see this, suppose that we have successfully constructed this chain and let M be the union. Then, by the definition of the union of a chain, M is also the union of both the chain $M_0 \subset M_1 \subset M_2 \subset \cdots$ and the chain $N_1 \subset N_2 \subset N_3 \subset \cdots$. Since the former chain is an elementary chain and M_0 models $\neg\varphi$, the union M models $\neg\varphi$ by Proposition 4.55. Since each N_i models φ and φ is preserved under unions of chains of models of T, M models φ. This is a contradiction and this contradiction proves the claim.

So we must describe how to construct a chain $M_0 \subset N_1 \subset M_1 \subset N_2 \subset \cdots$ possessing the above properties. We have already defined M_0. Suppose that, for some $i \in \mathbb{N}$ we have defined M_{i-1} so that $M_0 \prec M_{i-1}$. Then $M_{i-1} \models T \cup \{\neg\varphi\}$.

We must show that there exists an extension N_i of M_{i-1} that models $T \cup \{\varphi\}$. Let $\mathcal{ED}_\forall(M_{i-1})$ be the set of all universal sentences in prenex normal form that are in $\mathcal{ED}(M_{i-1})$.

Subclaim $\mathcal{ED}_\forall(M_{i-1}) \cup T \cup \{\varphi\}$ is consistent.

Contrarily, suppose that this set is inconsistent. Note that, for any φ_1 and φ_2 in $\mathcal{ED}_\forall(M_{i-1})$, there exists a sentence φ' in $\mathcal{ED}_\forall(M_{i-1})$ that is equivalent to $\varphi_1 \wedge \varphi_2$. So although $\mathcal{ED}_\forall(M_{i-1})$ is not closed under conjunction, the conclusion of Corollary 4.36 holds. If $\mathcal{ED}_\forall(M_{i-1}) \cup T \cup \{\varphi\}$ is not consistent, then $T \cup \{\varphi\} \vdash \neg \theta$ for some sentence $\theta \in \mathcal{D}(M_{i-1})$. As a sentence in $\mathcal{ED}_\forall(M_{i-1})$, the sentence θ has the form $\forall x_1 \ldots \forall x_n \psi(x_1, \ldots, x_n, c_1, \ldots, c_m)$ for some quantifier-free $\mathcal{V}$ formula $\psi(x_1, \ldots, x_n, y_1, \ldots, y_m)$ and constants c_i not in $\mathcal{V}$.

Since $T \cup \{\varphi\} \vdash \neg \forall x_1 \ldots \forall x_n \psi(x_1, \ldots, x_n, c_1, \ldots, c_m)$, we have

$T \cup \{\varphi\} \vdash \exists x_1 \ldots \exists x_n \neg \psi(x_1, \ldots, x_n, c_1, \ldots, c_m)$, and
$T \cup \{\varphi\} \vdash \forall y_1 \ldots \forall y_m \exists x_1 \ldots \exists x_n \neg \psi(x_1, \ldots, x_n, y_1, \ldots, y_m)$ by $\forall$-Introduction.

Since $\forall \bar{y} \exists \bar{x} \neg \psi(\bar{x}, \bar{y})$ is a $\forall_2$ sentence, it is in $\mathcal{C}$. Since $M_0 \models \mathcal{C}$ and $M_0 \prec M_{i-1}$, $M_{i-1} \models \forall \bar{y} \exists \bar{x} \neg \psi(\bar{x}, \bar{y})$. This contracts the assumption that $\forall \bar{x} \psi(\bar{x}, \bar{c}) \in \mathcal{D}(M_{i-1})$. This contradiction verifies the subclaim.

By Theorem 4.27, $\mathcal{ED}_\forall(M_{i-1}) \cup T \cup \{\varphi\}$ has a model N_i. Since $\mathcal{D}(M_{i-1}) \subset \mathcal{ED}_\forall(M_{i-1})$, we may assume that N_i is an extension of M_{i-1} by Proposition 2.79.

Next, we must show there exists an extension M_i of N_i that is an elementary extension of M_{i-1}. We apply Lemma 4.40 to the structures M_{i-1} and N_i. Since $N_i \models \mathcal{ED}_\forall(M_{i-1})$, N_i models every universal sentence that M_{i-1} models. It follows that M_{i-1} models every existential sentence that N_i models. By Lemma 4.40, there exists a structure M_i such that N can be embedded into M_i and M_{i-1} can be elementarily embedded into M_i. By Proposition 2.79, we may assume that M_i is an extension of both N_i and M_{i-1}.

Thus we construct the chain $M_0 \subset N_1 \subset M_1 \subset \cdots$ As we have shown, this construction proves the claim. By compactness, it follows from the claim that $T \vdash \varphi \leftrightarrow \psi$ for some sentence ψ in $\mathcal{C}$. This proves the proposition. $\square$

Theorem 4.61 Let T be a $\mathcal{V}$-theory. A $\mathcal{V}$-formula is T-equivalent to a $\forall_2$ formula if and only if it is preserved under unions of chains of models of T.

Proof This theorem follows from Proposition 4.60 in the same manner that Theorem 4.46 follows from Proposition 4.45. We leave the proof as Exercise 4.24. $\square$

Corollary 4.62 The formulas that are preserved under unions of chains are precisely those formulas that are equivalent to a $\forall_2$ formula.

Proof Take T to be empty in Theorem 4.61. $\square$

4.6 Amalgamation of vocabularies

In Section 4.4, we discussed various ways to amalgamate many structures into one. In each case, the given structures had the same vocabularies. For example, the Joint Embedding lemma 4.37 states that any two models M and N of a complete $\mathcal{V}$-theory T can be elementarily embedded into a single model of T. Here it is understood that M and N have the same vocabulary as T. In this section, we show that this remains true even if M is a $\mathcal{V}_1$-structure and N is a $\mathcal{V}_2$-structure where $\mathcal{V}_1$ and $\mathcal{V}_2$ are different expansions of $\mathcal{V}$. The primary result of this section is Robinson's Joint Consistency lemma. From this lemma, we are able to deduce results that are analogous to the amalgamation theorems of Section 4.4. We are also able to deduce the Craig Interpolation theorem and the Beth Definability theorem for first-order logic.

Lemma 4.63 (Robinson's Joint Consistency) Let T_1 be a $\mathcal{V}_1$-theory and let T_2 be a $\mathcal{V}_2$-theory. Let $\mathcal{V}_\cap = \mathcal{V}_1 \cap \mathcal{V}_2$ and let $\mathcal{V}_\cup = \mathcal{V}_1 \cup \mathcal{V}_2$. If $T_1 \cap T_2$ is a complete $\mathcal{V}_\cap$-theory, then $T_1 \cup T_2$ is a $\mathcal{V}_\cup$-theory.

Proof We show that there exists a $\mathcal{V}_\cap$-structure D that has expansions $D_1 \models T_1$ and $D_2 \models T_2$. If such a D exists, then we can define $D_\cup$ to be the $\mathcal{V}_\cup$-structure having the same underlying set as D that interprets $\mathcal{V}_1$ in the same manner as D_1 and $\mathcal{V}_2$ in the same manner as D_2. Since $D_\cup$ is a model of both T_1 and T_2, we can conclude that $T_1 \cup T_2$ is a theory as the lemma states.

To prove the existence of D, we construct elementary chains

$$M_0 \prec M_1 \prec M_2 \prec M_3 \prec \cdots \text{ of models of } T_1 \text{ and}$$
$$N_0 \prec N_1 \prec N_2 \prec N_3 \prec \cdots \text{ of models of } T_2.$$

Let $\tilde{M}_i$ and $\tilde{N}_i$ denote the reducts of M_i and N_i to the vocabulary $\mathcal{V}_\cap$. Since $T_\cap$ is a complete theory, $\tilde{M}_i \equiv \tilde{N}_j$ for any i and j. We want to construct the two chains in such a way that $\tilde{M}_i$ elementarily embeds into $\tilde{N}_i$ and $\tilde{N}_i$ elementarily embeds into $\tilde{M}_{i+1}$ for each i. We diagram the desired situation as follows:

$$\begin{array}{ccccccccc} M_0 & \prec & M_1 & \prec & M_2 & \prec & M_3 & \prec & \cdots \\ \downarrow & \nearrow & \downarrow & \nearrow & \downarrow & \nearrow & \downarrow & \nearrow & \\ N_0 & \prec & N_1 & \prec & N_2 & \prec & N_3 & \prec & \cdots \end{array}.$$

The arrows in this diagram represent embeddings that are elementary with respect to the vocabulary $\mathcal{V}_\cap$. Let $f_i : \tilde{M}_i \to \tilde{N}_i$ denote the embeddings represented by $\downarrow$ in the diagram, and let $g_i : \tilde{N}_i \to \tilde{M}_{i+1}$ denote the embeddings represented by $\nearrow$. We want to define these embeddings in such a way that $g_i(f_i(a)) = a$ for any a in the underlying set of M_i and $f_{i+1}(g_i(b)) = b$ for any b in the underlying set of N_i.

Before constructing these chains, we show how their existence proves the lemma. Suppose that we have successfully defined two chains as described in the previous paragraph. Our goal is to find a $\mathcal{V}_\cap$-structure D that has two expansions that model each of T_1 and T_2. Let M be the union of the chain $M_0 \prec M_1 \prec M_2 \prec \cdots$ and let D be the reduct of M to a $\mathcal{V}_\cap$-theory (so D is the union of the chain of $\tilde{M}_i$s). By Proposition 4.55, the expansion M of D

models T_1. To complete the proof of the lemma, we must define an expansion of D that models T_2.

Let N be the union of the chain $N_0 \prec N_1 \prec \cdots$. Then $N \models T_2$ by Proposition 4.55. Let $\tilde N$ be the reduct of N to a $\mathcal{V}_\cap$-theory. We claim that D and $\tilde N$ are isomorphic $\mathcal{V}_\cap$-structures. Let $f: D \to \tilde N$ be defined by $f(a) = b$ if and only if $f_i(a) = b$ for some i. Note that $f_i(a) = b$ implies $f_j(a) = b$ for all $j > i$ (since $g_i(f_i(a)) = a$ and $f_{i+1}(g_i(b)) = b$). So f is a well defined function. Since each f_i is elementary, so is f. Moreover, f is also one-to-one and onto. So f is an isomorphism as claimed. Let D_2 be the expansion of D to a $\mathcal{V}_2$-structure defined as follows. For any $\mathcal{V}_2$-formula $\varphi(x_1, \ldots, x_n)$,

$$D_2 \models \varphi(a_1, \ldots, a_n) \quad \text{if and only if} \quad N \models \varphi(f(a_1), \ldots, f(a_n))$$

for any n-tuple $(a_1, \ldots, a_n)$ of elements from the underlying set of D. Since N models T_2, so does D_2. So given two chains as described above, we can define the $\mathcal{V}_\cap$-structure D having expansions $M \models T_1$ and $D_2 \models T_2$ as was required to prove the lemma.

It remains to be shown that the two desired chains can be constructed. We carry out this construction by repeatedly applying Claim 2 below. As a stepping-stone toward Claim 2, we prove the following:

Claim 1 For any $M \models T_1$ and $N \models T_2$ there exists an elementary extension N^+ of N such that $\tilde M$ can be elementarily embedded into $\tilde N^+$ where the tildes denote the reduct to the vocabulary $\mathcal{V}_\cap$.

Proof We show that the set $\mathcal{ED}(\tilde M) \cup \mathcal{ED}(N)$ is consistent. We assume that the only constants occuring in both $\mathcal{ED}(\tilde M)$ and $\mathcal{ED}(N)$ are those constants in $\mathcal{V}_\cap$. If $\mathcal{ED}(\tilde M) \cup \mathcal{ED}(N)$ is not consistent, then, by compactness, $\mathcal{ED}(N) \vdash \neg \theta$ for some $\theta \in \mathcal{ED}(\tilde M)$. As a sentence in $\mathcal{ED}(\tilde M)$, θ has the form $\varphi(\bar a)$ for some $\mathcal{V}_\cap$-formula $\varphi(\bar x)$ and n-tuple $\bar a$ of constants not in $\mathcal{ED}(N)$. If, $\mathcal{ED}(N) \vdash \neg \varphi(\bar a)$, then $\mathcal{ED}(N) \vdash \forall \bar x \neg \varphi(\bar x)$ by $\forall$-Introduction. Since the theory T_2 contains the complete $\mathcal{V}_\cap$-theory T, the $\mathcal{V}_\cap$-sentence $\forall \bar x \neg \varphi(\bar x)$ must be in T. This contradicts the facts that $\tilde M \models T$ and $\varphi(\bar a) \in \mathcal{ED}(\tilde M)$. This contradiction proves the claim.

Now suppose that we are given $M \models T_1$ and $N \models T_2$ and an elementary embedding $g: \tilde N \to \tilde M$. By Claim 1, there exist elementary extension N^+ of N and elementary embedding $f: \tilde M \to \tilde N^+$. Moreover, we claim that we can find such N^+ and f so that $f(g(a)) = a$ for any a in the underlying set of N.

Claim 2 Suppose that $M \models T_1$, $N \models T_2$ and $g: \tilde N \to \tilde M$ is an elementary embedding (where the tildes again denote the reduct to the vocabulary $\mathcal{V}_\cap$). There exist elementary extension N^+ of N and elementary embedding $f: \tilde M \to \tilde N^+$ such that $f(g(a)) = a$ for any a in the underlying set of N.

Proof Let $C = \{c_a | a \in U_N\}$ be a set consisting of constants for each element a in the underlying set U_N of N. For any vocabulary $\mathcal{V}$, let $\mathcal{V}(C)$ denote the expansion

$\mathcal{V} \cup C$. Let $N(C)$ be the expansion of N to a $\mathcal{V}_1(C)$-structure that interprets each c_a as the element a. Let $\tilde{M}(C)$ be the expansion of $\tilde{M}$ to a $\mathcal{V}_\cap(C)$-structure that interprets each c_a as the element $g(a)$. Since $g : \tilde{N} \to \tilde{M}$ is elementary, $N(C) \models \varphi(\bar{c})$ if and only if $\tilde{M}(C) \models \varphi(\bar{c})$ for any $\mathcal{V}_\cap$-formula $\varphi(x_1, \ldots, x_n)$ and n-tuple $\bar{c}$ of constants from C. It follows that $N(C)$ and $\tilde{M}(C)$ are models of the same complete $\mathcal{V}_\cap(C)$-theory. By Claim 1, there exist elementary extension $N^+(C)$ of $N(C)$ and elementary embedding $f : \tilde{M}(C) \to \tilde{N}^+(C)$. Since embeddings must preserve constants, $f(g(a)) = a$ for each $a \in U_N$. Let N^+ be the reduct of $N^+(C)$ to a $\mathcal{V}_2$-structure.

Since T_1 and T_2 are theories, they have models. To begin the construction of the chains, we can use any models M_0 of T_1 and N_{-1} of T_2. By Claim 1, there exist elementary extension N_0 of N_{-1} and elementary embedding $f_0 : \tilde{M}_0 \to \tilde{N}_0$. Having successfully defined M_0, N_0, and f_0, we proceed to define the rest of the two chains inductively.

Suppose that, for some i, we have defined $M_i \models T_1$, $N_i \models T_2$ and elementary embedding $f_i : \tilde{M}_i \to \tilde{N}_i$. By claim 2, there exist elementary extension M_{i+1} of M_i and elementary embedding $g_i : \tilde{N}_i \to \tilde{M}_{i+1}$ such that $g_i(f_i(a)) = a$ for each a in the underlying set of M_i. (Here we have applied Claim 2 with M_i playing the role of N and N_i playing the role of M.) Applying Claim 2 yet again, there exist elementary extension N_{i+1} of N_i and elementary embedding $f_{i+1} : \tilde{M}_{i+1} \to \tilde{N}_{i+1}$ such that $f_{i+1}(g_i(b)) = b$ for any b in the underlying set of N_i. (Here N_i plays the role of N and M_{i+1} plays the role of M.) Repeating this process produces the two desired chains. □

The following generalization of the Joint Embedding lemma is an immediate consequence of Robinson's Joint Consistency lemma.

Corollary 4.64 Let M be a $\mathcal{V}_1$-structure and N be a $\mathcal{V}_2$-structure such that M and N are elementarily equivalent as $(\mathcal{V}_1 \cap \mathcal{V}_2)$-structures. There exists a $(\mathcal{V}_1 \cup \mathcal{V}_2)$-structure D such that M can be $\mathcal{V}_1$-elementarily embedded into D and N can be $\mathcal{V}_2$-elementarily embedded into D.

Proof Let T_1 be $\mathcal{ED}(M)$ and T_2 be $\mathcal{ED}(N)$. By Robinson's Joint Consistency lemma, there exists a model D of $T_1 \cup T_2$. □

Note that if $\mathcal{V}_1 = \mathcal{V}_2$, then the previous corollary is identical to the Joint Embedding lemma. Likewise, we can generalize the Elementary Amalgamation Over Structures theorem 4.38. We leave this as Exercise 4.34. We now turn our attention to two properties of first-order logic.

Theorem 4.65 (Craig Interpolation) Let φ be a $\mathcal{V}_1$ sentence and ψ be a $\mathcal{V}_2$-sentence. If $\models \varphi \to \psi$, then there exists a sentence θ that is both a $\mathcal{V}_1$-sentence and a $\mathcal{V}_2$-sentence such that $\models \varphi \to \theta$ and $\models \theta \to \psi$.

Proof Let $\mathcal{V}_\cap = \mathcal{V}_1 \cap \mathcal{V}_2$. Let $\mathcal{C}$ be the set of all $\mathcal{V}_\cap$ consequences of φ. That is $\mathcal{C}$ is the set of all $\mathcal{V}_\cap$-sentences θ such that $\models \varphi \to \theta$. We want to show that ψ is a consequence of $\mathcal{C}$.

Suppose for a contradiction that M is a $\mathcal{V}_1 \cup \mathcal{V}_2$-structure that models both $\mathcal{C}$ and $\neg\psi$. Let T be the $\mathcal{V}_\cap$-theory of M.

Claim $T \cup \{\varphi\}$ is consistent.

Proof Otherwise, $T \vdash \neg\varphi$. Since T is closed under conjunctions, $\{\theta\} \vdash \neg\varphi$ for some $\theta \in T$ (by compactness). So the contrapositive $\{\varphi\} \vdash \neg\theta$ also holds (by Example 1.34). So $\neg\theta$ is a consequence of φ and so $\neg\theta \in \mathcal{C}$. Since $\mathcal{C} \subset T$, we have both θ and $\neg\theta$ in T. Since M is a model of T, this is a contradiction.

The consistency of $T \cup \{\varphi\}$ leads to another contradiction. Let $T_1 = T \cup \{\varphi\}$ and $T_2 = T \cup \{\neg\psi\}$. If both T_1 and T_2 are consistent, then so is $T_1 \cup T_2$ by Robinson's Joint Consistency lemma. But since $\models \varphi \to \psi$, $T_1 \cup T_2$ cannot be consistent. The assumption that $\mathcal{C} \cup \{\neg\psi\}$ is satisfiable must be incorrect. It follows that $\mathcal{C} \vdash \psi$. Since $\mathcal{C}$ is closed under conjunctions, $\{\theta\} \vdash \psi$ for some $\theta \in \mathcal{C}$. It follows that $\models \theta \to \psi$ as was required. □

In order to state the Beth Definability theorem concisely, we introduce some terminology. We distinguish between two ostensibly different notions of definability. Beth's Definability theorem states that, for first-order logic, these two notions are the same.

Definition 4.66 Let T be a $\mathcal{V}$-theory and let R be an n-ary relation in $\mathcal{V}$. For any $\mathcal{V}' \subset \mathcal{V}$, we say that R is *explicitly defined by T in terms of $\mathcal{V}'$* if there exists a $\mathcal{V}'$-formula $\varphi(x_1, \ldots, x_n)$ such that

$$T \vdash \varphi(x_1, \ldots, x_n) \leftrightarrow R(x_1, \ldots, x_n).$$

Example 4.67 Let $\mathcal{V} = \{\leq, +, \cdot, 0, 1\}$ and let $\mathcal{V}' = \{+, \cdot, 0, 1\}$. Let $\mathbf{R}_{or} = (\mathbb{R}| \leq, +, \cdot, 0, 1)$ be the structure that interprets $\mathcal{V}$ in the usual way on the real numbers. Let $T = Th(\mathbf{R}_{or})$. Then the binary relation $\leq$ is explicitly defined by T in terms of $\mathcal{V}'$. To see this, let $\varphi(x, y)$ be the $\mathcal{V}'$-formula $\exists z(x + (z \cdot z) = y)$. Since $(z \cdot z) \geq 0$ for any real number z, $T \vdash \varphi(x, y) \leftrightarrow x \leq y$.

Definition 4.68 Let T be a $\mathcal{V}$-theory and let R be an n-ary relation in $\mathcal{V}$. For any $\mathcal{V}' \subset \mathcal{V}$, we say that R is *implicitly defined by T in terms of $\mathcal{V}'$* if the following holds. Given any $\mathcal{V}'$-structure M and two expansions N_1 and N_2 of M to $\mathcal{V}$-structures that model T,

$$N_1 \models R(\bar{a}) \text{ if and only if } N_2 \models R(\bar{a})$$

for any n-tuple $\bar{a}$ of elements from the underlying set of M.

If R is explicitly defined by T in terms of $\mathcal{V}$, then it is implicitly defined as well. So the binary relation $\leq$ from Example 4.67 is implicitly defined by T in terms of $\mathcal{V}'$. We now give a nonexample.

Example 4.69 Let M be the structure $(\mathbb{Z}|B)$ that interprets the binary relation B as a symmetric successor relation. By this we mean that $M \models B(a,b)$ if and only if $a = b+1$ or $b = a+1$. Let $N_1 = (\mathbb{Z}|B, <)$ be the expansion of M that interprets $<$ as the usual order on the integers. Let N_2 be the expansion of M that interprets $<$ backwards. That is, $N_2 \models a < b$ if and only if the integer a is greater than b.

Let $T = Th(N_1)$. Since N_1 and N_2 are distinct expansions of M that model T, the relation $<$ is not implicitly defined by T in terms of $\{B\}$.

If we replace the relation B with the successor relation S from Example 4.48, then the same conclusion holds. The relation $<$ is not implicitly defined by $Th(\mathbb{Z}|S, <)$ in terms of $\{S\}$. We leave the verification of this as Exercise 4.37.

Proposition 4.70 A relation R is implicitly defined by T in terms of $\mathcal{V}$ if and only if for any $\mathcal{V}$-structure M, there is at most one expansion N of M to a $\mathcal{V} \cup \{R\}$-structure such that $Th(N) \subset T$.

Proof Exercise 4.29. □

Example 4.71 Let T_Q be the theory of the rational numbers in the vocabulary $\mathcal{V}_{ar} = \{0, 1, +, \cdot\}$. Let $\mathcal{V} = \mathcal{V}_{ar} \cup \{R\}$ where R is a ternary relation. Let T be the theory $T_Q \cup \{\varphi\}$ for some $\mathcal{V}$-sentence φ.

If φ has the form $\forall x \forall y \forall z (\psi(x, y, z) \leftrightarrow R(x, y, z))$ for some $\mathcal{V}_{ar}$-formula $\psi(x, y, z)$, then, by definition, R is explicitly defined by T over $\mathcal{V}_{ar}$. In this case, there is exactly one way to expand a given model of T_Q to a model of T.

Conversely, suppose there is exactly one way to expand any given model of T_Q to a model of T. Then, by Proposition 4.70, R is implicitly defined by T over $\mathcal{V}_{ar}$. In this case, φ may not have the form $\forall x \forall y \forall z (\psi(x, y, z) \leftrightarrow R(x, y, z))$. For example, suppose that φ is the $\mathcal{V}$-sentence

$$\forall x \forall y \forall z \exists u \exists v \exists w (u \cdot v = 1 \wedge (R(x, y, z) \vee (x + x + y = z + v + u))$$
$$\wedge (w + y = 0 \wedge (R(x, y, z) \to x + x = z + w)).$$

There is exactly one way to expand a model of T_Q to a model of this sentence. So this sentence implicitly defines the ternary relation R.

Beth's Definability theorem states that R is defined implicitly if and only if it is defined explicitly. This means that the above sentence φ must be T_Q-equivalent to a sentence of the form $\forall x \forall y \forall z (\psi(x, y, z) \leftrightarrow R(x, y, z))$. Indeed, we can take $\psi(x, y, z)$ to be $2x + y = z$. We leave the verification of this to the reader.

Theorem 4.72 (Beth Definability) A relation is implicitly defined by a theory T in terms of $\mathcal{V}$ if and only if it is explicitly defined by T in terms of $\mathcal{V}$.

Proof Only one direction of this theorem requires proof. Suppose R is an n-ary relation that is implicitly defined by T in terms of $\mathcal{V}$. If $T \cup \{\exists \bar{x} R(\bar{x})\}$ is not consistent, then R is explicitly defined by the formula $\neg(x=x)$. So suppose that this is not the case.

Let $\mathcal{D}$ be the set of all $\mathcal{V}$-formulas ψ having free variables among $x_1, \ldots, x_n$ such that $T \vdash R(x_1, \ldots, x_n) \to \psi$. Let $\mathcal{V}(C) = \mathcal{V} \cup \{c_1, \ldots, c_n\}$, where $c_1, \ldots, c_n$ are constants that do not occur in $\mathcal{V}$. Let $\mathcal{D}(\bar{c})$ be the set of $\mathcal{V}(C)$-sentences obtained by replacing each occurrence of x_i in $\mathcal{D}$ with the constant c_i (for $i = 1, \ldots, n$).

Claim $T \cup \mathcal{D}(\bar{c}) \vdash R(\bar{c})$.

Proof Otherwise, $T \cup \mathcal{D}(\bar{c}) \cup \{\neg R(\bar{c})\}$ has a model M.

Let T_0 be the $\mathcal{V}(C)$-theory of M. We claim that $T_0 \cup R(\bar{c})$ is consistent. Otherwise, $\{R(\bar{c})\} \vdash \neg\psi(\bar{c})$ for some $\psi(\bar{c}) \in T_0$. But then $\neg\psi(\bar{c}) \in \mathcal{D}(C)$. This contradicts the facts that $\mathcal{D}(C) \subset T_0$ and T_0 are consistent.

Let $T_1 = T_0 \cup R(\bar{c})$ and let $T_2 = T \cup \mathcal{D}(\bar{c}) \cup \{\neg S(\bar{c})\}$ where S is an n-ary relation that does not occur in $\mathcal{V} \cup \{R\}$. Since $T_0 \cup \{\neg R(\bar{c})\}$ is consistent (M is a model), so is T_2. By Robinson's Joint Consistency lemma, $T_1 \cup T_2$ is consistent.

Let N be a model of $T_1 \cup T_2$. So N is a structure in the vocabulary $\mathcal{V} \cup \{R, S, c_1, \ldots, c_n\}$.

Let N_0 be the reduct of N to a $\mathcal{V}$-structure.

Let N_1 be the expansion of N_0 to a $\mathcal{V} \cup \{R\}$-structure that interprets R in the same manner as N.

Let N_2 be the expansion of N_0 to a $\mathcal{V} \cup \{R\}$-structure that interprets R as N interprets the relation S.

Let $\bar{a}$ be the n-tuple from the underlying set of N_0 that N interprets as the constants $\bar{c}$. Then $N_1 \models R(\bar{a})$ and $N_2 \models \neg R(\bar{a})$. This contradicts the assumption that R is implicitly defined by $T \subset T_0$ in terms of $\mathcal{V}$. This contradiction proves the claim.

Since $T \cup \mathcal{D}(\bar{c}) \vdash R(\bar{c})$, $T \vdash \varphi(\bar{c}) \to R(\bar{c})$ for some $\mathcal{V}$-formula $\varphi(\bar{x}) \in \mathcal{D}$. Since $\varphi(\bar{x}) \in \mathcal{D}$, we have

$$T \vdash (\varphi(\bar{x}) \leftrightarrow R(\bar{x})),$$

and so R is explicitly defined by T by the $\mathcal{V}$-formula φ. □

4.7 The expressive power of first-order logic

First-order logic, as any logic, is a language equipped with rules for deducing the truth of one sentence from that of another. These rules may be formulated as systems of deduction such as resolution and formal proofs discussed in Chapter 3. In this chapter, we have shown that the rules of deduction for first-order logic entail many nice properties. These properties give rise to the model theory of the next two chapters. Because of these desirable properties, the language of first-order logic is necessarily weak. In particular, the Compactness theorem and Downward Löwenhiem–Skolem theorem impose limitations on the expressive power of first-order logic.

We claim that every property of first-order logic discussed in this chapter is a consequence of the Compactness theorem and the Downward Löwenhiem–Skolem theorem. The completeness of first-order logic can be deduced from compactness in the same manner that this is done in Theorem 1.80 for propositional logic. The theorems of Section 4.4 stating that infinite structures M and N can be amalgamated in some manner into structure D are direct consequences of compactness. The Downward Löwenhiem–Skolem theorem guarantees that there exists such D having the same size as M or N. Inspecting the proofs, we see that Robinson's Joint Consistency lemma, the Beth Definability theorem, and the preservation theorems are consequences of compactness.

By compactness, there cannot exist a sentence of first-order logic that holds for infinite structures and only for infinite structures. By the Downward Löwenhiem–Skolem Theorem, there cannot exist a sentence of first-order logic that holds for uncountable structures and only uncountable structures. Because of these restrictions, there are basic concepts that first-order logic is incapable of expressing.

Example 4.73 In first-order logic, we cannot say that two definable subsets have the same size. To be precise, let $\mathcal{V}$ be a vocabulary that includes unary relations P and Q. For any $\mathcal{V}$-structure M having underlying set U,

$$\text{let } P(M) = \{a \in U | M \models P(a)\} \text{ and let } Q(M) = \{a \in U | M \models Q(a)\}.$$

There is no set of $\mathcal{V}$-sentences that says $P(M)$ and $Q(M)$ have the same size. In contrast, we can easily write sentences that say $P(M)$ and $Q(M)$ both have size n for any particular n. We can easily define a set of sentences that say $P(M)$ and $Q(M)$ are both infinite.

Note that $\mathcal{V}$ may contain symbols other than P and Q. For example, $\mathcal{V}$ may contain a unary function f. If this is the case, then we can write a $\mathcal{V}$-sentence φ_f that says f is a one-to-one correspondence between $P(M)$ and $Q(M)$. The existence of such a bijection is precisely what it means for $P(M)$ and $Q(M)$ to have the "same size." So if $M \models \varphi_f$, then $|P(M)| = |Q(M)|$. But the converse of this is not true. There exists $N \models \neg\varphi_f$ such that $|P(N)| = |Q(N)|$. Likewise, there is no $\mathcal{V}$-sentence (nor set of $\mathcal{V}$-sentences) that holds if and only if P and Q define subsets of equal size.

To verify this, let Γ be a set of $\mathcal{V}$-sentences. Suppose that $|P(M)| = |Q(M)|$ for any model M of Γ. We show that there necessarily exists a model N of Γ for which $P(N)$ and $Q(N)$ do not have the same size. Let N_0 be any $\mathcal{V}$-structure such that $P(N_0)$ is uncountable and $Q(N_0)$ is denumerable. Then $N_1 \models \neg \gamma$ for some $\gamma \in \Gamma$. Let X be a subset of the universe U of N_1 such that both $X \cap P(N_1)$ and $X \cap Q(N_1)$ are denumerable. By the Downward Löwenheim–Skolem Theorem, there exists a countable elementary substructure N of N_1 that contains X in its universe. Since $N \prec N_1$, we have $N \models \neg \gamma$. So N is a $\mathcal{V}$-structure that does not model Γ for which $|P(N)| = |Q(N)| = \aleph_0$.

Example 4.74 Let G be a graph. Recall that a *path* in G from vertex a to vertex b is a sequence of adjacent vertices beginning with a and ending with b. The length of the path is one less than the number of vertices in the sequence (that is, the number of edges in the path). By Exercise 2.13, there exist formulas $d_n(x, y)$ expressing the existence of a path between vertices x and y of length n. In contrast, we claim that the concept of a path cannot be expressed in first-order logic. Whereas we can say there is a path of some specified length, we cannot say there is a path of arbitrary length. Suppose to the contrary that we have a formula $\phi(x, y)$ that holds of any vertices x and y in any graph G if and only if there exists a path from x to y in G. Consider the following set of sentences in a vocabulary containing R and constants a and b:

$$\forall x \forall y \phi(x, y), \neg \varphi_1(a, b), \neg \varphi_2(a, b), \neg \varphi_3(a, b), \ldots$$

The first sentence says that there is a path between any two vertices. This sentence holds in a graph if and only if the graph is connected. Since the other sentences assert that there is no path between a and b, this set of sentences is contradictory. However, any finite subset of these sentences is satisfiable. This contradicts the Compactness theorem. We conclude that the formula $\phi(x, y)$ cannot exist.

So there is no first-order formula that defines the concept of a path. Likewise, there is no first-order sentence that holds in a graph if and only if it is connected. Another basic graph-theoretic property is k-colorability. A graph is said to be *k-colorable* if the vertices of the graph can be colored with k colors in such a way that no two vertices of the same color share an edge. There does not exist a first-order sentence φ_k such that $G \models \varphi_k$ if and only if G is a k-colorable graph. First-order logic cannot even say that there exists an even number of vertices in a finite graph. This is a consequence of the 0–1 law for first-order logic that is a subject of Section 5.4 of the next chapter.

This first-order impotence is by no means limited to graph theory. We list some of the many fundamental concepts from various areas of mathematics that first-order logic is incapable of expressing.

- Linear orders: there is no first-order sentence that holds for well ordered sets and only well ordered sets.
- Group theory: there is no first-order sentence that holds for simple groups and only simple groups.
- Ring theory: there is no first-order sentence that holds for Noetherian rings and only Noetherian rings.
- Metric spaces: there is no first-order sentence that holds for complete metric spaces and only complete metric spaces. In particular, the notion of a *Cauchy sequence* cannot be defined

To express these and other concepts, we must extend the logic. In Chapter 9, we consider extensions of first-order logic such as infinitary logics and second-order logic. Infinitary logics permit as formulas infinite conjunctions and disjunctions of first-order formulas. For example, consider the disjunction $\bigvee_{i \in \mathbb{N}} d_i(x, y)$ of the first-order formulas $d_i(x, y)$ from Example 4.74. This is a formula of the infinitary logic $\mathcal{L}_{\omega_1 \omega}$ as is the sentence $\forall x \forall y \bigvee_{i \in \mathbb{N}} d_i(x, y)$. This sentence holds in a graph if and only if it is connected. Now suppose that we want to say that two definable subsets have the same size as in Example 4.73. Second-order logic can express this. This logic allows quantification over subsets of the universe. Second-order logic is extremely powerful and can express each of the properties mentioned above.

Extending first-order logic comes at an expense. Since it can express the concept of a path, $\mathcal{L}_{\omega_1 \omega}$ must not have compactness. Likewise, since second-order logic can say that two definable sets have the same size, the Downward Löwenhiem–Skolem theorem must fail for this logic. Moreover, both compactness and completeness fail for second-order logic. Unlike first-order logic, we cannot list a set of rules from which we can deduce all truths of second-order logic. In this sense, the expressive power of second-order logic is too great.

The Compactness theorem and the Downward Löwenhiem–Skolem theorem make first-order logic the primary language of model theory. Model theory considers the relationship between a set of sentences T and the set of structures $Mod(T)$ that model T. Just as first-order logic can describe any finite structure up to isomorphism (by Proposition 2.81), infinitary logics and second-order logic can describe any countable structure up to isomorphism. This makes for an uninteresting model theory. If T is the second-order theory of a countable structure M, then M is the only structure in $Mod(T)$. Moreover, by the failure of completeness, we have no way to determine which sentences are in T.

Although there are basic concepts that cannot be defined in first-order logic, there are many concepts that can be defined. Moreover, we claim that those properties that are first-order definable form a natural class of mathematical objects. The language of first-order logic, containing ∃, ∀, ∧, and ¬ is a natural mathematical language to consider. First-order theories, which are the topic of the next two chapters, are natural objects of study. Since the Compactness and Downward Löwenhiem–Skolem theorems are central to model theory, we should consider the most powerful logic possessing these properties. By Lindström's theorem, which we shall prove in Section 9.4, first-order logic is this logic. This theorem states that any extension of first-order logic for which both the Compactness and Downward Löwenhiem–Skolem theorems hold must be equivalent to first-order logic itself. So in some precise sense, first-order logic is the most powerful logic that possesses the properties discussed in this chapter.

Exercises

4.1. Let T be an incomplete countable theory. For each of the following, either prove the statement or provide a counter example.
   (a) If T has an uncountable model, then T has a countable model.
   (b) If T has arbitrarily large finite models, then T has a denumerable model.
   (c) If T has finite models and a denumerable model, then T has arbitrarily large finite models.

4.2. Let T be an incomplete theory in an uncountable vocabulary. Repeat (a) and (b) from Exercise 4.1.

4.3. Let T_1 be a complete $\mathcal{V}_1$-theory and let T_2 be a complete $\mathcal{V}_2$-theory. Show that $T_1 \cup T_2$ is consistent if and only if $\varphi_1 \wedge \varphi_2$ is satisfiable for every $\varphi_1 \in T_1$ and $\varphi_2 \in T_2$.

4.4. Let φ be a first-order sentence that is not contained in any complete theory. Show that $\{\varphi\} \vdash \neg\varphi$.

4.5. Let $\varphi(x)$ be a quantifier-free $\mathcal{V}$-formula. Let $C = \{c_1, c_2, c_3, \ldots\}$ be a denumerable set of constants that do not occur in $\mathcal{V}$. Let $\mathcal{V}(C) = \mathcal{V} \cup C$. Show that the sentence $\exists x \varphi(x)$ is a tautology if and only if the sentence $\varphi(t_1) \vee \varphi(t_2) \vee \cdots \vee \varphi(t_n)$ is a tautology for some $n \in \mathbb{N}$ and $\mathcal{V}(C)$-terms $t_1, \ldots, t_n$.

4.6. Let $\mathcal{V}$ be a vocabulary containing denumerably many constants $\{c_1, c_2, c_3, \ldots\}$. Let T be a $\mathcal{V}$-theory having the following two properties.
   • If $T \models \exists x \theta(x)$, then $T \models \theta(c_i)$ for some $i \in \mathbb{N}$.
   • $T \models c_i \neq c_j$ for any $i, j \in \mathbb{N}$ with $i \neq j$.
   Show that T is complete.

4.7. Let T be an incomplete $\mathcal{V}$-theory and let θ be a $\mathcal{V}$-formula. Suppose that for each $M \models T$ there exists a $\mathcal{V}$-formula φ_M such that $M \models \theta \leftrightarrow \varphi_M$. Show that there exists finitely many $\mathcal{V}$-formulas $\varphi_1, \ldots, \varphi_n$ such that $T \vdash \bigvee_{i=1}^{n}(\theta \leftrightarrow \varphi_i)$.

4.8. Let $\mathcal{V}$ be a vocabulary that contains only constants (and neither functions nor relations). Let M and N be two infinite $\mathcal{V}$-structures. Using the Tarski-Vaught Criterion, show that if $M \subset N$, then $M \prec N$.

4.9. Let R be the structure $(\mathbb{R}|+,\cdot,0,1,<)$ having the real numbers as an underlying set that interprets the vocabulary in the usual manner.
   (a) Show that there exists an elementary extension M of R that has infinitesimals (an element c is an *infinitesimal* if $0 < c < 1/n$ for each $n \in \mathbb{N}$).
   (b) Let U_M be the underlying set of M. Show that the set of infinitesimals in U_M has the same size as the set of infinite elements in U_M (an element c is *infinite* if $n < c$ for each $n \in \mathbb{N}$).

4.10. Let $\mathbf{N}$ be the $\mathcal{V}$-structure $(\mathbb{N}|+,\cdot,1)$ from Exercise 2.7. By part (c) of Exercise 2.7, there exists a $\mathcal{V}$-formula $\lambda(x,y)$ such that, for any a and b in $\mathbb{N}$, $\mathbf{N} \models \lambda(a,b)$ if and only if $a < b$. By the Upward Löwenhiem–Skolem theorem, $\mathbf{N}$ has an elementary extension M of cardinality $\aleph_1$.
   (a) Let c be in the universe of M. Show that c is not in $\mathbb{N}$ if and only if $M \models \lambda(n,c)$ for each $n \in \mathbb{N}$. Call such an element c "infinite."
   (b) Show that there is no least infinite number in the universe of M. (That is, for every infinite c, there exists an infinite d such that $M \models \lambda(d,c)$.)
   (c) By part (b) of Exercise 2.7, there exists a $\mathcal{V}$-formula $\pi(x)$ such that, for any $n \in \mathbb{N}$, $\mathbf{N} \models \pi(n)$ if and only if n is prime. Show that $M \models \pi(c)$ for some infinite c. Call such a c an "infinite prime."
   (d) Show that there cannot be two consecutive infinite primes in the universe of M. (a and b are consecutive if $a + 1 = b$.)
   (e) Let $\varphi(x)$ be a $\mathcal{V}$-formula. Show that the following are equivalent:
      (i) $\mathbf{N} \models \varphi(n)$ for infinitely many $n \in \mathbb{N}$.
      (ii) $M \models \varphi(c)$ for some infinite c.
      (iii) There exists an elementary extension M_1 of M such that $M_1 \models \varphi(a)$ for $\aleph_{23}$ many elements a in its universe.

4.11. A graph is said to be *k-colorable* if the vertices can be colored with k different colors in such a way that no two vertices of the same color share an edge.

   A graph is said to be *planar* if it can be drawn on the Euclidian plane in such a way that no two edges cross each other. The *Four Color Theorem* states that any planar graph is four-colorable. This famous theorem was

proved by Appel and Haken in 1976. Assuming that this theorem is true for finite graphs, prove that it is true for infinite graphs.
(Hint: Given an infinite planar graph G, consider the union of $\mathcal{D}(G)$ and a suitable set of $\mathcal{V}'$-sentences where $\mathcal{V}'$ an expansion of $\mathcal{V}_R$ containing unary relations representing each of the colors.)

4.12. The relation $<$ is a *partial order* on a set A if
1. for all a and b in X, at most one of the following hold: either $a < b$, $b < a$, or $a = b$, and
2. for all a, b and c in X, if $a < b$ and $b < c$ then $a < c$.

If it is also true that either $a < b$ or $b < a$ for distinct a and b in A, then the partial order is a linear order. Using the compactness of first-order logic, show that any partial order on a set A can be extended to a linear order on A.
(Hint: First use induction to show that this is true for finite A.)

4.13. Let T be the set of all sentences in the vocabulary $\mathcal{V}_R$ that hold in every connected graph. Show that there exists a model G of T that is not a connected graph.

4.14. Derive the Compactness theorem from the Completeness theorem.

4.15. Let T be the set of all sentences in the vocabulary $\mathcal{V}_< = \{<\}$ that hold in every well ordered set. Show that there exists a model M of T that does not interpret $<$ as a well ordering of the underlying set of M.

4.16. Let M be a $\mathcal{V}$-structure having underlying set U. For any n-tuple $\bar{a}$ of elements from U, let $\langle \bar{a} \rangle$ be the substructure of M generated by $\bar{a}$ as defined in Exercise 2.34. Show that M can be embedded into a model of a theory T if and only if $\langle \bar{a} \rangle$ can be embedded into a model of T for every finite tuple $\bar{a}$ of elements from U.

4.17. Let $\mathcal{F}$ be a set of formulas having an infinite vocabulary $\mathcal{V}$. Show that $|\mathcal{F}| = |\mathcal{V}|$.

4.18. Show that the order $\triangleleft$ defined in the proof of Theorem 4.15 makes $\delta \times \delta$ a well ordered set.

4.19. For any set A of cardinals, let $sup A$ denote the least cardinal λ such that $\kappa \leq \lambda$ for each $\kappa \in A$. Let α be an infinite ordinal and let $\{\kappa_\iota \,|\, \iota < \alpha\}$ be a set of cardinals. Show that $\Sigma_{\iota<\alpha}\kappa_\iota = sup\{|\alpha|, \kappa_\iota | \iota < \alpha\}$.

4.20. Show that the following equalities hold for any ordinal α and any cardinal κ,
$\Sigma_{\iota<\alpha}\kappa = \kappa \cdot |\alpha|$, and
$\Pi_{\iota<\alpha}\kappa = \kappa^{|\alpha|}$.

4.21. Prove that there are uncountably many countable ordinals.

4.22. Let $\alpha_1 > \alpha_2 > \alpha_3 > \cdots$ be a descending sequence of ordinals. Show that there can be only finitely many ordinals in this sequence.

4.23. Let T be a complete theory. Let α be a nonzero ordinal. For each $\beta < \alpha$, let M_β be a model of T.
  (a) Show that there exists a model D of T such that each M_β can be elementarily embedded into D.
  (b) Show that we can find D in part (a) so that $|D| \leq |\alpha| \cdot |M_\beta|$ for each β.

4.24. Prove Theorem 4.61.

4.25. Let T be a $\mathcal{V}$-theory. Let $T_\forall$ be the set of all universal sentences ψ such that $T \vdash \psi$. Let M be a $\mathcal{V}$-structure that models $T_\forall$. Show that M can be embedded into a model of T.

4.26. Let T be a $\mathcal{V}$-theory and let $\varphi(x)$ and $\psi(x)$ be two $\mathcal{V}$-formulas. Suppose that, for any models M and N of T with $N \subset M$,

$$\text{if } M \models \varphi(a) \text{ then } N \models \psi(a)$$

for any element a in the universe of N.
Show that there exists a universal $\mathcal{V}$-formula $\theta(x)$ such that $T \vdash \varphi(x) \to \theta(x)$ and $T \vdash \theta(x) \to \psi(x)$.

4.27. Let T be an incomplete $\mathcal{V}$-theory and let $\varphi(\bar{x})$ be a $\mathcal{V}$-formula having n free variables (for $n \in \mathbb{N}$). Let M be a model of T having underlying set U_M.
  (a) Suppose that $M \models \varphi(\bar{a})$ if and only if $M \models \varphi(\bar{b})$ for any n-tuples $\bar{a}$ and $\bar{b}$ of elements of U_M that satisfy the same atomic $\mathcal{V}$-formulas in M. Show that $M \models \varphi(\bar{x}) \leftrightarrow \psi(\bar{x})$ for some quantifier-free $\mathcal{V}$-formula $\psi(\bar{x})$.
  (b) Show that $\varphi(\bar{x})$ is not necessarily T-equivalent to a quantifier-free formula by providing appropriate example.

4.28. Let T be a $\mathcal{V}$-theory and let $\varphi(x_1, \ldots, x_n)$ be a $\mathcal{V}$-formula. Prove that the following are equivalent:
  (i) $\varphi(x_1, \ldots, x_n)$ is T-equivalent to a quantifier-free formula.
  (ii) For any model M of T and any $\mathcal{V}$-structure C, if $f : C \to M$ and $g : C \to M$ are two embeddings of C into M, then

  $$M \models \varphi(f(c_1), \ldots, f(c_n)) \text{ if and only if } M \models \varphi(g(c_1), \ldots, g(c_n))$$

  for any n-tuple of elements from the underlying set of C.
  (Hint: see Exercise 2.34.)

4.29. Prove Proposition 4.70.

4.30. For any $\mathcal{V}$-theory T, let $T_{\forall\exists}$ be the set of $\forall_2$ $\mathcal{V}$-sentences that can be derived from T. Prove that the following are equivalent:
  (i) $T_{\forall\exists} \vdash T$.
  (ii) If M is the union of a chain of models of T, then $M \models T$.

(iii) Let M be a $\mathcal{V}$-structure having underlying set U. If for every $a \in U$, there exists $N \subset M$ such that a is in the universe of N and $N \models T$, then $M \models T$.

4.31. Let $\mathcal{V}$ be a vocabulary and let R be an n-ary relation not in $\mathcal{V}$. Let T be an incomplete theory in the vocabulary $\mathcal{V} \cup \{R\}$. Suppose that, for each $M \models T$, there exists a $\mathcal{V}$-formula $\varphi_M(\bar{x})$ such that $M \models R(\bar{x}) \leftrightarrow \varphi_M(\bar{x})$. Prove that R is explicitly defined by T in terms of $\mathcal{V}$. (Hint: see Exercise 4.7.)

4.32. (**Lyndon**) Refer to Exercise 2.33. A formula is said to be *positive* if it does not contain the symbols $\neg$, $\rightarrow$, nor $\leftrightarrow$. Let T be a $\mathcal{V}$-theory and let φ be a $\mathcal{V}$-formula. Show that the following are equivalent:
   (i) φ is T-equivalent to a positive formula.
   (ii) φ is preserved by every homomorphism $f : M \rightarrow N$ that is onto where both M and N are models of T.

4.33. (**Lyndon**) Let φ and ψ be $\mathcal{V}$-sentences in conjunctive prenex normal form. A relation R is said to occur *negatively* in φ if $\neg R$ occurs as subformula. Prove that if $\models \varphi \rightarrow \psi$ then there exists a $\mathcal{V}$-sentence θ in conjunctive prenex normal form such that $\models \varphi \rightarrow \theta$, $\models \theta \rightarrow \psi$, and every relation that occurs negatively in θ also occurs negatively in both φ and ψ. (Hint: Modify the proof of Theorem 4.65.)

4.34. Let $\mathcal{V}_1$ and $\mathcal{V}_2$ be two vocabularies. Let $\mathcal{V} = \mathcal{V}_1 \cap \mathcal{V}_2$. Let M be a $\mathcal{V}_1$-structure, N be a $\mathcal{V}_2$-structure, C be a $\mathcal{V}$-structure. Let $f_1 : C \rightarrow M$ and $f_2 : C \rightarrow N$ be $\mathcal{V}$-elementary embeddings. Show that there exist

   $(\mathcal{V}_1 \cup \mathcal{V}_2)$-structure D,
   $\mathcal{V}_1$-elementary embedding $g_1 : M \rightarrow D$, and
   $\mathcal{V}_2$-elementary embedding $g_2 : N \rightarrow D$

   such that $g_1(f_1(c)) = g_2(f_2(c))$ for each c in the underlying set of C.

4.35. Derive Robinson's Joint Consistency lemma from Compactness and Craig's Interpolation theorems.

4.36. Show that the Beth Definability theorem holds for functions as well as relations.

4.37. Let M be the structure $(\mathbb{Z}|S)$ that interprets the binary relation S as the successor relation on the integers. Let $N = (\mathbb{Z}|S, <)$ be the expansion of M that interprets the binary relation $<$ as the usual order. Let $T = Th(N)$.
   (a) Show that N is the only expansion of M to a the vocabulary $\{S, <\}$ that models T.
   (b) Show that $<$ is not explicitly defined by T in terms of $\{S\}$.

5 First-order theories

We continue our study of Model Theory. This is the branch of logic concerned with the interplay between sentences of a formal language and mathematical structures. Primarily, Model Theory studies the relationship between a set of first-order sentences T and the class $Mod(T)$ of structures that model T.

Basic results of Model Theory were proved in the previous chapter. For example, it was shown that, in first-order logic, every model has a theory and every theory has a model. Put another way, T is consistent if and only if $Mod(T)$ is nonempty. As a consequence of this, we proved the Completeness theorem. This theorem states that $T \vdash \varphi$ if and only if $M \models \varphi$ for each M in $Mod(T)$. So to study a theory T, we can avoid the concept of $\vdash$ and the methods of deduction introduced in Chapter 3, and instead work with the concept of $\models$ and analyze the class $Mod(T)$. More generally, we can go back and forth between the notions on the left side of the following table and their counterparts on the right.

Formal languages	Mathematical structures
Theory	Elementary class
T	$Mod(T)$
$Th(M)$	M
$\vdash$	$\models$
Sentences	Models
Formulas	Definable subsets
Consistent	Satisfiable
Syntax	Semantics

Progress in mathematics is often the result of having two or more points of view that are shown to be equivalent. A prime example is the relationship between the algebra of equations and the geometry of the graphs defined by the equations. Combining these two points of view yield concepts and results that would not be possible in either geometry or algebra alone. The Completeness theorem equates the two points of view exemplified in the above table. Model Theory exploits the relationship between these two points of view to investigate mathematical structures.

First-order theories 199

First-order theories serve as our objects of study in this chapter. A first-order theory may be viewed as a consistent set of sentences T or as an elementary class of structures $Mod(T)$. We shall present examples of theories and consider properties that the theories may or may not possess such as completeness, categoricity, quantifier-elimination, and model-completeness. The properties that a theory possesses shed light on the structures that model the theory. We analyze examples of first-order structures including linear orders, vector spaces, the random graph, and the complex numbers. In the final section, we use the model-theoretic properties of the theory of complex numbers to prove a fundamental result of algebraic geometry.

As in the previous chapter, all formulas are first-order unless stated otherwise. In particular, all theories are sets of first-order sentences.

5.1 Completeness and decidability

We demonstrate several examples of theories in this section. Variations of these theories are used throughout this chapter to illustrate the concepts to be introduced. Although any consistent set of sentences forms a theory, we typically restrict our attention to those theories that are deductively closed.

Definition 5.1 Let Γ be a set of sentences. The *deductive closure* of Γ is the set of all sentences that can be formally derived from Γ. If Γ equals its deductive closure, then Γ is said to be *deductively closed*.

Given a deductively closed theory, we consider the question of whether or not the theory is complete. To show that a $\mathcal{V}$-theory T is complete, we must show that, for every $\mathcal{V}$-sentence φ, either $\varphi \in T$ or $\neg\varphi \in T$. It is a much easier task to show that T is incomplete. To accomplish this, it suffices to produce only one sentence φ such that neither φ nor $\neg\varphi$ is in T. Instead of considering $\mathcal{V}$-sentences, we can consider $\mathcal{V}$-structures. To show that T is incomplete, it suffices to find two models of T that are not elementarily equivalent. This is also a necessary condition for T to be incomplete.

Proposition 5.2 Let T be a deductively closed theory. Then T is incomplete if and only if there exist models M and N of T that are not elementarily equivalent.

Proof First suppose that T is incomplete. Then there exists a sentence φ such that neither φ nor $\neg\varphi$ is in T. Since T is deductively closed, neither φ nor $\neg\varphi$ can be derived from T. This happens if and only if both $T \cup \{\varphi\}$ and $T \cup \{\neg\varphi\}$ are consistent. By Theorem 4.27, if these sets of sentences are consistent, then they are satisfiable. So if T is incomplete, then, for some $\mathcal{V}$-sentence φ, there exist models $M \models T \cup \{\varphi\}$ and $N \models T \cup \{\neg\varphi\}$. Clearly, such M and N are not elementarily equivalent.

Conversely, if there exist models M and N of T that are not elementarily equivalent, then there must be some sentence φ such that $M \models \varphi$ and $N \models \neg\varphi$. If this is the case, then T must be incomplete. □

Theories shall be presented in one of two ways. We may define T to be $Th(M)$ for some structure M. Such theories are necessarily complete by Proposition 2.86. Similarly, given a class of structures, we may define T to be the set of all sentences that hold in each structure in the set. In this case, T is complete if and only if the given structures are elementarily equivalent to one another. So if there are two or more structures in the class, then the theory T defined in this manner might be incomplete.

Example 5.3 Let $\mathcal{V}_R = \{R\}$ and $\mathcal{V}_E = \{E\}$ be vocabularies consisting of a single binary relation.

- Let T_G be the set of all $\mathcal{V}_R$-sentences that hold in every graph. This is the *theory of graphs*.
- Let T_E be the set of all $\mathcal{V}_E$-sentences that hold in every structure that interprets E as an equivalence relation. This is the *theory of equivalence relations*.

Since there exist finite models of T_G and T_E of different sizes, neither of these theories is complete.

Another way to define a theory T is to explicitly state which sentences are contained in T. Usually, T contains infinitely many sentences and we cannot simply list all of them. To present such a theory T, it suffices to provide a set of sentences Γ so that T is the deductive closure of Γ. That is, we *axiomatize* the theory.

Definition 5.4 Let T be a theory. An *axiomatization* of T is a subset Γ of T that has the same deductive closure of T (that is, $\Gamma \vdash \varphi$ for each $\varphi \in T$). We say that Γ *axiomatizes* T and that T is *axiomatized* by Γ.

Example 5.5 The theory of graphs T_G is the deductive closure of the two $\mathcal{V}_G$-sentences

$$\forall x \neg R(x,x) \quad \text{and} \quad \forall x \forall y (R(x,y) \leftrightarrow R(y,x)).$$

This agrees with our previous definition of T_G. These two definitions are equivalent because a "graph," by definition, is a structure that models these two sentences. Likewise, by the definition of "equivalence relation" the $\mathcal{V}_E$-theory

T_E is the deductive closure of the $\mathcal{V}_E$-sentences

$$\forall x E(x,x),$$
$$\forall x \forall y (E(x,y) \to E(y,x)), \quad \text{and}$$
$$\forall x \forall y \forall z ((E(x,y) \land E(y,z)) \leftrightarrow E(x,z)).$$

Of course, any theory T is an axiomatization of itself. This fact is neither interesting nor useful. An axiomatization is useful if it is somehow simpler than T. For example, whereas the theories T_G and T_E both contain infinitely many sentences, the axiomatizations of these theories are finite and easy to understand.

It is common practice in pure mathematics to define concepts by providing axiomatizations. However, not all axiomatizations are first-order axiomatizations. Our definition of *axiomatization* is more restrictive than the colloquial use of this word in mathematics. If we open a book on, say, real analysis, then we might see a set of *axioms* or *postulates* from which the theory is derived. For example, on page 17 of Ref. [42] we see the following axiom for the real numbers.

Completeness axiom. Every nonempty subset S of $\mathbb{R}$ that is bounded above has a least upper bound.

We mentioned this property of the real numbers in Section 2.4.3. Although it is a precise and formal statement, we cannot translate it to a sentence of first-order logic. To say "for all subsets S" we must quantify over subsets (as opposed to elements) of the set $\mathbb{R}$. We can do this in second-order logic, but not first-order logic.

Although not all axiomatizations can be translated to the language of first-order logic, there are many that can be. Of the plethora of possible examples in pure mathematics, we presently give three. These three examples are standard definitions of concepts that can be found in books on algebra, geometry, and logic, respectively.

Example 5.6 A *group* is defined as a set G equipped with a binary operation $\circ$, such that the following hold:

(Closure) If a and b are in G, then so is $a \circ b$.
(Associativity) For every a, b, and c in G, $a \circ (b \circ c) = (a \circ b) \circ c$
(Existence of identity) There is an element e in G such that $a \circ e = e \circ a = a$ for every a in G.
(Existence of inverses) For any a in G, there exists an element a^{-1} of G such that $a \circ a^{-1} = a^{-1} \circ a = e$.

These sentences can easily be expressed as first-order sentences in the vocabulary $\{\circ, e\}$ where $\circ$ is a binary function and e is a constant. In Exercise 2.5, they are expressed in the vocabulary $\mathcal{V}_{gp} = \{+, 0\}$. Let T_{gp} be the deductive closure of these $\mathcal{V}_{gp}$-sentences. This is the *theory of groups*. Note that we do not need to state the closure axiom, since, for any function f, the sentence $\forall \bar{x} \exists y (f(\bar{x}) = y)$ is a tautology of first-order logic.

Example 5.7 A *projective plane* is a set lines each of which is comprised of points in such a way that any two lines intersect in exactly one point and any two points are contained in exactly one line. Moreover, to rule out trivial examples, a projective plane must have at least four points and four lines.

We can translate this definition to a set of first-order sentences in the vocabulary $\mathcal{V}_{pg} = \{P, L, I\}$. This vocabulary contains two unary relations P (for "points") and L (for "lines") and one binary relation I (the "incidence relation"). The relation $I(x, y)$ is used to express "x is a point contained on the line y." We leave it to the reader to formalize the above definition as a set of $\mathcal{V}_{pg}$-sentences. Let T_{pg} denote the deductive closure of these sentences. This is the *theory of projective planes*. Note that the axiomatization of T_{pg} is symmetric with respect to P and L. That is, if we replace P with L and vice versa, then this set of sentences remains the same. It follows that for any sentence in T_{pg}, if we swap P and L we obtain another sentence of T_{pg}. This is the fundamental principle of duality for projective planes.

Example 5.8 In Section 4.2, we defined the concept of a *linearly ordered set* as follows. The relation $<$ is a linear order on structure M if M models the $\mathcal{V}_<$-sentences

$$\forall x \forall y ((x < y) \rightarrow \neg (y < x)),$$

$$\forall x (\neg (x < x)),$$

$$\forall x \forall y ((x < y) \vee (y < x) \vee (x = y)), \quad \text{and}$$

$$\forall x \forall y \forall z (((x < y) \wedge (y < z)) \rightarrow (x < z)).$$

Let T_{LO} be the deductive closure of these sentences. We refer to T_{LO} as the *theory of linear orders*.

So there are two ways to define a particular theory. It can be defined in terms of a class of structures or in terms of a set of sentences. We defined the theory of groups T_{gp} in terms of a set of sentences (an axiomatization). Equivalently, we could define T_{gp} as the set of all $\mathcal{V}_{gp}$-sentences that hold in all groups. Of course, this definition would not be helpful to a reader who is not previously familiar with groups. Another way that this latter definition is inferior is that it does not provide a method for determining precisely which sentences are in T_{gp}. If we are given an axiomatization Γ of a theory T, then (theoretically if not practically) we can determine whether or not a sentence is in T using the methods described in Chapter 3.

Definition 5.9 A $\mathcal{V}$-theory T is *decidable* if there exists an algorithm that will determine, in a finite number of steps, whether or not any given $\mathcal{V}$-sentence φ is in T.

Proposition 5.10 A complete countable theory is decidable if and only if it has an axiomatization that is decidable.

Proof Let T be a complete countable $\mathcal{V}$-theory. Since T is an axiomatization of itself, only one direction of this proposition requires proof. Suppose we are given a decidable axiomatization Γ of T. We want to show that T is decidable. Let φ be an arbitrary $\mathcal{V}$-sentence. We must describe a way to determine whether or not φ is in T.

Since T is countable, so is the set of all $\mathcal{V}$-sentences. So the set of all $\mathcal{V}$-sentences can be enumerated as $\{\psi_1, \psi_2, \psi_3, \ldots\}$. Moreover, we can find such an enumeration in a systematic way. For example, if $\mathcal{V}$ is finite, then we can list the finitely many sentences that have no more than 10 symbols followed by those that have no more that 20 symbols, and so forth. Since Γ is decidable, we can determine whether or not each ψ_i is in Γ in a finite number of steps. So there exists an enumeration $\{\gamma_1, \gamma_2, \gamma_3, \ldots\}$ of Γ and an algorithm that, for given $n \in \mathbb{N}$, produces the finite set $\{\gamma_1, \gamma_2, \ldots, \gamma_n\}$.

To determine whether or not φ is in T, we use the methods of Chapter 3 (either formal proofs, Herbrand's method, or resolution) to determine whether or not φ can be derived from Γ. For example, we can list every formal proof that has fewer than 1000 steps that can be derived from $\{\gamma_1, \ldots, \gamma_{10}\}$. There are only finitely many such proofs.

If $\Gamma \vdash \varphi$ occurs in one of these proofs, then we conclude "yes, φ is in T."

If $\Gamma \vdash \neg\varphi$ occurs in one of these finitely many proofs, then we conclude "no, φ is not in T."

Otherwise, if neither $\Gamma \vdash \varphi$ nor $\Gamma \vdash \neg\varphi$ occurs, then we proceed to check more formal proofs. We can list every formal proof that has fewer than 2000 steps that can be derived from $\{\gamma_1, \ldots, \gamma_{20}\}$. If that is not enough, we can then list every formal proof that has fewer than 3000 steps that can be derived from $\{\gamma_1, \ldots, \gamma_{30}\}$, and so forth. Since T is complete, either $\Gamma \vdash \varphi$ or $\Gamma \vdash \neg\varphi$. By compactness, the procedure we have described will eventually (in a finite number of steps) find a formal proof for either $\Gamma \vdash \varphi$ or $\Gamma \vdash \neg\varphi$.

This procedure is not practical, to say the least. We would not want to (nor be able to) actually list all of these formal proofs. However, the definition of "decidable" requires only the existence of an algorithm. It does not have to be a good algorithm. By this definition, T is decidable. $\square$

Of the two ways to define a theory, it is better to provide an axiomatization. If an axiomatization is not given, then it is desirable to find one. However, this is not always an easy task. In some cases, it may be difficult or impossible to provide an axiomatization for a theory.

Example 5.11 Let $T_{ar} = Th(\mathbf{A})$ where $\mathbf{A} = (\mathbb{Z}|+, \cdot, 0, 1)$ is as in Section 2.4.3. This is the theory of arithmetic. Although this is a perfectly well defined theory, we cannot provide a decidable axiomatization for it. The theory of arithmetic is undecidable. This is a consequence of Gödel's Incompleteness theorems that are the subject of Chapter 8.

Structures that have undecidable theories clearly do not lend themselves well to model-theoretic analysis. In the present chapter, we restrict our attention to first-order theories that are most accessible and do not consider undecidable theories.

Example 5.12 Let $\mathcal{V}_s = \{s\}$ where s is a unary function. Let $Z_s = (\mathbb{Z}|s)$ be the $\mathcal{V}_s$-structure that interprets s as the successor function on the integers. That is, for integers a and b, $Z_s \models s(a) = b$ if and only if $b = a + 1$. Let $T_s = Th(Z_s)$.

This is an unambiguous definition of T_s. There is only one $\mathcal{V}_s$-theory fitting this description. Now suppose that we want to provide an axiomatization for T_s. That is, from among the infinitely many sentences in T_s, we want to find a subset that succinctly describes this theory. One way to proceed is to ask: what are the salient features of the structure Z_s? If you were to describe this structure to someone who had no idea what the integers looked like, what would you say? There is no first element. There is no last element. The successor of any element is unique as is the predecessor. We can express these things with $\mathcal{V}_s$-sentences.

Let σ_1 be the sentence $\forall x \exists y (s(y) = x)$, and
let σ_2 be the sentence $\forall x \forall y (s(x) = s(y) \rightarrow x = y)$.

The first of these says that every element has a predecessor (there is no "first" element). The second of these sentences implies the uniqueness of the predecessor. We do not need to say that every element has a unique successor. Since any model interprets s as a function, the sentences

$$\forall x \exists y (s(x) = y) \quad \text{and} \quad \forall x \forall y (x = y \rightarrow s(x) = s(y))$$

are tautologies.

To axiomatize the $\mathcal{V}_s$-theory T_s we are merely listing some of the sentences that hold in the $\mathcal{V}_s$ structure Z_s. The problem is knowing when we are done. So far, we have listed the two sentences σ_1 and σ_2. Together these sentences say that s is one-to-one and onto. This is not enough. By Proposition 2.86, $T_s = Th(Z_s)$ is a complete theory. The set $\{\sigma_1, \sigma_2\}$ is not complete. There are finite models of these two sentences. An axiomatization of T_s must forbid finite cycles. That is, we must include sentences to say that for all x, $s(x) \neq x$, $s(s(x)) \neq x$, $s(s(s(x))) \neq x$, and so forth. For each $n \in \mathbb{N}$, let θ_n be the $\mathcal{V}_s$-sentence $\forall x \neg s^n(x) = x$ where $s^n(x)$ abbreviates $\underbrace{s(s(s \cdots s(x)))}_{n \text{ times}}$.

Let $\Gamma_s = \{\sigma_1, \sigma_2, \theta_n | n \in \mathbb{N}\}$. If the deductive closure of Γ_s is complete, then we are done. Otherwise, to obtain an axiomatization of T_s, we must proceed to add more sentences to Γ_s. We return to this example in Example 5.20 and show that Γ_s is indeed an axiomatization of T_s. It follows that T_s is decidable.

We need a way to verify that a given $\mathcal{V}$-theory T is complete. As we remarked at the outset, this is a more difficult task than showing that T is incomplete. It is not difficult to show that the theories T_G, T_E, T_{gp}, and T_{LO} are incomplete. Throughout this chapter, we will consider examples of complete theories that contain these theories as subsets. One of our goals in this chapter is to define various criteria that imply completeness.

5.2 Categoricity

A theory is complete if and only if all models of the theory are elementarily equivalent. This is a reformulation of Proposition 5.2. In particular, if all its models are isomorphic, then the theory must be complete. If this is the case, then we say that there is only one model *up to isomorphism* and that the theory is *categorical*.

Theories describe structures. We distinguish two types of descriptions that are desirable. A complete description describes its subject entirely. A categorical description describes its subject uniquely. Let us lift our restriction to first-order logic for the moment, and suppose that we want to describe an object using English sentences. Suppose we are in a crowded bar and I want to describe Dennis to you. If I tell you that Dennis is in the room, is over 2 m tall, has fuchsia hair, and is wearing sunglasses and a feather boa, then it is likely that there will be at most one person in the room fitting this description. If there is exactly one person fitting the description, then the description is *categorical*. A categorical description provides only enough information to single out its object and is not necessarily complete. We cannot deduce all there is to know about a person from a categorical description. Indeed, our categorical description leaves many unanswered questions about Dennis.

In English, a complete description is necessarily categorical, but not the other way around. In the language of first-order logic, since it is a weak language, this is reversed. A complete theory may not be categorical (it may have more than one model). But, as we pointed out in the opening paragraph, if a theory is categorical, then it must be complete.

Definition 5.13 A theory is *absolutely categorical* if it has only one model up to isomorphism.

Any complete theory having a finite model is absolutely categorical. This follows from Proposition 2.81 where it was shown that for any finite $\mathcal{V}$-structure M, there is a $\mathcal{V}$-sentence φ_M that describes M up to isomorphism. By the Upward Löwenhiem–Skolem theorem, these are the only examples of absolutely categorical theories. If a theory has an infinite model, then it has arbitrarily large models. In particular, any such theory has models of different cardinalities. Two structures of different cardinalities cannot possibly be isomorphic.

So absolutely categorical theories are nothing new. This is merely a new name for complete theories having a finite model. We extend the notion of categoricity so that it applies to theories having infinite models.

Definition 5.14 Let κ be a cardinal. A theory T is κ-**categorical** if T has exactly one model of size κ up to isomorphism.

This definition circumvents the Upward Löwenhiem–Skolem theorem. Let $N \models T$. If N is not the same size as M, then, of course, N cannot be isomorphic to M. If T is κ-categorical, then this is the only reason that N may not be isomorphic to a model M of size κ.

Among theories having infinite models, κ-categoricity is a very strong property. As we shall see, we can attain much information about a theory and the structure of its models merely by knowing for which cardinals κ the theory is κ-categorical. One basic result is the following:

Proposition 5.15 Let T be a deductively closed theory having only infinite models. If T is κ-categorical for some $\kappa \geq |T|$, then T is complete.

Proof We prove the contrapositive. Suppose T is not complete. By Proposition 5.2, there exist $M \models T$ and $N \models T$ such that $N \not\equiv M$. By Corollary 4.34 of the Löwenhiem–Skolem Theorems, there exist $M' \equiv M$ and $N' \equiv N$ such that $|M'| = |N'| = \kappa$. Since $M' \not\equiv N'$, M' and N' cannot be isomorphic and T is not κ-categorical. □

Definition 5.16 For any cardinal κ, we say that $\mathcal{V}$-structure M is κ-*categorical* if the $\mathcal{V}$-theory $Th(M)$ is κ-categorical.

Whereas all finite structures have theories that are absolutely categorical, relatively few infinite structures are κ-categorical for some κ. However, although it is rare, many important structures have this property. Structures that are

First-order theories 207

κ-categorical for infinite κ play a central role in Model Theory. Examples of these structures include the complex numbers, vector spaces, and the random graph. We shall investigate these structures and their theories later in this chapter. Presently, we provide some elementary examples.

Example 5.17 Recall from Section 2.4.1 that a *clique* is a graph that models the sentence $\forall x \forall y (\neg(x = y) \rightarrow R(x,y))$ saying that any two distinct vertices share an edge. Let T be the $\mathcal{V}_R$-theory axiomatized by this sentence together with the sentences that define a graph. Since any two cliques of the same size are isomorphic, T is κ-categorical for all cardinals κ. In particular, T is absolutely categorical. Since T has finite models of different sizes, it is not complete. Suppose that we add to this theory the sentences

$$\exists x_1 \exists x_2 \cdots \exists x_n \left(\bigwedge_{i \neq j} x_i \neq x_j \right)$$

for each $n \in \mathbb{N}$. These sentences express that the underlying set contains at least n elements for each $n \in \mathbb{N}$. That is, the universe is infinite. Let T_{clique} denote the set of $\mathcal{V}_<$-sentences that can be derived from the union of these sentences with the theory of cliques. Equivalently, T_{clique} is the set of all $\mathcal{V}_G$-sentences that are true in all infinite cliques. Since T_{clique} is κ-categorical for infinite κ and has only infinite models, it is complete by Proposition 5.15.

Example 5.18 Let T_E be the $\mathcal{V}_E$-theory of equivalence relations from Example 5.3. Each model of T_E is completely determined by the number and the sizes of its equivalence classes. We describe two models M_2 and N_2 of T_E.

Let M_2 have exactly two different equivalence classes each of which is denumerable.

Let N_2 have a denumerable number of equivalence classes each containing exactly two elements.

So both M_2 and N_2 have denumerable universes. Let $\{a_1, a_2, a_3, \ldots\}$ and $\{b_1, b_2, b_3, \ldots\}$ be the underlying sets of M_2 and N_2, respectively. We depict M_2 as tall and thin and N_2 as short and fat in Tables 5.1 and 5.2.

We claim that each of these structures is $\aleph_0$-categorical.

Consider first M_2. Let M be a countable $\mathcal{V}_E$-structure that is elementarily equivalent to M_2. The $\mathcal{V}_E$-sentence

$$\exists x \exists y (\neg E(x,y) \land \forall z (E(x,z) \lor E(y,z)))$$

expresses that there are exactly two equivalence classes. Since M_2 models this sentence, so does M. Also, M_2 models the sentences saying that each element has at least n elements in its equivalence class for each $n \in \mathbb{N}$. Since $M_2 \equiv M$, M also models these sentences. So M, like M_2, has two denumerable equivalence

Table 5.1 $\mathcal{V}_E$-structure M_2

.	.
.	.
.	.
a_5	a_6
a_3	a_4
a_1	a_2

Table 5.2 $\mathcal{V}_E$-structure N_2

| b_2 | b_4 | b_6 | b_8 | . | . | . |
| b_1 | b_3 | b_5 | b_7 | . | . | . |

classes. It follows that each equivalence class M can be put into one-to-one correspondence with either of the equivalence classes of M_2. Since they have the same number of equivalence classes, M_2 and M are isomorphic and M_2 is $\aleph_0$-categorical.

We now show that N_2 is κ-categorical for any infinite κ. Let κ be infinite and let N and N' be two $\mathcal{V}_E$-structures of size κ that are both elementarily equivalent to N_2. Then each equivalence class of either N or N' must contain exactly two elements (since this can be expressed with a first-order sentence). Since κ is infinite, both N and N' have κ many equivalence classes. So the equivalence classes of N can be put into one-to-one correspondence with the equivalence classes of N'. Since all equivalence classes have the same number of elements, N and N' are isomorphic and N_2 is κ-categorical as was claimed.

We return now to M_2 and show that this structure, unlike N_2, is not κ-categorical for uncountable κ. This follows from the fact that first-order logic cannot distinguish between one infinite cardinal and another. Whereas we can define a set of $\mathcal{V}_E$-sentences to say that each equivalence class is infinite, we cannot say that each equivalence class has size $\aleph_0$ or size $\aleph_{23}$ nor specify any other infinite cardinality. For any cardinals λ and κ, let $M_{\lambda\kappa}$ be the $\mathcal{V}_E$-structure having one equivalence class of size λ, one of size κ, and no other equivalence classes. Then $M_{\lambda\kappa} \equiv M_2$ for any infinite λ and κ. If $\lambda < \kappa$, then $M_{\lambda\kappa}$ is not isomorphic to $M_{\kappa\kappa}$. Moreover,

$$|M_{\kappa\kappa}| = 2 \cdot \kappa = \kappa = \kappa + \lambda = |M_{\kappa\lambda}|.$$

It follows that M_2 is not κ-categorical for uncountable κ as we claimed.

First-order theories 209

Definition 5.19 Let T be a theory having only infinite models.

T is *countably categorical* if it is $\aleph_0$-categorical.
T is *uncountably categorical* if it is κ-categorical for all uncountable κ.
T is *totally categorical* if it is κ-categorical for all infinite κ. That is, if it is both countably and uncountably categorical.

The theory $Th(M_2)$ from Example 5.18 is countably categorical but not uncountably categorical. The theory $Th(N_2)$ from that example is totally categorical as is the theory T_{clique} from Example 5.17. We now demonstrate an example of an uncountably categorical theory that is not countably categorical.

Example 5.20 Recall the $\mathcal{V}_s$-theory $T_s = Th(Z_s)$ from Example 5.12. Recall too the set Γ_s of $\mathcal{V}_s$-sentences expressing that s is a one-to-one and onto function having no finite cycles. We claim that Γ_s axiomatizes T_s. To do verify this, we show that every model of Γ_s is also a model of T_s. Let us consider some specific models of Γ_s.

Let Z_2 be a $\mathcal{V}_S$-structure having underlying set

$$\{\ldots, -3, -2, -1, 0, 1, 2, 3, \ldots\} \cup \{\ldots, a_{-3}, a_{-2}, a_{-1}, a_0, a_1, a_2, a_3, \ldots\}.$$

Let Z_2 interpret s the same way as Z_s on the integers. Further, suppose that $Z_2 \models s(a_i) = a_j$ if and only if $j = i + 1$. Then Z_2 interprets s as a one-to-one onto function having no finite cycles. So $M \models \Gamma_s$. We say that such M contains two copies of $\mathbb{Z}$. Likewise, we can define models of Γ_s having any number of copies of $\mathbb{Z}$. For any nonzero cardinal κ, let Z_κ be the $\mathcal{V}_s$-structure containing κ copies of $\mathbb{Z}$. (So Z_1 is Z_s.)

Let κ be an uncountable cardinal. Let N be a model of Γ_s of size κ. For any element a_0 in the universe U_N of N, there must exist a unique successor a_1 and predecessor a_{-1} in U_N. There must also exist successor a_2 of a_1 and predecessor a_{-2} of a_{-1}, and so forth. Since N has no finite cycles, each element $a_0 \in U_N$ is contained in a copy of $\mathbb{Z}$. Since $|N| = \kappa$, N must contain κ copies of $\mathbb{Z}$. It follows that $N \cong Z_\kappa$. So Z_κ is the only model of Γ_s of size κ up to isomorphism and Γ_s is κ-categorical for all uncountable κ. By Proposition 5.15, the deductive closure of Γ_s is the complete theory T_s. Since the nonisomorphic models $Z_1, Z_2, Z_3, \ldots, Z_{\aleph_0}$ are each countable, T_s is not $\aleph_0$-categorical.

We have demonstrated the existence of theories that are countably categorical and not uncountably categorical, theories that are uncountably categorical and not countably categorical, and theories that are totally categorical. We shall also see examples of theories that are not κ-categorical for any κ. For complete countable theories having infinite models, these are the only four possibilities. This is a consequence of Morley's theorem.

Theorem 5.21 (Morley) Let T be a countable theory. If T is κ-categorical for some uncountable κ, then T is κ-categorical for all uncountable κ.

Morley's proof of this theorem introduced methods and concepts to model theory that would bear fruit far beyond Morley's theorem itself. The proof gave rise to the subject of stability theory. We touch upon some of the ingredients of this proof in Chapter 6 (see Exercise 6.33). However, we do not prove Morley's theorem. Instead, we refer the reader to books devoted solely to model theory such as [29] and [39] and also to more advanced books on stability theory such as [1] and [6]. Also, for the serious student of model theory, Morley's original proof in [32] remains essential reading.

We conclude this section by stating without proof two results regarding categoricity and finite axiomatizability. Naturally, a theory is said to be *finitely axiomatizable* if it is axiomatized by a finite set of sentences. We have seen several examples of finitely axiomatizable theories including the theory of graphs, the theory of equivalence relations, the theory of groups, and others. All of these theories are incomplete. In the next section, we shall see examples of finitely axiomatizable complete theories having infinite models. Such theories necessarily contain a sentence that has only infinite models (see Exercises 2.37 and 2.38 for examples of such sentences). As a rule, most complete theories having infinite models are not finitely axiomatizable. If we restrict our attention to totally categorical theories, then we can be more precise.

Theorem 5.22 (Zil'ber) Totally categorical theories are not finitely axiomatizable.

Recall that the theory of cliques from Example 5.17 is finitely axiomatizable and κ-categorical for all κ. Since this theory has finite models, it is not totally categorical. In contrast, the theory T_{clique} of infinite cliques is totally categorical, but is not finitely axiomatizable. To axiomatize T_{clique} we must include sentences saying that there exist more than n elements for each $n \in \mathbb{N}$. Using counting quantifiers as defined in Exercise 2.20, we can express each of these sentences as $\exists^{\geq n} x(x=x)$. This is an example of a *quasi-finite axiomatization*.

Definition 5.23 A theory T is *quasi-finitely axiomatizable* if there exists a finite set F of formulas in one free variable such that T is axiomatized by sentences of the form $\exists^{\geq n} x \varphi(x)$ with $\varphi(x) \in F$.

Theorem 5.24 (Hrushovski) If T is totally categorical and has a finite vocabulary, then T is quasi-finitely axiomatizable.

Zil'ber's and Hrushovski's theorems are actually corollaries to results regarding the general structure of models of totally categorical theories. Their proofs of are beyond the scope of this book (these theorems are proved in [38]). We focus

First-order theories 211

on more elementary properties of these theories. In the next section, we prove a fundamental result regarding countably categorical theories.

5.3 Countably categorical theories

We investigate some specific countably categorical structures. We consider structures in the vocabulary $\mathcal{V}_<$ consisting of a single binary relation $<$. Each of the examples we consider interprets $<$ as a linear order. In the second part of this section, we prove a fundamental result that holds for all countably categorical theories.

5.3.1 Dense linear orders. Consider the closed unit interval of real numbers. Let $\mathbf{R}_{[0,1]}$ be the structure $\{[0,1]|<\}$ having the closed interval $[0,1]$ of real numbers as an underlying set and interpreting $<$ in the usual way. We list some $\mathcal{V}$-sentences that hold in $\mathbf{R}_{[0,1]}$. For reference, we label these sentences as δ_1–δ_7.

δ_1: $\forall x \forall y((x<y) \to \neg(y<x))$
δ_2: $\forall x(\neg(x<x))$
δ_3: $\forall x \forall y((x<y) \lor (y<x) \lor (x=y))$
δ_4: $\forall x \forall y \forall z(((x<y) \land (y<z)) \to (x<z))$
δ_5: $\forall x \forall y((x<y) \to \exists z(x<z \land z<y))$
δ_6: $\exists x \forall y((x=y) \lor (x<y))$
δ_7: $\exists x \forall y((x=y) \lor (y<x))$.

The first four of these sentences say that $<$ is a linear order. Recall from Example 5.8 that the theory T_{LO} is defined as the set of all consequences of these four sentences. The sentence δ_5 says that between any two elements, there exists another element. That is, the linear order is *dense* (see Section 2.4.3). Finally, δ_6 says that there exists a smallest element and δ_7 says that there exists a largest element.

Let T_{DLOE} denote the set of $\mathcal{V}_<$-sentences that can be derived from the above seven sentences. This is the theory of dense linear orders with endpoints. Clearly, $\mathbf{R}_{[0,1]} \models T_{DLOE}$. We claim that $Th(\mathbf{R}_{[0,1]}) = T_{DLOE}$. To show this, we must verify that T_{DLOE}, unlike T_{LO}, is a complete theory. We prove something stronger.

Proposition 5.25 T_{DLOE} is $\aleph_0$-categorical.
Proof Let M and N be two models of T_{DLOE} of size $\aleph_0$. We show that M and N are isomorphic. Let U_M and U_N denote the underlying sets of M and N

respectively. Enumerate these sets as follows:

$$U_M = \{a_1, a_2, a_3, \ldots\} \quad \text{and} \quad U_N = \{b_1, b_2, b_3, \ldots\}.$$

Since M and N model δ_6 and δ_7, these sets must contain a smallest and a largest element (with respect to the order $<$). We may assume that a_1 and b_1 are the smallest elements in each set and a_2 and b_2 are the largest elements.

We construct an isomorphism $f : M \to N$ step-by-step. In each step we define f for two elements of U_M:

Step 1: Let $f(a_1) = b_1$ and $f(a_2) = b_2$.

For $n > 1$, step n has two parts.

Step n: *Part a.* Let A_n be the set of all $a_i \in U_M$ for which $f(a_i)$ has been defined in some previous step. Let j be least such that a_j is not in A_n. We define $f(a_j)$. Since A_n is finite, we can find elements c and d of A_n so that no element of A_n is between these two elements and a_j is (that is, $c < a_j < d$ or $d < a_j < c$). In this case, let $f(a_j)$ be any element of U_N that lies between $f(c)$ and $f(d)$. Since $N \models \delta_5$, such $f(a_j)$ exists.

Part b. Let B_n be the set of all $b_i \in U_N$ for which $f^{-1}(b_i)$ has been defined in some previous step (including $f(a_j)$ from part a). Let j be least such that b_j is not in B_n. Since B_n is finite, we can find elements c and d of B_n so that no element of B_n is between these two elements and b_j is (that is, $c < b_j < d$ or $d < b_j < c$). In this case, let $f^{-1}(b_j)$ be any element of U_M that lies between $f^{-1}(c)$ and $f^{-1}(d)$. Since $M \models \delta_5$, such a $f^{-1}(d)$ exists.

After completing step n for all $n \in \mathbb{N}$, the function f is completely defined. Since $f(a_i)$ is defined in Step i (part a) if not before, each a_i is in the domain of f. Moreover, $f(a_i)$ is defined exactly once. So f has domain U_M and is one-to-one. Also, since $f^{-1}(b_i)$ is defined in Step i (part b) if not before, f is onto. By design, f preserves the order ($a_i < a_j$ implies $f(a_i) < f(a_j)$). So f is a literal embedding. By Proposition 2.57, f is an isomorphism as was desired. □

Corollary 5.26 T_{DLOE} *is complete.*

Proof By Proposition 5.15, any $\aleph_0$-categorical theory having only infinite models is complete. So to show that T_{DLOE} is complete, it suffices show that any dense linear order is necessarily infinite. If a linear order is finite, then it can be listed as $a_1 < a_2 < \cdots < a_n$ for some $n \in \mathbb{N}$. Such a linear order is not dense since there is no element between a_1 and a_2. So T_{DLOE} has only infinite models and is complete. □

It follows from this corollary that any model of T_{DLOE} is elementarily equivalent to $\mathbf{R}_{[0,1]}$. For example, suppose that we restrict the underlying set to the set of rational numbers in the interval $[0, 1]$. Let $\mathbf{Q}_{[0,1]}$ denote the $\mathcal{V}_<$-structure having this set of rationals as its underlying set and interpreting $<$ as the usual order. This is a countable model of T_{DLOE}. Since this theory is $\aleph_0$-categorical, it is essentially the only countable model of this theory. Any other countable model must be isomorphic to $\mathbf{Q}_{[0,1]}$. Moreover, since T_{DLOE} is complete, $\mathbf{Q}_{[0,1]}$ and $\mathbf{R}_{[0,1]}$, although not isomorphic, are elementarily equivalent.

Proposition 5.27 For any uncountable cardinal κ, T_{DLOE} is not κ-categorical.

Proof We define a model $\mathbf{H}_{[0,2]}$ of T_{DLOE} that has the same size as $\mathbf{R}_{[0,1]}$, but is not isomorphic to $\mathbf{R}_{[0,1]}$. The structure $\mathbf{H}_{[0,2]}$ is a hybrid of $\mathbf{Q}_{[0,1]}$ and $\mathbf{R}_{[0,1]}$. Its universe is the union of the set of all rational numbers in the interval $[0,1]$ and the set of all real numbers in the interval $[1,2]$. Again, this structure interprets $<$ in the usual way. This structure models T_{DLOE}. Since this theory is complete, $\mathbf{H}_{[0,2]} \equiv \mathbf{R}_{[0,1]}$. Moreover,

$$|\mathbf{H}_{[0,2]}| = |\mathbf{Q}_{[0,1]}| + |\mathbf{R}_{[0,1]}| = \aleph_0 + 2^{\aleph_0} = 2^{\aleph_0}.$$

So $\mathbf{H}_{[0,2]}$ has the same size as $\mathbf{R}_{[0,1]}$. To see that it is not isomorphic to $\mathbf{R}_{[0,1]}$, note that, in $\mathbf{R}_{[0,1]}$, there exist uncountably many elements between any two elements. This is not true in $\mathbf{H}_{[0,2]}$. So an isomorphism between these two models is impossible and T_{DLOE} is not $2^{\aleph_0}$-categorical. By Morley's theorem 5.21, T_{DLOE} is not κ-categorical for any uncountable κ. □

Recall the $\mathcal{V}_<$-structures $\mathbf{Q}_<$ and $\mathbf{R}_<$ from 2.4.3. Since they have no endpoints, these are not models of T_{DLOE}. We define the theory T_{DLO} of dense linear orders without endpoints as the set of all $\mathcal{V}_<$-sentences that can be derived from the sentences $\delta_1, \delta_2, \delta_3 \delta_4, \delta_5, \neg \delta_6$, and $\neg \delta_7$. We negate the sentences saying there exist a smallest and largest element. Both $\mathbf{Q}_<$ and $\mathbf{R}_<$ are models of this theory.

Corollary 5.28 T_{DLO} is $\aleph_0$-categorical.

Proof Any model of T_{DLO} can be extended to a model of T_{DLOE} by adding smallest and largest elements to the underlying set. If there were non-isomorphic countable models of T_{DLO}, then these could be extended to non-isomorphic countable models of T_{DLOE}. Since T_{DLOE} is $\aleph_0$-categorical, so is T_{DLO}. □

Corollary 5.29 T_{DLO} is complete.

Proof This is the same as the proof of Corollary 5.26. Since dense linear orders are necessarily infinite, T_{DLO} has no finite models. By Proposition 5.15, T_{DLO} is complete. □

Corollary 5.30 $\mathbf{Q}_< \equiv \mathbf{R}_<$.

Proof This follows immediately from the fact that $\mathbf{Q}_<$ and $\mathbf{R}_<$ are both models of the complete theory T_{DLO}. □

In light of these examples, and specifically of the proof of Proposition 5.25, we now investigate arbitrary countably categorical theories.

5.3.2 Ryll-Nardzewski et al.
Categoricity is a property of theories that is defined in terms of the models of the theory. A theory T is countably categorical if and only if there is exactly one countable model (up to isomorphism) in $Mod(T)$. As we shall prove, there is a purely syntactic characterization of these theories. We show that a theory T is $\aleph_0$-categorical if and only if there are only finitely many formulas in n free variables up to T-equivalence. Equivalently, a $\mathcal{V}$-structure M having universe U is $\aleph_0$-categorical if and only if, for each $n \in \mathbb{N}$, only finitely many subsets of U^n are $\mathcal{V}$-definable.

Proposition 5.31 Let T be a complete $\mathcal{V}$-theory. If there are only finitely many formulas in n free variables up to T-equivalence for each n, then T is $\aleph_0$-categorical.

We prove this proposition using a *back-and-forth* argument. This method of proof constructs an isomorphism between two structures by alternating back and forth between the elements of each of the two underlying sets. An example of a back-and-forth argument is provided by the proof of Proposition 5.25, where it was shown that any two countable models of T_{DLOE} are isomorphic. The proof of Proposition 5.31 resembles the proof of Proposition 5.25. In fact, Proposition 5.25 is a special case of this proposition.

Proof of Proposition 5.31 Suppose that there are only finitely many formulas in n free variables up to T-equivalence for each n. We show that T is $\aleph_0$-categorical.

Let M and N be two models of T of size $\aleph_0$. Let U_M and U_N denote the underlying sets of M and N respectively. Enumerate these sets as

$$U_M = \{a_1, a_2, a_3, \dots\} \quad \text{and} \quad U_N = \{b_1, b_2, b_3, \dots\}.$$

We construct an isomorphism $f : M \to N$ step-by-step. In each step, we define $f(a_i)$ for two elements a_i of U_M.

For $n \in \mathbb{N}$, let A_n be the set of all $a_i \in U_M$ for which $f(a_i)$ has been defined in some step prior to step n. Since A_n is finite, we may regard it as a tuple of elements of U_M. There are many ways to arrange the elements of a large finite set into a tuple. For any a_i and a_j in A_n, one of the two elements $f(a_i)$ and $f(a_j)$ of U_N must have been defined before the other. Let $\bar{a}_n$ be the tuple obtained by arranging the elements of A_n in the order in which f was defined. Likewise, let B_n be the corresponding set of all $f(a_i) \in U_N$ for $a_i \in A_n$. Let $\bar{b}_n$ be the tuple obtained by arranging the elements of B_n in the order in which f was defined.

Note that $A_1 = B_1 = \emptyset$. Note too that, since T is complete, $N \models \varphi$ if and only if $M \models \varphi$ for any $\mathcal{V}$-sentence φ.

For $n \in \mathbb{N}$, assume that A_n and B_n have been defined in such a way that $M \models \varphi(\bar{a}_n)$ if and only if $N \models \varphi(\bar{b}_n)$ for any $\mathcal{V}$-formula φ in $|A_n|$ free variables.

Step n: *Part a.* Let j be least such that a_j is not in A_n. We define $f(a_j)$.

Let $k = |A_n| = 2(n-1)$. By hypothesis, there exists a finite set $\mathcal{F}$ of $\mathcal{V}$-formulas in $k+1$ free variables so that every $\mathcal{V}$-formula having $k+1$ free variables is T-equivalent to a formula in $\mathcal{F}$.

Let $\Phi(\bar{x}, y)$ be the conjunction of those formulas $\varphi(\bar{x}, y)$ in $\mathcal{F}$ such that $M \models \varphi(\bar{a}_n, a_j)$. Then $M \models \exists y \Phi(\bar{a}_n, y)$.

This formula has $|A_n| = k$ free variables. By the definitions of A_n and B_n, $N \models \exists y \Phi(\bar{b}_n, y)$. It follows that $N \models \exists y \Phi(\bar{b}_n, b_i)$ for some $b_i \in U_N$.

Let $f(a_j) = b_i$.

Part b. Let $\bar{a}'_n$ be the tuple $(\bar{a}_n, a_j)$ and let $\bar{b}'_n$ be the tuple $(\bar{b}_n, b_i)$, where a_j and b_i are as defined in part (a). Let l be least such that b_l is not in $\bar{b}'_n$. As in part (a), we can find an element a_i of U_M so that $M \models \varphi(\bar{a}'_n, a_i)$ if and only if $N \models \varphi(\bar{b}'_n, b_l)$ for any φ in $2n$ free variables.

Define $f(a_i)$ to be b_l.

After completing step n for each $n \in \mathbb{N}$, the function f is completely defined. Since $f(a_i)$ is defined in Step i (part a) if not before, each a_i is in the domain of f. Moreover, $f(a_i)$ is defined exactly once. So f has domain U_M and is one-to-one. Also, since $f^{-1}(b_i)$ is defined in Step i (part b) if not before, f is onto. By design, f preserves all $\mathcal{V}$-formulas and is an isomorphism as was desired. □

We now have two ways to show that a given theory is countably categorical. We can give a back-and-forth argument as we did for T_{DLOE} in Proposition 5.25. Alternatively, we can show that there are only finitely many formulas in n free variables up to T-equivalence for each n. This may not seem practical. However, in Section 5.5 we discuss quantifier elimination and provide a systematic approach to understanding the definable subsets of certain structures.

The converse of Proposition 5.31 is also true. The countably categorical theories are precisely those theories that are complete and have few formulas (finitely many in n free variables for each n). Since this was proved by Ryll-Nardzewski in a 1959 paper, it is commonly referred to as the Ryll-Nardzewski theorem. Since it also appeared in 1959 in separate papers by Engeler and Svenonius, it is sometimes referred to as the Engeler–Ryll-Nardzewski–Svenonius theorem. We opt for brevity and refer to it as Theorem 5.32.

Theorem 5.32 A complete theory T is $\aleph_0$-categorical if and only if, for each $n \in \mathbb{N}$, there are only finitely many formulas in n free variables up to T-equivalence.

One direction of this theorem was proved as Proposition 5.31. We postpone the proof of the other direction until Chapter 6 where we shall see several equivalent characterizations of $\aleph_0$-categorical theories.

5.4 The Random graph and 0–1 laws

A random graph is a graph constructed by some random process such as rolling a die or flipping a coin. The idea of implementing random processes in graph theory was conceived by Paul Erdös and has served as a powerful tool for this and other areas of discrete mathematics. In this section, we discuss this idea and show how it gives rise to a complete first-order theory T_{RG} in the vocabulary of graphs. We prove that T_{RG} is $\aleph_0$-categorical. Whereas there are many possible finite random graphs, there is only one denumerable random graph. From this fact we deduce a 0–1 law for first-order logic.

We assume basic knowledge of probability.

Suppose that we have a set of vertices and want to build a graph. For example, suppose that we have five vertices v_1, v_2, v_3, v_4, and v_5. To define the graph, we must decide which pairs of vertices share an edge. Let us take the random approach and flip a coin to make our decisions. Given any two vertices (v_1 and v_2, say) we flip a coin. If the coin lands heads up, then v_1 and v_2 share an edge. If the coin lands tails up, then they do not share an edge. We repeat this for every pair of vertices. Since there are five vertices, there are $(5 \cdot 4)/2 = 10$ pairs of vertices to consider. After flipping the coin 10 times, we will have completed the graph.

Any graph having vertices v_1, v_2, v_3, v_4, and v_5 is a possible outcome of this process. Since each of the ten flips of the coin has two possible outcomes, there are 2^{10} possible graphs. If the coin is fair (landing heads up as frequently as tails up), then each of these 2^{10} graphs is equally likely. However, two or more of the outcomes may be isomorphic graphs. So, up to isomorphism, some graphs are more likely than others. For example, the 5-clique is an unlikely outcome. To obtain this result, each of our 10 flips of the coin must land heads up. The probability of this happening is $1/2^{10}$. It is more likely that the outcome will have exactly one edge. The probability of this happening is $10/2^{10}$ (so this will happen roughly 1% of the time).

There are two ways to compute the probabilities in the previous paragraph. Suppose we want to compute the probability that the resulting graph has exactly m edges for some $m \leq 10$. Using the formula for binomial probability distributions, this probability is $\binom{10}{m} (\frac{1}{2})^{10}$ (where $\binom{10}{m} = \frac{10!}{(10-m)!m!}$ is the number of ways that m of the 10 edges can be chosen). Alternatively, since each of the 2^{10} graphs are equally likely, this probability can be computed by counting the number of graphs having exactly m edges and dividing this number by 2^{10}. For example the 5-clique is the only one of the 2^{10} possible outcomes that has 10 edges. So the probability that this happens is $1/2^{10}$.

More generally, suppose that we randomly construct a graph having vertices $\{v_1, v_2, \ldots, v_n\}$ for some $n \in \mathbb{N}$. There are $n(n-1)/2$ pairs of vertices to consider. To ease notation, denote $n(n-1)/2$ by $e(n)$ (this is the number of edges in the n-clique). If we construct this graph by flipping a fair coin as before, then there are $2^{e(n)}$ possible outcomes each of which is equally likely. Let φ be a $\mathcal{V}_R$-sentence. Let $P_n(\varphi)$ be the probability that our randomly constructed graph models the sentence φ. This probability can be computed by counting the number of outcomes that model φ and dividing by the total number of possibilities $2^{e(n)}$.

Example 5.33 Let φ be the sentence $\forall x \forall y (x = y \lor R(x,y))$. For each $n \in \mathbb{N}$, this sentence holds in only one of the $2^{e(n)}$ graphs having vertices $\{v_1, \ldots, v_n\}$ (namely, the n-clique).

So $P_n(\varphi) = 1/2^{e(n)}$.

(In particular $P_5(\varphi) = 1/2^{e(5)} = 1/2^{10}$ as previously noted.)
If n is big, then this probability is close to zero.

Example 5.34 Let φ be the $\mathcal{V}_R$-sentence saying that the graph has exactly one edge. For each $n \in \mathbb{N}$, the number of graphs having vertices $\{v_1, \ldots, v_n\}$ that model this sentence is the number of possible edges $e(n)$.

So $P_n(\varphi) = e(n)/2^{e(n)}$.

If n is big, then this probability is close to zero.

For any $\mathcal{V}$-sentence φ and any $n \in \mathbb{N}$, $P_n(\varphi) + P_n(\neg \varphi) = 1$. This is because every graph either models φ or $\neg \varphi$ (and not both). So $P_n(\neg \varphi) = 1 - P_n(\varphi)$. In the previous two examples, since $P_n(\varphi)$ approaches zero, $P_n(\neg \varphi)$ approaches 1 as n gets large. We express this by using limit notation:

$$\lim_{n \to \infty} P_n(\varphi) = 0 \quad \text{and} \quad \lim_{n \to \infty} P_n(\neg \varphi) = 1.$$

The 0–1 *Law for Graphs* states that, for every $\mathcal{V}_R$-sentence θ, either $\lim_{n \to \infty} P_n(\theta) = 0$ or $\lim_{n \to \infty} P_n(\theta) = 1$. So either θ or its negation almost certainly holds in any large finite graph. This fact imposes limitations on what can be expressed by a $\mathcal{V}_R$-sentence. For example, the 0–1 Law for Graphs implies that there is no $\mathcal{V}_R$-sentence that holds only in those finite graphs having an even number of vertices.

We have verified the 0–1 Law for Graphs for a couple of particular sentences in the above examples. To prove this law, we must consider some other (more complicated) $\mathcal{V}_R$-sentences. For each $m \in \mathbb{N}$, let ρ_m be the $\mathcal{V}_R$-sentence

$$\forall x_1 \cdots \forall x_m \forall y_1 \cdots \forall y_m \left(\bigwedge_{i=1}^{m} \left(\bigwedge_{j=1}^{m} x_i \neq y_j \right) \right)$$

$$\to \exists z \left(\bigwedge_{i=1}^{m} R(x_i, z) \land \bigwedge_{j=1}^{m} \neg (R(z, y_j) \lor z = y_j) \right).$$

This sentence asserts that given any two sets of vertices $X = \{x_1, \ldots, x_m\}$ and $Y = \{y_1, \ldots, y_m\}$ such that $X \cap Y = \emptyset$, there exists a vertex z not in $X \cup Y$ that shares an edge with each vertex in X and with no vertex in Y. Note that $\rho_{m_2} \to \rho_{m_1}$ for $m_1 \leq m_2$. We examine ρ_m for various values of m.

$m = 1$: The sentence ρ_1 states that, for any vertices x and y there exists a vertex z that shares an edge with x but not y. Since x and z share an edge, they cannot be equal. Moreover, ρ_1 asserts there exists such a z that is not equal to y. So any model of ρ_1 must have at least three vertices corresponding to x, y, and z. Moreover, reversing the roles of x and y, there must exist a vertex w that shares an edge with y but not x. So any graph that models ρ_1 must have at least four vertices. In fact, the smallest example has five vertices (take the sides of a pentagon as edges).

$m = 2$: Let G be a graph that models ρ_2. Let a and b be two vertices of G. Then, there must exist a vertex c that shares and edge with a but not b and a vertex d that shares an edge with b but not a. Moreover, there must exist a vertex e that shares a vertex with both a and b and a vertex f that shares a vertex with neither a nor b. So there must be at least six vertices, but we are not done. There must also exist a vertex g that shares an edge with both e and d and with neither c nor d, and so forth. In contrast to ρ_1, it is not an easy task to draw a graph that models ρ_2 (nor is it easy to determine the minimal number of vertices for such a graph).

$m > 2$: It becomes increasingly difficult to demonstrate a finite graph that models ρ_m as m gets larger. We will not attempt to compute the precise value for $P_n(\rho_m)$ for given m and n. However, there are some things that we can say with certainty regarding this value. As a first observation, a graph that models ρ_m must have many vertices. In particular, $P_n(\rho_m) = 0$ for $n \leq m$. So if m is small, then so is $P_n(\rho_m)$. Less obvious is the fact that, for big m, $P_n(\rho_m)$ is close to 1. That is,

$$\lim_{n \to \infty} P_n(\rho_m) = 1.$$

So although it is hard to give a concrete demonstration of a finite graph that models, say, ρ_8, we have a process that will produce such a graph with high probability. If we construct a graph on the vertices $\{v_1, \ldots, v_n\}$ by flipping a coin, then we will most likely obtain a graph that models ρ_8 provided that n is sufficiently large. We prove this key fact as the following lemma.

Lemma 5.35 For each $m \in \mathbb{N}$, $\lim_{n \to \infty} P_n(\rho_m) = 1$.

Proof Fix $m \in \mathbb{N}$.

Given $N \in \mathbb{N}$, we compute $P_n(\neg \rho_m)$ where $n = N + 2m$.

Let G be a graph having n vertices. Let $X = \{x_1, \ldots, x_m\}$ and $Y = \{y_1, \ldots, y_m\}$ be two sets containing m vertices of G such that $X \cap Y = \emptyset$. If

$G \models \rho_m$, then there exists a vertex z not in X or Y such that z shares an edge with each vertex in X and with no vertex in Y. We want to compute the probability that this is not the case.

For any vertex z of G that is not in X or Y, say that z "works" for X and Y if z shares and edge with each vertex of X and no vertex of Y. For this to happen, each flip of the coin must land heads up for the m pairs of vertices (z, x_i) and tails up for the m pairs of vertices (z, y_i). The probability of this happening is $1/2^{2m}$. So given a particular z, it is unlikely that z works for X and Y. However, there are N possible vertices we may choose for z. Whereas the probability that any one of these does not work is $(1 - 1/2^{2m})$, the probability that all N of the vertices do not work is $(1 - 1/2^{2m})^N$.

Let $k = (1 - 1/2^{2m})$. Since k is between 0 and 1, $\lim_{n \to \infty} k^N = 0$. So if N is large, then it is likely that there exists a vertex z that works for X and Y even though the probability that any particular z works is small.

This indicates that it may be likely that a large graph will model ρ_m. However, we have not finished the computation. For the graph G to model ρ_m, there must exist a vertex z that works for X and Y *for all possible X and Y*. Since there are $n = N + 2m$ vertices in G, there are $\binom{n}{2m}$ ways to choose the vertices in $X \cup Y$. There are then $\binom{2m}{m}$ ways to choose m of these vertices for the set X (and the remaining m for set Y). In total, there are

$$\binom{n}{2m}\binom{2m}{m} = \frac{n!}{N!m!m!} < \frac{n^{2m}}{m!m!} \leq n^{2m}$$

possible choices for X and Y. For each choice, the probability that no z works for X and Y is only k^N. For G to model $\neg \rho_m$, this must happen for only one of these choices. Thus,

$$P_n(\neg \rho_m) \leq n^{2m} k^N = n^{2m} k^n / k^{2m}.$$

Since many of the possible choices for X and Y overlap, this is definitely an overestimate of this probability. However, this estimate serves our purpose.

Fact $\lim_{n \to \infty} n^{2m} k^n = 0$.

This fact follows solely from the fact that $k < 1$. Using calculus, it is easy to see that the function $x^{2m} k^x$ reaches a maximum at $x = -2m/\ln k$. To see that this function then decreases to zero, repeatedly apply L'Hopital's rule ($2m$ times) to the expression x^{2m}/k^{-x} having indeterminate form $\frac{\infty}{\infty}$.

Finally, since $\lim_{n \to \infty} P_n(\neg \rho_m) = 0$, $\lim_{n \to \infty} P_n(\rho_m) = 1$. □

We now define the $\mathcal{V}_R$-theory T_{RG}. This is the theory of infinite graphs that model ρ_m for all $m \in \mathbb{N}$. That is, T_{RG} is the $\mathcal{V}_R$-theory axiomatized by:

$$\forall x \neg R(x,x),$$

$$\forall x \forall y (R(x,y) \leftrightarrow R(y,x)),$$

$$\exists x_1 \ldots \exists x_n (\bigwedge_{i \neq j} x_i \neq x_j) \text{ for each } n \in \mathbb{N}, \text{ and}$$

$$\rho_m \text{ for each } m \in \mathbb{N}.$$

Let Δ be a finite subset of this infinite set of sentences. By Lemma 5.35, Δ is satisfied by a preponderance of the finite graphs of size n for sufficiently large n. Since every finite subset is satisfiable, T_{RG} is satisfiable by compactness. So T_{RG} is indeed a theory. We show that it is a complete theory.

Proposition 5.36 T_{RG} is $\aleph_0$-categorical.

Proof Let M and N be two denumerable models of T_{RG}. Let U_M and U_N denote the sets of vertices of M and N respectively. Enumerate these sets as follows:

$$U_M = \{a_1, a_2, a_3, \ldots\} \quad \text{and} \quad U_N = \{b_1, b_2, b_3, \ldots\}.$$

We construct an isomorphism $f : M \to N$ using a back-and-forth argument.

Step 1: Let $f(a_1) = b_1$.

Now suppose that we have defined $f(a_i)$ for n vertices in U_M. Let A_n be the set of vertices in U_M for which f has been defined. Let $B_n = \{f(a_i) | a_i \in A_n\}$.

Step $(n+1)$: *Part a.* Let j be least such that a_j is not in A_n. Since $N \models \rho_m$ for arbitrarily large m, there exists a vertex $f(a_j)$ such that $N \models R(f(a_i), f(a_j))$ if and only if $M \models R(a_i, a_j)$ for any $a_i \in A_n$.

Part b. Let j be least such that b_j is not in $B_n \cup \{f(a_j)\}$. By the same argument as in part a, we can find $f^{-1}(b_j)$ as desired.

The function f defined in this manner is a one-to-one function from M onto N that preserves the edge relation R. It follows that f is an isomorphism and T_{RG} is $\aleph_0$-categorical. □

Definition 5.37 The *random graph*, denoted by G_R, is the unique countable model of T_{RG}.

Proposition 5.38 Any finite graph can be embedded into any model of T_{RG}.

Proof Let G be a finite graph. Let M be an arbitrary model of T_{RG}. We show that G embeds into M by induction on $n = |G|$. Clearly this is true if $n = 1$. Suppose that any graph of size m embeds into M for some $m \in \mathbb{N}$. Let G be a graph having vertices $\{v_1, \ldots, v_{m+1}\}$. By our induction hypothesis, the substructure G' of G having vertices $\{v_1, \ldots, v_m\}$ can be embedded into M. Let

First-order theories 221

$f : G' \to M$ denote this embedding. Since M models ρ_k for arbitrarily large k, there exists a vertex $f(a_{n+1})$ of M such that $M \models R(f(a_i), f(a_{n+1}))$ if and only if $G \models R(a_i, a_{n+1})$ for $i = 1, \ldots, n$. Thus G is embedded into M. By induction, any finite graph can be embedded into M. □

Proposition 5.39 T_{RG} is complete.

Proof It follows from the previous proposition that T_{RG} has only infinite models. By Proposition 5.36, T_{RG} is $\aleph_0$-categorical. By Proposition 5.15, T_{RG} is complete. □

Theorem 5.40 (0–1 Law for Graphs) For every $\mathcal{V}_R$-sentence θ, either $\lim_{n\to\infty} P_n(\theta) = 0$ or $\lim_{n\to\infty} P_n(\theta) = 1$.

Proof Recall the axiomatization that was given for T_{RG}. By Lemma 5.35, $\lim_{n\to\infty} P_n(\varphi) = 1$ for each sentence φ in this axiomatization. It follows that $\lim_{n\to\infty} P_n(\varphi) = 1$ for every sentence φ in T_{RG}. Since T_{RG} is complete, either $T_{RG} \vdash \theta$ or $T_{RG} \vdash \neg\theta$ for every $\mathcal{V}_R$-sentence θ. It follows that either $\lim_{n\to\infty} P_n(\theta) = 1$ or $\lim_{n\to\infty} P_n(\neg\theta) = 1$. □

This result can be generalized. A vocabulary is *relational* if it contains no functions. So relational vocabularies may contain constants as well as relations. Let $\mathcal{M}_n$ be the set of all $\mathcal{V}$-structures having underlying set $\{1, 2, 3, \ldots, n\}$. Let $P_{\mathcal{V}}^n(\theta)$ be the number of structures in $\mathcal{M}_n$ that model θ divided by $|\mathcal{M}_n|$.

Theorem 5.41 (0–1 Law for Relations) Let $\mathcal{V}$ be a finite relational vocabulary. For any $\mathcal{V}$-sentence θ, either $\lim_{n\to\infty} P_{\mathcal{V}}^n(\theta) = 0$ or $\lim_{n\to\infty} P_{\mathcal{V}}^n(\theta) = 1$.

5.5 Quantifier elimination

Suppose that we want to analyze a given first-order structure M. We could begin by trying to find an axiomatization for $Th(M)$. Suppose we have accomplished this and $Th(M)$ is decidable. Then, for any sentence φ in the vocabulary of M, we can determine whether $M \models \varphi$ or $M \models \neg\varphi$. However, understanding the theory $Th(M)$ is only a first step toward understanding the structure M. To analyze M further, one must be familiar with the definable subsets of the structure.

For example, suppose that we are presented with a rather complicated graph G. We are given the set of vertices $\{v_1, v_2, v_3, \ldots\}$ along with the set of all pairs of vertices that share edges in G. Suppose too that we are given a decidable axiomatization of $Th(G)$. Then for any $\mathcal{V}_R$-sentence φ, we can determine whether or not φ holds in G. In some sense, this data represents all there is to know about the structure G. But suppose we want to determine which pairs of vertices (x_1, x_2) satisfy the $\mathcal{V}_R$-formula $\psi(x_1, x_2)$ defined by

$$\forall y \exists z \forall u \exists v (R(x_1, y) \wedge R(y, z) \wedge R(z, u) \wedge R(u, v) \wedge R(v, x_2)).$$

If neither $\forall x_1 \forall x_2 \psi(x_1, x_2)$ nor $\forall x_1 \forall x_2 \neg\psi(x_1, x_2)$ hold in G, then it may be a difficult task to determine whether or not a given pair of vertices satisfies this formula. In the terminology of Section 4.5.2, $\psi(x_1, x_2)$ is a $\forall_4$ formula. If you find the formula $\psi(x_1, x_2)$ easy to comprehend, then consider a $\forall_{23}$ formula or a $\exists_{45}$ formula. In Chapter 9, we shall introduce a technique that helps us get a handle on such complicated formulas (pebble games). In the present section, we study a property that allows us to utterly avoid them.

Definition 5.42 A $\mathcal{V}$-theory T has *quantifier elimination* if every $\mathcal{V}$-formula $\varphi(x_1, \ldots, x_n)$ (for $n \in \mathbb{N}$) there exists a quantifier-free $\mathcal{V}$-formula $\psi(x_1, \ldots, x_n)$ such that $T \vdash \varphi(x_1, \ldots, x_n) \leftrightarrow \psi(x_1, \ldots, x_n)$.

Quantifier elimination is a purely syntactic property that greatly facilitates the study of certain mathematical structures. If a $\mathcal{V}$-theory has this property, then every $\mathcal{V}$-definable subset of every model is defined by a quantifier-free formula.

For example, suppose G is a graph that has quantifier elimination. Since all vertices of a graph satisfy the same quantifier-free formulas (namely $x = x$ and $\neg R(x, x)$), any $\mathcal{V}_R$-formula in one free variable either holds for all vertices or no vertices of G. For a pair of distinct vertices x_1 and x_2 of G, there are two possibilities: either $R(x_1, x_2)$ or $\neg R(x_1, x_2)$ holds in G. In particular, we can determine whether or not the above $\forall_4$ formula $\psi(x_1, x_2)$ holds merely by checking whether or not x_1 and x_2 share an edge. One of the following two sentences must be in the $\mathcal{V}_R$-theory of G:

$$\forall x_1 \forall x_2 (R(x_1, x_2) \to \psi(x_1, x_2)) \text{ or } \forall x_1 \forall x_2 (R(x_1, x_2) \to \neg\psi(x_1, x_2)).$$

Likewise, any graph having quantifier elimination must model either

$$\forall x_1 \forall x_2 (\neg R(x_1, x_2) \to \psi(x_1, x_2)) \text{ or } \forall x_1 \forall x_2 (\neg R(x_1, x_2) \to \neg\psi(x_1, x_2)).$$

The goal of this section is to formulate methods that determine whether or not a given complete theory has quantifier elimination.

5.5.1 Finite relational vocabularies. Let T be a complete theory, and suppose that we want to determine whether or not T has quantifier elimination. We make two initial observations.

- Theorem 4.49 provides a sufficient criterion for a formula to be T-equivalent to a quantifier-free formula.
- To show that T has quantifier elimination, it suffices to check this criterion only for existential formulas having only one occurences of "$\exists$."

We elaborate and verify the latter point.

Let T be a $\mathcal{V}$-theory. To show that T has quantifier elimination, we must show that $\varphi(\bar{x})$ is T-equivalent to a quantifier-free formula for every $\mathcal{V}$-formula

$\varphi(\bar{x})$ having at least one free variable. One way to do this is to proceed by induction on the complexity of $\varphi(\bar{x})$. If $\varphi(\bar{x})$ is quantifier-free, then there is nothing to show. Suppose that both ψ and θ are T-equivalent to quantifier-free formulas. If $\varphi(\bar{x})$ is T-equivalent to either of these formulas, their negations, or their conjunction, then $\varphi(\bar{x})$ is T-equivalent to a quantifier-free formula. Now suppose $\varphi(\bar{x})$ is equivalent to $\exists y \psi(\bar{x}, y)$. To show that T has quantifier elimination, it suffices to show that this formula is T-equivalent to a quantifier-free formula. In this way, the problem of showing that T has quantifier elimination reduces to the problem of showing that formulas of the form $\exists y \psi(\bar{x}, y)$ are T-equivalent to quantifier-free formulas.

Proposition 5.43 A $\mathcal{V}$-theory T has quantifier elimination if and only if for every quantifier-free $\mathcal{V}$-formula $\varphi(x_1, \ldots, x_n, y)$ (for $n \in \mathbb{N}$), there exists a quantifier-free $\mathcal{V}$-formula $\psi(x_1, \ldots, x_n)$ such that $T \vdash \exists y \varphi(x_1, \ldots, x_n, y) \leftrightarrow \psi(x_1, \ldots, x_n)$.

Proof Suppose that $\exists y \varphi(\bar{x}, y)$ is T-equivalent to a quantifier-free formula for every quantifier-free $\mathcal{V}$-formula φ having at least two free variables. Then we can show that every $\mathcal{V}$-formula θ is T-equivalent to a quantifier-free formula by induction on the complexity of θ as in the preceeding paragraph. Conversely, if T has quantifier elimination, then $\exists y \varphi(\bar{x}, y)$, like every $\mathcal{V}$-formula, is T equivalent to a quantifier-free formula. □

So to eliminate all of the quantifiers from a formula like

$$\forall y \exists z \forall u \exists v (R(x_1, y) \wedge R(y, z) \wedge R(z, u) \wedge R(u, v) \wedge R(v, x_2)),$$

we need only be able to eliminate one occurrence of the quantifier $\exists$ at a time. For complete T, Theorem 4.49 gives us a criterion for determining whether a given formula is T-equivalent to a quantifier free formula. This yields a method for showing quantifier elimination. We first consider theories that have finite relational vocabularies. These vocabularies are particularly simple because of the following fact.

Proposition 5.44 The vocabulary $\mathcal{V}$ is finite and relational if and only if there are only finitely many atomic $\mathcal{V}$-formulas.

Proof If $\mathcal{V}$ contains a function f, then we have the atomic formulas $f(\bar{x}) = y$, $f(f(\bar{x})) = y$, $f(f(f(\bar{x}))) = y$, and so forth. □

Proposition 5.45 Let T be a complete theory in a finite relational vocabulary. The following are equivalent.

(i) T has quantifier-elimination.
(ii) For any model M of T and any $n \in \mathbb{N}$, if $(a_1, \ldots, a_n)$ and $(b_1, \ldots, b_n)$ are n-tuples of U_M that satisfy the same atomic formulas in M,

then for any $a_{n+1} \in U_M$ there exists $b_{n+1} \in U_M$ such that $(a_1, \ldots, a_n, a_{n+1})$, and $(b_1, \ldots, b_n, b_{n+1})$ satisfy the same atomic formulas in M (where U_M denotes the underlying set of M).

Proof Suppose first that T has quantifier elimination.

If $(a_1, \ldots, a_n)$ and $(b_1, \ldots, b_n)$ satisfy the same atomic formulas, then they satisfy the same quantifier-free formulas. This can be shown by induction on the complexity of a given quantifier-free formula. Given $a_{n+1} \in U_M$, let $\Phi(x_1, \ldots, x_{n+1})$ be the conjunction of all of the atomic and negated atomic formulas that hold of $(a_1, \ldots, a_n, a_{n+1})$ in M. Such a formula Φ exists since there are only finitely many atomic formulas. By quantifier elimination, $\exists y \Phi(x_1, \ldots, x_n, y)$ is T-equivalent to a quantifier-free formula $\theta(x_1, \ldots, x_n)$. Since M models $\theta(a_1, \ldots, a_n)$, M also models $\theta(b_1, \ldots, b_n)$. It follows that $M \models \exists y \Phi(b_1, \ldots, b_n, y)$. Let $b_{n+1} \in U_M$ be such that $M \models \Phi(b_1, \ldots, b_n, b_{n+1})$.

Conversely, suppose that (ii) holds. Let $\varphi(x_1, \ldots, x_n, x_{n+1})$ be a quantifier-free formula (for $n \geq N$). By (ii), if $(a_1, \ldots, a_n)$ and $(b_1, \ldots, b_n)$ satisfy the same atomic formulas, then $M \models \exists y \varphi(a_1, \ldots, a_n, y)$ if and only if $M \models \exists y \varphi(b_1, \ldots, b_n, y)$. By Theorem 4.49, $\exists y \varphi(x_1, \ldots, x_n, y)$ is T-equivalent to a quantifier-free formula. By Proposition 5.43, T has quantifier elimination. □

Example 5.46 Recall the $\mathcal{V}_S$-structure $Z_S = (\mathbb{Z}|S)$ from Example 4.48. This structure interprets the binary relation S as the successor relation on the integers. As was pointed out in Example 4.48, the ordered pairs $(0, 2)$ and $(4, 7)$ satisfy the same atomic $\mathcal{V}_S$-formulas. Let $T_S = Th(Z_S)$. If T_S had quantifier elimination, then, by condition (ii) of Proposition 5.45, for every integer x there would exist an integer y such that $(0, 2, x)$ and $(4, 7, y)$ satisfy the same atomic formulas. To show that this is not the case, let $x = 1$. Then both $S(0, x)$ and $S(x, 2)$ hold. Clearly, there is no y that bears these relations to 4 and 7. We conclude that T_S does not have quantifier elimination. In particular, the formula $\exists z(S(x, z) \wedge S(z, y))$ is not T_S-equivalent to a quantifier-free formula (this was also shown in Example 4.48).

Example 5.47 We show that T_{DLOE} does not have quantifier elimination. The only two atomic $\mathcal{V}_<$-formulas are $x < y$ and $x = y$. Any two n-tuples of elements listed in ascending order will satisfy the same atomic formulas. In particular, this is true if $n = 1$. Consider the two elements 0 and 0.01 from the underlying set of $\mathbf{Q}_{[0,1]}$. These two elements satisfy the same atomic $\mathcal{V}_<$-formulas in $\mathbf{Q}_{[0,1]}$. However, since 0 is the smallest element, there is no y so that $(0, y)$ satisfies the same atomic formulas as $(0.01, 0.001)$. By Proposition 5.45, T_{DLOE} does not have quantifier elimination. In particular, the formula $\exists y(y < x)$ is not T_{DLOE}-equivalent to a quantifier-free formula.

Using condition (ii) of Proposition 5.45, we can quickly show that some theories do not have quantifier elimination as in the previous examples. To show that a theory T does have quantifier elimination, (ii) requires us to consider all pairs of tuples from all models of T. If T is complete and has a finite relational vocabulary, then this condition can be simplified as the following corollary states.

Corollary 5.48 Let T be a complete theory. If T has a finite relational vocabulary, then T has quantifier elimination if and only if condition (ii) from Proposition 5.45 holds in some model M of T.

Proof This follows from the assumption that T is complete. Let Φ and θ be as in the proof of Proposition 5.45. Let M be some model of T. If M models $\exists y \Phi(x_1, \ldots, x_n, y) \leftrightarrow \theta(x_1, \ldots, x_n)$, then, since T is complete, so does every model of T. □

Proposition 5.49 The theory T_{DLO} of dense linear orders without endpoints has quantifier elimination.

Proof By the previous corollary, it suffices to verify condition (ii) of Proposition 5.45 for only one model. Let $\mathbb{Q}_<$ be the $\mathcal{V}_<$-structure that interprets $<$ as the usual order on the rational numbers.

Let $\bar{a} = (a_1, \ldots, a_n)$ and $\bar{b} = (b_1, \ldots, b_n)$ be two n-tuples of rational numbers. With no loss of generality, we may assume that $a_1 < a_2 < \cdots < a_n$. Suppose that $\bar{a}$ and $\bar{b}$ satisfy the same atomic $\mathcal{V}_<$-formulas. Then we have $b_1 < b_2 < \cdots < b_n$. Let a_{n+1} be any rational number. We must show that there exists a rational number b_{n+1} so that $(b_1, \ldots, b_n, b_{n+1})$ satisfies the same atomic $\mathcal{V}_<$-formulas as $(a_1, \ldots, a_n, a_{n+1})$.

There are four cases:

If $a_{n+1} = a_i$ for some $i = 1, \ldots, n$, then we can just let $b_{n+1} = b_i$.

If $a_{n+1} < a_i$ for each i, then, since $\mathbb{Q}$ has no smallest element, we can find $b_{n+1} \in \mathbb{Q}$ that is smaller than each b_{n+1}.

Likewise, if a_{n+1} is greater than each a_i, then we can find $b_{n+1} \in \mathbb{Q}$ that is greater than each b_i.

Otherwise, $a_i < a_{n+1} < a_{i+1}$ for some i. Since $\mathbb{Q}$ is dense, we can find $b_{n+1} \in \mathbb{Q}$ between b_i and b_{i+1}.

In any case, we can find b_{n+1} as desired. So T_{DLO} has quantifier elimination by Corollary 5.48. □

Proposition 5.50 The theory T_{RG} of the random graph has quantifier elimination.

Proof Since T_{RG} has a finite relational vocabulary, we can apply Corollary 5.48. Let U be the set of vertices of the random graph G_R. Let $(a_1, \ldots, a_n)$ and $(b_1, \ldots, b_n)$ be two n-tuples of U that satisfy the same atomic formulas in G_R. We must show that for any $x \in U$, there exists $y \in U$ such that $(a_1, \ldots, a_n, x)$ and $(b_1, \ldots, b_n, y)$ satisfy the same atomic formulas in G_R. This follows from the fact that G_R models the $\mathcal{V}_R$-sentence ρ_m for arbitrarily large m. □

Example 5.51 Let $\mathcal{V}^+ = \{<, P_s, P_b\}$ be an expansion of $T_<$ that includes two unary relations P_s and P_b. Let T^+_{DLOE} be the expansion of T_{DLOE} to a $\mathcal{V}^+$-theory that interprets $P_s(x)$ as the smallest element and $P_b(x)$ as the biggest element in the order. Then this theory has quantifier elimination. Recall from Example 5.47 that T_{DLOE} does not have quantifier elimination. It was shown in that example that $\exists y(x < y)$ is not T_{DLOE}-equivalent to a quantifier-free formula. This formula is T^+_{DLOE}-equivalent to the quantifier-free formula $\neg P_b(x)$. To show that T^+_{DLOE} has quantifier elimination, we can use an argument similar to the proof of Proposition 5.49.

Example 5.52 Let T be an $\mathcal{V}_E$-theory that says E is an equivalence relation having infinitely many infinite classes. We show that T has quantifier elimination. Let M be the $\mathcal{V}_E$-structure having denumerably many equivalence classes each of which is denumerable. Let $(a_1, \ldots, a_n)$ and $(b_1, \ldots, b_n)$ be two n-tuples of elements from the universe U_M of M that satisfy the same atomic $\mathcal{V}_E$-formulas in M. Let a_{n+1} be any element of U_M. Since there are infinitely many equivalence classes in U_M and each is infinite, we can surely find $b_{n+1} \in U_M$ such that $M \models E(b_i, b_{n+1})$ if and only if $M \models E(a_i, a_{n+1})$.

So T has quantifier elimination. Now suppose that we expand the vocabulary by adding a unary relation P. Let T_P be the expansion of T to the vocabulary $\{<, P\}$ that says that $P(x)$ holds for exactly one element x. Let $N_1 \models T_P$. Let a be the unique element such that $N_1 \models P(a)$. Let b be an element that is equivalent, but not equal, to a. Let c be an element that is not equivalent to a. Then b and c satisfy the same atomic formulas in N_1 (since $\neg P(b)$ and $\neg P(c)$ both hold). Since there is no element y so that (b, a) and (c, y) satisfy the same atomic formulas, T_P does not have quantifier elimination. In particular, the formula $\exists y(P(y) \wedge E(x, y))$ is not T_P-equivalent to a quantifier-free formula.

Now expand the vocabulary again to include the constant u. Let $T_{P(u)}$ be the expansion of T_P to the vocabulary $\{<, P, u\}$ that interprets u as the unique element for which $P(u)$ holds. Since we have merely provided a name for an element that was already uniquely defined, $T_{P(u)}$ is essentially the same as T_P (the models of T_P can easily be viewed as models of $T_{P(u)}$ and vice versa). However, in contrast to T_P, $T_{P(u)}$ does have quantifier elimination. The formula $\exists y(P(y) \wedge E(x, y))$ is $T_{P(u)}$-equivalent to the atomic formula $E(x, u)$.

The previous examples demonstrate that quantifier elimination is a purely syntactic property. The theories T_{DLOE} and T^+_{DLOE} (like the theories T_P and $T_{P(u)}$) have very similar models. Given any model M of one of these theories, we can find a model N of the other theory so that M and N have the same underlying set and the same definable subsets.

Definition 5.53 Let M_1 be a $\mathcal{V}_1$-structure and M_2 be a $\mathcal{V}_2$-structure having the same underlying set. If the $\mathcal{V}_1$-definable subsets of M_1 are the same as the $\mathcal{V}_2$-definable subsets of M_2 then M_1 and M_2 are said to be *bi-definable*.

Two theories T_1 and T_2 are *bi-definable* if every model of T_1 is bi-definable with some model of T_2 and vice versa.

The theories T_{DLOE} and T^+_{DLOE} are bi-definable as are T_P and $T_{P(u)}$. However, T^+_{DLOE} and $T_{P(u)}$ have quantifier elimination whereas T_{DLOE} and T_P do not. The following proposition states that any theory has a bi-definable theory with quantifier elimination.

Proposition 5.54 Let T be a $\mathcal{V}$-theory. There exists a theory T_m that is bi-definable with T and has quantifier elimination.

Proof For each $\mathcal{V}$-formula $\varphi(x_1,\ldots,x_n)$ (with $n \in \mathbb{N}$), let R_φ be an n-ary relation that is not in $\mathcal{V}$. Let $\mathcal{V}_m = \mathcal{V} \cup \{R_\varphi | \varphi \text{ is a } \mathcal{V}\text{-formula}\}$. Let T_m be the expansion of T to a $\mathcal{V}_m$-theory that contains the sentence $\varphi(\bar{x}) \leftrightarrow R_\varphi(\bar{x})$ for each $\mathcal{V}$-formula φ. Since each relation R_φ is explicitly defined by T_m in terms of $\mathcal{V}$, T, and T_m are bi-definable. Since every $\mathcal{V}_m$-formula is T_m-equivalent to a quantifier-free formula, T_m has quantifier elimination. □

The theory T_m in the previous proof is called the *Morleyization* of T. Morleyizations demonstrate that the property of quantifier elimination is not always useful. To analyze a structure M, we should choose an appropriate vocabulary. Ideally, we want to find a vocabulary $\mathcal{V}'$ so that M is bi-definable with some $\mathcal{V}'$-structure M' where $Th(M')$ has quantifier elimination. We must also require that the atomic $\mathcal{V}'$-formulas (and the relations between these formulas) are readily understood. The Morleyization of T_m is often of no use in this regard. If the atomic $\mathcal{V}'$-formulas are too complicated, then the quantifier elimination of $Th(M')$ does not lend insight into the structure M.

We have restricted our attention in this section to examples of theories that are particularly nice. With the exception of T_S from Example 5.46, each theory we have considered is bi-definable with a theory in a finite relational vocabulary that has quantifier elimination. This is a severe restriction. The astute reader may have anticipated the following fact.

Proposition 5.55 Let T be a complete theory in a finite relational vocabulary. If T has quantifier elimination, then T is $\aleph_0$-categorical.

228   First-order theories

Proof This follows immediately from Proposition 5.44 and Theorem 5.32. □

Corollary 5.48 provides a method for determining whether or not certain theories have quantifier elimination. By the previous proposition, this method can only be used to show that $\aleph_0$-categorical theories have quantifier elimination. To show that other theories have quantifier elimination, we must devise other methods.

5.5.2 The general case. Let T be a complete $\mathcal{V}$-theory, and suppose that we want to determine whether or not T has quantifier elimination. We have described a method that is useful for finite relational $\mathcal{V}$. If $\mathcal{V}$ is not both finite and relational, then this method may fail in one of two ways. We demonstrate these failures with two examples.

Example 5.56 Let $T_E = Th(M_2)$ where M_2 is the countable $\mathcal{V}_E$-structure defined in Example 5.18.

Let $T_s = Th(Z_s)$ where Z_s is the $\mathcal{V}_s$-structure defined in Example 5.12.

We define a theory T that contains both of these theories. Let $\mathcal{V}$ be the vocabulary $\{E, s\}$. Let T be the set of all $\mathcal{V}$-sentences that can be derived from the set

$$T_E \cup T_s \cup \{\forall x \forall y ((s(x) = y \to E(x, y))\}.$$

The models of T have two infinite equivalence classes. The sentence $\forall x \forall y ((s(x) = y \to E(x, y))$ implies that every element is in the same equivalence class as its successor, its successor's successor, and so forth. So each equivalence class contains copies of the structure $(\mathbb{Z}, s)$. As in Example 5.20, each equivalence class may contain any number of copies of $\mathbb{Z}$.

Let M be the model that has two copies of $\mathbb{Z}$ in one equivalence class and one copy in the other. This structure can be depicted as follows:

```
┌─────────────────────────────────────────────────────────┐
│                                                         │
│        (...a_{-2}, a_{-1}, a_0, a_1, a_2, ...)          │
│                                                         │
├─────────────────────────────────────────────────────────┤
│                                                         │
│   (..., b_{-2}, b_{-1}, b_0, b_1, b_2, ...)   (...c_{-2}, c_{-1}, c_0, c_1, c_2, ...)  │
│                                                         │
└─────────────────────────────────────────────────────────┘
```

The underlying set of M is $U_M = \{a_i, b_i, c_i | i \in \mathbb{Z}\}$. The two boxes represent the two E equivalence classes. The successor of a_i is a_{i+1}. Likewise for b_i and c_i.

The elements a_0 and b_0 satisfy the same atomic formulas in M (as does each element of U_M). However, there is no element y so that (a_0, y) and (b_0, c_0) satisfy the same atomic formulas.

If the vocabulary of T were finite and relational, then we could conclude that it does not have quantifier elimination. However, the vocabulary of T contains the function s. In fact, T is both complete and has quantifier elimination.

Example 5.57 Let $\mathcal{V} = \{E, P_i | i \in \mathbb{N}\}$ be the vocabulary consisting of a binary relation E and denumerably many unary relations P_i. Let M be a countable $\mathcal{V}$-structure that interprets E as an equivalence relation that has infinitely many equivalence classes of size 1, infinitely many equivalence classes of size 2, and no other equivalence classes. Moreover, each P_i holds for exactly two elements that are in the same equivalence class. So if a is in a class of size 1, then $\neg P_i(a)$ holds for each $i \in \mathbb{N}$. To complete our description of M, if $M \models E(b, c) \land \neg(b = c)$ then $M \models P_i(b) \land P_i(c)$ for exactly one i.

Let U_M be the underlying set of M. If two tuples of elements from U_M satisfy the same atomic $\mathcal{V}$-formulas in M, then they satisfy the same $\mathcal{V}$-formulas in M. However, $Th(M)$ does not have quantifier elimination. The formula $\exists y(\neg(x = y) \land E(x, y))$ is not $Th(M)$-equivalent to a quantifier-free formula.

Corollary 5.48 provides a necessary and sufficient condition for a complete theory T in a finite relational vocabulary to have quantifier elimination. We restate this condition.

(ii) For any model M of T and any $n \in \mathbb{N}$, if $(a_1, \ldots, a_n)$ and $(b_1, \ldots, b_n)$ are n-tuples of U_M that satisfy the same atomic formulas in M, then for any $a_{n+1} \in U_M$ there exists $b_{n+1} \in U_M$ such that $(a_1, \ldots, a_n, a_{n+1})$ and $(b_1, \ldots, b_n, b_{n+1})$ satisfy the same atomic formulas in M (where U_M denotes the underlying set of M).

Corollary 5.48 states that, if the vocabulary $\mathcal{V}$ of T is finite and relational, then T has quantifier elimination if and only if condition (ii) holds for some model M of T. If $\mathcal{V}$ is not finite and relational, then, as Example 5.57 demonstrates, we must verify (ii) for more than one model. So Corollary 5.48 fails for vocabularies that are not finite and relational. Example 5.56 demonstrates that one direction of Proposition 5.45 also fails for vocabularies that are not finite and relational. Condition (ii) does not necessarily hold for all theories that have quantifier elimination.

It is still true that (ii) implies quantifier elimination. The proof that (ii) implies (i) in Proposition 5.45 makes no use of the hypothesis that the vocabulary is finite and relational. So (ii) is a sufficient condition for quantifier elimination, but, if the vocabulary is not finite and relational, it is not a necessary condition. The following Proposition provides two necessary and sufficient conditions for an arbitrary theory to have quantifier elimination. Note that $(ii)'$ is a modified version of condition (ii).

Proposition 5.58 Let T be a complete $\mathcal{V}$-theory. The following are equivalent:

(i) T has quantifier elimination.

(ii)' For any model M of T and any $n \in \mathbb{N}$, if $\bar{a}$ and $\bar{b}$ satisfy the same atomic formulas in M, then for any $a_{n+1} \in U_M$ there exist an elementary extension N of M and an element b_{n+1} in U_N such that $(a_1, \ldots, a_n, a_{n+1})$ and $(b_1, \ldots, b_n, b_{n+1})$ satisfy the same atomic formulas in N
(where U_M and U_N denote the underlying sets of M and N).

(iii) For any $\mathcal{V}$-structure C, if $f : C \to M$ and $g : C \to M$ are two embeddings of C into a model M of T, then $M \models \exists y \varphi(f(c_1), \ldots, f(c_n), y)$ if and only if $M \models \exists y \varphi(g(c_1), \ldots, g(c_n), y)$
for any quantifier-free $\mathcal{V}$-formula $\varphi(x_1, \ldots, x_n, y)$ and any n-tuple $(c_1, \ldots, c_n)$ of elements from the underlying set of C.

Proof By modifying the proof that (i) implies (ii) in Proposition 5.45, it can be shown that (i) implies (ii)'. We leave this as Exercise 5.18.

That (ii)' implies (iii) follows from the fact that the tuples $(f(c_1), \ldots, f(c_n))$ and $(g(c_1), \ldots, g(c_n))$ satisfy the same quantifier-free formulas in M (by the definition of "embedding").

It remains to be shown that (iii) implies (i). We assume that (i) does not hold and show that (iii) does not hold.

Suppose that T does not have quantifier elimination. By Proposition 5.43, there exists a quantifier-free $\mathcal{V}$-formula $\varphi(x_1, \ldots, x_n, y)$ such that $\exists y \varphi(x_1, \ldots, x_n, y)$ is not T-equivalent to a quantifier-free formula. By Theorem 4.49, there exists a model M of T and n-tuples $\bar{a} = (a_1, \ldots, a_n)$ and $\bar{b} = (b_1, \ldots, b_n)$ from the universe U_M of M such that $\bar{a}$ and $\bar{b}$ satisfy the same atomic $\mathcal{V}$-formulas in M but

$$M \models \exists y \varphi(\bar{a}, y) \quad \text{and} \quad M \models \neg \exists y \varphi(\bar{b}, y).$$

Now if $\{a_1, \ldots, a_n\}$ happens to be the universe of a substructure of M, then we can take this to be C in (iii). Otherwise, we must consider the substructure $\langle \bar{a} \rangle$ of M *generated* by $\bar{a}$ (as defined in Exercise 2.34). This is the smallest substructure of M that contains $\{a_1, \ldots, a_n\}$. Likewise, let $\langle \bar{b} \rangle$ be the smallest substructure of M that contains $\{b_1, \ldots, b_n\}$.

Claim $\langle \bar{a} \rangle \cong \langle \bar{b} \rangle$.

First-order theories 231

Proof By Exercise 2.34, this claim follows from the fact that $\bar{a}$ and $\bar{b}$ satisfy the same atomic formulas (and, therefore, the same quantifier-free formulas) in M. For readers who have not completed this exercise, we sketch the idea.

Let A_0 be the union of $\{a_1, \ldots, a_n\}$ together with the set of all elements in U_M that interpret constants of $\mathcal{V}$. Let A be the closure of A_0 under all functions in $\mathcal{V}$. Then $\langle \bar{a} \rangle$ is the substructure of M that has A as an underlying set. The key point is that each element in $\langle \bar{a} \rangle$ can be represented by a quantifier-free $\mathcal{V}$-term having parameters among $\{a_1, \ldots, a_n\}$. Let f be the function defined by $g(a_i) = b_i$ for $i = 1, \ldots, n$. Since $\bar{a}$ and $\bar{b}$ satisfy the same quantifier-free formulas, and the elements of $\langle \bar{a} \rangle$ and $\langle \bar{b} \rangle$ can be expressed with quantifier-free terms, f can be extended to an isomorphism $g: \langle \bar{a} \rangle \to \langle \bar{b} \rangle$.

To see that (iii) does not hold, let $C = \langle \bar{a} \rangle$, let $f: C \to \langle \bar{a} \rangle$ be the identity function, and let $g: C \to \langle \bar{b} \rangle$ be as defined above. □

Let T be a complete theory. To determine whether or not T has quantifier elimination we can use either condition (ii)' or (iii) from the previous proposition. However, depending on how much information we have regarding T, verifying these conditions may or may not be practical. It may not be easy to consider arbitrary elementary extensions in (ii)' or arbitrary substructures in (iii).

In specific cases, when T is known to have certain properties, there are methods for determining quantifier elimination that are easier than (ii)' and (iii). For example, if T has a finite relational vocabulary, then, as we have discussed, it suffices to consider property (ii). If T is a small theory, then, regardless of whether the vocabulary is finite or relational, we only need to consider condition (ii) for the countable saturated model of T. However, this fact will not be immediately useful to those who are not reading this book backwards. *Small theories* and *saturated model* are defined and discussed in Chapter 6 (see Exercise 6.17). Another property that allows for a practical method for determining quantifier elimination is the *isomorphism property*.

Definition 5.59 We say that a structure M has the *isomorphism property* if any isomorphism between substructures of M can be extended to an isomorphism between submodels of M. If every model of T has the isomorphism property, then T is said to have the *isomorphism property*.

Example 5.60 Recall the $\mathcal{V}_S$-structure $\mathbb{Z}_S$ and the theory $T_S = Th(\mathbb{Z}_S)$ from Examples 4.48 and 5.46. Since $\mathcal{V}_S$ is relational, any subset of $\mathbb{Z}$ serves as the underlying set of a substructure of $\mathbb{Z}_S$. Let M_1 be the substructure having universe $\{0, 2\}$ and let M_2 be the substructure having universe $\{4, 7\}$. Since both of these structures model the sentence $\forall x \forall y \neg S(x, y)$, M_1 and M_2 are isomorphic. This isomorphism cannot be extended to submodels of $\mathbb{Z}_S$. So the theory T_S does not have the isomorphism property.

Example 5.61 Recall the $\mathcal{V}_s$-structure Z_s and the theory $T_s = Th(Z_s)$ from Examples 5.12 and 5.20. This theory is bi-definable with the $\mathcal{V}_S$-theory T_S from the previous example. We show that T_s, unlike T_S, has the isomorphism property. Let A and B be substructures of a model M of T_s. Since substructures must be closed under the function s, the underlying set of each of these substructures is a union of sets of the form $\{a, s(a), s(s(a)), \ldots\}$. With no loss of generality, we may assume that $\{a, s(a), s(s(a)), \ldots\}$ is the underlying set of A and $\{b, s(b), s(s(b)), \ldots\}$ is the underlying set of B. There is exactly one isomorphism between these $\mathcal{V}_s$-structures. This isomorphism can be extended to an isomorphism between submodels of M by mapping the predecessor of a to the predecessor of b, and so forth.

If T has the isomorphism property, then we can simplify condition (iii) of Proposition 5.62. Instead of dealing with substructures, we can focus on submodels of models of T.

Proposition 5.62 Let T be a complete $\mathcal{V}$-theory that has the isomorphism property. The following are equivalent:

(a) T has quantifier elimination.

(b) For any quantifier-free $\mathcal{V}$-formula $\varphi(x_1, \ldots, x_n, y)$ and any models M and N of T with $N \subset M$,

if $M \models \exists y \varphi(\bar{a}, y)$, then $N \models \exists y \varphi(\bar{a}, y)$

for any n-tuple $\bar{a}$ from the universe of N.

Proof If (a) holds, then, since quantifier-free formulas are preserved under submodels, (b) holds. We must prove the opposite direction. Suppose that (b) holds. We show that (iii) from Proposition 5.58 holds.

Let C be a $\mathcal{V}$-structure C and let $f : C \to M$ and $g : C \to M$ be two embeddings of C into a model M of T. Let C_1 be the range of f and let C_2 be the range of g. Then C_1 and C_2 are isomorphic substructures of M. Since T has the isomorphism property, the isomorphism $g(f^{-1}) : C_1 \to C_2$ extends to an isomorphism $h : N_1 \to N_2$ between submodels N_1 and N_2 of M. To verify (iii), we must show that for any quantifier-free $\mathcal{V}$-formula φ

$$M \models \exists y \varphi(f(c_1), \ldots, f(c_n), y) \text{ if and only if } M \models \exists y \varphi(g(c_1), \ldots, g(c_n), y)$$

for any n-tuple $(c_1, \ldots, c_n)$ of elements from the universe of C. Since existential formulas are preserved under supermodels,

if $N_1 \models \exists y \varphi(f(c_1), \ldots, f(c_n), y)$, then $M \models \exists y \varphi(f(c_1), \ldots, f(c_n), y)$.

Condition (b) provides the converse:

if $M \models \exists y \varphi(f(c_1), \ldots, f(c_n), y)$, then $N_1 \models \exists y \varphi(f(c_1), \ldots, f(c_n), y)$.

So we have

$$M \models \exists y \varphi(f(c_1), \ldots, f(c_n), y) \text{ if and only if } N_1 \models \exists y \varphi(f(c_1), \ldots, f(c_n), y).$$

Likewise,

$$M \models \exists y \varphi(g(c_1), \ldots, g(c_n), y) \text{ if and only if } N_2 \models \exists y \varphi(g(c_1), \ldots, g(c_n), y).$$

Since $h: N_1 \to N_2$ is an isomorphism that extends $g(f^{-1})$,

$$N_1 \models \exists y \varphi(f(c_1), \ldots, f(c_n), y) \text{ if and only if } N_2 \models \exists y \varphi(g(c_1), \ldots, g(c_n), y).$$

We conclude

$$M \models \exists y \varphi(f(c_1), \ldots, f(c_n), y) \text{ if and only if } M \models \exists y \varphi(g(c_1), \ldots, g(c_n), y)$$

as desired. □

Proposition 5.63 T_s has quantifier elimination.

Proof Since T_s has the isomorphism property, it suffices to verify condition (b) of Proposition 5.62. Let $\varphi(x_1, \ldots, x_n, y)$ be a quantifier-free $\mathcal{V}_s$-formula. Let $N \subset M$ be models of T_s and suppose $N \models \neg \exists y \varphi(\bar{a}, y)$ for some n-tuple $\bar{a}$ of elements from the underlying set of N. Any elementary extension of N also models $\neg \exists y \varphi(\bar{a}, y)$. Let N' be an elementary extension of N that contains many copies of $\mathbb{Z}$. Then $N' \models \neg \varphi(\bar{a}, b)$ for some b that is not in the same copy of $\mathbb{Z}$ as any of the a_is. Since any two such bs bear the same atomic relations to each a_i (namely $\neg s^m(a_i) = b$ and $\neg s^m(b) = a_i$ for $m \in \mathbb{N}$) any extension of N must model $\neg \exists y \varphi(\bar{a}, y)$. In particular, $M \models \neg \exists y \varphi(\bar{a}, y)$. This verifies condition (b). We conclude that T_s has quantifier elimination. □

5.6 Model-completeness

In this section, we discuss a property closely related to quantifier elimination. As we shall see, there are many equivalent ways to define this property. The following is the standard definition.

Definition 5.64 A theory T is *model-complete* if, for any models M and N of T, $N \subset M$ implies $N \prec M$.

Example 5.65 Let $N_s = \{\mathbb{N}|s\}$ be the $\mathcal{V}_s$-structure that interprets the binary relation s as the successor function on the natural numbers. Let N_4 be the substructure of N_s having underlying set $\{4, 5, 6, \ldots\}$. Then $N_4 \equiv N_s$ and $N_4 \subset N_s$. Since $N_s \models \exists x(s(x) = 4)$ and $N_4 \models \neg \exists x(s(x) = 4)$, N_4 is not an elementary substructure of N_s. It follows that $Th(N_s)$ is not model-complete. If we expand N_s to include a constant for the first element, then we obtain a structure that does have a model-complete theory.

The following proposition perhaps explains why this property it is called "model-complete."

Proposition 5.66 A theory T is model-complete if and only if, for any model M of T, $T \cup \mathcal{D}(M)$ is complete.

Proof Exercise 5.24. □

The Tarski–Vaught criterion 4.31 for elementary substructures yields the following criterion for model-completeness.

Proposition 5.67 Let T be a $\mathcal{V}$-theory. The following are equivalent:

(i) T is model-complete.

(ii) For any models M and N of T with $N \subset M$,

$$M \models \exists y \psi(\bar{a}, y) \text{ implies } N \models \exists y \psi(\bar{a}, y)$$

for any $\mathcal{V}$-formula $\psi(\bar{x}, y)$ and any tuple $\bar{a}$ of elements from the universe of N.

Proof It follows immediately from the definition of "model-complete" that (i) implies (ii). The converse follows from the Tarski–Vaught criterion 4.31. □

The following proposition shows that, in some sense, model-complete theories "almost" have quantifier elimination.

Proposition 5.68 Let T be a $\mathcal{V}$-theory. The following are equivalent:

(i) T is model-complete.

(ii) Every $\mathcal{V}$-formula is preserved under submodels and supermodels of T.

(iii) Every $\mathcal{V}$-formula is T-equivalent to an existential formula.

(iv) Every $\mathcal{V}$-formula is T-equivalent to a universal formula.

Proof

(i) implies (ii) by the definition of "model-complete."

(ii) implies (iii) by Proposition 4.45.

Suppose that (iii) holds. Let $\varphi(\bar{x})$ be a $\mathcal{V}$-formula. By (iii), $\neg \varphi(\bar{x})$ is T-equivalent to an existential formula. It follows that $\varphi(\bar{x})$ is T-equivalent to a universal formula. So (iii) implies (iv).

Finally, suppose that (iv) holds. Suppose that M and N are models of T and $N \subset M$. Let $\psi(\bar{x}, y)$ be a $\mathcal{V}$-formula and let $\bar{a}$ be a tuple of elements from the underlying set of N. Since $\exists y \psi(\bar{x}, y)$ is T-equivalent to a universal formula and universal formulas are preserved under submodels,

$$\text{if } M \models \exists y \psi(\bar{a}, y), \quad \text{then} \quad N \models \exists y \psi(\bar{a}, y).$$

By Proposition 5.67, T is model-complete. □

It follows from Proposition 5.68 that every theory with quantifier elimination is an example of a model-complete theory. We next demonstrate some examples that do not have quantifier elimination.

Example 5.69 Let T_S be the $\mathcal{V}_S$-theory from Examples 4.48 and 5.46. Recall that every $\mathcal{V}_S$-formula is preserved under submodels and supermodels of T_S. By Proposition 5.68, T_S is model-complete. As was shown in Example 5.46, T_S does not have quantifier elimination.

Example 5.70 Recall the $\mathcal{V}_<$-theory T_{DLOE} of dense linear orders with endpoints. As was shown in Example 5.47, T_{DLOE} does not have quantifier elimination. As was shown in Example 5.51, the expansion T_{DLOE}^+ of T_{DLOE} to the vocabulary $\{<, P_s, P_b\}$ does have quantifier elimination. Recall that T_{DLOE}^+ defines the unary relations P_s and P_b as follows:

$$T_{DLOE}^+ \models P_s(x) \leftrightarrow \forall y \neg(y < x) \quad \text{and} \quad T_{DLOE}^+ \models P_b(x) \leftrightarrow \forall y \neg(x < y).$$

Since both $P_s(x)$ and $P_b(x)$ are T_{DLOE}^+-equivalent to universal $\mathcal{V}_<$-formulas and T_{DLOE}^+ has quantifier elimination, it follows that every $\mathcal{V}$-formula is T_{DLOE}-equivalent to a universal formula. By Proposition 5.68, T_{DLOE} is model-complete.

Example 5.71 We state, but do not verify, some facts regarding the real numbers. We refer the reader to [16] or [29] for proofs of these facts. Let $\mathbf{R} = (\mathbb{R}|+, \cdot, 0, 1)$. Let $\mathbf{R}_{or}$ be the expansion of $\mathbf{R}$ to the vocabulary $\{<, +, \cdot, 0, 1\}$. Then $Th(\mathbf{R}_{or})$ has quantifier elimination and $Th(\mathbf{R})$ does not. As was shown in Example 4.67, the relation $<$ is explicitly defined by $Th(\mathbf{R}_{or})$ as $\exists z(x+(z \cdot z) = y)$. It follows that every formula in the vocabulary $\{+, \cdot, 0, 1\}$ is $Th(\mathbf{R})$-equivalent to an existential formula. By Proposition 5.68, $Th(\mathbf{R})$ is model-complete.

Definition 5.72 Let T be a $\mathcal{V}$-theory and let $M \models T$. We say that M is *existentially closed with respect to* T if, for any model N of T with $M \subset N$

$$\text{if } N \models \varphi(\bar{a}) \quad \text{then} \quad M \models \varphi(\bar{a})$$

for any existential $\mathcal{V}$-formula $\varphi(\bar{x})$ and any tuple $\bar{a}$ of elements from the universe of M.

Proposition 5.73 A theory T is model-complete if and only if every model of T is existentially closed with respect to T.

Proof It follows from the definitions that any model of a model-complete theory T is existentially closed with respect to T. We must prove the converse.

Suppose every model of T is existentially closed with respect to T. Let $\mathcal{V}$ be the vocabulary of T. To show that T is model-complete, we show that every

$\mathcal{V}$-formula is T-equivalent to a universal formula. It suffices to show that every existential $\mathcal{V}$-formula is T-equivalent to a universal formula (see Exercise 3.2).

Let $\varphi(\bar{x})$ be an existential $\mathcal{V}$-formula. If M and N are models of T with $M \subset N$, then, since M is existentially closed with respect to T, $N \models \varphi(\bar{a})$ implies $M \models \varphi(\bar{a})$ for any tuple $\bar{a}$ of elements from the universe of M. That is, $\varphi(\bar{x})$ is preserved under submodels of T. By Theorem 4.47, $\varphi(\bar{x})$ is T-equivalent to a universal $\mathcal{V}$-formula. By Exercise 3.2, every $\mathcal{V}$-formula is T-equivalent to a universal formula. By Proposition 5.68, T is model-complete. □

Proposition 5.73 improves Proposition 5.67. Instead of verifying condition (ii) of Proposition 5.67 for every $\mathcal{V}$-formula ψ, it suffices to verify this condition only for existential ψ.

Definition 5.74 A theory is said to be $\forall_2$-axiomatizable if it has an axiomatization consisting of $\forall_2$ sentences.

Proposition 5.75 If T is model-complete, then T is $\forall_2$-axiomatizable.

Proof Exercise 4.30. □

Lindström's theorem states that if T is κ-categorical for some κ, then the converse of Proposition 5.75 holds. To prove Lindström's theorem, we shall use the following result.

Proposition 5.76 If T is $\forall_2$-axiomatizable, then any model of T can be extended to a model that is existentially closed with respect to T.

Proof Let M be a model of T having underlying set U. We assume that both T and M are denumerable. For uncountable T or M, the proof is similar.

Let $\mathcal{V}$ be the vocabulary of T. Let $\mathcal{E}$ be the set of existential formulas having parameters from U and no free variables. That is, $\mathcal{E}$ consists of formulas of the form $\exists \bar{x} \varphi(\bar{x}, \bar{a})$ where φ is a quantifier-free $\mathcal{V}$-formula and $\bar{a}$ is a tuple of elements from U. Since both U and $\mathcal{V}$ are countable, so is $\mathcal{E}$. Enumerate $\mathcal{E}$ as $\{\theta_1, \theta_2, \theta_3, \ldots\}$. (In the case where M is uncountable, invoke the Well Ordering Principle.)

We inductively define a sequence of $\mathcal{V}$-structures as follows.

Let $A_0 = M$.

Suppose now that A_n has been defined. To define A_{n+1}, consider the formula θ_{n+1} in the enumeration of $\mathcal{E}$. Let A_{n+1} be any extension of A_n that models $T \cup \theta_{n+1}$. If no such extension exists, then just let $A_{n+1} = A_n$. (In the case where M is uncountable, let $A_\alpha = \bigcup_{\beta < \alpha} A_\beta$ for limit ordinals α.)

Let B_1 be the union of the chain $M = A_0 \subset A_1 \subset A_2 \subset \cdots$. Since each A_i models T and $\forall_2$ sentences are preserved under unions, B_1 is a model of T. We claim that, for any extention N of B_1 that models T, if $N \models \theta_i$, then $B_1 \models \theta_i$ for each $\theta_i \in \mathcal{E}$. If such an N exists, then $A_{i+1} \models \theta_i$. If this is the case, then $B_1 \models \theta_i$ since $A_{i+1} \subset B_1$ and existential formulas are preserved under supermodels.

To obtain an existentially closed extension of M, we must repeat this process. Given B_i, for some $i \in \mathbb{N}$, construct B_{i+1} in the same way that B_1 was constructed (with B_1 playing the role of M). Let $\varphi(\bar{x})$ be an existential $\mathcal{V}$-formula and let $\bar{b}$ be a tuple of elements from the universe of B_i. If B_i has an extension that models $T \cup \varphi(\bar{b})$, then B_{i+1} models $\varphi(\bar{b})$.

Let M_E be the union of the chain $B_1 \subset B_2 \subset \cdots$. Since T is $\forall_2$-axiomatizable, M_E models T. Let $\varphi(\bar{x})$ be an existential $\mathcal{V}$-formula and let $\bar{b}$ be a tuple of elements from the universe of M_E. Then $\bar{b}$ is a tuple from the universe of B_i for some i. If M_E has an extension that models $T \cup \varphi(\bar{b})$, then B_{i+1} models $\varphi(\bar{b})$. Since existential formulas are preserved under supermodels, $M_E \models \varphi(\bar{b})$. This shows that M_E is existentially closed with respect to T. □

Theorem 5.77 (Lindström) Let T be a κ-categorical for some $\kappa \geq |T|$. If T is $\forall_2$-axiomatizable, then T is model-complete.

Proof Suppose that T is $\forall_2$-axiomatizable and not model-complete. We show that T is not κ-categorical.

Let $\mathcal{V}$ be the vocabulary of T.

If T is not model-complete, then there exists a model M of T that is not existentially closed with respect to T (by Proposition 5.73). So there exists an extension N of M and an existential $\mathcal{V}$-formula $\varphi(\bar{x})$ such that $N \models T \cup \varphi(a_1, \ldots, a_n)$ and $M \models \neg\varphi(a_1, \ldots, a_n)$ for some n-tuple $(a_1, \ldots, a_n)$ of elements from the universe of M.

Let $\mathcal{V}' = \mathcal{V} \cup \{c_1, \ldots, c_n\}$ where each c_i is a constant not in $\mathcal{V}$. Let M' be the expansion of M to a $\mathcal{V}'$-structure that interprets each c_i as the element a_i. By Proposition 5.76, there exists an extension M'_1 of M' that is existentially closed with respect to $Th(M')$. By this same proposition, there exists an extension M_1 of M that is existentially closed with respect to T. Since it is existentially closed, $M_1 \models \varphi(\bar{a})$. Since $M'_1 \models Th(M')$, $M'_1 \models \neg\varphi(\bar{a})$.

By the Downward Löwenhiem–Skolem theorem, there exist $\mathcal{V}$-structure M_0 and $\mathcal{V}'$-structure M'_0 both of size $|T|$ such that $M_0 \prec M_1$ and $M'_0 \prec M'_1$. By the Upward Löwenhiem–Skolem theorem, there exist $\mathcal{V}$-structure M_2 and $\mathcal{V}'$-structure M'_2 both of size κ such that $M_0 \prec M_2$ and $M'_0 \prec M'_2$. Since M'_2 models $\neg\varphi(\bar{a})$, the reduct of M'_2 to $\mathcal{V}$ is not existentially closed. This reduct along with M_2 are two models of T of size κ. Since one is existentially closed and the other is not, they cannot be isomorphic and T is not κ-categorical as we wanted to show. □

Example 5.78 We demonstrate a countable complete theory T_L that is $\forall_2$-axiomatizable but not model-complete. By Lindström's theorem, T_L cannot be κ-categorical for any κ. To find such T_L, we expand upon Example 5.65. Recall that $Th(N_s)$ from Example 5.65 is not model-complete. This theory is also not $\forall_2$-axiomatizable. We cannot say that there exists an element with no predecessor using a $\forall_2$ $\mathcal{V}_s$-sentence. To express this with a $\forall_2$ sentence, we can expand to the vocabulary $\{s, 1\}$. Let $N_{s1} = \{\mathbb{N}|s, 1\}$ be the expansion of N_s that interprets 1 as 1. Then $Th(N_{s1})$ is $\forall_2$-axiomatizable. However, as was pointed out in Example 5.65, this theory is also model-complete.

We now define a structure N_L that contains infinitely many copies of N_{s1}. The vocabulary for N_L is $\mathcal{V}_L = \{r, u, c_i | i \in \mathbb{Z}\}$ containing two unary functions r and u and a denumerable set of constants. The underlying set of N_L is $\mathbb{N} \times \mathbb{Z}$. Each constant c_i is interpreted as $(0, i)$ in N_L. The functions are interpreted as follows. For each $(a, b) \in \mathbb{N} \times \mathbb{Z}$, N_L interprets $r(a, b)$ as $(a+1, b)$ and $u(a, b)$ as $(a, b+1)$. If we visualize N_L in a plane with $\mathbb{N}$ as a horizontal axis and $\mathbb{Z}$ as a vertical axis, then N_L interprets r as the "right-successor" and u as the "up-successor."

Let T_L be $Th(N_L)$.

There exist infinitely many elements of N_L that are not the right-successors of any element. Each is named by a constant. By compactness there exists an elementary extension N of N_L that has elements with no right-predecessor that are not named by constants. For the same reason that $Th(N_s)$ is not model-complete, T_L is not model-complete. Moreover, T_L is complete and $\forall_2$-axiomatizable. We leave the verification of these facts to the reader.

We conclude this section by providing methods for showing quantifier elimination that involve model-completeness.

Proposition 5.79 If T has the isomorphism property, then T is model-complete if and only if T has quantifier elimination.

Proof This follows immediately from Proposition 5.62 and the definition of "model-complete." □

For any theory T, let $T_\forall$ denote the set of universal sentences that can be derived from T. By Theorem 4.47, a sentence φ is T-equivalent to some sentence in $T_\forall$ if and only if φ is preserved under substructures of models of T. It follows that the models of $T_\forall$ are precisely the substructures of models of T (see Exercise 4.25).

Definition 5.80 A theory T has the *amalgamation property* if the following holds. For any models A, B, and C of T and embeddings $f_a : C \to A$ and $f_b : C \to B$, there exists a model D and embeddings $g_a : A \to D$ and $g_b : B \to D$ such that $g_a(f_a(c)) = g_b(f_b(c))$ for each c in the underlying set of C.

Proposition 5.81 If T is model-complete and $T_\forall$ has the amalgamation property, then T has quantifier elimination.

Proof To show that T has quantifier elimination we verify condition (iii) of Proposition 5.58. Let $M \models T$ and $C \models T_\forall$. Let $f : C \to M$ and $g : C \to M$

be two embeddings of C into M. Since $T_\forall$ has amalgamation, there exists a model D of $T_\forall$ and embeddings $f' : M \to D$ and $g' : M \to D$ such that $f'(f(c)) = g'(g(c))$ for each c in the underlying set of C. Since D models $T_\forall$, there exists an extension N of D that models T (by Exercise 4.25). Since T is model-complete, the embeddings $f' : M \to N$ and $g' : M \to N$ are elementary embeddings. Let $\varphi(x_1, \ldots, x_n)$ be any formula in the vocabulary of T. We have

$M \models \varphi(f(c_1), \ldots, f(c_n))$ if and only if
$N \models \varphi(f'(f(c_1)), \ldots, f'(f(c_n)))$ if and only if
$N \models \varphi(g'(g(c_1)), \ldots, g'(g(c_n)))$ if and only if
$M \models \exists y \varphi(g(c_1), \ldots, g(c_n), y)$,

and T satisfies condition (iii) of Proposition 5.58 as we wanted to show. □

5.7 Minimal theories

We define and discuss strongly minimal theories. In some sense, strongly minimal theories are the most simple of first-order theories. They are also among the most important and interesting theories. Strongly minimal theories have an intrinsic notion of independence that allows us to define, in an abstract setting, such concepts as basis and dimension. After discussing strongly minimal theories, we turn briefly to o-minimal theories. Like strong minimality, o-minimality is defined in terms of the definable subsets of models of a theory. Before giving these definitions, we must first define *definable*.

Let M be a $\mathcal{V}$-structure having underlying set U. Recall that "D is a $\mathcal{V}$-definable subset of M" means that D is a subset of U^n for some n and D is defined by some $\mathcal{V}$-formula $\varphi(x_1, \ldots, x_n)$. That is, $\bar{d} \in D$ if and only if $M \models \varphi(\bar{d})$. For $A \subset U$, we say that D is an "A-definable subset of M" if D is defined by some formula $\varphi(\bar{x}, \bar{a})$ having parameters $\bar{a} \in A^m$ for some m (where $\varphi(\bar{x}, \bar{y})$ is a $\mathcal{V}$-formula). We restate this important definition as follows.

Definition 5.82 Let A be a subset of the universe U of $\mathcal{V}$-structure M. Let $\mathcal{V}(A)$ be the expansion of $\mathcal{V}$ that contains a constant c_a for each $a \in A$. Let $M(A)$ be the expansion of M to a $\mathcal{V}(A)$-structure that interprets each c_a as the element $a \in U$. A $\mathcal{V}(A)$-definable subset of $M(A)$ is said to be an *A-definable* subset of M. A subset of U^n is said to be a *definable* subset of M if it is A-definable for some $A \subset U$.

When using this terminology, it is assumed that the vocabulary $\mathcal{V}$ is understood. Note that, for $\mathcal{V}$-structure M having universe U, $\emptyset$-*definable* means the same as $\mathcal{V}$-*definable* and U-*definable* means the same as *definable*.

Proposition 5.83 Let M be a $\mathcal{V}$-structure having underlying set U. Every finite subset of U is definable.

Proof Let $D = \{d_1, \ldots, d_k\}$ be a finite subset of U. Then D is definable since it is $\mathcal{V}(D)$-definable by the formula $\bigvee_{i=1}^{k}(x = d_i)$. □

A subset C of U is said to be *co-infinite* in U if there are infinitely many elements of U that are not in C. Likewise, C is said to be *co-finite* in U if its complement $U - C$ is finite. Since finite subsets of are definable, so are co-finite subsets (take the negation of the formula saying $x \in (U - C)$).

Definition 5.84 Let M be an infinite $\mathcal{V}$-structure having underlying set U. If the only definable subsets of U are finite or co-finite, then M is said to be a *minimal structure*.

Note that the definition of a minimal structure only considers definable subsets of U and not of U^n for $n > 1$. For any infinite structure M, the formula $(x = y)$ defines a subset of U^2 that is both infinite and co-infinite.

Example 5.85 Let $\mathcal{V}_< = \{<\}$. Let $\mathbb{Q}_<$ be the $\mathcal{V}_<$-structure that interprets $<$ as the usual order on the rationals. Every $\mathcal{V}$-formula $\varphi(x)$ either holds for all elements or no elements of the underlying set $\mathbb{Q}$. However, this is not true if we consider formulas having parameters from $\mathbb{Q}$. The formula $(x < 2)$ is clearly both infinite and co-infinite. So $\mathbb{Q}_<$ is not minimal. Likewise, no infinite model of the theory of linear orders T_{LO} is minimal.

Example 5.86 The random graph G_R is not minimal. Every $\mathcal{V}_R$-formula $\varphi(x)$ either holds for no vertices or all vertices of G_R. However, the formula $R(x, a)$ (having some vertex a of G_R as a parameter) defines a subset of G_R that is both infinite and co-infinite.

Definition 5.87 An infinite structure M is said to be *strongly minimal* if every structure N that is elementarily equivalent to M is minimal. A theory is said to be *strongly minimal* if all of its models are infinite and strongly minimal.

Strongly minimal structures (like minimal structures) are minimal in the sense that the definable subsets (definable by formulas in one free variable) are as few as possible. See Exercise 5.17 for an example of a minimal structure that is not strongly minimal.

The usual way to show that a given structure M is strongly minimal is to first show that the theory has quantifier elimination in an appropriate vocabulary. If this is the case, then it suffices to consider only atomic formulas.

First-order theories 241

Proposition 5.88 Let T be a $\mathcal{V}$-theory having quantifier elimination. The following are equivalent:

(i) T is strongly minimal.
(ii) For any model M of T and any atomic $\mathcal{V}$-formula $\varphi(x, y_1, \ldots, y_n)$, $\varphi(x, \bar{a})$ defines a finite or co-finite subset of the universe U of M for all $\bar{a} \in U^n$.

Proof It follows from the definition of "strongly minimal" that (i) implies (ii). We show that (ii) implies (i) by induction on the complexity of formulas. Condition (ii) provides the base step for the induction. Moreover, if both $\theta(x)$ and $\psi(x)$ define a finite or co-finite subset of U, then $\neg\theta(x)$ and $\theta(x) \land \psi(x)$ each define either a finite or co-finite subset of U. It follows by induction that every quantifier-free formula defines a finite or co-finite subset of U. Since T has quantifier elimination, this suffices to prove (i). □

Example 5.89 Recall the $\mathcal{V}_s$-theory T_s from Example 5.12. By Proposition 5.63, T_s has quantifier elimination. Each atomic $\mathcal{V}_s$-formula has the form $s^n(x) = y$ (where $s^n(x)$ denotes the nth-successor of x). Since each element of each model of T_s has a unique n^{th}-successor and a unique n^{th}-predecessor, T_s is strongly minimal by Proposition 5.88.

We use the following convenient notation. For any structure M and formula $\varphi(x)$, let $\varphi(M)$ denote the subset of the universe of M defined by $\varphi(x)$. That is, $\varphi(M) = \{a \in U | M \models \varphi(a)\}$ where U is the universe of M. This notation makes sense for any formula $\varphi(x)$ that is interpreted by the structure M. If M is a $\mathcal{V}$-structure having underlying set U, then $\varphi(M)$ is defined for any $\mathcal{V}(A)$-formula $\varphi(x)$ with $A \subset U$.

Definition 5.90 Let M be a structure and let $\varphi(x)$ be a formula in one free variable. If $\varphi(M)$ is finite, then $\varphi(x)$ is said to be *algebraic* in M.

Definition 5.91 Let M be a $\mathcal{V}$-structure having underlying set U. For any $A \subset U$ and $b \in U$, b is said to be *algebraic over* A in M if $b \in \varphi(M)$ for some algebraic $\mathcal{V}(A)$-formula $\varphi(x)$.

The set of all elements of U that are algebraic over A is called the *algebraic closure* of A in M and is denoted by $acl_M(A)$. We say that A is *algebraically closed* in M if $acl_M(A) = A$.

It is easy to see that $acl_M(A)$ is closed under all functions in $\mathcal{V}$ and contains all elements of U that interpret constants. For this reason, we regard $acl_M(A)$ as a substructure of M (provided $acl_M(A)$ is nonempty). If M is strongly minimal, then these substructures obey rules that justify the use of the word "closure." The following four rules are easily verified regardless of whether M is strongly minimal:

(Reflexivity) $A \subset acl_M(A)$.
(Monotonicity) If $A \subset B$, then $acl_M(A) \subset acl_M(B)$.
(Idempotency) $acl_M(acl_M(A)) = acl_M(A)$
(Finite character) If $a \in acl_M(A)$, then $a \in acl_M(A_0)$ for some finite subset A_0 of A.

If M happens to be strongly minimal, then we also have the *Exchange* rule.

Proposition 5.92 (Exchange) Let M be a strongly minimal structure. Let A be a subset of the universe of M and let b and c be elements from the universe of M. If $c \in acl_M(A \cup \{b\})$ and $c \notin acl_M(A)$, then $b \in acl_M(A \cup \{c\})$.

Proof Since $c \in acl_M(A \cup \{b\})$, there exists a formula $\varphi(\bar{x}, y, z)$ and parameters $\bar{a}$ from A such that $M \models \varphi(\bar{a}, b, c)$ and $M \models \exists^{=k} z \varphi(\bar{a}, b, z)$ for some $k \in \mathbb{N}$. ("$\exists^{=k} x \theta(x)$" is an abbreviation for the first-order formula saying that $\theta(x)$ holds for exactly k many elements.)

Claim Either $\varphi(\bar{a}, y, c)$ is algebraic (in which case $b \in acl_M(A \cup \{c\})$), or $\exists^{=k} z \varphi(\bar{a}, y, z)$ is algebraic (in which case $b \in acl_M(A) \subset acl_M(A \cup \{c\})$).

Proof If $\varphi(\bar{a}, y, c)$ is not algebraic, then it holds for all but finitely many elements y in U. So there exists $l \in \mathbb{N}$ such that $M \models \exists^{=l} y \neg \varphi(\bar{a}, y, c)$. Since $c \notin acl_M(A)$, the formula $\exists^{=l} y \neg \varphi(\bar{a}, y, z)$ holds for all but finitely many elements z in U. So for almost all z in U, the formula $\varphi(\bar{a}, y, z)$ holds for all but l elements y of U. It follows that, for all but at most l elements y in U, the formula $\varphi(\bar{a}, y, z)$ holds for almost all z in U. In particular, the formula $\exists^{=k} z \varphi(\bar{a}, y, z)$ does not hold for most choices of y. So this formula must be algebraic as we wanted to show. □

The exchange rule allows us to assign a dimension to subsets of the universe of a strongly minimal structure. Before defining this dimension, we first must define the notions of *independence* and *basis*.

Definition 5.93 Let A and C be subsets of the universe of M. We say that A is *independent over* C if, for every $a \in A$, a is not in $acl_M(A \cup C - \{a\})$. We say that A is *independent* if A is independent over $\emptyset$.

Definition 5.94 Let A and C be subsets of the universe of M. A *basis* for A is a subset $B \subset A$ such that B is independent and $acl_M(B) = acl_M(A)$. We say that B is a *basis for* A *over* C if B is independent over C and $acl_M(A \cup C) = acl_M(B \cup C)$.

The exchange rule entails that any two bases of a set have the same size. This allows us to define *dimension*.

First-order theories

Lemma 5.95 Let A and C be subsets of the universe U of a strongly minimal structure M. If A has a finite basis over C, then any two bases of A over C have the same size.

Proof We prove this for $C = \emptyset$. The proof is similar for $C \neq \emptyset$.

Claim Let E and F be finite independent subsets of U. If $|E| = |F|$ and $E \subset acl_M(F)$, then $F \subset acl_M(E)$.

Before proving the claim, we show that the claim implies the lemma. Let B_1 and B_2 be two bases for A (at least one of which is finite). With no loss of generality, we may assume that $|B_1| \leq |B_2|$. Let E be any subset of B_2 having the same size as B_1. Since $E \subset acl_M(A) = acl_M(B_1)$, the claim implies that $B_1 \subset acl_M(E)$. By monotonicity, $acl_M(B_1) \subset acl_M(acl_M(E))$. By idempotency $acl_M(B_1) \subset acl_M(E)$. Since $acl_M(B_1) = acl_M(A)$, E is a basis for A. Since B_2 is independent, E must be all of B_2. We conclude that B_2 has the same size as B_1 as we wanted to show.

Proof of Claim We prove the claim by induction on $n = |E|$. If $n = 1$, then $E = \{e\}$ and $F = \{f\}$. Since E is independent, $e \notin acl_M(\emptyset)$. By exchange, $e \in acl_M(f) - acl_M(\emptyset)$ implies $f \in acl(e)$.

Now suppose that $E = \{e_1, \ldots, e_{m+1}\}$ and $F = \{f_1, \ldots, f_{m+1}\}$ for some $m \in \mathbb{N}$. Our induction hypothesis is that the claim holds for any sets E and F with $|E| = |F| \leq m$. It follows that an independent set of size $m+1$ cannot be contained in the algebraic closure of a set of size m. In particular, E cannot be contained in $acl_M(f_2, \ldots, f_{m+1})$. So, for some i,

$$e_i \in acl_M(F) - acl_M(f_2 \ldots f_{m+1}).$$

With no loss of generality, we may assume $i = 1$. By exchange,

$$f_1 \in acl_M(e_1, f_2, \ldots, f_{m+1}).$$

Now suppose that, for some k, we have

$$\{f_1, \ldots, f_k\} \subset acl_M(e_1, \ldots, e_k, f_{k+1}, \ldots, f_{m+1}).$$

By our induction hypothesis, E is not in the algebraic closure of

$$\{e_1, \ldots, e_k, f_{k+1}, \ldots, f_{m+1}\} - \{f_{k+1}\}$$

(since this set has size m). So some $e_i \in E$ is not in this algebraic closure. Clearly, $i > k$. With no loss of generality, suppose $i = k+1$. Since

$$e_{k+1} \in acl_M(f_1, \ldots, f_{k+1}) \subset acl_M(e_1, \ldots, e_k, f_{k+1}, \ldots, f_{m+1}),$$

we have, by exchange,
$$f_{k+1} \in acl_M(e_1, \ldots, e_{k+1}, f_{k+2}, \ldots, f_{m+1}).$$
Continuing in this manner (for $m+1$ steps), we arrive at
$$\{f_1, \ldots, f_{m+1}\} \subset acl_M(e_1, \ldots, e_{m+1})$$
as we wanted to show. □

Proposition 5.96 Let A and C be subsets of the universe U of a strongly minimal structure M. If B_1 and B_2 are bases for A over C, then $|B_1| = |B_2|$.

Proof If B_1 or B_2 is finite, then this proposition is the same as the previous lemma. So suppose both bases are infinite. Let $\mathcal{P}_F(B_2)$ be the set of all finite subsets of B_2. By Exercise 2.36, $|B_2| = |\mathcal{P}_F(B_2)|$. By the finite character of algebraic closure, for each $b \in B_1$, there exists $F_b \in \mathcal{P}_F(B_2)$ such that $b \in acl_M(F_b)$. If $|B_1| > |B_2| = |\mathcal{P}_F(B_2)|$, then some $F \in \mathcal{P}_F(B_2)$ must equal F_b for infinitely many $b \in B_1$ (again by Exercise 2.36). Since F is a finite set, this is impossible by the previous lemma. We conclude that $|B_1| \leq |B_2|$. By the same argument, we have $|B_2| \leq |B_1|$, and so these two bases have the same size. □

Definition 5.97 Let A and C be subsets of the universe of a strongly minimal structure. The *dimension of A over C*, denoted $dim(A/C)$, is the cardinality of any basis for A over C. The *dimension* of A, denoted $dim(A)$, is the dimension of A over $\emptyset$.

The notion of *dimension* (as well as *basis* and *independence*) should be familiar to anyone who has studied linear algebra. As we shall see in the next section, infinite vector spaces (viewed in an appropriate vocabulary) provide examples of strongly minimal structures. The notion of *independence* that we have defined for strongly minimal theories corresponds exactly to the notion of *linear independence* in these examples. Likewise, the *dimension* of a subset of a vector space corresponds to the usual definition of dimension. The algebraic closure of a set of vectors corresponds to the span of the vectors. Vector spaces are completely determined by their dimension. The following lemma shows that this fact generalizes to arbitrary strongly minimal structures.

Lemma 5.98 Let M be a strongly minimal $\mathcal{V}$-structure. Let A and C be subsets of the universe U of M. If $dim_M(A) = dim_M(C)$, then $acl_M(A) \cong acl_M(C)$.

We use the following terminology in the proof of Lemma 5.98.

Definition 5.99 Let M be a $\mathcal{V}$-structure and let A be a subset of the universe U of M. A function $f : A \to U$ is said to be *M-elementary* if
$$M \models \varphi(a_1, \ldots, a_n) \quad \text{implies} \quad M \models \varphi(f(a_1), \ldots, f(a_n))$$
for any $\mathcal{V}$-formula $\varphi(x_1, \ldots, x_n)$ and tuple $(a_1, \ldots, a_n)$ of elements from A.

Proof of Lemma 5.98 Let α be an ordinal such that $|\alpha| = dim_M(A)$. Let $B_A = \{a_i | i < \alpha\}$ be a basis for A and let $B_C = \{c_i | i < \alpha\}$ be a basis for C. □

We first show that the function $f : B_A \to B_C$ defined by $f(a_i) = c_i$ is M-elementary. Second, we show that f can be extended to an isomorphism from $acl_M(A)$ onto $acl_M(C)$.

Claim 1 For any $\mathcal{V}$-formula $\varphi(\bar{x})$ and tuple $\bar{a}$ of B_A and corresponding tuple $f(\bar{a})$ of B_C, if $M \models \varphi(\bar{a})$ then $M \models \varphi(f(\bar{a}))$.

Proof We prove this by induction on the number of free variables in $\varphi(\bar{x})$. If there are zero free variables, then the claim asserts that $M \models \varphi$ implies $M \models \varphi$ for the $\mathcal{V}$-sentence φ. Suppose now that $\varphi(x_1, \ldots, x_{m+1})$ has $m+1$ free variables and the claim holds for any formula having fewer than $m+1$ free variables.

Suppose $M \models \varphi(a_1, \ldots, a_{m+1})$.
Since B_A is independent, the formula $\varphi(a_1, \ldots, a_m, y)$ is not algebraic.
Since M is strongly minimal, $\neg\varphi(a_1, \ldots, a_m, y)$ is algebraic.
So $M \models \exists^{=l} y \neg\varphi(a_1, \ldots, a_m, y)$ for some $l \in \mathbb{N}$ (where the counting quantifier $\exists^{=l}$ is as defined in the proof of Proposition 5.92).
By induction, $M \models \exists^{=l} y \neg\varphi(c_1, \ldots, c_m, y)$.
Since B_C is independent, $\models \varphi(c_1, \ldots, c_{m+1})$.

It follows that the claim holds for all $\varphi(\bar{x})$ and f is M-elementary.

Claim 2 If $E \subset acl_M(A)$ and $a \in acl_M(A) - E$, then any M-elementary function $g : E \to acl_M(C)$ extends to an M-elementary function $g' : (E \cup \{a\}) \to acl_M(C)$.

Proof Since $a \in acl_M(E)$, there exists a formula $\theta(x, \bar{e})$ having parameters from E such that $M \models \theta(a, \bar{e})$ and $M \models \exists^{=l} y \theta(y, \bar{e})$ for some $l \in \mathbb{N}$. Moreover, there exists such a θ so that l is as small as possible. This means that $M \models \theta(y, \bar{e}) \to \psi(y)$ for any $\mathcal{V}(E)$-formula $\psi(y)$ that holds for a (otherwise either $\theta(y, \bar{e}) \wedge \psi(y)$ would define a set smaller that l that contains a).

We want to show that g can be extended. Since g is M-elementary, $M \models \exists^{=l} y \theta(y, f(\bar{e}))$. So there exists $b \in acl_M(C)$ such that $M \models \theta(b, f(\bar{e}))$. We extend g to $E \cup \{a\}$ by defining $g'(a) = b$. For any $\mathcal{V}$-formula $\varphi(x_1, \ldots, x_n, y)$ and n-tuple $(d_1, \ldots, d_n)$ of elements from E,

$M \models \varphi(d_1, \ldots, d_2, a)$ implies
$M \models \theta(y, \bar{e}) \to \varphi(d_1, \ldots, d_2, y)$ which implies
$M \models \theta(y, g(\bar{e})) \to \varphi(g(d_1), \ldots, g(d_2), y)$ (since g is M-elementary)

which implies $M \models \varphi(g(d_1), \ldots, g(d_2), b)$ (since $M \models \theta(b, g(\bar{e}))$).
It follows that g' is M-elementary as we wanted to show.

Claim 2 shows that the M-elementary function defined in Claim 1 can be repeatedly extended. By induction (transfinite induction if $dim_M(A)$ is infinite), we can extend this to an M-elementary function from $acl_M(A)$ to $acl_M(B)$. Such a function must be onto (see Exercise 5.34), and is therefore an isomorphism. □

Theorem 5.100 Countable strongly minimal theories are uncountably categorical.

Proof Let T be a strongly minimal theory and let κ be an uncountable cardinal. Let M and N be two models of T of size κ. Let U_N and U_M be the underlying sets of N and M, respectively.

Claim $dim_M(U_M) = dim_N(U_N) = \kappa$.

Proof This follows from the assumption that the vocabulary $\mathcal{V}$ of T is countable. For any $A \subset U_M$, $|\mathcal{V}(A)| = |A| + \aleph_0$ implies $|acl_M(A)| \leq |A| + \aleph_0$. In particular, if $|A| < \kappa$, then A cannot be a basis for U_M.

By the Joint Embedding lemma 4.37, there exists a model D of T that is an elementary extension of both M and N. Since $dim_D(U_M) = dim_D(U_N) = \kappa$, we have $M = acl_D(U_M) \cong acl_D(U_N) = N$ by Lemma 5.98. □

Corollary 5.101 Strongly minimal theories are complete.

Proof This follows immediately from Proposition 5.15. □

We now turn to a variant of strong minimality. Let M be an infinite $\mathcal{V}$-structure. Suppose that $\mathcal{V}$ contains the binary relation $<$ and M interprets $<$ as a linear order on its underlying set U. An *interval* of M is a subset of U of the form

$$(a,b) = \{x \in U | a < x < b\}, (a, \infty) = \{x \in U | a < x\}, \text{ or}$$
$$(\infty, a) = \{x \in U | x < a\}$$

for some a and b in U. We also include singletons $\{a\} \subset U$ as (degenerate) intervals. Clearly, any interval is a definable subset of M. The structure M is said to be *o-minimal* if every definable subset of M is a finite union of intervals. A theory is *o-minimal* if its models are o-minimal.

As was demonstrated in Example 5.85, o-minimal theories are not strongly minimal. However, these two notions have much in common. The word "minimal" means the same for both. They are minimal in the sense that the definable subsets (definable by formulas in one free variable) are as few as possible. For o-minimal theories, "as few as possible" takes into account the presence of a linear order (o-minimal is short for "order-minimal"). Also, algebraically closed substructures of an o-minimal structure satisfy the exchange rule. So o-minimal structures, like strongly minimal structures, have an intrinsic notion of independence and dimension (however, o-minimal structures are not uncountably categorical).

Example 5.102 The following structures are o-minimal:

- $\mathbf{Q}_< = \{\mathbb{Q}| <\}$,
- $\mathbf{R}_{or} = \{\mathbb{R}| <, +, \cdot, 0, 1\}$, and
- $\mathbf{R}_{exp} = \{\mathbb{R}|exp, <, +, \cdot, 0, 1\}$,

where $\mathbf{R}_{exp}$ interprets the unary function $exp(x)$ as e^x and the other symbols are interpreted in the usual way.

That $\mathbf{Q}_<$ is o-minimal follows from the fact that T_{DLO} has quantifier elimination (Proposition 5.49). Likewise, the o-minimality of $\mathbf{R}_{or}$ can be deduced from *Tarski's theorem*. Tarski's theorem states that $T_{or} = Th(\mathbf{R}_{or})$ has quantifier elimination. (This fact was stated without proof in Example 5.71.) Not only did Alfred Tarski prove that T_{or} has quantifier elimination, he also provided an algorithm to carry out the quantifier elimination. Given any formula $\varphi(\bar{x})$ in the vocabulary $\mathcal{V}_{or}$ of T_{or}, Tarski's algorithm produces a quantifier-free $\mathcal{V}_{or}$-formula that is T_{or}-equivalent to $\varphi(\bar{x})$ (although this algorithm is far from efficient). Since Tarski's algorithm allows sentences as input, this also shows that T_{or} is decidable.

The question of whether $\mathbf{R}_{exp}$ has similar properties became known as Tarski's Problem (one of several problems by this name). This problem motivated the conception of o-minimality in the 1980s. Nearly half a century after Tarski's results regarding $\mathbf{R}_{or}$, Alex Wilkie proved in the 1990s that $\mathbf{R}_{exp}$ is o-minimal. This structure does not have a theory with quantifier elimination, but, as Wilkie proved, it is model-complete. Whether it is decidable remains unknown.

For more on o-minimal structures, the reader is referred to [10] written by Lou van den Dries, the mathematician who introduced the concept. We now end our brief discussion of o-minimality and return to strongly minimal structures. In this section, we have proved several facts regarding strongly minimal structures, but have provided a dearth of examples of such structures. We correct this deficiency in the next section by analyzing specific examples of strongly minimal theories.

5.8 Fields and vector spaces

We examine some basic algebraic structures that have strongly minimal theories. We consider vector spaces and the field of complex numbers. We show that these structures, viewed in appropriate vocabularies, have theories with quantifier elimination. From this we deduce strong minimality.

We use these examples to illustrate a fundamental trichotomy of strongly minimal theories. Strongly minimal theories are divided into those that are *trivial* and those that are *nontrivial*. They can also be divided into those that are *locally modular* and those that are *nonlocally modular*. Since trivial theories are necessarily locally modular (as we shall show), there are three possibilities: a strongly minimal theory is either nonlocally modular, trivial, or both nontrivial and locally modular. We shall define these concepts and provide examples of theories from each of these three categories. We begin with trivial strongly minimal theories.

Definition 5.103 A strongly minimal theory is *trivial* if for any $M \models T$ and any subset A of the universe of M, $acl_M(A) = \bigcup_{a \in A} acl_M(\{a\})$.

Example 5.104 Recall T_s from Example 5.12. This theory was shown to be strongly minimal in Example 5.89. Let M be a model of T_s having underlying set U. Recall from Example 5.20 that $M \cong Z_\kappa$ for some cardinal κ where Z_κ is the structure having κ copies of $\mathbb{Z}$ as its underlying set. For any $a \in U$, $acl_M(a)$ is the copy of $\mathbb{Z}$ that contains a. Likewise, for any $A \subset U$, $acl_M(A)$ consists of the copies of $\mathbb{Z}$ that contain some element of A. From this observation it follows that T_s is a trivial strongly minimal theory.

For examples of strongly minimal theories that are not trivial, recall the concept of a *group*. A group consists of a set together with a binary function that satisfies the axioms listed in Example 5.6. We can view any group as a first-order structure in the vocabulary $\mathcal{V}_{gp} = \{+, 0\}$ where $+$ is a binary function representing the group operation and 0 is a constant representing the identity of the group. Now suppose that T is a strongly minimal theory containing the $\mathcal{V}_{gp}$-theory T_{gp} of groups. Let $\{a, b\}$ be an independent set containing two elements from the universe of a model M of T. Then $a + b$ is an element that is in $acl_M(\{a, b\})$ but is contained in neither $acl_M(\{a\})$ nor $acl_M(\{b\})$. It follows that any such theory T is not trivial.

We shall demonstrate examples of strongly minimal groups in this section. Each of these examples happens to be an *Abelian* group. A group is *Abelian* if, in addition to the properties listed in Example 5.6, the following holds:

(Commutativity) For every a and b in G, $a \circ b = b \circ a$.

Here, as in Example 5.6, $\circ$ denotes the group's binary operation. This property can easily be expressed as a $\mathcal{V}_{gp}$-sentence. This sentence is consistent with, but not a consequence of, the theory of groups T_{gp} (see Exercise 2.5(c)).

Our choice of $\{+, 0\}$ as the vocabulary for groups is somewhat arbitrary. We can just as well use the vocabulary $\{\cdot, 1\}$ or any other vocabulary consisting of a binary function and a constant. An *additive group* is a group in the vocabulary

{+, 0}. A *multiplicative group* has {·, 1} as its vocabulary. A *field* is a structure with two binary operations each of which forms an Abelian group.

Definition 5.105 Let $\mathcal{V}_{ar}$ be the vocabulary $\{+, \cdot, 0, 1\}$ (the vocabulary of arithmetic). For any $\mathcal{V}_{ar}$-structure $F = (U|+, \cdot, 0, 1\}$, we say that F is a *field* if the following hold:

- The reduct $(U|+, 0)$ of F is an Abelian group.
- The substructure $(U - \{0\}|\cdot, 1)$ of the reduct $(U|\cdot, 1)$ of F is an Abelian group.
- $F \models \forall x \forall y \forall z (z \cdot (x + y) = z \cdot x + z \cdot y)$.
- $F \models \forall x \forall y \forall z ((x + y) \cdot z = x \cdot z + y \cdot z)$.

The *theory of fields*, denoted T_F, is the set of all $\mathcal{V}_{ar}$-sentences that hold in all fields.

So a field has both a multiplicative group structure and an additive group structure. The constant 0 necessarily has no multiplicative inverse and so must be excluded from the multiplicative group. The last two items in the above definition, called the *distributive rules*, dictate how the two operations interact.

Example 5.106 The rational numbers and the real numbers, viewed as structures in the vocabulary $\mathcal{V}_{ar}$, are examples of fields.

Example 5.107 The integers do not form a field. The structure $(\mathbb{Z} - \{0\}|\cdot, 1)$ is not a group since no element (other than 1) has a multiplicative inverse.

Suppose that we restrict our attention to the integers in the set $\mathbb{Z}_7 = \{0, 1, 2, 3, 4, 5, 6\}$. If we take the usual definition of addition and multiplication, then this set does not form a field since it is not closed under addition or multiplication. Let us instead consider addition and multiplication *modulo* 7. This means that we take the remainder of the sum or product when divided by 7. For example, $3 + 6 = 2$ (mod 7), $4 + 4 = 1$ (mod 7), $5 \cdot 4 = 6$ (mod 7), $5 \cdot 6 = 2$ (mod 7) and so forth. Let $\mathbf{F}_7 = (\mathbb{Z}_7|+, \cdot, 0, 1)$ be the $\mathcal{V}_{ar}$-structure that interprets + as addition modulo 7 and · as multiplication modulo 7 on the set $\mathbb{Z}_7$. Then $\mathbf{F}_7$ is an example of a finite field. For any positive integer a, $\mathbf{F}_a$ is defined analogously. This structure is a field if and only if a is prime. We leave the verification of these facts to the reader.

The examples of fields that we have given, namely $\mathbf{Q}$, $\mathbf{R}$, and $\mathbf{F}_7$, are not strongly minimal. To obtain a strongly minimal structure, we consider vector spaces over these fields.

First-order theories

Definition 5.108 Let F be a field. For each element a of F, let s_a denote a unary function. Let $\mathcal{V}_F = \{+, 0, s_a | a \in F\}$. A *vector space over F* is a structure M in the vocabulary $\mathcal{V}_F$ that satisfies the following:

- The reduct of M to $\{+, 0\}$ is an Abelian group.
- $M \models \forall x(s_1(x) = x)$.
- $M \models \forall x \forall y(s_a(x+y) = s_a(x) + s_a(y))$ for all $a \in F$.
- $M \models \forall x(s_{a+b}(x) = s_a(x) + s_b(x))$ for all a and b in F.
- $M \models \forall x(s_a(s_b(x)) = s_{a \cdot b}(x))$ for all a and b in F.

The *theory of vector spaces over F* is the set of $\mathcal{V}_F$-sentences that hold in each vector space over F.

Example 5.109 We consider various vector spaces over $\mathbb{R}$.

- Let $\mathbb{R}^n$ be the set of all ordered n-tuples $(a_1, \ldots, a_n)$ where each $a_i \in \mathbb{R}$.
- Let $\mathbb{R}[x_1, \ldots, x_n]$ be the set of all polynomials in n variables having coefficients in $\mathbb{R}$.
- Let $\mathbb{R}^{\leq 2}[x]$ be the set of all polynomials in $\mathbb{R}[x_1, \ldots, x_n]$ of degree at most 2.
- Let $M_{n \times n}(\mathbb{R})$ be the set of all $n \times n$ matrices having real numbers as entries.

There is a natural way to describe a vector space over $\mathbb{R}$ having any one of these sets as an underlying set. Each set carries a natural notion of addition and a zero element (either the matrix having all zero entries or the constant polynomial $p(x) = 0$). Moreover, we can define scalar multiplication for each. Given any element v from any one of these sets and any $r \in \mathbb{R}$, the product $r \cdot v$ is a well-defined element in the same set as v. Thus the unary function s_r has a natural interpretation.

We recall some facts about vector spaces from linear algebra. Let V be a vector space over a field F. Let $B = \{v_1, \ldots, v_n\}$ be a set of vectors in V. The *span* of B is the set of all linear combinations $a_1 \cdot + v_1 + \cdots a_n \cdot v_n$ where each a_i is in F. The set B is *linearly independent* if v_i is not in the span of $B - \{v_i\}$ for each $v_i \in B$. From this notion of independence, we can define *linear bases* and *linear dimension*. Two vector spaces having the same linear dimension over a field are necessarily isomorphic. We repeatedly use the adjective "linear" to distinguish these terms from their strongly minimal counterparts. However, we will show that these two notions are the same.

Proposition 5.110 The $\mathcal{V}_F$-theory T_V of a vector space over an infinite field F has quantifier elimination.

Proof Note that the theory of a vector space over F is $\forall_2$-axiomatizable. Also, any two uncountable models of the same size have the same linear dimension and, hence, are isomorphic. By Lindström's theorem, T_V is model-complete. To show that T_V has quantifier elimination, it suffices to show that it has the isomorphism property (by Proposition 5.79). Let $M \models T_V$.

Claim Every substructure of M is a submodel.

Proof A substructure is, by definition, closed under all functions in the vocabulary. Since $T_V \vdash \forall x(x + s_{-1}(x) = 0)$, every substructure contains the inverse for each element and also the constant 0. From this information it is easy to verify that any substructure of M is itself a vector space over F.

It follows from this claim that T_V has the isomorphism property and, hence, quantifier elimination as well. □

Proposition 5.111 *The $\mathcal{V}_F$-theory T_V of a vector space over an infinite field F is a nontrivial strongly minimal theory.*

Proof Let M be an arbitrary model of T_V. We must show that every $\mathcal{V}_F(M)$-formula $\theta(x)$ defines either a finite or co-finite subset of the underlying set of M. By the previous proposition, it suffices to consider only atomic $\theta(x)$ (by Proposition 5.88). Atomic $\mathcal{V}_F(M)$-formulas have the form $t_1 = t_2$ for some $\mathcal{V}_F(M)$-terms t_1 and t_2. If there is exactly one free variable x in the equation $t_1 = t_2$, then this formula is T_V-equivalent to a formula of the form $x = t$ for some quantifier-free $\mathcal{V}_F$-term t. That is, we can solve the equation for x (here we are using the fact that F is a field). Clearly, this formula defines a set of size 1. Since T_V has quantifier elimination and every atomic formula defines a finite subset of every model, T_V is strongly minimal. It is not trivial since $(a + b) \in acl_M(\{a, b\})$ for independent $\{a, b\}$. □

Corollary 5.112 *For any infinite field F, the $\mathcal{V}_F$-theory T_V of vector spaces over F is κ-categorical if and only if $\kappa > |F|$.*

Proof First note that T_V has no models smaller than $|F|$. If $\kappa > |F|$, then T_V is κ-categorical by Proposition 5.98. If $\kappa = |F|$, then T_V is not κ-categorical since any finite dimensional vector space over F has the same size as F. □

In particular, the theory of vector spaces over F is complete if F is infinite. This is not true for finite fields. Finite dimensional vector spaces over finite fields are finite. To obtain a complete theory, we must only consider vector spaces of infinite dimension over finite fields.

Proposition 5.113 *For any finite field F, the theory of infinite dimensional vector spaces over F is strongly minimal, nontrivial, and totally categorical.*

Proof This can be proved by repeating the arguments we gave for vector spaces over infinite fields. We leave the verification of this to the reader. □

Let M model the theory of vector spaces over a field F (either finite or infinite). For any algebraically closed subsets A and B of the universe of M, the following holds:

$$dim_M(A \cup B) = dim_M(A) + dim_M(B) - dim_M(A \cap B). \tag{5.1}$$

We state this fact from linear algebra without proof. We show that this is one property of vector spaces that does not generalize to all strongly minimal theories.

Example 5.114 Let M be an infinite dimensional vector space over a field F. Let $\mathcal{V}'_F = \mathcal{V}_F \cup \{f\}$ where f is a ternary function. Let M' be the expansion of M to a $\mathcal{V}'_F$-structure that interprets f as the function $f(x, y, z) = x + y - z$. This function is explicitly definable in terms of $\mathcal{V}_F$:

$$M' \models f(x, y, z) = u \text{ if and only if } M' \models \exists w(w + z = 0 \wedge x + y + w = u).$$

It follows that M and M' are bi-definable and M' is strongly minimal. Now let N be the reduct of M' to the vocabulary $\{f, s_a | a \in F\}$. That is, the vocabulary of N contains neither $+$ nor 0. Since M' is strongly minimal and every definable subset of N is also a definable subset of M', N is strongly minimal. We claim that Equation (5.1) does not hold for N. Let a, b, and c be elements from the underlying set such that $dim_M(a,b,c) = 3$. Then $dim_N(a,b,c) = 3$.

Clearly, $dim_N(a,b,c,f(a,b,c)) = dim_M(a,b,c,a+b-c) = 3$, $dim_N(a,b) = dim_M(a,b) = 2$, and $dim_N(c, f(a,b,c)) = dim_M(c, a+b-c) = 2$. If $A = acl_N(\{a,b\})$ and $B = acl_N(\{c, f(a,b,c)\})$, then $A \cap B = \emptyset$. Thus we have $3 = dim_N(A \cup B) \neq dim_N(A) + dim_N(B) - dim_N(A \cap B) = 2 + 2 - 0 = 4$ and Equation (5.1) fails.

Note that $acl_M(\{a,b\}) \cap acl_M(\{c, a+b-c\})$ is nonempty. It contains the constant 0 that was omitted from the vocabulary of N. This intersection also contains $a+b$ and all of its scalar multiples. So in M, this intersection has dimension 1 and Equation (5.1) holds (as it does in every vector space).

Definition 5.115 Let T be a strongly minimal theory.

If equation (5.1) holds for all $M \models T$, then T is said to be *modular*.

If Equation (5.1) holds whenever $A \cap B$ is nonempty, then T is said to be *locally modular*. We say that a strongly minimal structure is modular or locally modular if its theory is.

Equivalently, T is locally modular if and only if the expansion of T by a single constant is modular. Whereas the theory of a vector space over a field is modular, the theory $Th(N)$ from Example 5.114 is a locally modular strongly minimal theory that is not modular. If we expand N to include the constant 0, then the binary function $+$ can be recovered as $f(x, y, 0) = x + y$. So this expansion of N is bi-definable with the modular structure M.

Proposition 5.116 Let T be a strongly minimal theory. If T is trivial, then it is modular.

Proof Let A and C be subsets of the universe of a model M of T. Let B_0 be a basis for $A \cap C$. Let B_1 be a basis for $A - acl_M(A \cap C)$ and let B_2 be a basis for $C - acl_M(A \cap C)$. Consider $B_0 \cup B_1 \cup B_2$. For any elements a and b of this union, it is not the case that $a \in acl_M(\{b\})$ (by the definition of these three bases). It follows, since T is trivial, that $B_0 \cup B_1 \cup B_2$ is an independent set. So $B_0 \cup B_1$ is a basis for A, $B_0 \cup B_2$ is a basis for C, and $B_0 \cup B_1 \cup B_2$ is a basis for $A \cup C$. Equation (5.1) clearly holds. □

A strongly minimal theory is *nonlocally modular* if it is not locally modular. To demonstrate an example of a nonlocally modular strongly minimal theory, we consider the complex numbers. Recall that the set $\mathbb{C}$ of complex numbers consists of all numbers of the form $a + bi$ where a and b are real numbers an i is the square root of -1. Complex numbers are added and multiplied as follows:

$$(a + bi) + (c + di) = (a + c) + (b + d)i, \text{ and}$$
$$(a + bi) \cdot (c + di) = ac + adi + bci + bd(-1) = (ac - bd) + (ad + bc)i$$

In this way, we can view the complex numbers as a $\mathcal{V}_{ar}$-structure $\mathbf{C}$. This structure is a field (the multiplicative inverse of $a + bi$ is $a/(a^2 + b^2) - b/(a^2 + b^2)i$).

We axiomatize the theory $Th(\mathbb{C})$. We use without proof the *Fundamental Theorem of Algebra*. This theorem states that, for any nonconstant polynomial $p(x)$ having coefficients in $\mathbb{C}$, there exists a solution in $\mathbb{C}$ to the equation $p(x) = 0$. Moreover, there are no more than d such solutions where d is the degree of the polynomial.

Definition 5.117 The *theory of algebraically closed fields*, denoted T_{ACF}, is the $\mathcal{V}_{ar}$-theory axiomatized by:

- the axioms for the theory of fields T_F, and
- $\forall y_1 \cdots \forall y_n \exists x (x^n + y_1 \cdot x^{n-1} + \cdots + y_{n-1} \cdot x + y_n = 0)$ for each $n \in \mathbb{N}$
  (where x^n is an abbreviation for the $\mathcal{V}_{ar}$-term $x \cdot x \cdots x$).

Lemma 5.118 Let F be a field. There exists an extension $\tilde{F}$ of F that models T_{ACF}.

Proof Note that the axioms for the theory of fields T_f are each $\forall_2$-sentences. By Proposition 5.76, F has an extension that is existentially closed with respect to T_f. By definition, any existentially closed field is algebraically closed. □

The theory T_{ACF} of algebraically closed fields is not complete. By previous proposition, every field can be extended to a model of T_{ACF}. In particular, the field $\mathbf{F}_7$ from Example 5.107 has an extension $\tilde{\mathbf{F}}_7$ that models T_{ACF}. This structure is not elementarily equivalent to $\mathbf{C}$. To see this, let θ_7 be the sentence $(1+1+1+1+1+1+1 = 0)$. Then $\tilde{\mathbf{F}}_7 \models \theta_7$ and $\mathbf{C} \models \neg \theta_7$.

Definition 5.119 Let p be a prime number. Let θ_p be the V_{ar}-sentence saying that $p \cdot 1 = 0$. The *theory of algebraically closed fields of characteristic p*, denoted T_{ACFp}, is the deductive closure of $T_{ACF} \cup \{\theta_p\}$.

To axiomatize $\mathbf{C}$, we must include the negations of the θ_p.

Definition 5.120 The *theory of algebraically closed fields of characteristic 0*, denoted T_{ACF0}, is the deductive closure of $T_{ACF} \cup \{\neg \theta_p | p \text{ is prime}\}$.

We claim that T_{ACF0} is the complete V_{ar}-theory of $\mathbf{C}$.

Proposition 5.121 T_{ACF0} has quantifier elimination.

Proof We use condition (ii)′ of Proposition 5.58. Let $M \models T$ and let $(a_1, \ldots, a_n)$ and $(b_1, \ldots, b_n)$ be n-tuples from the universe U of M that satisfy the same atomic formulas in M. We must show that for any $a_{n+1} \in U$ there exists b_{n+1} in the universe of an elementary extension N of M such that $(a_1, \ldots, a_n, a_{n+1})$ and $(b_1, \ldots, b_n, b_{n+1})$ satisfy the same atomic formulas in N. We break the proof of this into two cases. In case 1, we are able to take N to be equal to M.

Case 1: a_{n+1} is a root of some polynomial having coefficients among $A = \{a_1, \ldots, a_n\}$. That is, $M \models p(a_{n+1}) = 0$ for some polynomial $p(x)$ having coefficients in A. We may assume that $p(x) = 0$ has the least number of solutions among all such polynomials (so $p(x)$ is the *minimal polynomial* over A). Let $q(x)$ be the polynomial obtained by replacing each occurrence of a_i in $p(x)$ with b_i (for each $i = 1, \ldots, n$). Since M is algebraically closed, $M \models q(b_{n+1}) = 0$ for some $b_{n+1} \in U$. Since $p(x)$ is minimal, $(a_1, \ldots, a_n, a_{n+1})$ and $(b_1, \ldots, b_n, b_{n+1})$ satisfy the same atomic formulas in M. This can be shown in the same way that Claim 2 was proved in the proof of Lemma 5.98.

Case 2: a_{n+1} is not a root of any polynomial having coefficients among $A = \{a_1, \ldots, a_n\}$. Let N be an elementary extension of M such that $|N| > |M|$. Since there are only countably many polynomials having coefficients in A and each has only finitely many roots, there must exist b_{n+1} in the universe of N that is not a root of any of them. Clearly $(a_1, \ldots, a_n, a_{n+1})$ and $(b_1, \ldots, b_n, b_{n+1})$ satisfy the same atomic formulas in N. □

First-order theories 255

Proposition 5.122 T_{ACF0} is a nonlocally modular strongly minimal theory.

Proof Let M be an arbitrary model of T_{ACF0}. To show that T_{ACF0} is strongly minimal, it suffices to show that every atomic $\mathcal{V}_{ar}(M)$-formula $\theta(x)$ defines either a finite or co-finite subset of the underlying set of M (by Proposition 5.88). Atomic $\mathcal{V}_{ar}(M)$-formulas are T_F-equivalent to formulas of the form $p(x) = 0$ where $p(x)$ is a polynomial having coefficients from the universe of M. Strong minimality follows from the fact that polynomials have finitely many roots.

It remains to be shown that T_{ACF0} is not locally modular. Let a, b, and c be elements from a model M of T_{ACF0} such that $dim_M(a, b, c) = 3$. Let $A = acl_M(\{a, b\})$ and let $B = acl_M(\{a + b \cdot c, c\})$. Then $dim_M(A) = dim_M(B) = 2$ and $dim_M(A \cup B) = 3$. We state without proof the following fact: if $d \in A \cap B$, then $d \in acl_M(\emptyset)$. From this we see that $dim_M(A \cap B) = 0$ and T_{ACF0} is not modular. □

Corollary 5.123 T_{ACF0} is complete and uncountably categorical.

It follows that T_{ACF0} is the complete theory of **C**. What does this fact tell us about the complex numbers? By quantifier elimination, we know that any $\mathcal{V}_{ar}$-formula φ is T_{ACF0}-equivalent to some quantifier-free $\mathcal{V}_{ar}$-formula ψ_φ. Let us consider some specific formulas φ.

For each $n \in \mathbb{N}$, let $p_n(x, y_0, y_1, \ldots, y_n)$ be the polynomial

$$y_0 + y_1 \cdot x + y_2 \cdot x^2 + \cdots + y_n \cdot x^n = 0.$$

Let $\varphi(y_0, \ldots, y_n)$ be the formula $\exists x p(x, y_0, y_1, \ldots, y_n) = 0$. Since T_{ACF0} has quantifier elimination, we know that this formula is T_{ACF0}-equivalent to a quantifier-free $\mathcal{V}_{ar}$-formula. However, T_{ACF0} implies every polynomial has a root. So the formula $\varphi(y_0, y_1, \ldots, y_n)$ holds for all $y_0, \ldots, y_n$ in any model of T_{ACF0}. It follows that $\varphi(y_0, y_1, \ldots, y_n)$ is T_{ACF0}-equivalent to the quantifier-free formula $1 = 1$. Do not try to impress your complex analysis professor with this fact.

Now, for any $n, m \in \mathbb{N}$, let $\theta_{n,m}(y_0, \ldots, y_n, z_0, \ldots, z_m)$ be the formula

$$\exists x (p_n(x, y_0, \ldots, y_n) = 0 \land p_m(x, z_0, \ldots, z_m) = 0).$$

This formula asserts that the two polynomials share a root. Whether or not this is true depends on the coefficients $(y_0, \ldots, y_n)$ and $(z_0, \ldots, z_m)$ of the two polynomials. Since T_{ACF0} has quantifier-elimination, there must exists a quantifier-free formula $\psi_\theta(y_0, \ldots, y_n, z_0, \ldots, z_m)$ that holds if and only if the two polynomials have a common root. This is not obvious. In fact, $\theta_{n,m}(y_0, \ldots, y_n, z_0, \ldots, z_m)$

holds if and only if the determinate of the following matrix is not zero:

$$\begin{pmatrix} y_0 & y_1 & \cdots & y_n & 0 & 0 & & \cdots & 0 \\ 0 & y_0 & y_1 & \cdots & y_0 & 0 & & \cdots & 0 \\ \cdot & & & \cdots & & & & & \cdot \\ \cdot & & & \cdots & & & & & \cdot \\ 0 & 0 & \cdots & 0 & y_0 & y_1 & & \cdots & y_m \\ z_0 & z_1 & & \cdots & & z_m & 0 & \cdots & 0 \\ 0 & z_0 & z_1 & & \cdots & & z_m & 0 & \cdots & 0 \\ \cdot & & & \cdots & & & & & \cdot \\ 0 & \cdots & 0 & z_0 & z_1 & & & \cdots & 0 \end{pmatrix}.$$

The determinant of this matrix is called the *resultant* of the two polynomials $p_n(x, y_0, \ldots, y_n)$ and $p_m(x, z_0, \ldots, z_m)$. Since the determinant is an algebraic expression in $(y_0, \ldots, y_n, z_0, \ldots, z_m)$, we can say that this determinate equals zero with a quantifier-free $\mathcal{V}_{ar}$-formula $\varphi_\theta(y_0, \ldots, y_n, z_0, \ldots, z_m)$.

Now suppose that we have k polynomials of the form $p_n(x, y_0, \ldots, y_n)$. Let $\bar{y}$ be the $k \cdot (n+1)$-tuple consisting of the coefficients of these polynomials. Suppose we want to determine whether there exists a number that is simultaneously the root of each of these k polynomials. Since T_{ACF0} has quantifier elimination, there exists some quantifier-free expression having $\bar{y}$ as variables that determines this. That is, there exist analogues for the resultant that work for each $k > 2$.

The perspective of model theory is somewhat askew compared to other branches of mathematics. The light shed by model theory will not fully illuminate a structure in all of its detail. However, it can bring to light certain features of a structure that are shaded by other approaches. As a basic example, we have the fact that there exist resultants for several polynomials in several variables. That is, there exists a polynomial $P(\bar{y})$ in the coefficients $\bar{y}$ of the given polynomials such that $P(\bar{y}) = 0$ if and only if the polynomials have a common zero. Model theory provides an immediate proof of this fact, but it does not provide a description of the polynomial $P(\bar{y})$.

Resultants provide a superficial example of the deep relationship between model theory and other branches of mathematics. Not only have model theoretic methods shed new light on various branches of mathematics, these methods have yielded results at the forefront of research. Most notable is Ehud Hrushovski's 1996 proof of the Mordell–Lang conjecture for function fields. Implementing model-theoretic tools (such as strong minimality), Hrushovski answered in the affirmative this long standing conjecture of algebraic geometry. The statement of this conjecture (not to mention the proof) is beyond the scope of this book. We consider an application of model theory to algebraic geometry that is far more fundamental.

5.9 Some algebraic geometry

The model-theoretic properties of **C** provide elementary proofs for some fundamental theorems of algebraic geometry. In this section, we give one prominent example known as Hilbert's Nullstellensatz.

Algebraic geometry arises from the interplay between the algebra of polynomial equations and the geometry of the solutions of these equations. Let $\mathbb{C}[x_1,\ldots,x_n]$ denote the set of all polynomials having variables $x_1,\ldots,x_n$ and coefficients in $\mathbb{C}$. Each $f(x_1,\ldots,x_n)$ in $\mathbb{C}[x_1,\ldots,x_n]$ defines a subset of $\mathbb{C}^n$, namely

$$V_f = \{(x_1,\ldots,x_n) \in \mathbf{C}^n | f(x_1,\ldots,x_n) = 0\}.$$

The set of solutions of a polynomial in two variables is called an *algebraic curve*. More specifically, if $f(x,y)$ is a polynomial having complex coefficients, then V_f is a *complex algebraic curve*.

Example 5.124 Consider the polynomials

$$f(x,y) = x^3 - xy + x^2y - y^2, \text{ and}$$
$$g(x,y) = x^4 + 2x^3y + x^2y^2 - x^2y + 2xy^2 + y^3.$$

These two polynomials define the same complex algebraic curves. This is because they factor as $f(x,y) = (x^2 - y)(x + y)$ and $g(x,y) = (x^2 - y)(x + y)^2$.

Since they have the same factors, they have the same curves. Whether we plot $f(x,y) = 0$ or $g(x,y) = 0$ in the real plane, we will see the union of the parabola defined by $y = x^2$ and the line $y = -x$. Likewise, the complex curves defined by these polynomials are identical.

Definition 5.125 A polynomial $f \in \mathbb{C}[x_1,\ldots,x_n]$ is *irreducible* if it cannot be factored as $f(x_1,\ldots,x_n) = p(x_1,\ldots,x_n) \cdot q(x_1,\ldots,x_n)$ for two nonconstant polynomials $p(x_1,\ldots,x_n)$ and $q(x_1,\ldots,x_n)$ in $\mathbb{C}[x_1,\ldots,x_n]$.

The polynomials $f(x,y)$ and $g(x,y)$ from the previous example are not irreducible. These polynomials have the two irreducible factors corresponding to the *irreducible curves* given by the line and the parabola. Hilbert's Nullstellensatz states that two polynomials in $\mathbb{C}[x,y]$ define the same curves if and only if they have the same irreducible factors. As the following example shows, this is not true when restricted to the real numbers.

Example 5.126 Let $h(x,y) = (x^2 + 1)(x^2 - y)(x + y)$. Since $(x^2 + 1)$ is not zero for any real numbers, $h(x,y)$ defines the same curve in $\mathbb{R}^2$ as the polynomials $f(x,y)$ and $g(x,y)$ from the previous example. In $\mathbb{C}^2$, however, $h(x,y)$ has the root $(i,0)$ that is not a root of $f(x,y)$. So the complex algebraic curve defined by $h(x,y)$ is not the same as the curve defined by $f(x,y)$.

Theorem 5.127 (Hilbert's Nullstellensatz) Let $g(x,y)$ and $h(x,y)$ be two polynomials having complex coefficients. The complex algebraic curves defined by $g(x,y)$ and $h(x,y)$ are the same if and only if $g(x,y)$ and $h(x,y)$ have the same irreducible factors.

Proof A point $(a,b) \in \mathbb{C}^2$ is on the curve defined by $g(x,y)$ if and only if $g(a,b) = 0$. This happens if and only if $p(x,y) = 0$ for some irreducible factor p of g. It follows that if $g(x,y)$ and $h(x,y)$ have the same irreducible factors, then $g(x,y)$ and $h(x,y)$ define the same curves.

Conversely, suppose that $g(x,y)$ and $f(x,y)$ do not have the same irreducible factors. Let $p(x,y)$ be an irreducible factor of $g(x,y)$ that is not a factor of $h(x,y)$. We show that there exists $(a,b) \in \mathbb{C}^2$ such that $p(a,b) = 0$ and $h(a,b) \neq 0$. If we show this, then we can conclude that the curves defined by $g(x,y)$ and $f(x,y)$ are not the same.

Let P be the set of all polynomials in $\mathbb{C}[x,y]$ that have $p(x,y)$ as a factor. Then $g(x,y) \in P$ and $h(x,y) \notin P$.

For each $f(x,y) \in \mathbb{C}[x,y]$, let $f(x,y) + P$ denote the set

$$\{f(x,y) + q(x,y) | q(x,y) \in P\}.$$

Note that $f_1(x,y) + P = f_2(x,y) + P$ if and only if the polynomial $f_1(x,y) - f_2(x,y)$ is in P. In particular, $f(x,y) + P = P$ if and only if $f(x,y)$ is in P.

Let $\mathbb{C}_P = \{f(x,y) + P | f(x,y) \in \mathbb{C}[x,y]\}$. So $\mathbb{C}_P$ is a set of sets.

We define a $\mathcal{V}_{ar}$-structure N having $\mathbb{C}_P$ as its underlying set. The $\mathcal{V}_{ar}$-structure N interprets the constants 0 and 1 as the elements P and $1 + P$, respectively. We next define addition and multiplication for this structure. For $f_1(x,y)$ and $f_2(x,y)$ in $\mathbb{C}[x,y]$ let:

$$(f_1 + P) + (f_2 + P) = (f_1 + f_2) + P, \text{ and } (f_1 + P) \cdot (f_2 + P) = (f_1 \cdot f_2) + P.$$

This completes our description of the $\mathcal{V}_{ar}$-structure $N = (\mathbb{C}_P | 0, 1, +, \cdot)$.

We claim that $e : \mathbf{C} \to N$ defined by $e(a) = a + P$ is an embedding. We leave the verification of this to the reader. The range $\mathbf{C}_e$ of e is a substructure of N that is isomorphic to $\mathbf{C}$. For any $f(x,y) \in \mathbb{C}(x,y)$, let $f_e(x,y)$ be the result of applying e to each coefficient of $f(x,y)$. By the definition of addition and multiplication in N, $f_e(x,y) = f(x,y) + P$.

For example, if

$$f(x,y) = 2x + 5xy^2,$$

then

$$f_e(x,y) = e(2)x + e(5)xy^2 = (2+P)x + (5+P)xy^2 = 2x + 5xy^2 + P.$$

Claim $N \models \exists w \exists z (p_e(w,z) = 0 \wedge h_e(w,z) \neq 0)$.

First-order theories 259

Proof This is witnessed by the elements $x + P$ and $y + P$ of $\mathbb{C}_P[x,y]$. We have $p_e(x + P, y + P) = p(x + P, y + P) + P$ (by the definition of p_e), and $p(x + P, y + P) = p(x, y) + P$ (by the definition of $+$ and $\cdot$ in N). So we have $p_e(x + P, y + P) = (p(x, y) + P) + P = (p(x, y) + P) + (0 + P) = p(x, y) + P$. Since $p(x, y) \in P$, $p(x, y) + P = P$. Since N interprets 0 as the element P,

$$N \models p_e(x + P, y + P) = 0.$$

Likewise, $h_e(x+P, y+P) = h(x, y) + P$. Since $h(x, y)$ is not in P, $h(x, y) + P \neq P$, and $N \models h_e(x + P, y + P) \neq 0$. Thus the claim is verified.

We further claim that N is a field. The axioms T_F are easily verified. We leave this verification to the reader. By Proposition 5.118, there exists an extension $\mathbb{C}_P$ of N that models T_{ACF0}. Since $\mathbb{C}_P$ is an extension of the model $\mathbf{C}_e$ and T_{ACF0} is model-complete, $\mathbb{C}_P$ is an elementary extension of $\mathbf{C}_e$. We have

$\mathbf{C}_e \models \exists w \exists z (p_e(w, z) = 0 \wedge h_e(w, z) \neq 0)$ (since $\mathbf{C}_e \prec N$), and
$\mathbf{C} \models \exists w \exists z (p(w, z) = 0 \wedge h(w, z) \neq 0)$ (since $e : \mathbf{C} \to \mathbf{C}_e$ is an isomorphism).

By the semantics of $\exists$, $\mathbf{C} \models (p(a, b) = 0 \wedge h(a, b) \neq 0)$ for some $(a, b) \in \mathbb{C}^2$ as we wanted to show. □

Exercises

5.1. A theory T is $\forall_1$-*axiomatizable* if it has an axiomatization consisting of universal sentences.
   (a) Prove that T is $\forall_1$-axiomatizable if and only if for every $M \models T$ and every $A \subset M$, A is a model of T.
   (b) Find an example of a complete $\forall_1$-axiomatizable theory or show that no such theory exists.

5.2. A theory T is $\exists_1$-*axiomatizable* if it has an axiomatization consisting of existential sentences.
   (a) Prove that T is $\exists_1$-axiomatizable if and only if for any model M of T and any embedding $f : M \to N$, N is also a model of T.
   (b) Find an example of a complete $\exists_1$-axiomatizable theory or show that no such theory exists.

5.3. Show that the following are equivalent:
   (i) T is finitely axiomatizable.
   (ii) T is axiomatized by a single sentence.
   (iii) Any axiomatization of T has a finite subset that axiomatizes T.

5.4. Show that the following theories are not finitely axiomatizable:
  (a) The theory T_s of the integers with a successor function.
  (b) The theory T_{RG} of the random graph.
  (c) The theory T_{ACF0} of algebraically closed fields of characteristic 0.

5.5. Let T be a complete $\mathcal{V}_E$-theory that contains the theory of equivalence relations T_E. Show that T is finitely axiomatizable if and only if T has a finite model.

5.6. Let Γ_1 be the set of $\mathcal{V}_<$-sentences that hold in every finite model of T_{LO}. Let Γ_2 be the set of sentences saying that there exist at least n elements for each $n \in \mathbb{N}$. Let T_{FLO} be the set $\mathcal{V}_<$-sentences that can be derived from $\Gamma_1 \cup \Gamma_2$.
  (a) Show that T_{FLO} is a theory.
  (b) Show that T_{FLO} is quasi-finitely axiomatizable.
  (c) Show that T_{FLO} is not κ-categorical for any κ.

5.7. Let T_1 and T_2 be bi-definable theories each having finite vocabularies.
  (a) Show that T_1 is complete if and only if T_2 is.
  (b) Show that T_1 is finitely axiomatizable if and only if T_2 is.
  (c) Show that T_1 is quasi-finitely axiomatizable if and only if T_2 is.
  (d) Show that T_1 is κ-categorical if and only if T_2 is.
  (e) Show that T_1 is strongly minimal if and only if T_2 is.

5.8. Let $\mathcal{V}_P$ be the vocabulary consisting of a single unary relation P. Let T be a complete $\mathcal{V}_P$-theory having infinite models.
  (a) Show that T is countable categorical.
  (b) Give examples showing that T may or may not be totally categorical.

5.9. Show that there exists a complete quasi-finitely axiomatizable $\mathcal{V}$-theory having infinite models for every finite vocabulary $\mathcal{V}$.

5.10. For any first-order sentence φ, let $Spec(\varphi)$ denote the finite spectrum of φ (as defined in Exercise 2.3). Show that either $Spec(\varphi)$ or $Spec(\neg\varphi)$ is cofinite. (Hint: Use the previous exercise.)

5.11. Let $\mathcal{V}_E$ be the vocabulary consisting of a single binary relation E. Let M be an infinite $\mathcal{V}_E$-structure that interprets E as an equivalence relation. Suppose that each equivalence class of M has the same size.
  (a) Show that $Th(M)$ is countably categorical.
  (b) Show that $Th(M)$ is uncountably categorical if and only if the equivalence classes are finite.
  (c) Show that $Th(M)$ has quantifier elimination.

5.12. Let $\mathcal{V}_E$ be the vocabulary consisting of a single binary relation E. Let T be the $\mathcal{V}_E$-theory saying that E is an equivalence relation having infinitely many equivalence classes of size 3, infinitely many equivalence classes of size 5, and no other equivalence classes.
   (a) Axiomatize T.
   (b) How many models of size $\aleph_0$ does T have up to isomorphism?
   (c) How many models of size $\aleph_1$ does T have up to isomorphism?
   (d) Show that T does not have quantifier elimination.
   (e) Show that T is model-complete.

5.13. Show that T_{DLO} has $2^{\aleph_0}$ nonisomorphic models of size $2^{\aleph_0}$.

5.14. Let $\mathcal{V}_<^+ = \{<, P_b, P_s\}$ be the vocabulary consisting of a single binary relation $<$ and two unary relations P_b and P_s. Let T_{DLOE}^- be the $\mathcal{V}_<^+$-theory axiomatized by the $\mathcal{V}_<$-sentences δ_1–δ_5 in Section 5.3 together with the following two $\mathcal{V}_<^+$-sentences:

$$\delta_6' : \forall x \forall y (P_s(x) \to \neg(y < x))$$
$$\delta_7' : \forall x \forall y (P_b(x) \to \neg(x < y)).$$

So $P_s(x)$ means x is small and $P_b(X)$ means x is big.
   (a) Show that T_{DLOE}^- is incomplete.
   (b) Show that T_{DLOE}^- has exactly four countable models up to isomorphism.
   (c) Show that T_{DLOE}^- has quantifier elimination.

5.15. Let $\mathcal{V}_<(C)$ be the vocabulary $\{<, c_1, c_2, c_3, \ldots\}$ consisting of a binary relation $<$ and a denumerable set of constants. Let T_{CDLO} be the complete expansion of T_{DLO} to a $\mathcal{V}^+(C)$-theory that says $c_i < c_j$ if and only if $i < j$.
   (a) Show that T_{CDLO} has exactly three countable models up to isomorphism.
   (b) Show that T_{CDLO} is complete.
   (c) Show that T_{CDLO} has quantifier elimination.

5.16. Let $\mathcal{V}_E$ be the vocabulary consisting of a single binary relation E. Let T_E be the $\mathcal{V}_E$-theory that says E is an equivalence relation. Let M be a model of T_E that has exactly one equivalence class of size n for each $n \in \mathbb{N}$ and no other equivalence classes.
   (a) Axiomatize $Th(M)$.
   (b) Show that $Th(M)$ is not finitely axiomatizable.
   (c) Show that M is not κ-categorical for any κ.
Let $\mathcal{V}^+ = \{E, f\}$ where f is a unary function. Let φ^+ be the $\mathcal{V}^+$-sentence saying for each x there exists a unique y such that both $E(x, y)$ and

$\forall z(\neg f(z) = y)$. Let M^+ be an expansion of M to a $\mathcal{V}^+$-structure that interprets f as a one-to-one and onto function and models φ^+.
    (d) Show that $Th(M^+)$ is finitely axiomatizable.
    (e) Show that $Th(M^+)$ is not κ-categorical for any κ.

5.17. Let M and M^+ be as in Exercise 5.16.
    (a) Show that M is minimal but not strongly minimal.
    (b) Show that M^+ is not minimal.

5.18. Complete the proof of Proposition 5.58 by showing that T has quantifier elimination if and only if condition (ii)' holds.

5.19. Let T be a countable complete theory. Show that T has quantifier elimination if and only if condition (ii)' from Proposition 5.58 holds for all countable models M of T.

5.20. Let $\mathcal{B}$ be the set of all finite sequences of 0s and 1s (including the empty sequence). Let $M = (\mathcal{B}|S)$ be the structure in the vocabulary of a single binary relation S that interprets S as follows. For sequences s_1 and s_2 in $\mathcal{B}$, $M \models S(s_1, s_2)$ if and only if s_2 is obtained by adding a 0 or a 1 to the end of s_1. So S is a successor relation and every element of $\mathcal{B}$ has exactly two successors and at most one predecessor.
    (a) Show that $Th(\mathcal{B})$ is bi-definable with a model-complete theory that has a finite relational vocabulary. (Include a constant for the element having no predecessor.)
    (b) Show that any theory in a finite relational vocabulary that is bi-definable with $Th(\mathcal{B})$ cannot have quantifier elimination.
    (c) Show that $\mathcal{B}$ is a strongly minimal structure.

5.21. Let $\mathcal{V}_{Ps}$ be the vocabulary consisting of denumerably many unary relations $P_1, P_2, P_3, \ldots$ and let I and O be disjoint finite subsets of $\mathbb{N}$. Let $\varphi_{I,O}(x)$ be the $\mathcal{V}_{Ps}$-formula $\bigwedge_{i \in I} P_i(x) \wedge \bigwedge_{i \in O} \neg P_i(x)$.

This formula says that x is in each of the sets defined by P_i for $i \in I$ and outside each of the sets defined by P_i for $i \in O$. Let T_P be the $\mathcal{V}_{Ps}$-theory axiomatized by the sentences saying that there exist at least n elements satisfying $\varphi_{I,O}$ for each n in $\mathbb{N}$ and any finite disjoint subsets I and O of $\mathbb{N}$.
    (a) Show that T_P has quantifier elimination.
    (b) Show that T_P is not κ-categorical for any κ.

5.22. An *automorphism* of a structure M is an isomorphism $f : M \rightarrow M$ from M onto itself. Let T be a countable complete theory.
    (a) Suppose that, for any $M \models T$ and tuples $(a_1, \ldots, a_n)$ and $(b_1, \ldots, b_n)$ satisfying the same atomic formulas in M, there is an automorphism

f of M with $f(a_i) = b_i$ for $i = 1,\ldots,n$. Show that T has quantifier elimination.

(b) Suppose that T has quantifier elimination. Show that, for any $M \models T$ and tuples $(a_1,\ldots,a_n)$ and $(b_1,\ldots,b_n)$ satisfying the same atomic formulas in M, there exist an elementary extension N of M and an automorphism f of N with $f(a_i) = b_i$ for $i = 1,\ldots,n$.

5.23. Let $T_E = Th(M_2)$ where M_2 is the countable $\mathcal{V}_E$-structure defined in Example 5.18. Let $T_s = Th(Z_s)$ where Z_s is the $\mathcal{V}_s$-structure defined in Example 5.20.

We define a theory T that contains both of these theories. Let $\mathcal{V}$ be the vocabulary $\{E, s\}$. Let T be the set of all $\mathcal{V}$-sentences that can be derived from the set $T_E \cup T_s \cup \{\forall x \forall y (s(x) = y \to E(x, y))\}$.

(a) Show that T is complete.

(b) Refer to Exercise 5.22. Demonstrate a model M of T and tuples $(a_1,\ldots,a_n)$ and $(b_1,\ldots,b_n)$ from the universe of M such that

- $(a_1,\ldots,a_n)$ and $(b_1,\ldots,b_n)$ satisfy the same atomic formulas in M, and
- there is no automorphism f of M for which $f(a_i) = b_i$ for $i = 1,\ldots,n$.

(c) Show that T has quantifier elimination.

5.24. Let T be a theory. Prove that T is model-complete if and only if, for any model M of T, $T \cup D(M)$ is complete.

5.25. Let T be a theory. Let M be a model of T that can be embedded into any model of T. Show that if T is model-complete, then T is complete.

5.26. Show that T is model-complete if and only if, for any models M and N of T with $M \subset N$, there exists an elementary extension M' of M such that $M \subset N \subset M'$.

5.27. Let T be a model-complete theory. Let $T_{\forall\exists}$ be the set of all $\forall_2$-sentences φ such that $T \vdash \varphi$. Show that $M \models T_{\forall\exists}$ if and only if $M \models T$.
(Hint: Show that every model of $T_{\forall\exists}$ has an elementary extension that is the union of a chain of models of T.)

5.28. Let T be a theory and let M be a model of T. Show that M is existentially closed with respect to T if and only if M is existentially closed with respect to $T_\forall$.

5.29. Show that the following theories have the amalgamation property:
(a) The theory of graphs T_G.
(b) The theory of linear orders T_{LO}.
(c) The theory of fields T_F.

5.30. Let T be a theory. A theory T' is the *model-companion* of T if $T'_\forall = T_\forall$ and T' is model-complete.
  (a) Show that T_{RG} is the model-companion of T_G.
  (b) Show that T_{DLO} is the model-companion of T_{LO}.
  (c) Show that T_{ACF} is the model-companion of T_F.

5.31. Refer to the previous two exercises. Prove that if T has the amalgamation property and T' is the model-companion of T, then T' has quantifier elimination.

5.32. Verify that $acl_M(acl_M(A)) = acl_M(A)$ for any structure M and any subset A of the underlying set of M.

5.33. Let M be a strongly minimal $\mathcal{V}$-structure having underlying set U. Let $\varphi(x,y)$ be a $\mathcal{V}$-formula having two free variables. Show that there exists $n \in \mathbb{N}$ such that, for all $a \in U$: $|\varphi(a,M)|$ is infinite if and only if $|\varphi(a,M)| \geq n$. Show that this is not true for the minimal structure M from Exercise 5.17.

5.34. Let M be a structure and let $f: A \to B$ be an M-elementary function between subsets A and B of M. Show that A is algebraically closed if and only if B is algebraically closed.

5.35. Let G be a graph having a strongly minimal theory. Let a and b be vertices of G such that $dim_G(a,b) = 2$. Let $d_G(a,b)$ be the length of the shortest path (in G) from a to b if such a path exists and ∞ otherwise. Prove that there are exactly three possible values for $d_G(a,b)$ (including ∞).

5.36. Let T be a strongly minimal theory. Show that the following are equivalent.
  (i) T is locally modular.
  (ii) If T is expanded by adding one constant to the vocabulary, then the result is modular.
  (iii) Some expansion of T by constants is modular.

5.37. Let T be a strongly minimal theory. Show that the following are equivalent.
  (i) T is modular.
  (ii) If $c \in acl_M(A \cup \{b\})$, then $c \in acl_M(\{a,b\})$ for some $a \in A$ for any model M of T and any subset $A \cup \{b\}$ of the underlying set of M with $acl_M(A) = A$.
  (Hint: To show (ii) implies (i) use induction on $n = dim_M(A)$.)

First-order theories 265

5.38. Let $\mathbf{R}_f = \{\mathbb{R}|f, <, +, \cdot, 0, 1\}$ be an expansion of $\mathbf{R}_{or}$ where f is a unary function.
   (a) Show that if $\mathbf{R}_f$ interprets $f(x)$ as a polynomial, then $\mathbf{R}_f$ is o-minimal. (Use the fact that $\mathbf{R}_{or}$ is o-minimal.)
   (b) Show that if $\mathbf{R}_f$ interprets $f(x)$ as $\sin(x)$, then $\mathbf{R}_f$ is not o-minimal.
   (c) For any real number x, the *floor of x*, denoted $\lfloor x \rfloor$, is greatest integer less than or equal to x. Show that if $\mathbf{R}_f$ interprets $f(x)$ as $\lfloor x \rfloor$, then $\mathbf{R}_f$ is not o-minimal.

5.39. Let M be a $\mathcal{V}$-structure and let A be a subset of the universe U of M. The *definable closure* of A in M, denoted $dcl_M(A)$, is the set of all $d \in U$ such that $M \models \forall x(x = d \leftrightarrow \varphi(x))$ for some $\mathcal{V}(A)$-formula $\varphi(x)$. (The formula $\varphi(x)$ is said to *define* the unique element d over A.) Show that if M is o-minimal, then $dcl_M(A) = acl_M(A)$ for all $A \subset U$. Show that this is not necessarily true if M is strongly minimal.

5.40. Show that T_{RG} is not uncountably categorical.

5.41. We randomly construct a graph having vertices $V = \{v_1, v_2, v_3, \ldots\}$. For each pair of vertices v_i and v_j, we flip a coin. If the coin lands heads up, v_i and v_j share an edge. Otherwise, they do not share an edge. Suppose that our coin is unfair. Say that our coin lands heads up only 1 out of 1000 times. Show that (after flipping the coin infinitely many times) the resulting random graph will be isomorphic to G_R (with probability 1).

5.42. We define a graph having $\mathbb{N}$ as vertices. Any natural number n can be uniquely factored as $p_1^{a_1} \cdot p_2^{a_2} \cdot p_3^{a_3} \cdots p_m^{a_m}$ where the p_is are distinct primes. We say that each of the exponents a_i in this factorization are "involved in n." We now define our graph: two natural numbers a and b share an edge if and only if either a is involved in b or b is involved in a. Show that the resulting graph is isomorphic to the random graph.

5.43. Show that for every substructure A of the random graph G_R, either $G_R \cong A$ or $G_R \cong (G_R - A)$ (where $(G_R - A)$ is the substructure having the vertices that are not in A as an underlying set).

5.44. Show that the 0–1 law fails for vocabularies that are not relational. (Hint: Consider the sentence $\exists x f(x) = x$.)

5.45. Let T_{ACFp} be the $\mathcal{V}_{ar}$-theory of algebraically closed fields of characteristic p (for prime p). Prove that, for any $\mathcal{V}_{ar}$-sentence φ, the following are equivalent:
   (i) $T_{ACF0} \models \varphi$,

(ii) $T_{ACFp} \models \varphi$ for sufficiently large primes p, and

(iii) $T_{ACFp} \models \varphi$ for arbitrarily large primes p.

5.46. Algebraically closed fields of any characteristic are necessarily infinite. However, every finite subset of T_{ACFp} has arbitrarily large finite models for any prime p. Using this fact (and the previous exercise) prove Ax's theorem.

Ax's theorem: Let $f(x)$ be a polynomial having complex coefficients. If $f : \mathbb{C} \to \mathbb{C}$ is one-to-one, then f is onto.

6 Models of countable theories

We define and study *types* of a complete first-order theory T. This concept allows us to refine our analysis of $Mod(T)$. If T has few types, then $Mod(T)$ contains a uniquely defined smallest model that can be elementarily embedded into any structure of $Mod(T)$. We investigate the various properties of these small models in Section 6.3. In Section 6.4, we consider the "big" models of $Mod(T)$. For any theory, the number of types is related to the number of models of the theory. For any cardinal κ, $I(T, \kappa)$ denotes the number of models in $Mod(T)$ of size κ. We prove two basic facts regarding this cardinal function. In Section 6.5, we show that if T has many types, then $I(T, \kappa)$ takes on its maximal possible value of 2^κ for each infinite κ. In Section 6.6, we prove Vaught's theorem stating that $I(T, \aleph_0)$ cannot equal 2.

All formulas are first-order formulas. All theories are sets of first-order sentences. For any structure M, we conveniently refer to an n-tuple of elements from the underlying set of M as an "n-tuple of M."

6.1 Types

The notion of a *type* extends the notion of a *theory* to include formulas and not just sentences. Whereas theories describe structures, types describe elements within a structure.

Definition 6.1 Let M be a $\mathcal{V}$-structure and let $\bar{a} = (a_1, \ldots, a_n)$ be an n-tuple of M. The type of $\bar{a}$ in M, denoted $tp_M(\bar{a})$, is the set of all $\mathcal{V}$-formulas $\varphi(\bar{x})$ having free variables among $x_1, \ldots, x_n$ that hold in M when each x_i in $\bar{x}$ is replaced by a_i. More concisely, but less precisely: $tp_M(\bar{a}) = \{\varphi(\bar{x}) | M \models \varphi(\bar{a})\}$.

If $\bar{a}$ is an n-tuple, then each formula in $tp_M(\bar{a})$ contains at most n free variables but may contain fewer. In particular, the type of an n-tuple contains sentences. For any structure M and tuple $\bar{a}$ of M, $tp_M(\bar{a})$ contains $Th(M)$ as a subset. The set $tp_M(\bar{a})$ provides the complete first-order description of the tuple $\bar{a}$ and how it sits in M. This description is not necessarily categorical; many tuples within the same structure may have the same type.

Example 6.2 Let $\mathbf{Q}_<$ be the structure $(\mathbb{Q}| <)$ that interprets $<$ as the usual order on the rational numbers. This structure is a model of the theory T_{DLO} of dense linear orders discussed in Section 5.4. Consider the four-tuple $(-2, -1, 1, 2)$. The type $tp_{\mathbf{Q}_<}(-2,-1,1,2)$ contains the formulas $x_1 < x_2$, $x_2 < x_3$, and $x_3 < x_4$. Since T_{DLO} has quantifier elimination, for any four-tuple $\bar{a} = (a_1, a_2, a_3, a_4)$ of rational numbers, if $a_1 < a_2 < a_3 < a_4$ then $tp_{\mathbf{Q}_<}(\bar{a})$ is the same as $tp_{\mathbf{Q}_<}(-2,-1,1,2)$.

Definition 6.3 Let Γ be a set of formulas having free variables among $x_1, \ldots, x_n$. A structure M *realizes* Γ if Γ is a nonempty subset of $tp_M(\bar{a})$ for some tuple $\bar{a}$ of M. Otherwise, M is said to *omit* Γ. The set Γ is *realizable* if it is realized in some structure.

Note the distinction between the terms *realizable* and *satisfiable*. The set $tp_M(\bar{a})$ is realizable by definition, but rarely is $tp_M(\bar{a})$ satisfiable (see Exercise 6.4). This is because $tp_M(\bar{a})$ contains formulas that are not sentences. Recall that a formula $\varphi(\bar{x})$ is equivalent to the sentence $\forall \bar{x} \varphi(\bar{x})$. So when we say that a formula $\varphi(x_1, \ldots, x_n)$ is satisfiable, we mean that it holds for all n-tuples of a structure. When we say that $\varphi(x_1, \ldots, x_n)$ is realizable, we mean that it holds for some n-tuple of a structure.

We now define the key concept of this chapter.

Definition 6.4 An *n-type* is a realizable set of formulas having free variables among $x_1, \ldots, x_n$. A *type* is an n-type for some n.

The sets $tp_M(\bar{a})$ are examples of types. Moreover, these are the only examples we need to consider. Every type is a subset of $tp_M(\bar{a})$ for some M and $\bar{a}$. The types $tp_M(\bar{a})$ are called *complete types*. Types that are not complete are called *partial types*. We typically use p, q, and r to denote types (Γ is used to denote arbitrary sets of formulas). We often write a type with its free variables as $p(x_1, \ldots, x_n)$. The notation $p(t_1, \ldots, t_n)$ represents the set of formulas obtained by replacing each x_i with the term t_i.

Since types are generally not satisfiable, they are not consistent. This is unfortunate. Much of the previous chapters has been devoted to consistent sets of formulas. We can recover results from the previous chapters and apply them to types by making the following observation: the formula $\varphi(x)$ is realizable if and only if the sentence $\varphi(c)$ is satisfiable for some constant c. We state this more generally as the following proposition.

Proposition 6.5 Let $\Gamma(x_1, \ldots, x_n)$ be a set of formulas having free variables among $x_1, \ldots, x_n$. Let $c_1, \ldots, c_n$ be constants not in the vocabulary of Γ. Then $\Gamma(x_1, \ldots, x_n)$ is realizable if and only if $\Gamma(c_1, \ldots, c_n)$ is satisfiable.

Proof $\Gamma(x_1, \ldots, x_n)$ is realizable if and only if

$\Gamma(x_1, \ldots, x_n)$ is a subset of $tp_M(\bar{a})$ for some M and $\bar{a}$.

This happens if and only if $M' \models \Gamma(c_1,\ldots,c_n)$ where M' is an expansion of M that interprets the constants $c_1,\ldots,c_n$ as the tuple $\bar{a}$ of M. □

So for any realizable set of formulas, there is a closely related set of sentences that is satisfiable. This allows us to apply properties regarding satisfiability to types that are not satisfiable. In particular, the Compactness theorem remains true when "satisfiable" is replaced with "realizable."

Proposition 6.6 Let $\Gamma(x_1,\ldots,x_n)$ be a set of formulas having free variables among $x_1,\ldots,x_n$. Every finite subset of Γ is realizable if and only if Γ is realizable.

Proof Let $c_1,\ldots,c_n$ be constants not in the vocabulary of Γ.

By Proposition 6.5, $\Gamma(x_1,\ldots,x_n)$ is realizable if and only if $\Gamma(c_1,\ldots,c_n)$ is satisfiable.

By the Compactness theorem, $\Gamma(c_1,\ldots,c_n)$ is satisfiable if and only in every finite subset of $\Gamma(c_1,\ldots,c_n)$ is satisfiable.

Finally, again by Proposition 6.5, every finite subset of $\Gamma(c_1,\ldots,c_n)$ is satisfiable if and only if every finite subset of $\Gamma(x_1,\ldots,x_n)$ is realizable. □

Let T be a complete theory. Any type that is realized in a model of T, whether it is partial or complete, is called a *type of T*. The set of all complete types of T is denoted $S(T)$. Equivalently, $S(T)$ is the set of all complete types that contain T as a subset. We denote by $S_n(T)$ the set of all n-types in $S(T)$.

Corollary 6.7 Let T be a complete theory and let Γ be a set of formulas having free variables among $x_1,\ldots,x_n$. If each finite subset of Γ is a type of T, then Γ is an type of T.

Proof Apply Proposition 6.6 to the set $\Gamma \cup T$. □

Example 6.8 Let $\mathcal{V}_E$ be the vocabulary consisting of a single binary relation E. Let M be the $\mathcal{V}$-structure that interprets E as an equivalence relation having exactly one equivalence class of size n for each $n \in \mathbb{N}$ and no other equivalence classes. Let $T = Th(M)$. We depict M as follows:

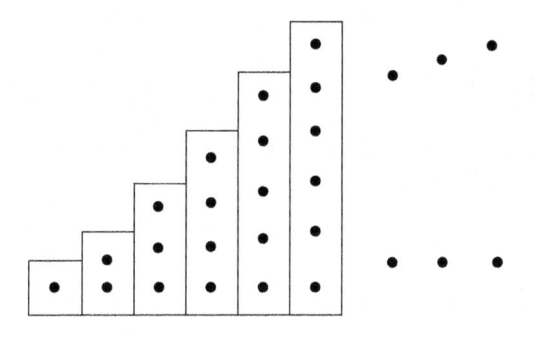

Each box represents an equivalence class. Each of these equivalence classes determines a unique type in $S_1(T)$. For any $m \in \mathbb{N}$, let p_m be the type of an element in the equivalence class containing exactly m elements. Any two such elements have the same type (we cannot distinguish between two elements in the same equivalence class using the vocabulary $\mathcal{V}_E$). Let φ_m be the $\mathcal{V}_E$-formula saying that there are exactly m elements equivalent to x_1. Then $\varphi_m(x_1) \in p_m$ and p_m is the only type in $S_1(T)$ that contains φ_m.

The set $S_1(T)$ contains the types p_1, p_2, p_3, and so forth. Given any element a in the universe of M, $tp_M(a)$ equals p_m, where m is the number of elements in the equivalence class containing a. So M realizes each of the types p_m for $m \in \mathbb{N}$ and no other types. However, there does exist another type in $S_1(T)$.

Consider the set of $\mathcal{V}$-formulas $\Gamma(x_1) = \{\neg\varphi_m(x_1) | m \in \mathbb{N}\}$. These formulas say that, for each $m \in \mathbb{N}$, there are not exactly m elements equivalent to x_1. Given any finite subset Δ of Γ, $\Delta \subset p_m$ for sufficiently large $m \in \mathbb{N}$. By Corollary 6.7, $\Gamma(x_1)$ is a type of T. Let $p_\infty \in S_1(T)$ be a complete type containing Γ as a subset. This type says that there exist infinitely many elements equivalent to x_1. This type is not realized in M, but it is realized in an elementary extension of M. Let N be the model of T having M as a substructure and also having one denumerable equivalence class and no other infinite equivalence classes. Then N realizes the type p_∞ as well as the types $p_1, p_2, p_3, \ldots$ These are all of the types in $S_1(T)$.

Now consider $S_2(T)$. This is the set of all 2-types realized in some model of T. Each 2-type contains formulas having at most two free variables (namely x_1 and x_2). For any m and l in $\mathbb{N}$, let $p_{m,l}$ be the complete 2-type that says there are exactly m elements equivalent to x_1 and l elements equivalent to x_2. Then $p_{m,l}$ is the unique 2-type containing the two 1-types $p_m(x_1)$ and $p_l(x_2)$ as subsets. Each formula in $p_{m,l}$ can be derived from $T \cup p_m(x_1) \cup p_l(x_2)$. In particular, the formula $E(x_1, x_2)$ is in $p_{m,l}$ if and only if $l = m$.

Consider now the partial 2-type $\Gamma_{\infty,\infty} = p_\infty(x_1) \cup p_\infty(x_2)$. This 2-type is the union of two complete 1-types, but it is not complete. To obtain a complete 2-type, we must say whether or not x_1 is equivalent to x_2. Let $p_{\infty,\infty}$ be the complete 2-type that contains $\Gamma_{\infty,\infty}$ and the formula $\neg E(x_1, x_2)$. This type is not realized in N_1, but it is realized in the elementary extension N_2 of N_1 that contains exactly two denumerable equivalence classes.

Let T be a complete theory and let p be in $S(T)$. By definition, p is realized in some model of T. As the previous example indicates, something stronger is true. Given any $M \models T$ there exists an elementary extension N of M that realizes p.

Proposition 6.9 Let T be a complete theory and let $M \models T$. Each type in $S(T)$ is realized in some elementary extension of M.

Proof Let p be an n-type in $S(T)$. Let $\mathcal{ED}(M)$ be the elementary diagram of M. Let $c_1, \ldots, c_n$ be constants that do not occur in $\mathcal{ED}(M)$. By the Joint Consistency lemma 4.63, $\mathcal{ED}(M) \cup p(c_1, \ldots, c_n)$ is consistent. By Proposition 4.27, there exists a model N of $\mathcal{ED}(M) \cup p(c_1, \ldots, c_n)$. This model realizes p and is an elementary extension of M. □

6.2 Isolated types

A type in $S(T)$ may be realized in some models of T and omitted in others. In this section, we focus on those types in $S(T)$ that are realized in every model of T.

Definition 6.10 Let T be a complete theory and let p be an n-type in $S_n(T)$. If there exists some formula $\theta \in p$ such that p is the only type in $S_n(T)$ containing θ, then p is said to be *isolated* in $S_n(T)$ and the formula θ is said to *isolate* p in $S_n(T)$. A partial n-type is an *isolated* type of T if it is contained in a complete type that is isolated in $S_n(T)$. Otherwise, it is a *nonisolated* type of T.

Our goal for this section is to show that the isolated types of T are realized in every model of T and that, for countable T, these are the only types realized in every model of T.

Example 6.11 In Example 6.8, the type p_m is isolated by the formula φ_m for each m in $\mathbb{N}$. These are the only isolated types in $S_1(T)$. Likewise, the 2-types $p_{m,l}$ for m and l in $\mathbb{N}$ are the only isolated types of $S_2(T)$. The isolated types are precisely the types that are realized in M. These types are also realized in every model of $Th(M)$.

Proposition 6.12 Let T be a complete theory and let p be a type of T. If p is isolated, then p is realized in every model of T.

Proof Let q be an isolated type in $S_n(T)$ that contains p. Let $\theta(\bar{x})$ be a formula that isolates q in $S_n(T)$. There exists a model M of T that realizes q. In particular, $M \models \exists \bar{x} \theta(\bar{x})$. Since T is complete, the sentence $\exists \bar{x} \theta(\bar{x})$ is in T.

Let N be an arbitrary model of T. By the semantics of $\exists$, $N \models \theta(\bar{a})$ for some tuple $\bar{a}$ of N. Since $tp_N(\bar{a})$ is in $S_n(T)$ and contains $\theta(\bar{x})$, this type must be q. Since N was an arbitrary model of T, every model realizes q, and, hence, p as well. □

If the vocabulary is countable, then the converse of Corollary 6.12 holds. If a type of T is realized in every model of a countable theory T, then that type must be isolated. Put another way, every nonisolated type of T is omitted by some model of T. To prove this, we use a Henkin construction to obtain a model that omits a given nonisolated type. The proof of Theorem 4.2 serves a precedent for such a construction. The reader may want to refer to that proof. Essentially, the following theorem is proved by adding one step to the proof of Theorem 4.2.

Theorem 6.13 (Omitting Types) Let T be a complete theory in a countable vocabulary $\mathcal{V}$. If p is a nonisolated type in $S(T)$, then there exists a model M of T that omits p.

Proof We want to demonstrate a structure that models T and omits p.

Let $\mathcal{V}$ be the vocabulary of Γ. Let $\mathcal{V}^+ = \mathcal{V} \cup \{c_1, c_2, c_3, \ldots\}$ where each c_i is a constant that does not occur in $\mathcal{V}$. Let C denote the set $\{c_1, c_2, c_3, \ldots\}$. Let D denote the set of $\mathcal{V}^+$-terms and let D^n be the set of n-tuples of elements from D.

We shall define a complete $\mathcal{V}^+$-theory T^+ with the following three properties.

Property 1 Every sentence of T is in T^+.

Property 2 For every $\mathcal{V}^+$-sentence in T^+ of the form $\exists x \theta(x)$, the sentence $\theta(c_i)$ is also in T^+ for some $c_i \in C$.

Property 3 For each $\bar{d}$ in D^n, there exists a formula $\varphi(\bar{x})$ in p such that the sentence $\neg \varphi(\bar{d})$ is in T^+.

As in the proof of Theorem 4.2, Property 2 allows us to find a model M^+ of T^+. By Property 1, M^+ is a model of T. Property 3 ensures that p is not the type of any tuple of $\mathcal{V}^+$-terms in D^n. Recall from the proof of Theorem 4.2 that the underlying set of M^+ is a set of $\mathcal{V}^+$-terms. It follows that M^+ is a model of T that omits p as was required. So if we can successfully define T^+ having the above three properties, then this will prove the theorem.

We define T^+ in stages. Let T_0 be T.

Since the vocabulary $\mathcal{V}$ is countable, $\mathcal{V}^+$ is denumerable. By Proposition 2.47, there are denumerably many $\mathcal{V}^+$-formulas. It follows that there are denumerably many $\mathcal{V}^+$-terms and $\mathcal{V}^+$-sentences. Let $\{\bar{d}_1, \bar{d}_2, \bar{d}_3, \ldots\}$ be an enumeration of D^n and let $\{\varphi_1, \varphi_2, \varphi_3 \ldots\}$ enumerate the set of all $\mathcal{V}^+$-sentences.

Suppose that T_m has been defined in such a way that T_m is consistent and only finitely many sentences of T_m contain constants in C. We define T_{m+1} in two steps. First, we define T'_{m+1} in the same way that T_{m+1} was defined in both Theorems 4.2 and 4.27.

Step 1:
(a) If $T_m \cup \{\neg \varphi_{m+1}\}$ is consistent, then define T'_{m+1} to be $T_m \cup \{\neg \varphi_{m+1}\}$.
(b) If $T_m \cup \{\neg \varphi_{m+1}\}$ is not consistent, then $T_m \cup \{\varphi_{m+1}\}$ is consistent. We divide this case into two subcases.
   (i) If φ_{m+1} does not have the form $\exists x \theta(x)$ for some formula $\theta(x)$, then just let T'_{m+1} be $T_m \cup \{\varphi_{m+1}\}$.
   (ii) Otherwise φ_{m+1} has the form $\exists x \theta(x)$.

In this case let T'_{m+1} be $T_m \cup \{\varphi_{m+1}\} \cup \{\theta(c_i)\}$, where i is such that c_i does not occur in $T_m \cup \{\varphi_{m+1}\}$.

We know from the proof of Theorem 4.2 that if T_m is consistent, then so is T'_{m+1}. This was the first claim of that proof. Also, if T_m contains only finitely many sentences that use constants from C, then so does T'_{m+1}. This is because T'_{m+1} is obtained by adding only a sentence or two to T_m.

In Step 2, we ensure that the tuple $\bar{d}_{m+1}$ in the enumeration of D^n does not realize p.

Step 2: Let $T_{m+1} = T'_{m+1} \cup \{\neg\psi(\bar{d}_{m+1})\}$, where $\psi(\bar{x})$ is any formula in p such that T_{m+1} is consistent.

We must verify that such a formula $\psi(\bar{x}) \in p$ exists. Let $\Theta(\bar{c}, \bar{d}_{m+1})$ be the conjunction of the finitely many sentences in T'_{m+1} that contain constants from C. Then T'_{m+1} is equivalent to $T \cup \{\Theta(\bar{c}, \bar{d}_{m+1})\}$. In particular, $T'_{m+1} \vdash \Theta(\bar{c}, \bar{d}_{m+1})$ (by $\wedge$-Introduction). The tuple $\bar{c}$ contains all the constants that occur in T'_{m+1} and do not occur in $\bar{d}_{m+1}$. The sentence $\Theta(\bar{c}, \bar{d}_{m+1})$ may not contain all (or any) of the constants in $\bar{d}_{m+1} = (d_1, d_2, \ldots, d_n)$. Let $\Theta(\bar{c}, \bar{x})$ be the formula obtained by replacing each occurrence of d_i in Θ with x_i (for $i = 1, \ldots, n$).

Consider the formula $\exists \bar{y} \Theta(\bar{y}, \bar{x})$. Since $T'_{m+1} \vdash \Theta(\bar{c}, \bar{d}_{m+1})$, this formula is realized in every model of T'_{m+1}.

If this formula is not in p, then its negation is (since p is a complete type). In this case, let $\psi(\bar{x})$ be $\neg \exists \bar{y} \Theta(\bar{y}, \bar{x})$. Since $\neg\psi(\bar{x})$ is equivalent to $\exists \bar{y} \Theta(\bar{y}, \bar{x})$, $T'_{m+1} \vdash \neg\psi(\bar{d}_{m+1})$. In particular, $T_{m+1} = T'_{m+1} \cup \{\neg\psi(\bar{d}_{m+1})\}$ is consistent.

So we may assume that the formula $\exists \bar{y} \Theta(\bar{y}, \bar{x})$ is in p. In this case, we cannot let ψ be this formula (since we want $T'_{m+1} \cup \{\neg\psi(\bar{d}_{m+1})\}$ to be consistent) nor its negation (since it is not in p). To find a formula ψ that works we use the fact that p is not isolated.

Since p is nonisolated, it is not the only type in $S(T)$ containing the formula $\exists \bar{y} \Theta(\bar{y}, \bar{x})$. Let q be another type in $S(T)$ that contains this formula. Since p and q are different types, there must be a formula in p that is not in q. Let $\psi(\bar{x})$ be any such formula. Then $T \cup \{\exists \bar{y} \Theta(\bar{y}, \bar{x}), \neg\psi(\bar{x})\}$ is realizable (since it is a subset of $q \in S(T)$). It follows that $T \cup \{\Theta(\bar{y}, \bar{x}), \neg\psi(\bar{x})\}$ is also realizable. By Proposition 6.5, $T \cup \{\Theta(\bar{c}, \bar{d}_{m+1}), \neg\psi(\bar{d}_{m+1})\}$ is satisfiable. Note that $T \cup \{\Theta(\bar{c}, \bar{d}_{m+1}), \neg\psi(\bar{d}_{m+1})\}$ is $T_{m+1} = T'_{m+1} \cup \{\neg\psi(\bar{d}_{m+1})\}$. Since it is satisfiable, it is consistent as we wanted to show.

So given a consistent $\mathcal{V}^+$-theory T_m containing T and only finitely many other sentences, we have defined the consistent $\mathcal{V}^+$-theory T_{m+1} by adding a few sentences to T_m. Starting with $T_0 = T$, this iterative process generates $\mathcal{V}^+$-theories T_0, T_1, T_2, and so forth. Let T^+ be the union of these theories.

Since each T_m is consistent, so is T^+. Also, by Step 1 of the definition of T_{m+1}, either φ_{m+1} or $\neg\varphi_{m+1}$ is in T_{m+1}. Since this is true for each φ_{m+1} in the enumeration of all $\mathcal{V}^+$-sentences, T^+ is a complete theory. Since $T = T_0 \subset T^+$, T^+ has Property 1. Part (b)(ii) of Step 1 guarantees that T^+ has Property 2. Step 2 guarantees Property 3.

So T^+ has all of the desired properties and a model M^+ of T^+ can be defined as in the proof of Theorem 4.2. The underlying set of M^+ is a subset of D. Property 3 of T^+ ensures that no n-tuple of elements in D satisfies all formulas of p. It follows that M^+ is a model of $T \subset T^+$ that omits p. □

So, if T is countable, then a type in $S(T)$ is realized in every model of T if and only if it is isolated. This remains true when restricted to countable models.

Corollary 6.14 Let T be a countable complete theory. A type $p \in S(T)$ is isolated if and only if it is realized in every countable model of T.

Proof By Proposition 6.12, if $p \in S(T)$ is isolated, then it is realized in every model of T. In particular, it is realized in every countable model of T. Conversely, if p is nonisolated, then p is omitted from a model M^+ of T by the Omitting Types theorem. Moreover, the model M^+ constructed in the proof of that theorem is countable. □

In particular, every type realized in the countable model of an $\aleph_0$-categorical theory must be an isolated type. This yields characterizations of $\aleph_0$-categorical theories in terms of $S(T)$.

Theorem 6.15 Let T be a complete theory having infinite models. The following are equivalent.

(i) T is $\aleph_0$-categorical.
(ii) $S(T)$ is countable and every type in $S(T)$ is isolated.
(iii) $S_n(T)$ is finite for each $n \in \mathbb{N}$.
(iv) There are finitely many formulas in n free variables up to T-equivalence for each $n \in \mathbb{N}$.

Proof Note that "(i) if and only if (iv)" is a restatement of Theorem 5.32. We proved as Proposition 5.31 that (iv) implies (i). It remains to be shown that (i) implies (iv).

First, we show that (i) implies (ii). Suppose that T is $\aleph_0$-categorical and let p be any type in $S(T)$. Then p is realized as $tp_N(\bar{a})$ in some model N of T. By the Downward Löwenhiem–Skolem theorem, there exists a countable elementary substructure M of N containing $\bar{a}$ in its universe. So p is realized in a countable model of T. Since T is $\aleph_0$-categorical, p is realized in every countable model of T. By Corollary 6.14 of the Omitting Types theorem, p is isolated. Moreover, since every type in $S(T)$ is realized in the countable model M, there are only countably many types in $S(T)$.

Next we show that (ii) implies (iii). Suppose that (ii) holds. Then $S(T)$ is countable. Suppose for a contradiction that $S_n(T)$ is denumerable. Let $\{p_1, p_2, p_3, \ldots\}$ be an enumeration of $S_n(T)$. For each $i \in \mathbb{N}$, there is a formula φ_i that isolates p_i. Let $\Gamma = \{\neg\varphi_1, \neg\varphi_2, \neg\varphi_3, \ldots\}$. Every finite subset of Γ is contained in infinitely many types in $S_n(T)$. By Corollary 6.7, Γ is an n-type of T. So Γ is a subset of some p_i in $S_n(T)$. Since Γ contains the formula $\neg\varphi_i$, this is a contradiction. This contradiction proves that $S_n(T)$ must be finite.

It remains to be shown that (iii) implies (iv). Suppose that there are only finitely many types in $S_n(T)$. Let $\{p_1, \ldots, p_k\}$ enumerate $S_n(T)$. If p_i and p_j are distinct types in this set, then there is some formula φ_{ij} that is in p_i and not in p_j. By taking the conjunction of the formulas φ_{ij} for various js, we obtain a formula Φ_i that is contained in p_i and no other type of $S_n(T)$. We see that each type in $S_n(T)$ is isolated (we have incidently shown that (iii) implies (ii)).

Now let φ be any formula in the vocabulary of T having n free variables. We claim that φ is T-equivalent to a disjunction of the formulas Φ_i for various values of i. For example, if φ is contained in p_1 and p_1 and no other type in $S_n(T)$, then φ is T-equivalent to the formula $\Phi_1 \vee \Phi_2$. Since there are only finitely many possible disjunctions of this form, there are finitely many formulas having n free variables up to T-equivalence. □

The proof of the Omitting Types theorem is similar to the proof of Theorem 4.2. Unlike Theorem 4.2, however, the Omitting Types theorem does not hold for theories having uncountable vocabularies (see Exercise 6.7). The Omitting Types theorem can be extended in another way.

Theorem 6.16 (Countable Omitting Types) Let T be a complete theory in a countable vocabulary $\mathcal{V}$. Let $p_1, p_2, p_3, \ldots$ be countably many types in $S(T)$ each of which is not isolated. There exists a model M of T that omits each p_i.

Proof This can be proved by modifying the proof of the Omitting Types theorem. See Exercise 6.9. □

6.3 Small models of small theories

Let T be a countable complete theory having infinite models. If $S(T)$ is countable, then T is said to be a *small theory*. In this and the next sections, we investigate some of the countable structures in $Mod(T)$. We show that if T is small, then $Mod(T)$ contains a smallest countable model and a biggest countable model. Of course, any two countable structures have the same size. When we refer to the smallest or biggest countable model, we are referring to the types realized in the model.

At minimum, any model of T must realize the isolated types in $S(T)$. Countable *atomic* models realize only these types. When they exist, these are the smallest countable models in $Mod(T)$. At the other extreme, countable *saturated* models realizes all types in $S(T)$ (although this is not the full definition). When they exist, these are the biggest countable models in $Mod(T)$. We show that such models do exist if T is small. Moreover, we show that atomic and saturated models posses many useful properties.

We deal with atomic models in this section and saturated models in the next. We begin with some examples (and nonexamples) of small theories.

Example 6.17 By Proposition 6.15, any $\aleph_0$-categorical theory is small.

Example 6.18 Let M be the $\mathcal{V}_E$-structure defined in Example 6.8. Let T be $Th(M)$. It follows from the discussion in Example 6.8 that $S_n(T)$ is countable for each n. So T is small.

Example 6.19 Let $\mathbf{R}_<$ be the $\mathcal{V}_<$-structure $(\mathbb{R}|<)$. The $\mathcal{V}_<$-theory of $\mathbf{R}_<$ is T_{DLO}. Since T_{DLO} is $\aleph_0$-categorical, it is small.

Let $C_{\mathbb{Q}} = (c_a | a \in \mathbb{Q})$ be a countable set of constants. Let $\mathcal{V}_{\mathbb{Q}}$ be $\mathcal{V}_< \cup C_{\mathbb{Q}}$. Let $R_{\mathbb{Q}}$ be the expansion of $\mathbf{R}_<$ to a $\mathcal{V}_{\mathbb{Q}}$-theory that interprets each constant c_a as the number $a \in \mathbb{Q}$. Let $T = Th(R_{\mathbb{Q}})$. Then T is countable but not small.

For each real number r, let p_r denote $tp_{R_{\mathbb{Q}}}(r)$. To see that T is not small, consider the set $P = \{p_r | r \in \mathbb{R}\}$. Given any two distinct real numbers b and c, there exists some rational number a between b and c. With no loss of generality, we may assume that b is smaller than c. Then $x_1 < c_a$ is a $\mathcal{V}_{\mathbb{Q}}$-formula in p_b that is not in p_c. We see that no two types in P are the same. It follows that $|S_1(T)| \geq 2^{\aleph_0}$ and T is not small.

6.3.1 Atomic models. If T is small, then $S(T)$ contains countably many non-isolated types. By the Countable Omitting Types theorem 6.16, there exists a model of T that omits all of them. Such a model is said to be *atomic*.

Definition 6.20 A structure M is *atomic* if $tp_M(\bar{a})$ is an isolated type of $Th(M)$ for every tuple $\bar{a}$ of M.

Example 6.21 The $\mathcal{V}_E$-structure M from Examples 6.8 and 6.18 is atomic. Since M has no infinite equivalence class, each type realized in M is isolated by the formula φ_m for some $m \in \mathbb{N}$ where φ_m is as defined in Example 6.8.

Proposition 6.22 Let T be a countable complete theory. If T is small, then there exists a countable atomic model M of T.

Proof By the Countable Omitting Types theorem 6.16 there exists a model that omits all nonisolated types. By the Downward Löwenhiem–Skolem theorem, we can find such a model that is countable. □

As the following example shows, the converse of this proposition does not hold.

Example 6.23 Let T be the $\mathcal{V}_Q$-theory from Example 6.19. This theory is not small, but it does have a countable atomic model. Let $\mathbf{Q}_Q$ be the $\mathcal{V}_Q$-structure having $\mathbb{Q}$ as an underlying set and interpreting $<$ as the usual order and each constant c_a as the rational number a. This is a model of T. Each 1-type realized in $\mathbf{Q}_Q$ is isolated by the formula $x_1 = c_a$ for some $a \in \mathbb{Q}$. It follows that every type realized in $\mathbf{Q}_Q$ is isolated.

Next, we give an example of a countable theory that does not have a countable atomic model. By Proposition 6.22, such a theory necessarily has uncountably many n-types for some n.

Example 6.24 Let $\mathcal{V}_{Ps}$ be the vocabulary consisting of denumerably many unary predicates P_i for $i \in \mathbb{N}$. Let T_P be the $\mathcal{V}_{Ps}$-theory defined in Exercise 5.21.

For any subset A of $\mathbb{N}$, let Γ_A be the set of formulas containing $P_i(x_1)$ for each $i \in A$ and $\neg P_i(x_1)$ for each $i \notin A$. By Corollary 6.7, this is a type of T_P. Since T_P has quantifier elimination (by Exercise 5.21(a)), there is exactly one type p_A in $S_1(T_P)$ containing Γ_A. Moreover, each type p in $S_1(T_P)$ is p_A for some $A \subset \mathbb{N}$. It follows that $|S_1(T_P)| = |\mathcal{P}(\mathbb{N})| = 2^{\aleph_0}$.

For any $p \in S_1(T_P)$, we claim that p is not isolated. Given any finite subset Δ of p, there are infinitely many relations P_i that do not occur in Δ. By the axioms of T_{Ps}, both $\Delta \cup \{P_i(x_1)\}$ and $\Delta \cup \{\neg P_i(x_1)\}$ are realizable. So p is not isolated by any formula. Since $S_1(T_P)$ has no isolated types, T_P cannot possibly have an atomic model.

6.3.2 Homogeneity. Having established in Proposition 6.22 the existence of countable atomic models for small T, we now investigate some of the properties of these models. We show that countable atomic models are unique, prime, and homogeneous.

Definition 6.25 A countable structure M is said to be *homogeneous* if, given n-tuples $\bar{a}$ and $\bar{b}$ of M with $tp_M(\bar{a}) = tp_M(\bar{b})$, for any c of M, there exists d such that $tp_M(\bar{a}, c) = tp_M(\bar{b}, d)$.

Most of the countable structures we have discussed have been homogeneous. An example of a nonhomogeneous structure is provided by the structure M in Example 5.56. Referring to that example, let $\bar{a} = (a_1, \ldots, a_n)$ and let $\bar{b} = (b_1, \ldots, b_n)$. These tuples share the same type in M even though they are different in an obvious way. Whereas $\bar{b}$ has equivalent elements that are far away (in terms of s), there are no such elements equivalent to $\bar{a}$. The structure is not homogeneous because this distinction cannot be expressed by a $\mathcal{V}$-formula. A structure is homogeneous if any two tuples having the same type are indistinguishable (unlike $\bar{a}$ and $\bar{b}$ in our example). This intuitive idea is made precise in Exercise 6.19.

We now develop some properties of homogeneous structures that will be useful for our investigation of atomic and saturated structures.

Proposition 6.26 Let T be a countable complete theory. Let M be a countable model of T and let N be a homogeneous model of T. Suppose that every type in $S(T)$ that is realized in M is also realized in N. There exists an elementary embedding $f: M \to N$.

Proof If a complete theory has a finite model, then all of its models are isomorphic (by Proposition 2.81). So we may assume that M is denumerable. Enumerate the underlying set of M as $U_M = \{a_1, a_2, a_3, \ldots\}$. Let U_N be the underlying set of N.

We construct an elementary embedding $f: M \to N$ step-by-step. In step n we define $b_n = f(a_n)$.

Step 1: Since N realizes every type realized in M, there exists $b_1 \in U_N$ such that $tp_M(a_1) = tp_N(b_1)$. Let $f(a_1) = b_1$.

Let $\bar{a}_n$ denote the n-tuple $(a_1, \ldots, a_n)$. Suppose that, for some $n \in \mathbb{N}$, $\bar{b}_n = (b_1, \ldots, b_n)$ has been defined such that $tp_M(\bar{a}_n) = tp_N(\bar{b}_n)$.

Step $n+1$: We want to define $b_{n+1} = f(a_{n+1})$ so that $tp_M(\bar{a}_n, a_{n+1}) = tp_N(\bar{b}_n, b_{n+1})$.

Since N realizes every type realized in M, there exists an $(n+1)$-tuple $(c_1, \ldots, c_{n+1})$ of N such that $tp_N(c_1, \ldots, c_{n+1}) = tp_M(a_1, \ldots, a_{n+1})$. Let $\bar{c}_n$ be $(c_1, \ldots, c_n)$.

Since $tp_N(\bar{c}_n)$ and $tp_N(\bar{b}_n)$ both equal $tp_M(\bar{a}_n)$, these two types equal each other. Since N is homogeneous, there exists b_{n+1} such that $tp_N(\bar{b}_n, b_{n+1}) = tp_N(\bar{c}_n, c_{n+1})$. Since $tp_N(\bar{c}_n, c_{n+1}) = tp_M(\bar{a}_{n+1})$ we have $tp_N(\bar{b}_n, b_{n+1}) = tp_M(\bar{a}_n, a_{n+1})$ as desired. Let $f(a_{n+1}) = b_{n+1}$.

In this manner we construct an infinite sequence $b_1, b_2, b_3, \ldots$ of elements of U_N and define $f: M \to N$ by $f(a_i) = b_i$ for each $i \in \mathbb{N}$. Since $tp_M(\bar{a}_n) = tp_N(\bar{b}_n)$ for each $n \in \mathbb{N}$, this function is an elementary embedding. □

If two countable homogeneous models M and N realize the same types, then they can be elementarily embedded into each other by the previous proposition. Moreover, expanding on the proof of this proposition, we can construct and isomorphism between M and N.

Proposition 6.27 Let T be a countable complete theory. Let M and N be two countable homogeneous models of T that realize the same types of $S(T)$. Then $M \cong N$.

Proof We may assume that M and N are denumerable and enumerate the underlying sets as

$$U_M = \{a_1, a_2, a_3, \ldots\} \quad \text{and} \quad U_N = \{b_1, b_2, b_3, \ldots\}.$$

As in Proposition 5.31, we can give a back-and-forth argument to construct an isomorphism $f: M \to N$. In step $n+1$ of this construction, we must define both $f(a_{n+1})$ and $f^{-1}(b_{n+1})$. We can define both of these in the same manner that $f(a_{n+1})$ was defined in the proof in the previous proposition. We leave the details as Exercise 6.21. □

We return to our discussion of small models. We show that countable atomic models are homogeneous. As we shall see in the next section, the property of being homogeneous is by no means restricted to countable atomic models.

Proposition 6.28 Countable atomic structures are homogeneous.

Proof Let M be a countable atomic structure and let $\bar{a}$ and $\bar{b}$ be two n-tuples of M realizing the same type p in $S_n(Th(M))$. Since M is atomic, this type is isolated by a formula $\theta(\bar{x})$. Let c be any element of M. Let $\psi(\bar{x}, x_{n+1})$ isolate $tp_M(\bar{a}, c)$. Then $\exists y \psi(\bar{x}, y)$ is in $tp_M(\bar{a}) = p$. Since p is also the type of $\bar{b}$ in M, $M \models \exists y \psi(\bar{b}, y)$. So $M \models \psi(\bar{b}, d)$ for some d in M. So $\psi(\bar{x}, y)$ is in $tp_M(\bar{b}, d)$. Since this formula isolates $tp_M(\bar{a}, c)$, we have $tp_M(\bar{a}, c) = tp_M(\bar{b}, d)$. By the definition of *homogeneous*, M is homogeneous. □

Corollary 6.29 Countable atomic models are unique up to isomorphism.

Proof Let M and N be two countable atomic models of a complete theory T. Then M and N are homogeneous by the previous proposition. Since M and N each realize only the isolated types in $S(T)$, M and N are isomorphic by Proposition 6.27. □

6.3.3 Prime models. At the outset of this section, we said that atomic countable models are in some sense the "smallest" countable models. We justify this terminology by showing that the atomic countable model of a theory, if it exists, can be elementarily embedded into any other model of that theory.

Definition 6.30 Let T be a theory and let M be a model of T. If M can be elementarily embedded into every model of T, then M is said to be a *prime model* of T.

Proposition 6.31 Let T be a countable complete theory. A model M of T is a prime model of T if and only if M is countable and atomic.

Proof Suppose M is prime. By the Downward Löwenhiem–Skolem theorem, there exists a countable model of T. Since M can be elementarily embedded into this model, M must be countable. It remains to be shown that M is atomic. Let p be a nonisolated type in $S(T)$. By the Omitting Types theorem 6.13, there exists a model N of T that omits p. Since M can be elementarily embedded into N, M must also omit p (see Exercise 6.1). So M realizes only the isolated types in $S(T)$ and is atomic.

Now suppose that M is countable and atomic. Then every type realized in M is realized in every model of T. By Proposition 6.26, M can be elementarily embedded into any homogeneous model of T. In a similar manner, we show that, since M is atomic, it can be elementarily embedded into any model (homogeneous or not).

Since M is countable, we can enumerate the underlying set of M as $U_M = \{a_1, a_2, a_3, \ldots\}$. (As usual, if M is finite, then this proposition is trivial.) For each $n \in \mathbb{N}$, let $\bar{a}_n$ denote the n-tuple $(a_1, \ldots, a_n)$. Let N be an arbitrary model of T. Let b_1 be an element of N that realizes the isolated type $tp_M(a_1)$.

Suppose that, for some $n \in \mathbb{N}$, we have defined an n-tuple $\bar{b} = (b_1, \ldots, b_n)$ of N so that $tp_N(\bar{b}_n) = tp_M(\bar{a}_n)$. Let $\theta(x_1, \ldots, x_{n+1})$ be a formula that isolates $tp_M(\bar{a}, a_{n+1})$. Since $\bar{b}_n$ has the same type as $\bar{a}_n$, $N \models \exists y \theta(\bar{b}_n, y)$. So $N \models \theta(\bar{b}_n, b_{n+1})$ for some element b_{n+1} of N. Since there is only one type in $S_{n+1}(T)$ containing θ, $tp_N(b_1, \ldots, b_{n+1}) = tp_M(a_1, \ldots, a_{n+1})$.

In this manner we can construct a sequence $b_1, b_2, b_3, \ldots$ as in the proof of Proposition 6.26. Let function f defined by $f(a_i) = b_i$ is an elementary embedding of M into N.

Since N was arbitrary, M is prime. □

We summarize the results of this section. If T is a small theory, then there exists an atomic countable model M of T. This model is unique up to isomorphism, is homogeneous, and can be elementarily embedded into any model of T. In this sense, M is the smallest model of T. Countable atomic models also exist for theories that are not small (recall Example 6.23 and see Exercise 6.14). We now turn our attention to big countable models.

6.4 Big models of small theories

We define and investigate countable saturated models of a countable complete theory. We show that countable saturated models, like countable atomic models, are homogeneous and unique up to isomorphism. We also show that every countable model of a theory can be elementarily embedded into the countable saturated model (provided it exists). So countable saturated models are the largest countable models in the same sense that countable atomic models are

Models of countable theories 281

the smallest models. In the second part of this section, we extend the notion of saturation to apply to uncountable structures.

6.4.1 Countable saturated models. Before defining countable saturated models, we must introduce the concept of a type *over* a set. Let M be a $\mathcal{V}$-structure having underlying set U_M. Let A be a subset of U_M. A *type over A* is a type that allows parameters from A. More specifically, an n-*type over* A is a set of $\mathcal{V}(A)$-formulas in n free variables that is realized in some elementary extension of M.

Example 6.32 Consider the structure $\mathbf{Q}_< = \{\mathbb{Q}| <\}$.

Let A be the set $\{1, 2, 3\}$. The three numbers in A break $\mathbb{Q}$ into four intervals. Each of these intervals correspond to a 1-type over A. These types are isolated by the formulas

$$(x_1 < 1), \quad \neg(x_1 < 1) \wedge \neg(x_1 = 1) \wedge (x_1 < 2),$$
$$\neg(x_1 < 2) \wedge \neg(x_1 = 2) \wedge (x_1 < 3), \text{ and } \neg(x_1 < 3) \wedge \neg(x_1 = 3).$$

In addition, there are the three types over A isolated by the formulas $x_1 = 1$, $x_1 = 2$, and $x_1 = 3$. So there are seven isolated types over A.

Let B be the natural numbers and let C be the set of all rational numbers. Then there are denumerably many types over B exactly one of which is nonisolated. The nonisolated type contains the formulas $\neg(x_1 < n)$ for each $n \in B$. This type is not realized in $\mathbf{Q}_<$ but is realized in an elementary extension of $\mathbf{Q}_<$. As was shown in Example 6.19, there are $2^{\aleph_0}$ types over C.

We make formal the definition of a type over a set and introduce notation for this concept.

Definition 6.33 Let M be a $\mathcal{V}$-structure having underlying set U_M. For any subset A of U_M, and for any tuple $\bar{b} = (b_1, \ldots, b_n)$ of elements of U_M, the *type of $\bar{b}$ over A in M*, denoted $tp_M(\bar{b}/A)$, is the set of all $\mathcal{V}(A)$-formulas having free variables among $x_1, \ldots, x_n$ that hold in M when each x_i in $\bar{x}$ is replaced by b_i.

The types $tp_M(\bar{b}/A)$ are called *complete types over A*. Let $S(A)$ denote the set of all complete types over A. The subset of n-types in $S(A)$ is denoted by $S_n(A)$. Since the theory T is not mentioned in this notation, $S(A)$ is ambiguous when taken out of context. In Example 6.32, we said that $S(A)$ contains seven types when $A = \{1, 2, 3\}$. If T is the theory of the rational numbers with addition and multiplication, then this is not true. We shall only use the notation $S(A)$ when T is understood.

Definition 6.34 Let T be a complete theory. A countable model M of T is *saturated* if, for every finite subset A of the underlying set of M, every 1-type in $S(A)$ is realized in M.

Example 6.35 Let T be the $\mathcal{V}_E$-theory defined in Example 6.8. Recall that M is the model of T having exactly one equivalence class of size n for each $n \in \mathbb{N}$. It was shown that there exists a type in $S_1(T)$ that is not realized in M. So this structure is not saturated. Let N_m be the model of T having exactly m infinite equivalence classes. Let $A = \{a_1, \ldots, a_m\}$ be a set of elements from each of these infinite classes. The type over A saying that x_1 has an infinite class but is not equivalent to a_i for each i is not realized in N_m. The only countable saturated model of T is the countable model containing denumerably many infinite equivalence classes.

Example 6.36 Let T be the theory defined in Example 5.56. The model containing countably many copies of $\mathbb{Z}$ in each equivalence class is the only saturated model of T.

As with atomic models, countable saturated models exist for small theories. Unlike the atomic models, these are the only theories having countable saturated models.

Proposition 6.37 A complete theory T is small if and only if it has a countable saturated model.

Proof Suppose first that T has a countable saturated model M. Then every type in $S(T)$ is realized by some tuple of M. Since M is countable, $S(T)$ must be countable also and T is small.

Conversely, suppose that T is small. Let M_1 be a countable model of T. We define an elementary chain of countable models $M_1 \prec M_2 \prec M_3 \ldots$

Suppose that countable M_n has been defined. Let A be a finite subset of the underlying set of M_n. If $S_1(A)$ is uncountable, then so is $S_{k+1}(T)$ where $k = |A|$. Since T is small, this is not the case. So we can enumerate $S_1(A)$ as the possibly finite set $\{p_1, p_2, \ldots\}$. For each p_i in this set, there exists an elementary extension of M_n realizing p_i. By the Downward Löwenhiem–Skolem theorem, there exists a countable elementary extension N_i of M_n that realizes p_i. By Proposition 4.37, there exists a countable model M_A of T such that M_n and each N_i can be elementarily embedded into M_A. Since M_n is countable, there are countably many finite subsets of M_n. Again applying Proposition 4.37, there exists a countable model M_{n+1} of T such that M_A can be elementariliy embedded into M_{n+1} for each finite subset A of M_n.

Let M be the limit of the elementary chain $M_1 \prec M_2 \prec M_3 \cdots$. Then M is a countable model of M. Any finite subset A of the universe of M is in the universe of M_n for some $n \in \mathbb{N}$. By the definition of M_{n+1}, every type in $S_1(A)$ is realized in M_{n+1}. Since $M_{n+1} \prec M$, every type in $S_1(A)$ is realized in M and M is saturated. □

Proposition 6.38 Countable saturated models are homogeneous.

Proof Let M be a countable saturated model of a complete theory T and let $\bar{a} = (a_1, \ldots, a_n)$ and $\bar{b} = (b_1, \ldots, b_n)$ be two n-tuples of M that realize the same type in $S_n(T)$. Let c be an element of M. Let $p_1(x_1) = tp_M(c/\bar{a})$. Let p_2 be the type over $\bar{b}$ obtained by replacing each occurrence of a_i in p_1 with b_i (for $i = 1, \ldots, n$).

Claim $p_2(x_1)$ is realizable.

Proof Let $\Phi(x_1, \bar{b})$ be the conjunction of a given finite set of formulas in $p_2(x_1)$. Then $\Phi(x_1, \bar{a}) \in p_1(x_1, \bar{a})$. By the definition of p_1, $M \models \Phi(c, \bar{a})$. So $M \models \exists y \Phi(x_1, \bar{a})$. Since $\bar{a}$ and $\bar{b}$ have the same type in M, $M \models \exists y \Phi(y, \bar{b})$. This shows that any finite subset of $p_2(x_1)$ is realizable. The claim then follows from Proposition 6.6.

Since M is saturated, $p_2(x_1)$ is realizable in M. Let d be an element of M that realizes this type. Then $tp_M(\bar{b}, d) = tp_M(\bar{a}, c)$ and M is homogeneous. □

So countable saturated models, like atomic models, are homogeneous. From this fact we can immediately deduce two more properties of countable saturated models. They are universal and unique.

Definition 6.39 Let T be a theory and let M be a countable model of T. If every countable model of T can be elementarily embedded into M, then M is said to be a *universal* model of T.

Corollary 6.40 Countable saturated models are universal.

Proof This follows from Propositions 6.38 and 6.26. □

The converse of Corollary 6.40 does not hold. Exercise 6.26 provides an example of a countable universal model that is not saturated. For the universal model to be saturated, it must be homogeneous.

Proposition 6.41 Let T be a small theory. A countable model M of T is saturated if and only if it is universal and homogeneous.

Proof A countable saturated model is universal and homogeneous by Corollary 6.40 and Proposition 6.38. We must prove the converse. Suppose that M is a countable model of T that is both universal and homogeneous. Let A be a finite subset of M and let p be a type in $S_1(A)$. We must show that there exists an element d so that $tp_M(d/A) = p$.

The type p is realized in some elementary extension of M. By the Downward Löwenhiem–Skolem theorem, p is realized in some countable model N containing A. So $tp_N(c/A) = p$ for some element c of N. Since M is universal, N can be elementarily embedded into M. Let $f: N \to M$ be elementary. Let B be

$\{f(a)|a \in A\}$. Note that B does not necessarily equal A, but it does have the same type as A in M. That is, $tp_M(\bar{a}) = tp_M(\bar{b})$, where $(a_1,\ldots,a_k)$ is some enumeration of A and $\bar{b} = (f(a_1),\ldots,f(a_k))$. Since M is homogeneous, there exists d so that $tp_M(d,\bar{a}) = tp_M(f(c),\bar{b})$. For all formulas $\varphi(x_1)$ in p, since $N \models \varphi(c)$ and $f: N \to M$ is elementary, we have $M \models \varphi(d)$. So $tp_M(d/A) = p$ and p is realized in M. Since p is arbitrary, M is saturated. □

Next we show that the saturated model of a theory is unique up to isomorphism. This fact, like Corollary 6.40, is an immediate consequence of Proposition 6.38.

Corollary 6.42 Let T be a complete small theory. Any two countable saturated models of T are isomorphic.

Proof This follows immediately from Propositions 6.38 and 6.27. □

We summarize. Let T be a small theory. Then T possesses both a countable atomic model M and a countable saturated model N. Each of these is unique up to isomorphism. The countable atomic model is the smallest model in the sense that it can be elementarily embedded into any model of T. The countable saturated model M is the biggest countable model of T in the sense that every countable model of T can be elementarily embedded into M. Countable saturated models are characterized by this property together with homogeneity. Likewise, countable atomic models are characterized as prime models (which are necessarily homogeneous).

We turn to theories that are not small in the next section. We close the present subsection by extracting the following characterization of $\aleph_0$-categorical theories from the above results.

Proposition 6.43 A theory is $\aleph_0$-categorical if and only if it has an atomic model and a countable saturated model that are isomorphic.

Proof Suppose T is $\aleph_0$-categorical. Since $\aleph_0$-categorical theories are small, T possesses a countable atomic model and a countable saturated model. These models must be isomorphic since T only has one countable model up to isomorphism.

Conversely, suppose that T has an atomic model N and a countable saturated model M with $N \cong M$. Since M is universal, any countable model of T can be elementarily embedded into M. Since $N \cong M$, any countable model of T can be elementarily embedded into the atomic model. It follows that every countable model of T realizes only isolated types. So every type in $S(T)$ must be isolated and T is $\aleph_0$-categorical by Proposition 6.15. □

6.4.2 Monster models. By definition, countable saturated models, as well as homogenous models and universal models, are countable. The following definitions extend these notions to uncountable structures.

Definition 6.44 Let M be a $\mathcal{V}$-structure having universe U and theory $T = Th(M)$. Let κ be an infinite cardinal.

We say that M is κ-*saturated* if, for each $A \subset U$ with $|A| < \kappa$, every type in $S_1(A)$ is realized in M. We simply say that M is *saturated* if M is $|M|$-saturated.

We say that M is κ-*universal* if every model N of T with $|N| < \kappa$ can be elementarily embedded into M. We simply say that M is *universal* if M is $|M|$-universal.

To extend the notion of notion of a *homogeneous* model, recall from Section 5.7 the definition of an M-elementary function. Note that a countable model M is homogeneous if and only every finite M-elementary function can be extended.

Definition 6.45 Let M be a structure and let κ be an infinite cardinal. We say that M is κ-*homogeneous* if, for each $A \subset U$ with $|A| < \kappa$ and each $a \in U$, every M-elementary function $f : A \to U$ extends to an M-elementary function $g : A \cup \{a\} \to U$. We simply say that M is *homogeneous* if M is $|M|$-homogeneous.

Proposition 6.46 A structure M is κ-saturated if and only if M is both κ-homogeneous and κ-universal.

Proof This can be proved in the same manner as Proposition 6.41. We leave this as Exercise 6.30. □

In particular, a model is saturated if and only if it is both homogeneous and universal. As with countable saturated models, we can use the homogeneity of saturated models to show that any two elementarily equivalent saturated models of the same cardinality must be isomorphic (see Exercise 6.31).

Now let T be a theory and suppose we wish to analyze the collection $Mod(T)$ of all models of T. Suppose that we only care to consider models in $Mod(T)$ of size less than κ. Since κ may be a ridiculously large cardinal, this is a reasonable assumption. If M is a saturated model of T of size κ, then we can replace the collection $Mod(T)$ with the model M. Every structure in $Mod(T)$ that we care to consider is an elementary substructure of M (since M is κ-universal). Moreover, any isomorphism between substructures of these models extends to an automorphism of M (by homogeneity and Exercise 6.20). Rather than considering the elements of $Mod(T)$ as separate entities, the saturated model M allows us the convenience of working within a single model. Such a model is referred to as a *monster model*. Model theorists often use the preamble "we work inside of a monster model $M \dots$."

Unfortunately, saturated models of large cardinalities may not exist. To guarantee the existence of arbitrarily large saturated models, we must assume set theoretic hypotheses beyond ZFC such as the General Continuum Hypothesis or (less severely) the existence of inaccessible cardinals. If we want to avoid such considerations, then we must settle for κ-saturated models instead of saturated models. Since they are both κ-universal and κ-homogeneous, κ-saturated models may serve as monster models. Although they are not necessarily homogeneous, κ-saturated models possess the fortunate property of existence. We prove this as the following proposition. The proof of this proposition also shows why saturated models may not exist. The κ-saturated model we construct is much larger than κ and so is not necessarily saturated.

Proposition 6.47 Let T be a complete theory having infinite models. Let κ be a cardinal. There exists a κ-saturated model of T.

Proof As in the proof of Proposition 6.37, we define an elementary chain of models $M_1 \prec M_2 \prec M_3 \cdots$. To begin, let M_1 be any model of T. Given M_i, let M_{i+1} be a model of T that realizes every type over every subset of the universe of M_i. If δ is a limit ordinal, let M_δ be the union of the chain of M_β for $\beta < \delta$. Consider the model M_α where $|\alpha| = \kappa$. Any subset of size κ of the universe of M_α must also be a subset of M_β for some $\beta < \alpha$. Every type over A is realized in $M_{\beta+1} \prec M_\alpha$. □

6.5 Theories with many types

Let T be a theory that is not small. By definition, $|S(T)|$ is uncountable. We show that, in fact, $|S(T)| = 2^{\aleph_0}$. We use this fact to show that T has the maximal number of nonisomorphic countable models.

Lemma 6.48 Let T be a countable complete theory. Let P be an uncountable subset of $S(T)$. There exists a formula ψ such that both ψ and $\neg\psi$ are contained in uncountably many types of P.

Proof Let $\mathcal{V}$ be the vocabulary of T. Let $F(T)$ denote the set of all $\mathcal{V}$-formulas that occur in some p in $S(T)$. That is, $F(T)$ is the set of formulas that are realized in some model of T. Each formula in $F(T)$ is either contained in uncountably many types of P or countably many (possibly zero) types of P. Let $\{\varphi_1, \varphi_2, \varphi_3, \ldots\}$ be the (possibly finite) set of those formulas in $F(T)$ that occur in only countably many types of P.

Let P_i be the set of all types in P that contain the formula φ_i. Then P_i is countable. Let P_φ be the union of all the P_is. Since it is a countable union of countable sets, P_φ is countable (by Proposition 2.43).

Now, P_φ is the set of all types in P that contain φ_i for some i. Since P is uncountable, there must be uncountably many types in P that are not in P_φ. Suppose it were the case that, for every formula $\psi \in F(T)$, either $\psi \in P_\varphi$ or $\neg\psi \in P_\varphi$. Then there would be at most one type of in P not contained in P_φ (namely, the type consisting of $\neg\varphi_i$ for each i). So this cannot be the case and there must exist some formula ψ such that neither ψ nor $\neg\psi$ is in P_φ. By the definition of P_φ, both ψ and $\neg\psi$ are contained in uncountably many types of P. □

Proposition 6.49 Let T be a countable complete theory. If T is not small, then $|S(T)| = 2^{\aleph_0}$.

Proof First, we show that $|S(T)| \leq 2^{\aleph_0}$. Since T is countable, the set of all formulas in the vocabulary of T can be placed into one-to-one correspondence with $\mathbb{N}$ (by Proposition 2.47). Since each type is a set of formulas, $|S(T)| \leq |\mathcal{P}(\mathbb{N})| = 2^{\aleph_0}$.

Now suppose that T is not small. We show that $|S(T)| \geq 2^{\aleph_0}$.

By definition, $2^{\aleph_0}$ is the cardinality of the set of all functions from $\mathbb{N}$ to the set $\{0, 1\}$. Each such function can be viewed as a denumerable sequence of 0s and 1s. For each of these sequences, we define a distinct type in $S(T)$.

Since T is not small, $S(T)$ is uncountable. By Lemma 6.48, there exists a formula ψ such that both ψ and $\neg\psi$ are contained in uncountably many types of $S(T)$.

Let χ_0 be $\neg\psi$ and χ_1 be ψ.

Let s be a finite sequence of 0s and 1s. For either $i = 0$ or $i = 1$, let $s \smallfrown i$ be the sequence obtained by adding an i to the end of sequence s.

Suppose that we have defined a formula χ_s that is contained in uncountably many types in $S(T)$. Let P_s be the set of types in $S(T)$ that contain χ_s. By Lemma 6.48, there exists a formula ψ such that both ψ and $\neg\psi$ are contained in uncountably many types of P_s.

Let $\chi_{s \smallfrown 0}$ be $\chi_s \wedge \neg\psi$ and $\chi_{s \smallfrown 1}$ be $\chi_s \wedge \psi$.

In this manner we define a formula χ_s for each finite sequence s of 0s and 1s. By design, we have both $T \vdash \chi_{s \smallfrown 0} \to \chi_s$ and $T \vdash \chi_{s \smallfrown 1} \to \chi_s$. Moreover, $\chi_{s \smallfrown 0}$ and $\chi_{s \smallfrown 1}$ cannot both be realized in a model of T since one formula implies ψ and the other implies $\neg\psi$.

Let $\{0,1\}^\omega$ denote the set of all denumerable sequences of 0s and 1s. For $t \in \{0,1\}^\omega$ and $n \in \mathbb{N}$, let $t|n$ denote the first n terms of the sequence t. Let Γ_t be the set of all formulas χ_s such that $s = t|n$ for some $n \in \mathbb{N}$.

Claim For each $t \in \{0,1\}^\omega$, Γ_t is realizable.

Proof By Proposition 6.6, it suffices to show that any finite subset of Γ_t is realizable. If Δ is a finite subset of Γ_t, then, for some $m \in \mathbb{N}$ and every χ_s in Δ, the sequence s has length less than m. Then $T \vdash \chi_{t|m} \to \chi_s$ for each χ_s in Δ. By definition, $\chi_{t|m}$ is contained in uncountably many types of $S(T)$. It follows that Δ is contained in uncountably many types of $S(T)$. In particular, Δ is realizable.

Let p_t be a type in $S(T)$ containing Γ_t. If t_1 and t_2 are distinct sequences in $\{0,1\}^\omega$, then p_{t_1} and p_{t_2} are distinct types in $S(T)$ since there exists a formula ψ such that ψ is contained in one of these types and $\neg\psi$ is contained in the other. It follows that $|S(T)| \geq |\{0,1\}^\omega| = 2^{\aleph_0}$. □

So, for any countable T, there are only two possibilities for $|S(T)|$. Either $|S(T)| = \aleph_0$ or $|S(T)| = 2^{\aleph_0}$. This is true even if the continuum hypothesis is false. Even if there exists cardinal numbers between $\aleph_0$ and $2^{\aleph_0}$, the set $S(T)$ is forbidden from having such cardinalities.

We now show that if T is not small, then T has the maximal number of nonisomorphic models of size $\aleph_0$. First we compute this maximal number.

Proposition 6.50 Let T be a countable $\mathcal{V}$-theory and let κ be an infinite cardinal. There exist at most 2^κ nonisomorphic models of T of size κ.

Proof Let U be any set of size κ. Let us count the number of $\mathcal{V}$-structures having U as an underlying set. Suppose we wish to define such a $\mathcal{V}$-structure.

Given any constant c in $\mathcal{V}$, we may interpret c as any element of U. There are $|U| = \kappa$ many possibilities.

We may interpret each n-ary relation in $\mathcal{V}$ as any subset of U^n. There are $|\mathcal{P}(U^n)| = 2^\kappa$ possible choices.

Finally, there are $\kappa^\kappa = 2^\kappa$ functions from U^n to U. We may interpret each n-ary function in $\mathcal{V}$ as any one of these functions.

So each symbol in $\mathcal{V}$ can be interpreted in at most 2^κ different ways on the set U. Since $\mathcal{V}$ is countable, there are at most $\aleph_0 \cdot 2^\kappa = 2^\kappa$ ways to interpret this vocabulary on U. That is, there are at most 2^κ $\mathcal{V}$-structures having underlying set U.

Let M be a model of T of size κ. Then there is a one-to-one correspondence between U and the underlying set of M. So, with no loss of generality, we may assume that M has underlying set U. It follows that there are at most 2^κ models of T of size κ. □

So a countable theory T can have at most $2^{\aleph_0}$ countable models. If T is not small, then it attains this maximal number.

Proposition 6.51 Let T be a countable theory. If $|S(T)| = 2^{\aleph_0}$ then the number of nonisomorphic countable models is $2^{\aleph_0}$.

Proof Suppose $|S(T)| = 2^{\aleph_0}$. Since each type in $S(T)$ is realized in some countable model (by the Downward Löwenhiem–Skolem theorem), there must be at least $2^{\aleph_0}$ countable models. By the previous proposition there are also at most this many countable models of T. $\square$

6.6 The number of nonisomorphic models

Let T be a theory. For any infinite cardinal κ, let $I(T,\kappa)$ denote the number of nonisomorphic models of T of size κ. The function $I(T,x) = y$, restricted to infinite cardinals x, is called the *spectrum* of T. When restricted to uncountable cardinals, this function is called the *uncountable spectrum* of T. The spectra provide a natural classification of the class of first-order theories. For example, totally categorical theories are the theories having the constant function $I(T,x) = 1$ as a spectrum. We have also seen uncountably categorical theories T having spectrum

$$I(T,x) = \begin{cases} \aleph_0, & x = \aleph_0 \\ 1, & x > \aleph_0. \end{cases}$$

The Baldwin–Lachlan theorem states that every uncountably categorical theory that is not totally categorical has this function as a spectrum.

Of course, there are many possible spectra. Let T be a countable complete theory that has infinite models. For any infinite cardinal κ, $1 \leq I(T,\kappa) \leq 2^\kappa$. The lower bound of 1 follows from the Löwenhiem–Skolem theorems and the upper bound is from Proposition 6.50. It may seem that the possibilities for $I(T,\kappa)$ are endless. It is a most remarkable fact that we can list the possible uncountable spectra for T.

Largely due to the work of Shelah, the uncountable spectra for the seemingly boundless and unmanageable class of countable first-order theories have been determined. Moreover, the work of Shelah shows that the spectrum of a given theory has structural consequences for the models of the theory. If a theory T has an uncountable spectrum other than the maximal $I(T,\kappa) = 2^\kappa$, then the models of T have an inherent notion of independence. We defined "independence" for strongly minimal structures in Section 5.7. By Theorem 5.100, any strongly minimal T has uncountable spectrum $I(T,\kappa) = 1$. The notion of independence for strongly minimal theories generalizes to a class of theories known as the *simple theories* that includes all theories having uncountable spectra other than $I(T,\kappa) = 2^\kappa$. For these theories, the notion of independence give rise to a system of invariants (analogous to dimension) that determine up to isomorphism the models of the theory. As humbling as it sounds, simple theories are beyond the scope of this book (as are supersimple theories). In the next section, we shall say a little more about simple theories and other classes of theories and provide references.

Whereas the possible uncountable spectra for countable theories have been determined, there remain open questions regarding the possible values of $I(T, \aleph_0)$. We will discuss these unanswered questions in the next section. In the present section, we prove one notable fact regarding $I(T, \aleph_0)$. We prove that $I(T, \aleph_0)$ cannot equal 2. In contrast, there exist theories T for which $I(T, \aleph_0) = n$ for every natural number n other than 2. The theory T_{CDLO} from Exercise 5.15 has exactly three countable models. For any $n > 3$, there exist expansions of T_{CDLO} that have exactly n countable models up to isomorphism. The fact that $I(T, \aleph_0)$ cannot be 2 was proved by Vaught in [49].

Theorem 6.52 (Vaught) A complete theory cannot have exactly two countable models.

Proof Let T be a complete theory. Suppose that T has two distinct countable models. We show that there exists a third countable model for T.

Since T is complete and has more than one model, every model of T is infinite.

If T is not small, then by Proposition 6.51, T has uncountably many countable models. So we may assume that T is small.

By Proposition 6.22, there exists an atomic model M_1 of T.

By Proposition 6.37, there exists a countable saturated model M_2 of T.

Since T is not $\aleph_0$-categorical, there exists a nonisolated type $p \in S_m(T)$ for some $m \in \mathbb{N}$ (by Theorem 6.15). Let $\bar{a}$ be an m-tuple from the saturated model M_2 that realizes the m-type p. Expand the vocabulary $\mathcal{V}$ of T by adding constants $c_1, \ldots, c_m$. Let N be the expansion of M_2 to this vocabulary interpreting the constants as the m-tuple $\bar{a}$.

Let $T' = Th(N)$. Since T is not $\aleph_0$-categorical, $S_m(T)$ is infinite for some n (again, by Theorem 6.15). If follows that $S_m(T')$ is also infinite, and so T', too, is not $\aleph_0$-categorical. If T' has uncountably many nonisomorphic models, then so does T. Since we are assuming that this is not the case, T', like T, has both an atomic model N_1 and a saturated model N_2. Moreover, by Proposition 6.43, N_1 and N_2 are not isomorphic. So N_1 is not saturated. Let M_3 denote the reduct of N_1 to the vocabulary of T. Since N_1 is not saturated, neither is M_3. Since M_3 realizes the nonisolated type p ($\bar{a}$ is an m-tuple of M_3), M_3 is not atomic. It follows that M_3 is a model of T that is isomorphic to neither M_1 nor M_2. □

6.7 A touch of stability

In this final and all too brief section, we give an overview of some of the concepts that have shaped model theory during the nearly 40 years since the proof of Morley's theorem. We state without proof many nontrivial facts. References are provided at the end of the section.

Definition 6.53 Fix a countable complete theory T and an infinite cardinal κ. We say that T is κ-*stable* if for every subset A of a model of T, if $|A| \leq \kappa$, then $|S(A)| \leq \kappa$.

This notion divides all first-order theories into four classes.

Definition 6.54 Let T be a countable complete theory.

T is *stable* if it is κ-stable for arbitrarily large κ.
T is *superstable* if it is κ-stable for sufficiently large κ.
T is ω-*stable* if it is κ-stable for all infinite κ.

We say that T is *strictly stable* if it is stable but not superstable. Likewise, we say that T is *strictly superstable* if it is superstable but not ω-stable.

Alternatively, stable theories can be defined as those theories that are κ-stable for some κ. If a theory is κ-stable for some κ, then it must be κ-stable for arbitrarily large κ. Countable superstable theories are characterized as those theories that are κ-stable for each $\kappa \geq 2^{\aleph_0}$. Also, "ω-stable" is synonymous with "$\aleph_0$-stable." From these facts we see that every theory is ω-stable, strictly superstable, strictly stable, or unstable.

Stability is a robust notion that has several equivalent formulations. The set of stable theories is often defined in terms of definable orderings of the underlying sets of models.

Definition 6.55 T has the *order property* if there is a formula $\varphi(\bar{x}, \bar{y})$ such that $M \models \varphi(\bar{a}_i, \bar{b}_j)$ if and only if $i < j$ for some $M \models T$ and sequences $(\bar{a}_i | i < \omega)$ and $(\bar{b}_j | j < \omega)$ of tuples of M.

Theorem 6.56 A theory is unstable if and only if it has the order property.

In particular, the theory of any infinite linear order is not stable. Note that unstable theories, like T_{DLO}, can be small. Like theories that are not small, theories possessing the order property necessarily have the maximal uncountable spectra.

Corollary 6.57 If T is unstable, then $I(T, \kappa) = 2^\kappa$ for all uncountable κ.

The proof of Morley's theorem in [32] utilizes the notion of ω-stability, although Morley did not use this terminology. A key component of Morley's proof shows that if T is κ-categorical for uncountable κ, then T is ω-stable. The terminology "κ-stable" is due to Rowbottom. In the 1970s, the properties of stable theories were developed (primarily in the work of Shelah). These theories possess an intrinsic notion of independence, called *forking independence*. By a "notion of independence" we mean that forking independence satisfies some basic properties such as symmetry: if $\bar{a}$ is forking independent from $\bar{b}$, then $\bar{b}$ is forking independent from $\bar{a}$.

A primary motivation for the study of stable theories was Shelah's program of classifying theories according to their uncountable spectra (as discussed in the previous section). By the previous corollary, unstable theories have the maximal uncountable spectra. Shelah proved that this is also true for theories that are not superstable. So, for the classification problem, one may focus on the superstable theories and further refine this class (into DOP and NDOP). However, the tools developed for this study apply to a much wider class of theories.

In his 1996 PhD thesis, Byunghan Kim proved that the notion of forking independence extends beyond the stable theories to a class known as the *simple theories*. Introduced by Shelah in a 1980 paper, the simple theories contain several important unstable first-order theories such as the theory of the random graph (from Section 5.4). The term "simple" is used to indicate that unstable simple theories share some of the properties of stable theories. Among the unstable theories, these are the simplest to understand (from the point of view of stability theory). By Theorem 6.56, unstable simple theories must possess the order property. However, they avoid the following strict version of this property.

Definition 6.58 A theory T has the *strict order property* if there is a formula $\varphi(x_1,\ldots,x_n,y_1,\ldots,y_n)$ such that
$M \models \varphi(\bar{a}_i, \bar{a}_j)$ if and only if $i < j$ for some $M \models T$ and sequences $(\bar{a}_i | i < \omega)$ of n-tuples of M.

Simple theories, like stable theories, do not have the strict order property. Moreover, the notion of "forking" can be defined for these theories. Kim's theorem states that forking gives rise to a notion of independence for simple theories just as it does for stable theories. In particular, forking independence is symmetric. This discovery, 16 years after Shelah introduced simple theories, has made this an active area of current research in model theory. The notion of *simplicity* is now viewed as an extension of *stability*. The simple theories encompass the stable theories just as the stable theories encompass the superstable theories, and so forth.

Thus the world of all first-order theories is divided into several classes. We list these classes below and provide at least one example from each. The class of *tame theories* is not precisely defined. A theory is "tame" if it lends itself to model-theoretic analysis. This class certainly includes all simple theories. It also includes the o-minimal theories discussed in Section 5.7. Whereas these theories share some of the model-theoretic features of strongly minimal theories, the o-minimal theories by definition possess the strict order property and so are not simple. Theories that are not tame are called *wild*.

A hierarchy of first-order theories:

- The *wild theories* contain undecidable theories such as T_N discussed in Chapter 8.

- The *tame theories* include well-behaved theories such as the o-minimal theories and each of the theories below.

- The *simple theories* include the theory of the random graph and each of the theories below.

- The *stable theories* include the superstable theories below and also strictly stable theories such as the following.
  Let $\mathcal{V}_\omega = \{E_i | i < \omega\}$ be the vocabulary containing countably many binary relations.
  Let T_{st} be the $\mathcal{V}$-theory saying that
  - each E_i is an equivalence relation,
  - $\forall x \forall y (E_{i+1}(x,y) \to E_i(x,y))$ for each i,
  - each equivalence class of E_i contains infinitely many equivalence classes of E_{i+1} for each i.

  This theory is stable but not superstable.

- The *supertable theories* include the ω-stable theories below and also strictly superstable theories such as the following $\mathcal{V}_\omega$-theory.
  Let T_{sst} be the $\mathcal{V}_\omega$-theory saying that
  - each E_i is an equivalence relation,
  - there are exactly two E_i equivalence classes for each i,
  - the equivalence relation defined by $E_i(x,y) \wedge E_j(x,y)$ has exactly four infinite classes for $i \neq j$.

  This theory is superstable but not ω-stable.

- The ω-*stable theories* include all uncountably categorical theories as well as some noncategorical theories such as the theory of a single equivalence relation having two infinite classes.

- The *uncountably categorical theories* include the strongly minimal theories by Proposition 5.100. Each uncountably categorical structure is closely linked to a strongly minimal structure. For example, recall the strongly minimal structure $Z_s = (\mathbb{Z}|s)$ (the integers with a successor function). Let $Z_P = (\mathbb{Z}|s, P)$ be the expansion of Z_s containing a unary relation P that holds for every other integer. That is, Z_P models $\forall x (P(x) \leftrightarrow \neg P(s(x)))$. This structure is closely linked to Z_s. Like Z_s, Z_P is an uncountably categorical structure. Since $P(x)$ defines an infinite and co-infinite subset of Z_P, the theory of Z_P is not strongly minimal.

- The *strongly minimal theories* include the theory of Z_s as well as the theories of vector spaces and algebraically closed fields discussed in Section 5.7.

Note that the countably categorical theories are omitted from the above list. Whereas uncountably categorical theories are necessarily ω-stable, countably categorical theories provide a cross-section of the above classification. There exist countably categorical theories that are strongly minimal (such as T_{clique}), unstable (the random graph), and unsimple (dense linear orders). Lachlan's Theorem states that countably categorical theories cannot be strictly superstable. Whether the same is true for strictly stable theories is an open question.

A related open question is the following:

Does there exist a countable simple theory T with $1 < I(T, \aleph_0) < \aleph_0$?

The theory T_{CDLO} (defined in Exercise 5.15) is an example of such a theory that is not simple. The question regarding simple theories has been resolved for certain cases. The Baldwin–Lachlan theorem states that there are no uncountably categorical examples. Byunghan Kim has extended this result to a class of simple theories known as the supersimple theories (this class contains all superstable theories as well as some unstable theories).

A most famous open question in this area is known as *Vaught's Conjecture*. This conjecture asserts that the number of countable models of a complete countable theory is either countable or $2^{\aleph_0}$. If the continuum hypothesis holds and $2^{\aleph_0} = \aleph_1$, then this conjecture is an immediate consequence of Proposition 6.50. If the continuum hypothesis does not hold, then there exist cardinals between $\aleph_0$ and $2^{\aleph_0}$. Vaught's Conjecture asserts that $I(T, \aleph_0)$ cannot take on such values. Robin Knight has recently proposed a counter example to this conjecture. However, Knight's example is not simple. So the following question remains:

Does there exist a countable simple theory T with $\aleph_0 < I(T, \aleph_0) < 2^{\aleph_0}$?

As with the former question, partial results have been obtained. Shelah, Harrington, and Makkai showed that the answer is "no" for ω-stable theories. Buechler proved that Vaught's Conjecture holds for a class of superstable theories.

For more on stability theory, I recommend to the beginner both Poizat [39] and Buechler [6]. The later chapters of [39] are devoted to stability theory and serve as an excellent introduction. As the title suggests, [6] is entirely dedicated to the subject. Many of Shelah's important results are contained in his epic [43]. Although this book is the well spring of stability theory, I do not recommend it as a source for learning stability theory. I also do not recommend studying the sun by staring directly at it. Secondary sources are preferable. Baldwin [1] is more comprehensive than both [39] and [6] and far more accessible than [43]. Both [26] and [37] are also recommended, but are unfortunately currently out of print. For simple theories, Wagner [50] is recommended.

Exercises

6.1. Let M and N be $\mathcal{V}$-structures and let $f : M \to N$ be an elementary embedding. Show that N realizes every type that M realizes, but N does not omit every type that M omits.

6.2. Let T be a complete $\mathcal{V}$-theory. For any $\mathcal{V}$-formula φ, let S_φ denote the set of all types in $S(T)$ that contain φ. The set $\{\varphi_1, \varphi_2, \ldots\}$ of $\mathcal{V}$-formulas is said to be a *cover* of $S_n(T)$ if every type in $S_n(T)$ is in S_{φ_i} for some i. Show that, for any $n \in \mathbb{N}$, any cover of $S_n(T)$ has a finite subset that is also a cover of $S_n(T)$.

6.3. Let T be a complete theory that has a finite model. Show that every type in $S(T)$ is isolated.

6.4. Let T be a complete theory having infinite models.
   (a) Show that any type in $S_n(T)$ for $n > 1$ is not satisfiable.
   (b) Give an example of a type $p \in S_1(T)$ that is satisfiable.

6.5. Let T be a complete theory and let p be a type of T. Show that the formula θ isolates p over T if and only if θ is realized in some model of T and, for each formula $\varphi \in p$, $T \vdash \theta \to \varphi$.

6.6. Let T be a complete $\mathcal{V}$-theory. Suppose that $S_n(T)$ is finite for some $n \in \mathbb{N}$.
   (a) Show that there are only finitely many $\mathcal{V}$-formulas in n free variables up to T-equivalence.
   (b) Show that T is not necessarily $\aleph_0$-categorical by providing an appropriate example.

6.7. We demonstrate that the Omitting Types theorem fails for theories in uncountable vocabularies. Let M be the $\mathcal{V}_E$-structure defined in Example 6.8 (and Exercise 5.16). Let T be $Th(M)$. Let $C = \{c_i | i < \omega_1\}$ be an uncountable set of constants. Let $\mathcal{V}'$ be the expansion $\mathcal{V}_E \cup C$. Let T' be any expansion of T to a complete $\mathcal{V}'$-theory that says each of the constants in C are distinct. Consider the partial type $q = \{\neg\varphi_1(x_1), \neg\varphi_2(x_1), \neg\varphi_3(x_1), \ldots\}$. Recall that $\varphi_m(x_1)$ is the $\mathcal{V}_E$-formula saying that there exists exactly m elements equivalent to x_1.
   (a) Show that q is not isolated.
   (b) Show that q is realized in every model of T'.
   (c) Show that for each complete type p in $S_1(T')$ containing q, either p is isolated or p is omitted from some model of T.

6.8. Suppose we attempt to generalize the proof of the Omitting Types theorem 6.13 to theories having uncountable theories in the same manner that Theorem 4.2 was generalized to Theorem 4.27. Specifically what goes wrong?

6.9. Let T be a complete theory in a countable vocabulary $\mathcal{V}$.
   (a) Let p_1 and p_2 be nonisolated types in $S_n(T)$. Following the proof of the Omitting Types theorem 6.13, prove that there exists a countable model of T that omits both p_1 and p_2.
   (b) Let $p_1, p_2, \ldots, p_k$ be nonisolated types in $S_n(T)$ (for $k \in \mathbb{N}$). Describe how the proof of the Omitting Types theorem 6.13 can be modified to construct a model of T that omits each p_i.
   (c) Let $p_1, p_2, p_3, \ldots$ be countably many types in $S_n(T)$ each of which is not isolated. Describe how the proof of the Omitting Types theorem 6.13 can be modified to construct a model of T that omits each p_i.

6.10. Give an example of a complete countable theory T and a nonisolated type $p \in S(T)$ such that p is not omitted in any uncountable model of T. That is, show that Corollary 6.14 fails for uncountable models.

6.11. Let T be a small theory. Prove that there exists an isolated type in $S(T)$.

6.12. Find an example of a theory T in a finite vocabulary so that $S_n(T)$ contains no isolated types for some $n \in \mathbb{N}$.

6.13. Let T be a countable complete theory and let p be a type in $S_1(T)$.
   (a) Suppose that, for any model M of T, there are only finitely many elements a such that $p = tp_M(a)$. Show that p is isolated.
   (b) Suppose that, for some model M of T, there is exactly one element a such that $p = tp_M(a)$. Show that p is not necessarily isolated by demonstrating an appropriate example.

6.14. Let T be a complete countable $\mathcal{V}$-theory. The isolated types are said to be *dense* in $S(T)$ if every $\mathcal{V}$-formula contained in a type of $S(T)$ is contained in an isolated type of $S(T)$. Prove that T has a countable atomic model if and only if the isolated types are dense in $S(T)$.

6.15. Let T be a countable complete theory. Show that M is a prime model of T if and only if T can be elementarily embedded into every countable model of T.

6.16. For any n-tuple $\bar{a}$ of elements from an $\mathcal{V}$-structure M, let $qftp_M(\bar{a})$ denote the set of quantifier-free $\mathcal{V}$-formulas in $tp_M(\bar{a})$. Prove that the following are equivalent statements regarding a complete theory T.
   (i) For any $M \models T$, $n \in \mathbb{N}$, and any n-tuples $\bar{a}$ and $\bar{b}$ of M, if $qftp_M(\bar{a}) = qftp_M(\bar{b})$ then $tp_M(\bar{a}) = tp_M(\bar{b})$.

(ii) For any $n \in \mathbb{N}$, models N and M of T, and any n-tuple $\bar{a}$ of M and $\bar{b}$ of N, if $qftp_M(\bar{a}) = qftp_N(\bar{b})$ then $tp_M(\bar{a}) = tp_N(\bar{b})$.

(iii) T has quantifier elimination.

6.17. Let T be a small theory and let M be the countable saturated model of T. Prove that T has quantifier elimination if and only if the following holds. For any $n \in \mathbb{N}$, if $(a_1, \ldots, a_n)$ and $(b_1, \ldots, b_n)$ are n-tuples of M that satisfy the same atomic formulas in M, then for any a_{n+1} there exists b_{n+1} such that $(a_1, \ldots, a_n, a_{n+1})$ and $(b_1, \ldots, b_n, b_{n+1})$ satisfy the same atomic formulas in M.

6.18. Let T be a theory. Suppose that there exists a unique nonisolated type in $S_1(A)$ for each subset A of each model of T. Prove that T is uncountably categorical but not countably categorical. (Hint: Show that T is strongly minimal.)

6.19. Recall that an isomorphism $f : M \to M$ from a structure M onto itself is called an *automorphism* of M. Say that two elements a and b from the universe of M are *indistinguishable* in M if there exists an automorphism of M such that $f(a) = b$. Suppose that M is a countable homogeneous structure. Prove that a and b are indistinguishable if and only if $tp_M(a) = tp_M(b)$.

6.20. We show that the notion of a *homogeneous structure* can be defined in terms of automorphisms (see the previous exercise).

(a) Let M be a countable structure. Prove that M is homogeneous if and only if, given any two tuples $(a_1, \ldots, a_n)$ and $(b_1, \ldots, b_n)$ of M having the same type in M, there exists an automorphism f of M such that $f(a_i) = b_i$ for each $i = 1, \ldots n$.

(b) Let M be a structure (not necessarily countable). Prove that M is homogeneous if and only if every M-elementary function can be extended to an automorphism.

6.21. Prove Proposition 6.27.

6.22. Let T be an incomplete theory. Show that T may have a countable atomic model, but it cannot have a prime model.

6.23. Let M be a countable saturated structure. Show that, for any subset A of the underlying set of M, every type in $S(A)$ is realized in M. (That is, if every 1-type is realized, then every type is realized.)

6.24. Let M be a countable model of a complete theory T. Prove that M is the countable saturated model of T if and only if M is homogeneous and realizes every type in $S(T)$.

6.25. Let T be the theory in the vocabulary $\{E, s\}$ defined in Example 5.56 of the previous chapter. Prove that T is complete by showing that T has a countable atomic model.

6.26. Let $\mathcal{V}$ be the vocabulary consisting of a binary relation E and a unary function s. Let T_∞ be the $\mathcal{V}$-theory that says E is an equivalence relation having infinitely many equivalence classes and each class contains a copy of $(\mathbb{Z}|s)$. (So T_∞ is a modified version of T from the previous exercise.) Show that T_∞ has a countable universal model that realizes every type in $S(T)$ but is not saturated.

6.27. A set of sets $\mathcal{S}$ is said to have the *finite intersection property* if the intersection of any finite number of sets in $\mathcal{S}$ is nonempty.
   (a) Let $\mathcal{Z}_s = (\mathbb{Z}|s)$ be the $\mathcal{V}_s$-structure that interprets the unary function s as the successor function on the integers. Let $\mathcal{D}$ be the set of all infinite subsets of $\mathbb{Z}$ that are definable (by some $\mathcal{V}_s$ with parameters from $\mathbb{Z}$). Show that $\mathcal{D}$ has the finite intersection property but the intersection of all sets in $\mathcal{D}$ is empty. (For the first part, use the fact that $\mathcal{Z}_s$ is strongly minimal.)
   (b) Let M be a countable $\mathcal{V}$-structure having underlying set U. Show that M is saturated if and only if the intersection of all sets in $\mathcal{D}$ is nonempty for any set $\mathcal{D}$ of definable subsets of U having the finite intersection property.

6.28. Show that if a structure in $\aleph_n$-saturated, then it is $\aleph_{n+1}$-universal.

6.29. Show that, for $m \leq n$, if a structure is $\aleph_n$-homogeneous and $\aleph_m$-universal, then it is $\aleph_{m+1}$-universal.

6.30. Show that a structure is κ-saturated if and only if it is both κ-homogeneous and κ-universal.

6.31. Let T be a complete theory. Let M and N be two uncountable homogeneous models of T that realize the same types of $S(T)$. Show that $M \cong N$ using transfinite induction. (Compare this with Proposition 6.27.)

6.32. Assuming that the general continuum hypothesis it true, show that every theory has arbitrarily large saturated models.

6.33. Prove Morley's theorem using the following facts.
   - Let κ be an uncountable cardinal. If T is ω-stable and has an uncountable model that is not saturated, then T has a model of size κ that is not $\aleph_1$-saturated.
   - If T is κ-categorical for some uncountable κ, then T is ω-stable.

7 Computability and complexity

In this chapter we study two related areas of theoretical computer science: computability theory and computational complexity. Each of these subjects take mathematical problems as objects of study. The aim is not to solve these problems, but rather to classify them by level of difficulty. Time complexity classifies a given problem according to the length of time required for a computer to solve the problem. The polynomial-time problems **P** and the nondeterministic polynomial-time problems **NP** are the two most prominent classes of time complexity. Some problems cannot be solved by the algorithmic process of a computer. We refer to problems as *decidable* or *undecidable* according to whether or not there exists an algorithm that solves the problem. Computability theory considers undecidable problems and the brink between the undecidable and the decidable.

There are only countably many algorithms and uncountably many problems to solve. From this fact we deduce that most problems are not decidable. To proceed beyond this fact, we must state precisely what we mean by an "algorithm" and a "problem." One of the aims of this chapter is to provide a formal definition for the notion of an *algorithm*. The types of problems we shall consider are represented by the following examples.

- *The even problem:* Given an $n \in \mathbb{N}$, determine whether or not n is even.
- *The 10-clique problem:* Given finite graph, determine whether or not there exists a subgraph that is isomorphic to the 10-clique.
- *The satisfiability problem for first-order logic:* Given a sentence of first-order logic, determine whether or not it is satisfiable.

The first problem is quite easy. To determine whether a given number is even, we simply check whether the last digit of the number is 0, 2, 4, 6 or 8. The second problem is harder. If the given graph is large and does contain a 10-clique as a subgraph, then we may have to check many subsets of the graph before we find it. Time complexity gives precise meaning to the ostensibly subjective idea of one problem being "harder" than another. The third problem is the most difficult of the three problems. This problem is not decidable (it is semi-decidable). Whereas time complexity categorizes the first two problems, the third problem falls into the realm of computability theory.

We still have not defined what constitutes a "problem." We have merely provided a few examples. Note that each of these examples may be viewed as a function. Given some input, each problem requires a certain output. In each of the examples we have cited, the output is a "yes" or "no" answer. This type of problem is called a *decision problem*. Other problems may require various output: the problem of finding the largest prime that divides a given number, for example. More generally, let $f(\bar x)$ be a k-ary function on the non-negative integers. Consider the following problem:

- *The f-computability problem:* Given $\bar x$, determine the value of $f(\bar x)$.

If this problem is decidable, then the function f is said to be *computable*. We claim that the f-computability problems are the only problems we need to consider. In fact, we take this as our definition of a *problem*.

> Every *problem* corresponds to a function
> on the non-negative integers.

A *decision problem* corresponds to a function that only takes on the values 0 and 1 (we interpret 0 as "no" and 1 as "yes"). Let A be the set of all k-tuples $\bar x$ for which $f(\bar x) = 1$. The problem represented by f corresponds to the following decision problem: given $\bar x$, determine whether or not $\bar x$ is in A. In this way, every decision problem can be viewed as a subset of $(\mathbb{N} \cup \{0\})^k$ for some k.

> Every *decision problem* corresponds to a relation
> on the non-negative integers.

The Even Problem corresponds to the set of non-negative even numbers. For the 10-Clique Problem, we can code each finite graph as a k-tuple of non-negative integers. Likewise, every formula of first-order logic can be coded as a natural number. We shall discuss the required coding procedures in this chapter and the next.

The goal of this chapter is to give precise meaning to notions such as *algorithm*, *decidable*, and *polynomial-time*. These notions have arisen in earlier chapters. In Chapter 5, we defined *decidable theories* in terms of algorithms. Given a $\mathcal{V}$-theory T and a $\mathcal{V}$-sentence φ, we may ask whether or not φ is a consequence of T. This decision problem is decidable if and only if the theory T is decidable. The Satisfiability Problem for First-order Logic (which we shall abbreviate FOSAT), was the central topic of Chapter 3. The corresponding problem for propositional logic (denoted PSAT), was discussed in Chapter 1.

The definitions and results of the present chapter allow us to prove in the final chapters some of the claims that were made in the previous chapters. In Chapter 8, we prove that the theory of arithmetic and other theories are undecidable. In chapter 10, we prove that FOSAT is undecidable. In contrast, PSAT is decidable. However, there is no known algorithm that decides PSAT quickly.

This problem is **NP**-complete. Again, we do not prove this until the final chapter. In the present chapter, we define the concept of **NP**-completeness and discuss the relationship between PSAT and the **P** = **NP** question.

The topics of computability and complexity do not flow from the stream of ideas embodied in the model theory of the previous chapters. In fact, Sections 7.2 and 7.6.2 are the only sections of this chapter that require logic. However, there are strong connections between the topics of the present chapter and the logic of the previous chapters. The formal nature of computability has historically and philosophically linked this subject with the formal languages of logic. In addition to the connections mentioned in the previous paragraphs, we have the following fact: many of the classes of problems that naturally arise in computability and complexity can be defined in terms of logic. We shall see evidence of this fundamental fact in Sections 7.2, 7.6.2, 8.3, and 10.4.

7.1 Computable functions and Church's thesis

We consider functions on the non-negative integers. Whenever we refer to *a function* in this chapter, we always mean a function of the form $f : D \to \mathbb{N} \cup \{0\}$ where the domain D of f is a subset of $(\mathbb{N} \cup 0)^k$ for some k. If the domain D happens to be all of $(\mathbb{N} \cup \{0\})^k$, then the function f is said to be a *total* function. Otherwise, f is a *partial* function.

Definition 7.1 Let f be an k-ary function on the non-negative integers. Then f is *computable* if there exists an algorithm which, given an k-tuple $\bar{a}$ of non-negative integers,

- outputs $f(\bar{a})$ if $\bar{a}$ is in the domain of f, and
- yields no output if $\bar{a}$ is not in the domain of f.

To make this definition precise, we must state what constitutes an algorithm. Informally, an *algorithm* is a computer program. We claim that the definition of a *computable function* is invariant under our choice of programming language. If we have a computer program in C++ that computes the function $f(x)$, then we can translate this to a program in Fortran or any other language.

There are countably many computer programs and uncountably many functions (Propositions 2.47 and 2.48). So most functions are not computable. This fact flies in the face of empirical evidence. Nearly every function that naturally arises in mathematics (every function one encounters in, say, calculus) is computable. We shall demonstrate several functions that are not computable in Section 7.6.1.

The set of computable functions is our primary object of study in this chapter. We take two approaches to circumscribing this set from among the uncountably many functions that are not computable. In Section 7.3, we make definite the notion of an algorithm by specifying a simplified computer language. This facilitates the analysis of the set of computable functions. In the present section, we take a different approach. We consider some basic computable functions and show how more complex computable functions can be derived from them. We define a set of computable functions known as the *recursive* functions. This set is generated from a few functions by applying a few rules. We prove in Sections 7.3 and 7.4 that the two approaches yield the same functions: the recursive functions of the present section are precisely those that are computable by an algorithm in the sense of Section 7.3. Moreover, we claim that these functions are precisely those functions that can be computed by a C++ program, a Maple program, a Pascal program, or any other computer program. Thus, we demonstrate that *computability* is not a vague notion. It is a robust concept suitable for mathematical analysis.

7.1.1 Primitive recursive functions.

We begin our analysis of computable functions with some examples.

Example 7.2 The following functions are unquestionably computable:

the zero function $Z(x) = 0$,
the successor function $s(x) = x + 1$, and
the projection functions $p_i^k(x_1, x_2, \ldots, x_k) = x_i$.

The projection functions are defined for any k and i in $\mathbb{N}$ with $i \leq k$.

Definition 7.3 The functions $s(x)$, $Z(x)$, and $p_i^k(\overline{x})$ from the previous example are called the *basic functions*.

From two given unary functions f and g we can define various new functions. One of these is the composition $h(x) = f(g(x))$. If we can compute both $f(x)$ and $g(x)$, then we can compute $h(x)$. We generalize this idea to k-ary functions.

Definition 7.4 Let S be a set of functions on the non-negative integers. The set is said to be *closed under compositions* if given any m-ary function h in S and k-ary functions $g_1, g_2, \ldots, g_m$ in S (for any k and m in $\mathbb{N}$) the k-ary function f defined by

$$f(x_1, \ldots, x_k) = h(g_1(x_1, \ldots, x_k), \ldots, g_m(x_1, \ldots, x_k))$$

is also in S.

The set of computable functions is closed under compositions. In particular, since the basic functions $Z(x)$ and $s(x)$ are both computable, so is the function $s(Z(x))$. This is the constant function $c_1(x) = 1$. Likewise, each of the constant functions $c_n(x) = n$ is a computable function.

Functions on the non-negative integers can also be defined inductively. That is, we can define a function $f(x)$ by first stating the value of $f(0)$ and then describing how to compute $f(x+1)$ given the value of $f(x)$. For example, we define the function $f(x) = x!$ inductively as follows:

$$f(0) = 1 \quad \text{and} \quad f(x+1) = f(x) \cdot (x+1).$$

There are numerous ways to extend this idea to k-ary functions for $k > 1$. One way is provided by *primitive recursion*. Primitive recursion is a method for defining an k-ary function in terms of two given functions: one of these is $(k-1)$-ary and the other is $(k+1)$-ary. For the case where $k = 1$, we define the *0-ary functions* to be the unary constant functions $Z(x)$ and $c_n(x)$ for $n \in \mathbb{N}$. Before defining primitive recursion we demonstrate how it works with an example.

Example 7.5 Let $h(x, y) = 1$ and let $g(x, y, z, w) = w \cdot z^{xy}$. From these 2-ary and 4-ary functions we define, using primitive recursion, the 3-ary function f.

Let $f(0, y, z) = h(y, z)$, and $f(x+1, y, z) = g(x, y, z, f(x, y, z))$. For fixed values y_0 and z_0 of y and z, the above definition inductively defines the unary function given by $f(x, y_0, z_0)$. We have

$$f(0, y_0, z_0) = 1 \quad \text{and} \quad f(x+1, y_0, z_0) = f(x, y_0, z_0) \cdot z_0^{xy_0}.$$

From this definition, $f(x, y, z)$ can be computed for any tuple (x, y, z) of non-negative integers. Specifically, for any y and z, we have:

$$\begin{aligned} f(1, y, z) &= z^y & f(4, y, z) &= z^{6y} z^{4y} = z^{10y} \\ f(2, y, z) &= z^y z^{2y} = z^{3y} & f(5, y, z) &= z^{10y} z^{5y} = z^{15y} \quad , \text{ and} \\ f(3, y, z) &= z^{3y} z^{3y} = z^{6y} & f(6, y, z) &= z^{15y} z^{6y} = z^{21y}. \end{aligned}$$

We can see from this sequence that $f(x, y, z)$ is explicitly defined as follows:

$$f(x, y, z) = z^{y(x^2+x)/2}.$$

Definition 7.6 Let $\mathcal{S}$ be a set of functions on the non-negative integers. The set is *closed under primitive recursion* if, for any $k \in \mathbb{N}$, any $(k-1)$-ary function h in $\mathcal{S}$, and any $(k+1)$-ary function g in $\mathcal{S}$, the k-ary function f defined as follows

is also in $\mathcal{S}$:
$$f(0, x_2, \ldots, x_k) = h(x_2, \ldots, x_k),$$
$$f(x_1 + 1, x_2, \ldots, x_k) = g(x_1, \ldots, x_k, f(x_1, \ldots, x_k)).$$

If $k = 1$, then we allow h to be any constant function.

If both h and g are computable, then so is any function defined from h and g be primitive recursion. So the set of computable functions is closed under primitive recursion as well as composition. Also, the computable functions include the basic functions $Z(x)$, $s(x)$, and p_i^k from Example 7.2.

Definition 7.7 The set of *primitive recursive functions* is the smallest set containing the basic functions and closed under both composition and primitive recursion.

Whereas every primitive recursive function is computable, it is not true that every computable function is primitive recursive. The primitive recursive functions are only part of the set of computable functions. They are, however, a sizable part.

Proposition 7.8 The addition function defined by $a(x, y) = x + y$ is primitive recursive.

Proof Let $h(y) = p_1^1(y) = y$ and $g(x, y, z) = s(p_3^3(x, y, z)) = s(z)$. Since $h(x)$ is a basic function, it is primitive recursive. Since $g(x, y, z)$ is the composition of basic functions, it is also primitive recursive. We define $a(x, y)$ using primitive recursion as follows:
$$a(0, y) = h(y) = y, \text{ and}$$
$$a(x + 1, y) = g(x, y, a(x, y)) = s(a(x, y)) = a(x, y) + 1.$$

It follows that $a(x, y)$ is primitive recursive. □

Unlike addition, subtraction does not determine a total binary function on the non-negative integers. We let $\dotdiv$ denote a modified subtraction operation that is total. This function is defined as follows:
$$a \dotdiv b = \begin{cases} a - b & \text{if } b \leq a \\ 0 & \text{otherwise.} \end{cases}$$

Proposition 7.9 The function $sub(x, y) = x \dotdiv y$ is primitive recursive.

Proof First we show that the predecessor function $pred(x) = x \dotdiv 1$ is primitive recursive. Let $h_1(x) = Z(x)$ and $g_1(x, y) = p_1^2(x, y)$. Let $pred(0) = h_1(0) = 0$, and $pred(x + 1) = g_1(x, pred(x)) = x$.

To define $sub(x,y)$ by primitive recursion, let $h_2(x) = p_1^1(x)$ and $g_2(x,y,z) = pred(z)$. Then $sub(x,0) = h_2(x) = x$ and $sub(x,y+1) = g_2(x,y,sub(x,y)) = pred(sub(x,y))$ and so $sub(x,y)$ is primitive recursive. □

Proposition 7.10 The multiplication function defined by $m(x,y) = xy$ is primitive recursive.

Proof Let $h(y) = Z(y)$ and $g(x,y,z) = a(y,z)$. By Proposition 7.8, $a(y,z)$ is primitive recursive. We define $m(x,y)$ using primitive recursion as follows:

$$m(0,y) = h(y) = 0, \quad \text{and}$$
$$m(x+1,y) = g(x,y,m(x,y)) = a(y,m(x,y)) = y + m(x,y).$$

It follows that $m(x,y)$ is primitive recursive. □

Consider the exponential function $exp(x,y) = y^x$. Since this function is not total, it cannot be primitive recursive. If we define $exp(0,0)$ to be 1 (or any other number), then the resulting function is primitive recursive.

Proposition 7.11 The function $exp(x,y)$ is primitive recursive.

Proof Let $h(y) = s(0)$ and $g(x,y,z) = m(y,z)$. By the previous proposition, $m(y,z)$ is primitive recursive. We define $exp(x,y)$ using primitive recursion as follows:

$$exp(0,y) = h(y) = s(0) = 1, \quad \text{and}$$
$$exp(x+1,y) = g(x,y,exp(x,y)) = m(y,exp(x,y)) = y \cdot exp(x,y).$$

It follows that $exp(x,y)$ is primitive recursive. □

Proposition 7.12 Let $p(x)$ be any polynomial having natural numbers as coefficients. Then $p(x)$ is primitive recursive.

Proof This follows from the fact that $p(x)$ can be written as a composition of constant functions, $a(x,y)$, $m(x,y)$, and $exp(x,y)$. □

By definition, the set of primitive recursive functions is closed under the operations of composition and primitive recursion. By repeatedly using is fact, we can generate more and more primitive recursive functions from the basic functions. We next show that the set of primitive recursive functions is necessarily closed under operations other than composition and primitive recursion.

Definition 7.13 Let S be a set of functions on the natural numbers. The set is *closed under bounded sums* if, for any k-ary function $f(x_1,\ldots,x_k)$ in S, the function $sum_f(y,x_2,\ldots,x_k) = \sum_{z<y} f(z,x_2,\ldots,x_k)$ is also in S.

Proposition 7.14 The set of primitive recursive functions is closed under bounded sums.

Proof Let $f(x_1, \ldots, x_k)$ be a primitive recursive function.

Let $h(x_2, \ldots, x_k) = 0$ and let $g(x_1, \ldots, x_k, x_{k+1}) = f(x_1, \ldots, x_k) + x_{n+1}$. Both h and g are primitive recursive. Moreover, the function $sum_f(y, x_2, \ldots, x_k)$ can be defined from h and g by primitive recursion:

$$sum_f(0, x_2, \ldots, x_k) = 0,$$
$$sum_f(k+1, x_2, \ldots, x_k) = g(k, x_2, \ldots, x_k, sum_f(k, x_2, \ldots, x_k))$$
$$= f(k, x_2, \ldots, x_k) + sum_f(k, x_2, \ldots, x_k). \qquad \square$$

Definition 7.15 Let S be a set of functions on the natural numbers. The set is *closed under bounded products* if, for any k-ary function $f(x_1, \ldots, x_k)$ in S, the function $prod_f(y, x_2, \ldots, x_k) = \Pi_{z<y} f(z, x_2, \ldots, x_k)$ is also in S.

Proposition 7.16 The set of primitive recursive functions is closed under bounded products.

Proof The function $prod_f(y, x_2, \ldots, x_k)$ can be defined using primitive recursion in the same manner that $sum_f(y, x_2, \ldots, x_k)$ is defined in Proposition 7.18. $\square$

Definition 7.17 Let S be a set of functions on the non-negative integers. The set is *closed under bounded search* if the following holds. If $f(\bar{x}, y)$ is in S, then so is the function $bs_f(\bar{x}, y)$ defined as follows

$$bs_f(\bar{x}, y) = \begin{cases} \text{the least } z \text{ less than } y \text{ such that } f(\bar{x}, z) = 0 \\ y \text{ if no such } z \text{ exists.} \end{cases}$$

Proposition 7.18 The set of primitive recursive functions is closed under bounded search.

Proof Let $f(\bar{x}, y)$ be a primitive recursive function. We show that $bs_f(\bar{x}, y)$ can be written as a composition of primitive recursive functions.

The predecessor function $pred(x)$ was shown to be primitive recursive in the proof of Proposition 7.9. Let $pos(x)$ be the function $x \dotminus pred(x)$. This function determines whether or not x is positive: $pos(x) = 1$ for nonzero x and $pos(0) = 0$. We claim that

$$bs_f(y, \bar{x}) = \sum_{z<y} (\Pi_{w<z} pos(f(w, \bar{x}))).$$

We leave the verification of this claim to the reader. By the previous propositions, $bs_f(y, \bar{x})$ is a composition of primitive recursive functions and is therefore primitive recursive. $\square$

Thus, we see that the primitive recursive functions form a vast set of computable functions. In fact, it may seem difficult to demonstrate a total computable function that is not primitive recursive. We now give one well known example of such a function.

7.1.2 The Ackermann function. We define a total binary function $A(n, x)$ that is computable but not primitive recursive. We refer to this function as the *Ackermann function*. It is one of several variations of a function first introduced by Wilhelm Ackermann in 1928. The function $A(n, x)$ is defined as follows. For $(n, x) \in \mathbb{N}^2$,

$$A(0, x) = x + 1,$$
$$A(n, 0) = A(n - 1, 1), \text{ and}$$
$$A(n, x) = A(n - 1, A(n, x - 1)).$$

This function is computable. Let us compute some specific values of $A(n, x)$. Since $A(0, x) = x + 1$, we have $A(0, 1) = 2$, $A(0, 2) = 3$, and so forth. We also have: $A(1, 0) = A(0, 1) = 2$, $A(2, 0) = A(1, 1) = A(0, A(1, 0)) = A(0, 2) = 3$, $A(1, 2) = A(0, A(1, 1)) = A(0, 3) = 4$, and $A(3, 0) = A(2, 1) = A(1, A(2, 0)) = A(1, 3) = A(0, A(1, 2)) = A(0, 4) = 5$. We record these and other values of $A(n, x)$ in Table 7.1.

Table 7.1 Values for $A(n, x)$

	$n = 0$	1	2	3
$x = 0$	1	2	3	5
1	2	3	5	13
2	3	4	7	29
3	4	5	9	61

Although these computations may seem innocuous, it is practically impossible to extend the above table much further to the right. The column for $n = 4$ begins $A(4, 0) = 13$, $A(4, 1) = 65534$, and then the numbers get really big. The column for $n = 5$ begins

A(5,0) = A(4,1) = 65533.

The computation of $A(5, 1)$ is beyond the capabilities of any computer.

For each $n \in \mathbb{N}$, let $a_n(x)$ denote the unary function $A(n, x)$. From the above computations, we see that

$$a_0(x) = A(0, x) = x + 1,$$
$$a_1(x) = A(1, x) = x + 2,$$

$$a_2(x) = A(2, x) = 2x + 3, \text{ and}$$
$$a_3(x) = A(3, x) = 2^{x+3} - 3.$$

The function $a_4(x)$ is defined by $A(4, x) = \underbrace{2^{2^{\cdot^{\cdot^{2}}}}}_{x+3 \text{ times}} - 3.$

The functions $a_n(x)$ for $n > 4$ cannot be expressed using conventional notation. We make two observations regarding these functions. First, note that the function $a_n(x)$ can be defined inductively from the function $a_{n-1}(x)$. This follows from the rules $A(n, 0) = A(n-1, 1)$ and $A(n, x) = A(n-1, A(n, x-1))$ in the definition of $A(n, x)$. It follows that each $a_n(x)$ is a primitive recursive function (since $a_0(x)$ is). The second observation is that the function $a_{n-1}(x)$ is smaller than $a_n(x)$ for each $n \in \mathbb{N}$.

Definition 7.19 We say that $g(x_1, \ldots, x_k)$ is smaller than $f(x)$, and write $g < f$, if $g(x_1, \ldots, x_k) < f(m)$ for all k-tuples $\bar{x}$ having maximum entry $x_i = m$. If g happens to be unary, then $g < f$ simply means that $g(n) < f(n)$ for all n.

Let $a_\infty(x) = A(x, x)$. This function is not smaller than any of the functions $a_n(x)$.

Proposition 7.20 The function $a_\infty(x) = A(x, x)$ is not primitive recursive.

Proof We show that every primitive recursive function is smaller than $a_m(x)$ for some $m \in \mathbb{N}$. The basic functions $Z(x)$ and $p_i^k(\bar{x})$ are smaller than $a_1(x)$ and $s(x)$ is smaller than $a_2(x)$.

Suppose now that f is the composition of g and h and $g < a_i$ and $h < a_j$ for some i and j in $\mathbb{N}$. Further suppose that g and h are unary functions (we claim that what follows generalizes to higher arities). We show that $f(x) = g(h(x))$ is smaller than a_{m+1} where m is larger than both i and j. To show this, we make two observations: for any $x \in \mathbb{N}$ and $m \in \mathbb{N} \cup \{0\}$:

1. $x < a_m^{(x-1)}(1)$ (for $x > 0$ and $m > 0$), and
2. $a_m^{(x+1)}(1) = a_{m+1}(x)$,

where $a_m^{(x)}(1)$ denotes the composition $\underbrace{a_m(a_m(\cdots(a_m(1))))}_{x \text{ times}}$. We regard $a_m^{(x-1)}(1)$ as a function of x.

To see that the first observation holds, note that $a_0^{(x-1)}(1) = x$ for $x > 0$.

For the second observation, recall from the definition of the Ackermann function that
$$a_{m+1}(x) = a_m(a_{m+1}(x-1)) = a_m(a_m(a_{m+1}(x-2))) = \cdots.$$

We see that $a_{m+1}(x) = a_m^{(x)}(a_{m+1}(0))$. Again by the definition of the Ackermann function: $a_{m+1}(0) = a_m(1)$. This establishes the second observation.

We now show that the composition of h and g is smaller than $a_{m+1}(x)$. We have

$g(h(x)) < a_i(a_j(x)) < a_m(a_m(x))$.
By observation 1, $a_m(a_m(x)) < a_m(a_m(a_m^{(x-1)}(1)))$.
By observation 2, $a_m(a_m(a_m^{(x-1)}(1))) = a_m^{(x+1)}(1) = a_{m+1}(x)$.

Now suppose that f is defined from h and g by primitive recursion. Again, we suppose for simplicity that f is unary. In this case, $h(x) = c$ for some constant c and g is a binary function. The function f is defined for each $n \in \mathbb{N}$ as follows:

$$f(0) = c \quad \text{and} \quad f(n+1) = g(n, f(n)).$$

Suppose that $g < a_i$. We show that $f < a_m$ for some m. Since $a_j < a_{j+1}$ for each j, there exists j such that $a_j(0)$ is bigger than the constant c. Let m be greater than both j and $i + 1$.

We may assume that $f(n) > n$ for each n (since f is smaller than some function with this property). Likewise, we may also assume that g is an increasing function. In particular, we use the inequality $g(n, f(n)) < g(f(n), f(n))$.

Claim $f(n) < a_m(n)$ for all n.

Proof We prove this by induction on n.

$$f(0) = c < a_m(0) \text{ by our choice of } m.$$
$$f(n+1) = g(n, f(n)) < g(f(n), f(n)) \text{ (by our previous remarks)}$$
$$< a_i(f(n)) < a_i(a_m(n)) \text{ (by induction)}$$
$$< a_{m-1}(a_m(n)) \text{ (since } m > i+1\text{)}$$
$$= a_m(n+1) \text{ (since } A(m, n+1) = A(m-1, A(m, n))).$$

This completes the proof. □

We conclude that the inductive method used to define $A(n, x)$ is stronger than primitive recursion. To generate the set of computable functions from the basic functions, we must include closure operations beyond composition and primitive recursion. We claim that, in addition these two operations, we need only one more operation to obtain the entire set of computable functions. As we shall show, it suffices to modify the operation of bounded search.

7.1.3 Recursive functions. We now define the *recursive functions*. In addition to being closed under composition and primitive recursion, the set of recursive functions is also closed under *unbounded search*.

Definition 7.21 Let $f(\bar{x}, y)$ be a $(k+1)$-ary function on the non-negative integers. We denote by $us_y f(\bar{x}, y)$ the (k)-ary function defined as follows. For a given k-tuple $\bar{x}$, $us_y f(\bar{x}, y) = z$ if and only if $f(\bar{x}, y)$ is defined for each $y \leq z$ and z is least such that $f(\bar{x}, z) = 0$. If no such z exists, then $\bar{x}$ is not in the domain of $us_y f(\bar{x}, y)$.

Definition 7.22 Let $\mathcal{S}$ be a set of functions on the non-negative integers. The set is *closed under unbounded search* if the following holds. If $f(\bar{x}, y)$ is in $\mathcal{S}$, then so is the function $us_y(\bar{x}, y)$.

Definition 7.23 The set of *recursive functions* is the smallest set containing the basic functions and closed under composition, primitive recursion, and unbounded search.

Every recursive function is computable. To verify this, consider the unbounded search process. Suppose that $f(\bar{x}, y)$ is computable. To compute $us_y f(\bar{a}, y)$ for a given input $\bar{a}$, follow these steps:

(1) Let $n = 0$
(2) Compute $f(\bar{a}, n)$
(3) If $f(\bar{a}, n) = 0$, then $us_y f(\bar{a}, y) = n$.
 Otherwise add 1 to the value of n and go to (2).

Thus, $us_y f(\bar{a}, y)$ is computed in a finite number of steps provided that $\bar{a}$ is in the domain of $us_y f(\bar{x}, y)$. If $\bar{a}$ is not in the domain of $us_y f(\bar{x}, y)$, then these three steps will produce no output (the algorithm will either be unable to compute step (2) for some n or it will form an infinite loop).

Conversely, we claim that every computable function is recursive. In particular, the Ackermann function is recursive. We prove this as a consequence of Kleene's Recursion theorem 7.38 of Section 7.4. Thus, we show that the Ackermann function is recursive without providing an explicit definition of this function in terms of unbounded search. We leave this explicit definition as Exercise 7.8.

The assertion that every computable function is recursive is known as *Church's thesis*. Section 7.4 provides evidence for Church's thesis. Not only do we show that the Ackermann function is recursive, but we also prove that all functions computable by programs in a specific programming language are recursive. We do not prove Church's thesis. Such a proof would have to consider every conceivable programming language. This is not feasible. Moreover, even if we were to somehow succeed in this task, then this still would not suffice as a proof of Church's thesis.

To fully understand Church's thesis, we must make the distinction between functions that are computable by a computer program and those functions that are computable in the intuitive sense. A function is "programmable" if there is a computer program (in some conventional programming language) that computes the function. It need not be feasible to execute the program in a reasonable amount of time or space. A function is "intuitively computable" if it can somehow be computed by some algorithm. Whereas the notion of a recursive function is precise, the notions of "programmable" and "intuitively computable" are increasingly more vague. From these definitions, we have the following two inclusions:

$$\left\{\begin{matrix}\text{Recursive}\\ \text{Functions}\end{matrix}\right\} \subset \left\{\begin{matrix}\text{Programmable}\\ \text{Functions}\end{matrix}\right\} \subset \left\{\begin{matrix}\text{Intuitively Computable}\\ \text{Functions}\end{matrix}\right\}.$$

Church's thesis implies that each of these three sets contain the same functions. There are two possible scenarios in which this thesis is false. One possibility is that there exists a clever computer capable of computing a nonrecursive function. We claim that this is not the case. That is, we claim that the first of the above inclusions is not proper. We provide evidence for this following Theorem 7.38 of Section 7.4. The other scenario involves the second inclusion.

Let us temporarily entertain the possibility that there exist functions that are computable by some algorithm that cannot be implemented into a computer program. For example, consider the function $\pi(n) =$ the nth digit in the decimal expansion of π. This function is intuitively computable since we have the following algorithm: construct a really big circle, measure the circumference and diameter, and divide ("really big" depends on n). Likewise, consider the function $f(n) = |1000 \cdot \sin n|$ (rounded to the nearest integer). We can compute this function by constructing a really big triangle having an angle of n radians and taking the ratio of two sides. If we are trying to program a computer to evaluate these functions, then these algorithms do not lend much insight. There do, of course, exist computer algorithms that compute $\pi(n)$ and $f(n)$ (for example, we can use the Taylor series to compute the sine). However, it seems plausible that there exist other algorithms that cannot be carried out by a computer. Church's thesis states that this is not the case: any function that is computable in the intuitive sense is recursive. From this point of view, Church's thesis is not a statement of mathematics, but a statement of faith that precludes the possibility of proof (although disproof is possible).

We make no assumptions regarding the veracity of Church's thesis. We focus our attention on the set of recursive functions. We show in Section 7.4 that these functions are equivalent to the functions that are computable in a certain computer language. Moreover, we claim that these are precisely those functions that are computable by a program in C++ (or Basic or Cobol or Maple). This provides evidence in support of Church's thesis, but this not our aim. Regardless of whether or not Church's thesis is true, the set of recursive functions is of natural interest. Henceforth, when we refer to the *computable functions* we mean a function in this precisely defined set. Our aim in this chapter is to investigate this set of computable functions and the corresponding concept of decidability.

7.2 Computable sets and relations

Let A be a subset of the non-negative integers. The *characteristic function* of A (denoted $\chi_A(x)$) is defined as follows:

$$\chi_A(x) = \begin{cases} 1 & x \in A \\ 0 & x \notin A. \end{cases}$$

The set A is said to be *recursive* if its characteristic function is recursive. Likewise, A is said to be *primitive recursive* if its characteristic function is. These definitions can be extended to any relation R on $\mathbb{N} \cup \{0\}$. That is, the *recursive* (and *primitive recursive*) subsets of $(\mathbb{N} \cup \{0\})^k$ for $k > 1$ are defined in the same manner that we have defined the recursive and primitive recursive subsets of $\mathbb{N} \cup \{0\}$.

To each relation R on the non-negative integers there is an associated decision problem: namely, the problem of determining whether or not a given tuple $\bar{x}$ is in the set R. This decision problem is said to be *decidable* if and only if R is a recursive relation. By definition, every decision problem corresponds to some relation. So to study recursive relations is to study the set of all decidable decision problems. In this section, we restrict our attention to the primitive recursive relations. In Section 7.5 we consider the wider set of recursive relations and demonstrate (in Section 7.6) various subsets of $\mathbb{N} \cup \{0\}$ that are not recursive.

Example 7.24 Consider the binary relation $x < y$. The characteristic function for this relation is defined by $\chi_<(x, y) = 1$ if $x < y$ and $\chi_<(x, y) = 0$ otherwise. This function can be defined as a composition of primitive recursive functions:

$$\chi_<(x, y) = 1 \mathbin{\dot{-}} (x \mathbin{\dot{-}} y).$$

It was shown in Proposition 7.9 that $sub(x, y) = x \mathbin{\dot{-}} y$ is primitive recursive. It follows that the relation $x < y$ is also primitive recursive.

Primitive recursive relations allow us to define primitive recursive functions *by cases*. For example, consider the function $f(x)$ defined by cases as follows.

$$f(x) = \begin{cases} g(x) & \text{if } x < 10 \\ h(x) & \text{otherwise.} \end{cases}$$

If $g(x)$ and $h(x)$ are both primitive recursive, then so is $f(x)$. More generally, we have the following.

Proposition 7.25 Let $R_1, \ldots, R_n$ be disjoint k-ary relations and let $g_1(\bar x), \ldots, g_n(\bar x)$ be k-ary functions. If each of these functions and relations is primitive recursive, then so is the function defined by cases as follows

$$h(\bar x) = \begin{cases} g_1(\bar x) & \text{if } R_1(\bar x) \\ g_2(\bar x) & \text{if } R_2(\bar x) \\ \ldots & \ldots \\ g_k(\bar x) & \text{if } R_k(\bar x). \end{cases}$$

Proof We have

$$f(\bar x) = g_1(\bar x) \cdot \chi_1(\bar x) + g_2(\bar x) \cdot \chi_2(\bar x) + \cdots + g_k(\bar x) \cdot \chi_k(\bar x),$$

where $\chi_i(\bar x)$ denotes the characteristic function for the relation $R_i(\bar x)$ (for $i = 1, \ldots, k$). □

In this section, we show that several familiar relations (in addition to $a < b$) are primitive recursive. We also show that, given two primitive recursive relations A and B on the non-negative integers, the relations $A \cup B$, $A \cap B$, $A \times B$, and other relations are also primitive recursive. Rather than considering these relations one-by-one and proving that each is primitive recursive, we instead take advantage of our background in first-order logic. We show that if a relation is definable by a quantifier-free formula in the vocabulary of arithmetic, then that relation is primitive recursive.

Let $\mathbf{N_0} = (\mathbb{N} \cup \{0\} | +, \cdot, 0, 1)$ be the structure having the non-negative integers as an underlying set that interprets the vocabulary $\mathcal{V}_{ar} = \{+, \cdot, 0, 1\}$ in the usual way.

Proposition 7.26 Let A be a definable subset of $\mathbf{N_0}$. If A is definable by a quantifier-free $\mathcal{V}_{ar}$-formula, then A is primitive recursive.

Proof Let $\varphi_A(x_1, \ldots, x_n)$ be a $\mathcal{V}_{ar}$-formula that defines the k-ary relation A. We show that A is primitive recursive by induction on the complexity of φ_A. However, we do not proceed in the usual order. We first prove the induction step and lastly consider the case where φ_A is atomic.

Suppose that $\theta(\bar x)$ and $\psi(\bar x)$ are $\mathcal{V}_{ar}$-formulas that define primitive recursive subsets of $\mathbf{N_0}$. Let B be the relation defined by θ and let C be the relation defined by ψ. Then the characteristic functions χ_B and χ_C are both primitive recursive. We must show that the characteristic function of A is also primitive recursive.

If $\varphi_A(\bar x) \equiv \theta(\bar x)$, then χ_A and χ_B are the same primitive recursive function. If $\varphi_A(\bar x)$ is the formula $\neg \theta(\bar x)$, then $\chi_A(\bar x)$ is 1 if and only if $\chi_B(\bar x)$ is 0.

So $\chi_A(\bar x) = 1 \dot - \chi_B(\bar x)$. Since both $\chi_B(\bar x)$ and $sub(x,y) = x \dot - y$ are primitive recursive, so is $\chi_A(\bar x)$.

Now suppose $\varphi_A(\bar{x})$ is the formula $\theta(\bar{x}) \wedge \psi(\bar{x})$. Then $\chi_A(\bar{x})$ equals the composition $m(\chi_B(\bar{x}), \chi_C(\bar{x})) = \chi_B(\bar{x}) \cdot \chi_C(\bar{x})$.

In any of these cases, $\chi_A(\bar{x})$ is primitive recursive. This concludes the induction step of the proof. It remains to be shown that $\chi_A(\bar{x})$ is primitive recursive in the case where $\varphi_A(\bar{X})$ is an atomic $\mathcal{V}_{ar}$-formula.

Since there are no relations in the vocabulary, atomic $\mathcal{V}_{ar}$-formulas have the form $t_1 = t_2$ for $\mathcal{V}_{ar}$-terms t_1 and t_2. Each term is a composition of the functions $+$ and $\cdot$ applied to the constants and variables. So each $\mathcal{V}_{ar}$-term may be regarded as a polynomial $p(\bar{x})$ having natural numbers as coefficients. If $\varphi_A(\bar{x})$ is atomic, then it must have the form $p_1(\bar{x}) = p_2(\bar{x})$. By Proposition 7.12, each of the polynomials $p_1(\bar{x})$ and $p_2(\bar{x})$ are primitive recursive functions. We must show that the relation of equality is also primitive recursive.

We previously demonstrated that the relation $x < y$ is primitive recursive. It follows that $y \leq x$ (the negation of $x < y$) is also primitive recursive. Likewise, $x \leq y$ is a primitive recursive relation. Finally, the relation $x = y$, defined as $x \leq y \wedge y \leq x$, is primitive recursive. (We are using the fact that the primitive recursive relations are closed under negations and conjunctions. This was proved as part of the induction step.)

If $\varphi_A(\bar{x})$ is atomic, then $\chi_A(\bar{x})$ is the composition $\chi_{eq}(p_1(\bar{x}), p_2(\bar{x}))$, where $\chi_{eq}(x, y) = 1$ if $x = y$ and is otherwise zero. Since it is the composition of primitive recursive functions, $\chi_A(\bar{x})$ is primitive recursive.

This completes the base step for the induction. We conclude that every quantifier-free $\mathcal{V}_{ar}$-formula defines a primitive recursive subset of $\mathbf{N}_0$. □

The converse of Proposition 7.26 does not hold. In the next chapter we prove that every primitive recursive relation is definable (see Corollary 8.15). However, not every primitive recursive relation is definable by a quantifier-free formula. The formula that defines a primitive recursive relation may require quantifiers.

Definition 7.27 Let $\mathcal{F}$ be a set of $\mathcal{V}_{ar}$-formulas. We say that $\mathcal{F}$ is *closed under bounded quantifiers* if for any $\varphi(x, y) \in \mathcal{F}$, the formula $\exists y(y < x \wedge \varphi(x, y))$ is also in $\mathcal{F}$ where $y < x$ is an abbreviation for the formula $\exists z(y + z = x)$. (The formula $\varphi(x, y)$ may have free variables other than x and y.)

Let Δ_0 be the smallest set of $\mathcal{V}_{ar}$-formulas containing the atomic formulas that is closed under equivalence, negation, conjunction, and bounded quantifiers. Note that the negation of the formula $\exists y(y < x \wedge \varphi(x, y))$ is equivalent to the formula $\forall y(y < x \rightarrow \neg \varphi(x, y))$. So in any Δ_0 formula, each variable y that is quantified by either $\exists$ or $\forall$ is bounded by another variable as $y < x$. That is, the bound variables are bounded by free variables. In particular, each Δ_0 sentence must be quantifier-free.

Computability and complexity   315

A relation is primitive recursive if and only if it is definable by a Δ_0 formula. We presently prove one direction of this fact. The other direction shall become apparent after Section 8.3 of the next chapter and is left as Exercise 8.6.

Proposition 7.28 Let A be a definable subset of $\mathbf{N}_0$. If A is definable by a Δ_0 formula, then A is primitive recursive.

Proof We must add one step to the proof of the previous proposition. Suppose that $\varphi(x, y)$ defines a primitive recursive subset A of $\mathbf{N}_0$. We must show the formula $\exists y (y < x \land \varphi(x, y))$ also defines a primitive recursive subset. For convenience, we assume that x and y are the only free variables of $\varphi(x, y)$ (this assumption does not alter the essence of the proof).

Let $\chi_A(x, y)$ be the characteristic function for A. Since this function is primitive recursive, so is the function $sum\chi(x, y) = \sum_{z < y} \chi_A(x, z)$ by Proposition 7.18. It follows that the function $g(x) = sum\chi(x, x)$ is also primitive recursive. Note that $1 \dot{-} g(x)$ equals 0 if $\chi_A(x, z) = 1$ for some $z < x$ and otherwise $1 \dot{-} g(x)$ equals 1. From this observation, we see that the function $1 \dot{-} (1 \dot{-} g(x))$ is the characteristic function for the set defined by $\exists y(y < x \land \varphi(x, y))$. It follows that this is a primitive recursive set. □

Propositions 7.26 and 7.28 allow us to succinctly show that certain functions and relations are primitive recursive. The aim for the remainder of this section is twofold. One aim is to demonstrate some of the many familiar functions and relations that are primitive recursive. The other aim is to show that a specific binary function, namely $pf(x, i)$, is primitive recursive. The name "pf" bestowed to this function is an abbreviation for "prime factorization." We shall make use of this function and the fact that it is primitive recursive in Section 7.4.

Prior to defining the function $pf(x, i)$, we define the relations $div(x, y)$ and $prime(x)$. For any pair (x, y) of non-negative integers, the relation $div(x, y)$ says that x divides y and $prime(x)$ says that x is prime. The relation $div(x, y)$ holds if and only if there exists a z such that $x \cdot z = y$. Clearly, if such a z exists, then z is at most y. So $div(x, y)$ is definable by the Δ_0 formula

$$\exists z(z < y \land x \cdot z = y) \lor x = 1 \lor (y = 0 \land \neg x = 0).$$

Likewise, $prime(x)$ is defined by the formula

$$\forall z(z < x \to (z = 1 \lor \neg div(z, x))) \land (\neg x = 1).$$

Since these formulas are Δ_0, the relations $div(x, y)$ and $prime(x)$ are primitive recursive by Proposition 7.28.

There are infinitely many primes. Let $p_1, p_2, p_3, \ldots$ represent the enumeration of the primes in increasing order. So $p_1 = 2$, $p_2 = 3$, $p_3 = 5$, and so forth. We claim that the function $pr(i) = p_i$ is primitive recursive. To make this function total, let us set $pr(0) = 0$. This function can be defined by primitive recursion. Let $h(x) = Z(x)$ and let $g(x, y)$ be the least prime number greater than y. To verify that $g(x, y)$ is primitive recursive, note that we can define this function using bounded search. The least prime number greater than y must be less than $2y$. ("Chebychev proved it, and Erdös proved it again, there is always a prime between n and $2n$.") The function $pr(i)$ is defined as follows:

$$pr(0) = h(0) = 0,$$
$$pr(n+1) = g(n, pr(n)).$$

We now define the prime factorization function $pf(x, i)$. Every natural number x can be factored as

$$x = p_1^{a_1} p_2^{a_2} \ldots p_k^{a_k},$$

where p_i denotes the ith prime number. Moreover, the exponents a_i are uniquely determined by n. This is the *Fundamental Theorem of Arithmetic*. We define $pf(x, i)$ to be the exponent a_i that occurs on the ith prime in the prime factorization of x. To make this a total function, we define $pf(x, i)$ to be 0 if $x = 0$ or $i = 0$.

Proposition 7.29 The function $pf(x, i)$ is primitive recursive.

Proof We sketch the proof. The function $pf(x, i)$ equals y if and only if $div(pr(i)^y, x)$ and $\neg div(pr(i)^{(y+1)}, x)$ both hold. Such a number y must be less than x (since $p^x > x$ for all primes p and integers x). So we can define $pf(x, i)$ in terms of the primitive recursive function $pr(i)$, the primitive recursive relation $div(x, y)$, and the primitive recursive operation of bounded search. It follows that $pf(x, i)$ is primitive recursive. □

7.3 Computing machines

In the 1930s, Alan Turing described a theoretical computing machine to capture the notion of computability. *Turing's thesis* states that every function that is intuitively computable can be computed by a Turing machine. Modern computers may be viewed as crude approximations of Turing's machine (crude since they do not have infinite memory). Variations of Turing's machine known as *register machines* were developed in the 1960s by Shepherdson, Sturgis, and others. It was shown that each of these theoretical machines have the same computing power. The functions computable by either a Turing machine or a register machine are precisely the recursive functions. In light of these results, Turing's thesis is equivalent to Church's thesis. Henceforth, we refer to this as the Church–turing thesis.

In this section, we describe a variation of the register machines. This machine executes programs written in a specific programming language that we shall describe. Functions computable by programs in this language are called *T-computable*. At the conclusion of this section, we prove that every recursive function is T-computable. We prove the converse of this in the next section. So the computing machine we describe has the same computing power as any register machine or Turing machine.

We now describe our programming language. As we have previously indicated, it does not matter which programming language we choose. If a function is computable by a program in PASCAL, then this program can be translated to a program in C++ or any other programming language. For convenience and definiteness, we use a simplified programming language we call T^{++}. This language is convenient because it has only four types of commands. Of course, if we actually wanted to program a computer to perform a complicated task, then this language would not be so convenient. For each $i \in \mathbb{N}$, T^{++} has the following commands:

Add i, Rmv i, RmvP i, and GOTO i.

A *program* in T^{++} is a finite sequence of commands.

We now describe a machine that runs a given T^{++} program P. This is called a *turnip machine* or, more simply, a T-*machine*. The machine consists of an enumerated row of bins. Each bin contains turnips. Let B_i denote the ith bin. We assume there are enough bins (and turnips) to carry out the program. Some of these bins may be empty. Whereas the bins are enumerated B_1, B_2, $B_3, \ldots$, the commands that constitute the program are enumerated $(1), (2), \ldots$ (the latter sequence is finite).

To run program P, the T-machine begins by reading the first command. We now describe how the T-machine executes each of the four possible commands. Suppose that the machine is reading command (10) of program P.

- If this command is Add i, then the machine puts one turnip in bin B_i and then proceeds to the next command (namely, command (11)).
- If the tenth command is RmvP i, then there are two possibilities. If bin B_i is empty, then the machine does nothing and proceeds to the next command. Otherwise, the turnip machine removes one turnip from bin B_i and then goes to the previous command (namely, command (9)).
- If the tenth command is Rmv i, then the T-machine removes a turnip from bin B_i (if there is one) and then, regardless of whether or not there was a turnip to be removed, proceeds to the next command (namely, (11)).
- Finally, the command GOTO i causes the turnip machine to go to command (i) of program P.

The T-machine continues to operate until it comes to a line of the program that does not exist. For example, the following program causes the T-machine to halt immediately without doing anything:

(1) GOTO 12.

It is possible that the T-machine will never halt as the following T^{++} program demonstrates:

(1) Add 4
(2) RmvP 4.

By adding one line the beginning of the previous program we obtain:

(1) RmvP 4
(2) Add 4
(3) RmvP 4.

If there is a turnip in bin B_4 when we run this program, then the T-machine removes a turnip and halts. Otherwise, if B_4 is empty, the T-machine will never halt.

The number of turnips in each bin when the T-machine halts (if it halts) depends on how many turnips were in the bins at the outset. Thus, each T^{++} program determines a function. In fact, each program P determines many functions. For each $k \in \mathbb{N}$, we describe a k-ary function $P^{(k)}$ on the non-negative integers. Given $(x_1, \ldots, x_k)$ as input, put x_i turnips in bin B_i for $i = 1, \ldots, k$ and leave the bins B_j empty for $j > k$. Run program P. We define $P^{(k)}(x_1, \ldots, x_k)$ to be the number of turnips in bin B_1 when the machine halts. If the T-machine does not halt, then $P^{(k)}(x_1, \ldots, x_k)$ is undefined.

Definition 7.30 Let f be a partial or total k-ary function on the non-negative integers. We say that f is T-computable if f is $P^{(k)}$ for some T^{++} program P. That is, f and $P^{(k)}$ have the same domain and $f(x_1, \ldots, x_k) = P^{(k)}(x_1, \ldots, x_k)$ for any $(x_1, \ldots, x_k)$ in this domain.

Of course, the actual hardware for the T-machine is irrelevant. We could use cabbage instead of turnips. In fact, the concept of a T-machine does not require any vegetables. Modern computers can be used to simulate turnip machines. From now on, we assume that a T-machine is a computer that has been programmed to carry out the above commands. We view each B_i as a program variable that may take on any non-negative integer value.

Although they may seem primitive, T-machines are capable of computing any recursive function.

Proposition 7.31 Every recursive function is T-computable.

Proof We first show that the basic functions are T-computable. The successor function $s(x)$ corresponds to the one-lined program: (1) Add 1. The zero function is computed by the following T^{++} program.

(1) Rmv 1
(2) RmvP 1.

Now consider the projection function $p_i^k(x_1, x_2, \ldots, x_k) = x_i$ (for $i \leq k$). If $i = 1$, then this function is computed by the program (1) GOTO 12 or any other program that causes the T-machine to do nothing. For $i > 1$, consider the following program.

(1) Rmv 1
(2) RmvP 1
(3) Add 1
(4) RmvP i
(5) Rmv 1.

This program moves the contents of B_i to B_1. The first two lines set B_1 to zero. Lines (3) and (4) successively increase B_1 while decreasing B_i. When B_i reaches zero, we will have increased B_1 one too many times. The final line of the program corrects this.

We claim that the set of T-computable functions is closed under both composition and primitive recursion. We leave the verification of this as Exercise 7.8. It follows that every primitive recursive function is T-computable.

It remains to be shown that the T-computable functions are closed under unbounded search. Suppose that the function $h(x_1, \ldots, x_k, y)$ is T-computable. We describe a T^{++} program that computes the least value of y for which $h(x_1, \ldots, x_k, y) = 0$.

(1) ZERO B_{n+1}
(2) COMPUTE $h(B_1, \ldots, B_n, B_{n+1})$ STORE IN B_{n+2}
(3) GOTO 5
(3) GOTO 8
(5) RmvP $n + 2$
(6) MOVE B_{n+1} TO B_1
(7) GOTO 10
(8) Add $n + 1$
(9) GOTO 2.

The command ZERO B_{n+1} is an abbreviation for the commands that set B_{n+1} to zero. Line (6) moves the contents of B_{n+1} to B_1. We previously described programs for each of these operations. Likewise, if h is computable, then we can replace line (2) with a series of T^{++} commands (see Exercise 7.8).

We conclude that every recursive function is T-computable. □

7.4 Codes

We describe a process for coding and decoding T^{++} programs as natural numbers. To each T^{++} program P we assign a natural number called the *code* for P. Given the natural number, we can recover the entire program. Codes provide an invaluable tool for analyzing the set of T-computable functions. Using these codes, we shall be able to prove that every T-computable function is recursive. In light of this fact, the codes also lend insight into the recursive functions. The codes allow us to show, among other things, that the Ackermann function is recursive.

Prior to assigning codes to programs, we assign codes to individual commands. To each command we assign a natural number as follows:

Command	Number
Add i	$4i$
Rmv i	$4i - 1$
RmvP i	$4i - 2$
GOTO i	$4i - 3$

Each command corresponds to exactly one natural number and each natural number corresponds to exactly one command. In particular, 0 is the only nonnegative integer that does not correspond to some T^{++} command. Let P be a T^{++} program. We may view P as a finite sequence of natural numbers. Suppose that P corresponds to the sequence $(n_1, n_2, \ldots, n_k)$. That is, n_i is the number corresponding to command (i) of P (for $i \leq k =$ the length of P). Let $e = 2^{n_1} 3^{n_2} 5^{n_3} \cdots p_k^{n_k}$, where p_k denotes the kth prime number. The program P uniquely determines the number $e \in \mathbb{N}$. We refer to e as the *code* for P.

A given natural number e is the code for some T^{++} program if and only if e is divisible by each of the first k primes (for some k) and no other primes. If e is the code for a program P, then we can recover this program by factoring e. This follows from the Fundamental Theorem of Arithmetic which states that every natural number can be factored into primes in a unique manner. For example, the number 12 factors as $2^2 3^1$. This number corresponds to the T^{++} program

(1) RmvP 1

(2) GOTO 1

having sequence $(n_1, n_2) = (2, 1)$. The number $42 = 2 \cdot 3 \cdot 7$ does not correspond to a program since it is divisible by 7 but not 5.

We assign a program P_e to each e in $\mathbb{N} \cup \{0\}$. If e is the code for a T^{++} program, then let P_e denote this program. For those numbers that do not code a program, we assign a "default program." We arbitrarily choose the one-lined program (1) GOTO 12 to be this program. So if e does not code a program, then, by default, P_e is (1) GOTO 12. Consider the list of programs $P_0, P_1, P_2, P_3, \ldots$. Since every program has a code, this list includes every T^{++} program.

For each $k \in \mathbb{N}$, let φ_e^k denote the function $P_e^{(k)}$ (this notation helps distinguish the computable function from the program that computes the function). Consider the list of k-ary functions $\varphi_0^k, \varphi_1^k, \varphi_2^k, \varphi_3^k, \ldots$. By Proposition 7.31, this list includes every recursive k-ary function on the non-negative integers.

With the notable exception of the program (1) GOTO 12, every T^{++} program occurs exactly once in the list $P_0, P_1, P_2, \ldots$. The same cannot be said of the list of k-ary T-computable functions. Let $f(\bar{x})$ be a k-ary T-computable function. We show that $f(\bar{x})$ occurs as φ_e^k for infinitely many e.

Notation 2 Let $f(\bar{x})$ and $g(\bar{x})$ be partial functions. We write $f(\bar{x}) \simeq g(\bar{x})$ if the two functions have the same domain and $f(\bar{x}) = g(\bar{x})$ for any $\bar{x}$ in this domain.

Proposition 7.32 If $f(\bar{x})$ is a T-computable k-ary function, then $f(\bar{x}) \simeq \varphi_e^k(\bar{x})$ for infinitely many e.

Proof To any program that computes $f(\bar{x})$, we can add extraneous commands to obtain another program that computes $f(\bar{x})$. $\square$

In particular, each recursive k-ary function occurs infinitely many times in the list $\varphi_0^k, \varphi_1^k, \varphi_2^k, \varphi_3^k \ldots$. We next show that the recursive functions expend this list.

Theorem 7.33 Every T-computable function is recursive.

Proof Let $f(\bar{x})$ be a T-computable k-ary function. Then $f(\bar{x})$ is $\varphi_e^k(\bar{x})$ for some $e \in \mathbb{N}$. Our goal is to show that φ_e^k is recursive.

For convenience, suppose that $k = 1$. (This assumption does not alter the essence of the proof.)

To compute $\varphi_e^1(x)$, we set B_1 equal to x and B_j equal to 0 for $j > 1$ and then run the program P_e. The T-machine executes the commands of P_e one-by-one in the order determined by the program. We regard each executed command as a "step" of the computation. Suppose that the T-machine has completed n steps of the computation for some non-negative integer n.

- Let $bin(e, x, n, j)$ denote the value of bin B_j at this stage of the computation, and
- let $line(e, x, n)$ denote the line of the program that is to be executed next by the T-machine according to the program P_e.

We claim that the functions $line(e, x, n)$ and $bin(e, x, n, j)$ are primitive recursive. To verify this, we define these functions in a primitive recursive manner. For fixed values of e and x, we define the functions $line(e, x, n)$ and $bin(e, x, n, j)$ by induction on n. For each value of n, the function $bin(e, x, n, j)$ is defined for all j.

When $n = 0$, the T-machine has not yet begun the computation. We have

$$line(e, x, 0) = 1, \text{ and}$$

$$bin(e, x, 0, j) = \begin{cases} x & \text{if } j = 1 \\ 0 & \text{otherwise} \end{cases}$$

To determine $line(e, x, n + 1)$ and $bin(e, x, n + 1, j)$, we consider the line of the program previously executed, namely $(line(e, x, n))$, and examine the current contents $bin(e, x, n, j)$ of bin B_j. Let $L_n = line(e, x, n)$. Since there are four types of T^{++} commands, there are four possibilities for L_n.

If line (L_n) of P_e is the command GOTO 12, then we set $line(e, x, n+1) = 12$ and $bin(e, x, n+1, j) = bin(e, x, n, j)$. Note that "line (L_n) of P_e is the command GOTO 12" means that the exponent on the L_n^{th} prime in the prime factorization of e is the number $4 \cdot 12 - 3 = 45$ that corresponds to the command GOTO 12. Another way to express this is $pf(e, L_n) = 45$.

More generally, if $pf(e, L_n) = 4i - 3$ (corresponding to GOTO i), then

$$line(e, x, n + 1) = i \quad \text{and} \quad bin(e, x, n + 1, j) = bin(e, x, n, j) \quad \text{(for all } j\text{)}.$$

If $pf(e, L_n) = 4i$ (corresponding to Add i), then

$$line(e, x, n + 1) = line(e, x, n) + 1, \text{ and}$$

$$bin(e, x, n + 1, j) = \begin{cases} bin(e, x, n, j) + 1 & \text{if } j = i \\ bin(e, x, n, j) & \text{if } j \neq i. \end{cases}$$

If $pf(e, L_n) = 4i - 1$ (corresponding to Rmv i), then
$$line(e, x, n+1) = line(e, x, n) + 1, \text{ and}$$
$$bin(e, x, n+1, j) = \begin{cases} bin(e, x, n, j) \dot{-} 1 & \text{if } j = i \\ bin(e, x, n, j) & \text{if } j \neq i. \end{cases}$$

Finally, if $pf(e, L_n) = 4i - 2$ (corresponding to RmvP i), then
$$bin(e, x, n+1, j) = \begin{cases} bin(e, x, n, j) \dot{-} 1 & \text{if } j = i \\ bin(e, x, n, j) & \text{if } j \neq i \end{cases}, \text{ and}$$
$$line(e, x, n+1) = \begin{cases} line(e, x, n) \dot{-} 1 & \text{if } bin(e, x, n, i) \neq 0 \\ line(e, x, n) + 1 & \text{if } bin(e, x, n, i) = 0 \end{cases}.$$

Thus, we define the functions $bin(e, x, n, j)$ and $line(e, x, n)$. To see that this definition is primitive recursive, we make three observations.

- By Proposition 7.29, $pf(e, L_n)$ is primitive recursive.
- By Proposition 7.25, the above definitions by cases (including the cases based on the values of $pf(e, L_n)$) are primitive recursive.
- The process of inductively defining the two functions simultaneously is primitive recursive. We leave this as Exercise 7.8.

We conclude that the functions $bin(e, x, n, j)$ and $line(e, x, n)$ are primitive recursive functions as claimed.

Our goal is to show that the function $\varphi_e^1(x)$ is recursive. If the computation terminates, then $\varphi_e^1(x)$ equals the value of $bin(e, x, n, 1)$ for any n beyond the final step of the computation. Moreover, the computation terminates precisely when $lin(e, x, n)$ refers to a line of the program that does not exist. If $L_n = lin(e, x, n)$ is not a line of the program P_e, then $pf(e, L_n) = 0$. So the program terminates at step n if n is least such that $pf(e, L_n) = 0$. So we can define $\varphi_e^1(x)$ from $bin(e, x, n, j)$ and $line(e, x, n)$ using unbounded search. Explicitly, for any $e \in \mathbb{N}$:
$$\varphi_e^1(x) \simeq bin(e, x, y, 1), \quad \text{where } y = us_n pf(e, L_n); L_n = lin(e, x, n).$$

That is, $\varphi_e^1(x)$ is the composition $bin(e, x, us_n pf(e, lin(e, x, n)), 1)$. Since this function is defined from primitive recursive functions using unbounded search, it is a recursive function. Since e was arbitrary, we conclude that every T-computable function is recursive. □

Theorem 7.33 frees us from our restrictive programming language T^{++}. Whereas this choice of programming language was somewhat arbitrary, the resulting set of T-computable functions is not arbitrary. If we upgrade the T-machine so that it recognizes commands for adding and multiplying B_i and

B_j, then this will not provide any new computable functions. We may assume, without altering our concept of computability, that our programming language contains any number of commands for various recursive operations. This assumption may alter the concept of complexity. To define complexity classes in Section 7.7, we choose a particular extension of T^{++}.

The proof of Theorem 7.33 yields more than the statement of the theorem. Suppose we add a truly new feature to T^{++}. Consider the command $Copy(i, B_j)$ that sets B_c equal to B_i where c represents the contents of B_j. For example, if B_1 equals 9 and B_2 equals 5, then $Copy(1, B_2)$ sets B_5 equal to 9 (the contents of B_1 are "copied" to B_5). This command offers a versatility in writing programs that is found in virtually every programming language other than our contrived T^{++}. For example, this command allows us to write a program that, given input n in bin B_1, sets bin B_n equal to 1 and then halts. This simple task cannot be performed by a T-machine operating on T^{++} commands (try it).

Corollary 7.34 Suppose that T-machine (version 7.4) is an upgraded version of the T-machine that recognizes the command $Copy(i, B_j)$ as defined above for each i and j in $\mathbb{N}$. The functions computable by this machine are precisely the T-computable functions.

Proof This can be proved in the same manner as Theorem 7.33. The coding must be changed to accommodate the new commands. The crux of the proof shows that the functions $line(e, x, n)$ and $bin(e, x, n, j)$ are primitive recursive. We can define these functions inductively as in the proof of Theorem 7.33 with the following addition.

If $pf(e, L_n)$ is the code for the command $Copy(i, B_j)$, then

$$line(e, x, n+1) = line(e, x, n) + 1, \text{ and}$$
$$bin(e, x, n+1, k) = \begin{cases} bin(e, x, n, i) & \text{if } k = bin(e, x, n, j) \\ bin(e, x, n, k) & \text{otherwise} \end{cases}$$

□

The Church–Turing thesis implies that the computing power of T^{++} cannot be improved upon. The previous corollary corroborates this. We next provide stronger evidence by showing that any function defined from recursive functions in an inductive manner (such as the Ackermann function) is itself recursive. We prove this as a consequence of Kleene's Recursion theorem at the end of this section. Prior to proving this, we extract two further results from the proof of Theorem 7.33.

Corollary 7.35 (Kleene Normal Form) For every recursive function $\varphi_e^k(\bar{x})$, there exist two $k+2$-ary primitive recursive functions f and g such that $\varphi_e^k(\bar{x}) \simeq f(e, \bar{x}, us_n g(e, \bar{x}, n))$.

Proof Let $f(e, \bar{x}, n) = bin(e, \bar{x}, n, 1)$ and $g(e, \bar{x}, n) = pf(e, lin(e, \bar{x}, n))$. □

So not only is every computable function recursive, every computable function is a recursive function having a certain form. The definition of $\varphi_e^k(\bar{x})$ has only one occurrence of the unbounded search process. Since every recursive function is T-computable, every recursive function can be defined from the basic functions using primitive recursion, composition, and at most one application of unbounded search.

Corollary 7.36 The $(k+1)$-ary function U_k defined by $U_k(e, \bar{x}) \simeq \varphi_e^k(\bar{x})$ is recursive.

Proof By the proof of Theorem 7.33, U_k is the recursive function

$$bin(e, x, us_n pf(e, lin(e, x, n)), 1).$$

□

To prove Kleene's Recursion theorem, we make use of the following lemma.

Lemma 7.37 For all natural numbers n and m, there exists a binary primitive recursive function S_n^m such that

$$S_m^n(e, x_1, \ldots, x_m) = z \text{ implies } \varphi_z^n(y_1, \ldots, y_n) \simeq \varphi(m+n)_e(x_1, \ldots, x_m, y_1, \ldots, y_n).$$

Let us consider the content of this lemma prior to proving it. Suppose for simplicity that $m = n = 1$. Let $f(x, y)$ be a recursive binary function. Then $f(x, y)$ is the function $\varphi_e^2(x, y)$ for some e. For each number a, let $f_a(x)$ denote the unary function defined by $f_a(y) \simeq f(a, y)$. Since $f(x, y)$ is recursive, so is $f_a(y)$. So $f_a(y)$ is the function $\varphi_z^1(y)$ for some z. The lemma states that there exists a function S_1^1 that produces a code z for $f_a(y)$ given (e, a) as input. Moreover, this function is a primitive recursive function. If we replace x with an m-tuple $\bar{x}$ and y with an n-tuple $\bar{y}$, then we obtain the statement of the lemma in its full generality. This lemma is commonly referred to as the S–m–n Theorem.

Proof of Lemma 7.37 We prove this theorem for $m = n = 1$. The proof is the same for arbitrary m and n.

Let $f(x, y)$ denote $\varphi_e^2(x, y)$. To compute this function, we set bin B_1 equal to x, bin B_2 equal to y, and run the T^{++} program P_e.

We now describe a T^{++} program P_z that computes the function $f_a(y) \simeq f(a, y)$ for given a.

(1) MOVE B_1 to B_2
(2) Add 1

...

(a+1) Add 1
(a+2) P_e.

This program moves the contents of B_1 to B_2, then sets B_1 equal to a, and then runs the program P_e that computes $f(a, y)$. So this program computes the function $f_a(x)$.

Given a and e, the function $S_1^1(a, e)$ computes the code z for the above program. Clearly, this can be done for any given a and e and so $S_1^1(a, e)$ is a total function. We must show that it is primitive recursive.

Let $E(a)$ be the code for the program represented by the first $a + 1$ "lines" of the program P_z. Since MOVE B_1 to B_2 is itself a program, it is more than one line. It is a subroutine. Let w be the number of lines in this subroutine and let $E(0)$ be its code. We define $E(a)$ inductively by $E(a+1) = E(a)p_{(a+w+1)}^4$ (the exponent 4 corresponds to the command Add 1). This is a primitive recursive definition of the function $E(a)$.

We now describe how to compute z from a and e. Since e is a code for a program, e factors as $e = p_1^{a_1} p_2^{a_2} \cdots p_k^{a_k}$ for some k and nonzero $a_1 \cdots a_k$. The code for the above program P_z is the following product:

$$z = E(a) p_{1+w}^{\hat{a}_1} p_{2+w}^{\hat{a}_2} \cdots p_{k+w}^{\hat{a}_k},$$

where $\hat{a}_j = a_j + 4w$ if a_j has the form $4i - 3$ and $\hat{a}_j = a_j$ otherwise (for $j = 1, \ldots, k$). This represents "shifting" the lines of the program P_e. This program constitutes lines $(1 + w)$ through $(k + w)$ of the program P_z. Because of this shift, any occurrence of the command GOTO i (having code $4i - 3$) in P_e must be changed to GOTO $i + w$ in P_z.

The definition of $\hat{a}_j$ by cases is primitive recursive by Proposition 7.25. The prime factorization of e is primitive recursive by Proposition 7.29. Moreover, we have shown the function $E(a)$ to be primitive recursive. We conclude that the function $S_1^1(e, a) = z$, where z is as defined above, is primitive recursive. $\square$

Theorem 7.38 (Kleene's Recursion theorem) Let $f(y, x_1, \ldots, x_k)$ be a $(k+1)$-ary recursive function. For some number e, the k-ary function defined by $f(e, x_1, \ldots, x_k)$ is the same function as $\varphi_e^k(x_1, \ldots, x_k)$.

Proof Consider the $(k + 1)$-ary function h defined as

$$h(y, x_1, \ldots, x_k) \simeq f(s_n^1(y, y), x_1, \ldots, x_k),$$

where S_n^1 is as in Lemma 7.37. Since it is the composition of recursive functions, h is recursive. So $h \simeq \varphi_d^{k+1}$ for some d. Let $e = S_n^1(d, d)$.

We have $f(e, x_1, \ldots, x_k) \simeq f(S_n^1(d, d), x_1, \ldots, x_k)$ (by our choice of e) $\simeq h(d, x_1, \ldots, x_k) \simeq \varphi_d^{k+1}(d, x_1, \ldots, x_k)$ (by our definition of h and d). By Lemma 7.37, $\varphi_d^{k+1}(d, x_1, \ldots, x_k) \simeq \varphi_z^k(x_1, \ldots, x_k)$, where $z = S_k^1(d, d)$. By our definition of e, $e = z$ and $f(e, x_1, \ldots, x_k) \simeq \varphi_e^k(x_1, \ldots, x_k)$ as was required to show. $\square$

Corollary 7.39 The Ackermann function $A(x, y)$ is recursive.

Proof Let U_2 be the ternary function defined by $U_2(e, x, y) \simeq \varphi_e^2(x, y)$. This function was shown to be recursive in Proposition 7.36. Using this function, we define another ternary function f as follows:

$$f(y, n, x) = \begin{cases} x + 1 & \text{if } n = 0 \\ U_2(y, n-1, 1) & \text{if } x = 0 \text{ and } n > 0 \\ U_2(y, n-1, U_2(y, n, x-1)) & \text{otherwise.} \end{cases}$$

By Kleene's Recursion theorem, there exists $e \in \mathbb{N}$ such that $f(e, n, x) \simeq \varphi_e^2(n, x)$. It follows that

$$\varphi_e^2(n, x) = \begin{cases} x + 1 & \text{if } n = 0 \\ \varphi_e^2(n-1, 1) & \text{if } x = 0 \text{ and } n > 0 \\ \varphi_e^2(n-1, \varphi_e^2(n, x-1)) & \text{otherwise.} \end{cases}$$

Comparing this with the definition of $A(, n, x)$ (Section 7.1.2), we see that $A(n, x) = \varphi_e^2(n, x)$ for all n and x. Since $\varphi_e^2(n, x)$ is recursive, so is $A(n, x)$. □

In a similar way, we can show that any given function defined from recursive functions in an inductive manner is itself recursive. This gives credence to our claim that every programmable function is recursive. For any specified programming language, we could prove this claim. We have done this for the contrived language T^{++}. To prove that the set of C++ computable functions is the same as the set of recursive functions, we would have to delve into the grammar of C++. The skeptical reader may pursue the details regarding this or any other programming language, but we do not. We accept our claim as fact and use the terms *computable* and *recursive* interchangeably.

7.5 Semi-decidable decision problems

We further study the subsets of $\mathbb{N} \cup \{0\}$. In Section 7.2, we defined the recursive subsets of $\mathbb{N} \cup \{0\}$. In the present section we consider the *recursively enumerable* sets. The recursive sets are computable in the sense that they have computable characteristic functions. The recursively enumerable sets are *computably generated* in the following sense.

Definition 7.40 Let A be a set of non-negative integers. We say that A is *recursively enumerable* if there exists a total recursive function f such that $A = \{f(0), f(1), f(2), f(3), \ldots\}$.

So a set is *recursively enumerable* if it is the range of some total recursive function. Recall that every subset of $(\mathbb{N} \cup \{0\})^k$ corresponds to a decision problem. Whereas the recursive subsets correspond to decidable decision problems, recursively enumerable subsets correspond to *semi-decidable* decision problems.

Definition 7.41 Let R be a subset of $(\mathbb{N} \cup \{0\})^k$. The decision problem corresponding to R is *semi-decidable* if the following k-ary function is computable:

$$h(\bar{x}) = \begin{cases} 1 & \text{if } \bar{x} \in R \\ \text{undefined} & \text{otherwise.} \end{cases}$$

Example 7.42 Consider the Validity Problem for First-Order Logic (FOVAL). Given a first-order sentence φ, we must determine whether or not φ is valid. We claim that this problem is semi-decidable. We describe an algorithm that determines the correct answer given valid φ. This algorithm lists each of the countably many formal proofs and checks them one-by-one. If a formal proof for $\emptyset \vdash \varphi$ is found, then the algorithm stops and outputs "yes, φ is valid." Otherwise, the algorithm produces no output. Intuitively, this is what is meant by "semi-decidable." Whereas this algorithm correctly determines whether a given sentence is valid, it will not tell us whether a given sentence is not valid.

Formally, a decision problem is a relation on the non-negative integers. The algorithm from the previous example is stated informally. To prove that FOVAL is semi-decidable but not decidable, we code FOVAL as a subset of $\mathbb{N}$. In Section 8.4, we describe a procedure for coding sentences of first-order logic. That the set of codes for valid sentences is recursively enumerable follows from the completeness of first-order logic.

Examples of recursively enumerable sets that are not recursive are given in the next section. The codes from the previous section are used to define these and other noncomputable subsets of $\mathbb{N} \cup \{0\}$. In the present section, we discuss some of the numerous equivalent ways to define the concept of recursively enumerable sets.

Proposition 7.43 Let A be a proper subset of the non-negative integers. The following are equivalent:

(1) A is recursively enumerable
(2) A is the domain of a partial recursive function
(3) the decision problem of determining whether or not a given number x is in A is semi-decidable.

Proof Suppose first that A is recursively enumerable. Then A is the range of a total recursive function $f(x)$.

The binary function $g(x,y) = (f(x) \dot{-} y) + (y \dot{-} f(x))$ is also total recursive. This function equals 0 if and only if $y = f(x)$. The set A is the domain of the function $ub_x g(x,y)$. Since this function is defined from a recursive function by unbounded search, $ub_x g(x,y)$ is recursive. So (1) implies (2).

Suppose now that (2) holds. Suppose that A is the domain of a recursive function $f(x)$. By definition, the decision problem corresponding to A is semi-decidable if and only if the following function is computable:

$$h(x) = \begin{cases} 1 & \text{if } x \in A \\ \text{undefined} & \text{otherwise.} \end{cases}$$

Since $f(x)$ is recursive, so is the composition $c_1(f(x)) \simeq h(x)$ (where c_1 is the constant function $c_1(x) = 1$). So (2) implies (3).

Finally, suppose that (3) holds. Then $h(x)$ (as defined above) is recursive. So $h(x) \simeq \varphi_e^1(x)$ for some e. To compute $h(x)$ we run program P_e with input x in bin B_1. Recall the primitive recursive functions $bin(e, x, n, 1)$ and $lin(e, x, n)$ from the proof of Theorem 7.33. If we run program P_e with input x, then the computation terminates when $pf(e, lin(e, x, n)) = 0$. Let a be any element of A. Let

$$g(x, n) = \begin{cases} x & \text{if } pf(e, lin(e, x, n)) = 0 \\ a & \text{otherwise.} \end{cases}$$

The range of $g(x, n)$ is A. To prove (1) we must find a unary function having range A. Let

$$f(x) = \begin{cases} g(n, m) & \text{if } x = 2^n 3^m \\ a & \text{otherwise.} \end{cases}$$

Clearly $f(x)$ is a unary function having the same range as $g(x, y)$. So A is recursively enumerated by the function $f(x)$ and (1) holds. □

The characterization of recursively enumerable sets as the domains of partial recursive functions yields the following characterization of the recursive sets.

Proposition 7.44 A set A is recursive if and only if both A and $\bar{A}$ are recursively enumerable (where $\bar{A}$ is the set of non-negative integers not in A).

Proof First, we show that *recursive* implies *recursively enumerable*. Let $g(x)$ be any recursive function having 1 in its domain, but not 0. If A is a recursive set, then the composition $g(\chi_A(x))$ is a recursive function. This function has domain A. By the previous proposition, A is recursively enumerable. Moreover, if A is recursive, then so is $\bar{A}$, whence both A and $\bar{A}$ are recursively enumerable.

Conversely, suppose that both A and $\bar{A}$ are recursively enumerable. Then A is the domain of $\varphi_e^1(x)$ and $\bar{A}$ is the domain of $\varphi_d^1(x)$ for some e and d. Let $f(x) = us_n(pf(e, lin(e, x, n)) \cdot pf(d, lin(e, x, n)))$. That is $f(x)$ is the number of steps it takes for either P_e or P_d to halt on input x. Since each x is either in A or $\bar{A}$, the function $f(x)$ is total.

We have $\chi_A(x) = 1 \dotminus pf(e, lin(e, x, f(x)))$. Since this function is recursive, so is the set A. □

We next state without proof a most remarkable characterization of the recursively enumerable sets.

Definition 7.45 A set A of non-negative integers is said to be *Diophantine* if there exists a polynomial $p(x, y_1, \ldots, y_n)$ having integer coefficients such that $A = \{x |$ there exist non-negative integers $a_1, \ldots, a_n$ such that $p(x, a_1, \ldots, a_n) = 0\}$.

Theorem 7.46 (Matiyasevich) A set is recursively enumerable if and only if it is Diophantine.

Prior to its proof, this theorem was known as Davis' conjecture. Yuri Matiyasevich proved this theorem in 1970 following the work on this conjecture by Martin Davis, Hilary Putnam, and Julia Robinson. To understand why this theorem is remarkable, let us consider some examples of Diophantine sets:

- The set of even numbers is Diophantine (let $p(x, y) = x - 2y$).
- The set of perfect squares is Diophantine (let $p(x, y) = x - y^2$).
- The set of composite numbers (numbers that are not prime) is Diophantine (let $p(x, y_1, y_2) = x - (y_1 + 2)(y_2 + 2)$).

It is far more difficult to show that the following sets are Diophantine:

- The set $\{2, 4, 8, 16, 32, \ldots\}$ of powers of 2.
- The set of prime numbers.
- The set of palindromes $\{1221, 343, 11, 82628, 1010101, 8, 747, \ldots\}$.

There is no obvious polynomial $p(x, y_1, \ldots, y_n)$ that works for any of these three sets. Prior to Matiyasevich's proof, the question of whether or not these sets are Diophantine was an open question. In particular, Alfred Tarski posed the question of whether or not the powers of 2 form a Diophantine set. Matiyasevich's theorem answers this and many other questions in the affirmative. Since the function $f(n) = 2^n$ is easily shown to be primitive recursive, the range of this function is recursively enumerable and, therefore, Diophantine. Likewise, since the prime numbers form a primitive recursive set, this set, too, is Diophantine. A number written in base 10 is a palindrome if it represents the same number when written backwards. It is not difficult to show that this set is recursive. By Matiyasevich's theorem the set of palindromes and countless other sets are necessarily Diophantine.

Matiyasevich's theorem equates the number theoretic concept of Diophantine set with the concept of a computability generated set. In proving this theorem, Matiyasevich showed that the class of Diophantine sets is far more extensive than its definition suggests. This theorem also resolved a famous open problem of mathematics known as Hilbert's 10th Problem. This is one of the 23 problems selected by David Hilbert as the most important mathematical problems at the turn of the twentieth century. The 10th problem is one of the more succinctly stated of Hilbert's problems:

Hilbert's 10th Problem Given a Diophantine equation with any number of unknown quantities and with rational integral numerical coefficients: *to devise a process according to which it can be determined by a finite number of operations whether the equation is solvable in rational integers.*

There is no loss of generality in replacing the "rational integers" in this problem with integers. Phrased this way, the problem is to find a method for determining whether $p(x_1, \ldots, x_n) = 0$ has integer solutions where $p(\bar{x})$ is a given polynomial having integer coefficients. As stated, the problem is not to determine whether such a process exists, but rather to find such a process. The implied optimism of this statement underestimates the complexity of the integers and reflects misconceptions that were commonly held at the time. These misconceptions were dispelled by Gödel's Incompleteness theorems that are the subject of our next chapter. The First Incompleteness theorem shows that the integers are extraordinarily complex in the following sense: the first-order theory of the natural numbers (in a vocabulary containing both multiplication and addition) is undecidable.

The subject of computability began with Gödel's results and the work of Kleene, Post, Turing, and others that followed. This subject made Hilbert's 10th Problem precise by providing a formal definition for the "process" described in the problem. It became apparent that Hilbert's 10th Problem may be unsolvable. This idea motivated the work of Davis, Putnam, Robinson, and others that culminated in Matiyasevich's theorem. Matiyasevich's theorem resolves Hilbert's 10th Problem by showing that an algorithmic process as described in the problem cannot exist. This follows from Matiyasevich's theorem because there exist well known recursively enumerable sets that are not computable. These sets are the topic of the next section. For more on Matiyasevich's theorem, the reader is referred to Matiyasevich's book [31].

7.6 Undecidable decision problems

In this section, we view some sets that lie beyond the brink of computability.

7.6.1 Nonrecursive sets. Let W_i be the domain of the function $\varphi_e^1(x)$. By Proposition 7.43, the list

$$W_0, W_1, W_2, W_3, W_4, \ldots.$$

includes every recursively enumerable set. Let $J = \{x \, | \, x \notin W_x\}$. If this set is recursively enumerable, then $J = W_e$ for some e. But then we have $e \in W_e$ if and only if $e \notin W_e$ (by the definition of J). This is contradictory. We conclude that J must be different from W_e for each e and so J is not recursively enumerable. It follows that the characteristic function of J, although it is a well-defined function, is not computable.

Now consider the set $K = \{x \, | \, x \in W_x\}$.

Proposition 7.47 K is recursively enumerable, but not recursive.

Proof If K is recursive, then so is the complement of K in the non-negative integers. The complement of K is J. Since, J is not recursive, neither is K. Moreover, K is recursively enumerable since it is the range of the function $U_2(x, x)$ from Proposition 7.36. □

Let H_1 be the set of ordered pairs (e, x) such that $x \in W_e$. That is, (e, x) is in H_1 if and only if x is in the domain of the function φ_e^1 computed by the program P_e. To determine whether or not a given pair (e, x) is in H_1 is to determine whether or not the program P_e halts given input x. Likewise, we define H_k as the set of $(k+1)$-tuples $(e, \bar{x})$ such that P_e halts given input $\bar{x}$. Let $H = \{e \, | (e, 0) \in H_1\}$. To determine whether or not e is in H is to determine

Computability and complexity

whether or not P_e halts on input 0. This decision problem is known as the *Halting Problem*. The problem corresponding to H_k is the *Halting Problem on k-tuples*. These problems are undecidable.

Proposition 7.48 H is not recursive.

Proof We first show that H_1 is not recursive. Note that $K = \{x \mid (x,x) \in H_1\}$. If we could determine whether or not a given pair is in H_1, then we could determine whether or not a given element is in K. Since K is not recursive, neither is H_1.

We now show that H is not recursive. Given program P_e and input x, let P_d be the program obtained by adding as a prefix x copies of the command Add B_1 to the program P_e. Running P_d with input 0 has the same outcome as running P_e with input x. So $d \in H$ if and only if $(e,x) \in H_1$. Since H_1 is not recursive, neither is H. $\square$

We have now demonstrated that J, K, and H are three examples of nonrecursive sets. We have also demonstrated our primary tool for showing that a given set is not recursive. To show that K is not recursive, we *reduced* K to J. That is, we showed that if K is recursive, then so is J. Similarly, we reduced H to K (by way of H_1).

Definition 7.49 Let A and B be two sets of non-negative integers. We say that A is *recursively reducible* to B, denoted $A \leq_r B$, if there exists a recursive unary function f such that $x \in A$ if and only if $f(x) \in B$.

Proposition 7.50 If B is recursive and $A \leq_r B$, then A is also recursive.

Proof If $A \leq_r B$, then $x \in A$ if and only if $f(x) \in B$ for some recursive function f. It follows that the characteristic function of A is the composition $\chi_B(f(x))$. $\square$

Conversely, if $A \leq_r B$ and A is not recursive, then B is not recursive. We exploit this fact to provide many examples of nonrecursive sets. The set J is not recursive by its definition. Each of the other nonrecursive sets we define can be reduced to J. Rather than considering each set one-by-one, we prove Rice's theorem. This theorem provides a plethora of nonrecursive sets. Rice's theorem states that any nontrivial index set is not recursive.

Definition 7.51 A set of non-negative integers A is said to be an *index set* if the following holds. If $x \in A$ and $\varphi_x^1 \simeq \varphi_y^1$, then $y \in A$.

For example, both J and K are index sets (they are defined in terms of the indices i of the sets W_i). Likewise, the set of all x such that W_x contains the number 5 is an index set. For a nonexample, consider the set

GOTO 12 = $\{x : P_x$ is the program (1) GOTO 12 $\}$.

Recall that, by default, any number such as 42 that does not code a T^{++} program is in this set. Let y be the code for the program (1) GOTO 23. Then φ_y^1, like φ_{42}^1, is the identity function. But whereas $42 \in$ GOTO 12, $y \notin$ GOTO 12. So this set is not an index set. Note that this set is primitive recursive and, therefore, decidable. The following theorem states that this is not the case for any nontrivial index set.

Theorem 7.52 (Rice) Let A be an index set. If A is neither $\emptyset$ nor $\mathbb{N} \cup \{0\}$, then A is not recursive.

Proof Let A be a proper subset of $\mathbb{N} \cup \{0\}$. Let c be the code for the program:

(1) Add 1
(2) RmvP 1.

Since this program never halts $\varphi_c^1(x)$ is undefined for all x.

Claim If $c \in A$, then $K \leq_r \bar{A}$.

Proof By Proposition 7.47, K is recursively enumerable. It follows that the following function is recursive:

$$h_K(x) = \begin{cases} 1 & x \in K \\ \text{undefined} & \text{otherwise.} \end{cases}$$

Since A is not $\mathbb{N}$, there exists $e \in \bar{A}$.
  Since $h_K(x)$ is recursive, so is the function

$$g(x, y) = \begin{cases} \varphi_e(y) & h_K(x) = 1 \\ \text{undefined} & \text{otherwise.} \end{cases}$$

By Lemma 7.37, there exists a recursive function $f(x)$ such that, for each x, $\varphi_{f(x)}(y) \simeq g(x, y)$.
  Note that if $x \in K$, then $\varphi_{f(x)}(y) \simeq \varphi_e$. Otherwise, $\varphi_{f(x)} \simeq \varphi_c$. Since A is an index set and $c \in A$ and $e \notin A$, we have $x \in K$ if and only if $f(x) \in \bar{A}$. So $K \leq_r \bar{A}$ as claimed.

So if $c \in A$, then $\bar{A}$ is not recursive by the claim. If $\bar{A}$ is not recursive, then neither is A. If $c \notin A$, then applying the claim to $\bar{A}$ yields $K \leq_r A$. Either way, A is not recursive. □

Rice's theorem provides an uncountable supply of nonrecursive sets. For example, consider the following:

$$ID = \{x \,|\, \varphi_x^1 \text{ is the identity function}\}$$
$$SQUARE = \{x \,|\, \varphi_x^1(y) \simeq y^2\}$$
$$FIN = \{x \,|\, W_x \text{ is finite}\}$$
$$INF = \{x \,|\, W_x \text{ is infinite}\}$$
$$COF = \{x \,|\, W_x \text{ is co-infinite}\}$$
$$TOT = \{x : \varphi_x \text{ is total}\}$$
$$REC = \{x : W_x \text{ is recursive}\}.$$

Since each of these is a nontrivial index set, each is nonrecursive by Rice's theorem. In fact, none of these sets is recursively enumerable. Whereas each of these sets is recursively reducible to K (and, therefore, to J as well), K is not reducible to any of these sets. In this sense, each of the above sets is more complicated than K.

The notion of recursive reducibility, as the notation $\leq_r$ suggests, imposes an order on the subsets of $\mathbb{N} \cup \{0\}$. Each set is ranked in a hierarchy according to this order. For example, $INF \leq_r COF$ (Exercise 7.8). Intuitively, this means that the decision problem corresponding to COF is at least as difficult as the decision problem for INF. If we had some way of determining whether or not a given number is in COF (which we do not), then we could use this procedure to determine whether or not a given number is in INF. Since both of these problems are undecidable, this may seem like hairsplitting. There are two reasons that we consider the hierarchy of undecidable decision problems. One reason is that it serves as a precursor to the classification of decidable decision problems that is the topic of the final two sections of this chapter. Another reason is that this hierarchy relates concepts from computability to the first-order logic of the previous chapters.

7.6.2 The arithmetic hierarchy. The definable subsets of $\mathbb{N}_0$ are called *arithmetic sets*. The *arithmetic hierarchy* is the classification of these sets according to the syntax of the formulas that define the set. Recall the definition of a Δ_0 formula from Section 7.2.

Definition 7.53 A $\mathcal{V}_{ar}$-formula is said to be Π_1 if it has the form $\forall y \varphi(\bar{x}, y)$ for some Δ_0 formula $\varphi(\bar{x}, y)$.

A $\mathcal{V}_{ar}$-formula is said to be Σ_1 if it has the form $\exists y \varphi(\bar{x}, y)$ for some Δ_0 formula $\varphi(\bar{x}, y)$.

An arithmetic set is said to be Π_1 if there exists a Π_1 formula that defines the set. The Σ_1 sets are defined analogously.

At the bottom of the arithmetic hierarchy we have the primitive recursive sets. These are precisely the sets definable by Δ_0 formulas. The Σ_1 sets correspond to recursively enumerable sets.

Proposition 7.54 If a set is Σ_1, then it is recursively enumerable.

Proof Suppose that A is Σ_1. Let $\exists y \varphi(x, y)$ be a $\mathcal{V}_{ar}$-formula that defines A, where $\varphi(x, y)$ is Δ_0. By Proposition 7.28, the set $B = \{(x, y) | \mathbf{N}_0 \models \varphi(x, y)\}$ is primitive recursive. Let $f(x) = us_y(1 \dot{-} \chi_B(x, y))$. Then f is a recursive function having domain A and A is recursively enumerable. □

Corollary 7.55 If a set is both Σ_1 and Π_1, then it is recursive.

Proof The negation of a Π_1 formula is a Σ_1 formula. The corollary follows from Propositions 7.44 and 7.54. □

The converses of these statements also hold. The Σ_1 sets are precisely the recursively enumerable sets and the recursive sets are those in both Σ_1 and Π_1. This is proved as Corollary 8.15 of the next chapter.

Likewise, we classify every definable subset of $\mathbf{N}_0$.

Definition 7.56 Let n be a natural number.

- A $\mathcal{V}_{ar}$-formula is said to be Π_{n+1} if it has the form $\forall y \varphi(\bar{x}, y)$ for some Δ_n formula $\varphi(\bar{x}, y)$.
- A $\mathcal{V}_{ar}$-formula is said to be Σ_{n+1} if it has the form $\exists y \varphi(\bar{x}, y)$ for some Δ_n formula $\varphi(\bar{x}, y)$.
- A $\mathcal{V}_{ar}$-formula is said to be Δ_{n+1} if it both a Π_{n+1} formula and a Σ_{n+1} formula.

An arithmetic set A is said to be Π_n, Σ_n, or Δ_n according to the formulas that define A.

Note that every Π_n set is Σ_{n+1} and every Σ_n set is Π_{n+1} for each $n \in \mathbb{N}$. Since each first-order formula is equivalent to a formula in prenex normal form, the arithmetic hierarchy includes every arithmetic set. Moreover, there exist arithmetic sets at each of the denumerably many levels of the hierarchy. For each $n \in \mathbb{N}$, there exist sets that are Σ_n but not Π_n and vice versa. This is the Hierarchy theorem. We leave the proof of this as Exercises 7.29 and 7.30.

In this section, not only have we demonstrated examples of nonrecursive sets, but we have also presented a vast hierarchy of such sets. Each of these sets corresponds to both an undecidable decision problem and a noncomputable function (namely, its characteristic function). Based on the results of Section 2.5, we made the initial observation that, since there are uncountably many sets and countably many recursive sets, most sets are not recursive. Likewise, most decision problems are not decidable and most functions are not computable. In this section, we have shown more than this. We have shown that, even among the countably many definable subsets of $\mathbf{N}_0$, the recursive sets are the exceptions and not the rule. Just as viewing the heavens puts earth into perspective, viewing the plethora of nonrecursive sets puts the recursive sets and decidable decision problems into proper perspective. We now return to earth and consider decidable decision problems. For those who would like to pursue the study of nonrecursive sets further, [48] is recommended.

7.7 Decidable decision problems

The previous sections of this chapter have concerned the distinction between decision problems that are decidable and those that are not. We now focus on the distinction between those problems that are decidable and those that are *really* decidable. By definition, a decision problem is decidable if it can be resolved by some algorithm. There is nothing in this definition that requires the algorithm to be practical. A decision problem is said to be *feasible* if it can be resolved by an algorithm using a reasonable amount of time and space. This is an intuitive notion that will not be precisely defined. This notion depends not only on our perception of "reasonable," but also on our technological capabilities. Algorithms that are not feasible today may become feasible with quantum computers or other potential technologies of the future.

Rather than considering the vague notion of feasibility, we focus on precisely defined complexity classes. In particular, we consider the class of polynomial-time decision problems **P** and the class of nondeterministic polynomial-time problems **NP** that contains **P**. The class **P** was defined in the Preliminaries prior to Chapter 1. We repeat the definition.

Definition 7.57 An algorithm is *polynomial-time* if there exists a $k \in \mathbb{N}$ such that, given any input of length $n > 2$, the algorithm halts in fewer than n^k steps. A decision problem is *polynomial-time* if it can be decided by a polynomial-time algorithm. The set of polynomial-time decision problems is denoted **P**.

If a decision problem is not in **P**, then it is certainly not feasible. The converse is not true. If an algorithm halts in fewer than $n^{1,000,000}$ steps (given input of length n), then it is polynomial-time but not necessarily feasible. So the set of polynomial-time algorithms contains the set of feasible algorithms as a proper subset.

To make the above definition of polynomial-time precise, we must specify both how the "length" of the input is to be measured and what constitutes a "step" of an algorithm. The *length* of the input is the number of digits. For example, the length of 8427 is 4. The length of the ordered triple $(17, 8, 109)$ is 6.

More generally, the length of a natural number x is $\lfloor log(x) \rfloor + 1$, where $\lfloor log(x) \rfloor$ denotes the greatest integer less than $log(x)$. The length of zero is 1. The length of a k-tuple of numbers is the sum of each of the k lengths. We assume that numbers are presented in the usual base 10 notation, in which case log is the common base 10 logarithm.

To define a "step" of an algorithm, we must first define the notion of an algorithm. We provide this definition in the second part of this section. The formal definition of an algorithm yields natural measures of computational time and space. In addition to the class of polynomial-time decision problems **P**, we also define the classes of polynomial-space (**PSPACE**) and logarithmic space (**L**). In the third and final part of this section, we define the notion of a *nondeterministic algorithm* and make precise the class **NP**. We discuss the relationship between these various complexity classes. We begin with some examples.

7.7.1 Examples. We present several examples of decidable decision problems. For each problem, we informally describe an algorithm (using either English prose or pseudo-code) to verify that the problem is decidable. For some problems, we also provide (again, informally) a *nondeterministic* algorithm. An algorithm is a step-by-step procedure. At any stage of the algorithm, the next step is completely determined. In contrast, a *nondeterministic* algorithm may have more than one possible "next step" at a given stage. Essentially, in a nondeterministic algorithm, we are allowed to guess. One purpose of this subsection is to illustrate the complexity classes **P**, **NP**, and **coNP**. The "N" in **NP** and **coNP** indicates that these classes are defined in terms of nondeterministic algorithms. Both **NP** and **coNP** contain **P** as a subset.

Although we have not yet defined a "step" of an algorithm, we can verify that certain problems are in **P**. We use the fact that any feasible algorithm is polynomial-time. So tasks such as multiplying or dividing numbers that are clearly feasible must be polynomial-time. We assume that if the input has length n, then it can be read in at most n steps. So, in polynomial-time, we can repeatedly read the input 1000 times, n times, or $3n^5$ times, but not 2^n times. We also use the fortunate fact that the composition of polynomials is again a polynomial. If we write an algorithm that uses a certain polynomial-time algorithm as a subroutine a polynomial number of times, then the resulting algorithm, too, is polynomial-time.

For future reference, each decision problem is given a short name. To avoid ambiguity, this name shall be written in capital letters.

The evens problem (EVENS)

The evens problem corresponds to the set of even natural numbers. Since we can determine whether or not a given number is even merely by looking at its last digit, EVENS is not only decidable, it is feasible. In contrast, consider the problem PRIMES of determining whether or not a given natural number is prime.

The primes problem (PRIMES)

The prime problem corresponds to the set of prime numbers. Since this set is primitive recursive, we know that PRIMES, like the EVENS, is decidable. The following algorithm, written in pseudo-code, resolves this problem:

Given: natural number n

if $n = 1$ then output "not prime" and halt

if $n = 2$ then output "prime" and halt

else for k = 2,...,n−1 do:

    divide n by k

    if remainder is 0 then output "not prime" and halt

end for

output "prime"

halt

This algorithm outputs "prime" if and only if the input is a prime number. Given input n, the algorithm checks each k between 1 and n to see if k divides n. In fact, we only need to check this for k between 1 and $\sqrt{n}$ (if $n = a \cdot b$ and $a > \sqrt{n}$, then $b < \sqrt{n}$). This observation offers a more efficient algorithm. Now, suppose we actually want to use this algorithm to determine whether or not a given number n is prime. If n is a three-digit number, then $100 \leq n \leq 999$. To execute our algorithm, we must divide n by at most 32 numbers (since $32 > \sqrt{999}$). We can easily do this on a computer. But suppose n is 23 digits long. Then n is between $10^{22} - 1$ and 10^{23}. If n happens to be prime, then it will take at least $\sqrt{10^{22}}$ computations for the algorithm to arrive at the output "prime." If your computer can do this, then take a prime number 45 digits long. The algorithm will take 10^{22} steps. Compared to the length of the input, the algorithm takes exponentially long. This algorithm is not polynomial-time.

The composites problem (COMP)

A natural number n is *composite* if $n = a \cdot b$ for natural numbers a and b both smaller than n. Put another way n is composite if it is neither 1 nor prime. So the above algorithm for *PRIMES* can be altered slightly to produce an algorithm deciding the decision problem $COMP$ corresponding to the set of composite numbers. Consider now the following nondeterministic algorithm.

Given: natural number n
if $n > 2$ then choose i between 1 and n
divide n by i
if remainder is 0 then output "composite" and halt
halt

This algorithm is nondeterministic because of the command "choose i between 1 and n." If n is big, then there are many possible values for i. So there are more than one way to proceed in the next step of the algorithm. If we are lucky and choose a value of i that divides n, then the algorithm quickly concludes that n is composite. So this algorithm, when its chooses correctly, determines whether a number is composite in polynomial-time.

Definition 7.58 We define the class **NP** of *nondeterministic polynomial time* decision problems. Let PROB be an arbitrary decision problem. Given certain input, PROB produces an output of either "yes" or "no." Let Y be the set of all inputs for which PROB produces the output of "yes" and let N be the analogous set of inputs that produce output "no."

- If there exists a nondeterministic algorithm which, given input x, can produce the output "yes" in polynomial-time if and only if $x \in Y$, then PROB is in **NP**.
- If there exists a nondeterministic algorithm which, given input x, can produce the output "no" in polynomial-time if and only if $x \in N$, then PROB is in **coNP**.

The nondeterministic algorithm we gave for $COMP$ demonstrates that this decision problem is in **NP**. Since a number is not prime if and only if it is 1 or composite, $PRIMES$ is in **coNP**. It can also be shown that $PRIMES$ is in **NP**. This is not apparent from the above algorithms. To show that $PRIMES$ is in **NP**, we must come up with another algorithm. In fact, something much stronger is true. In 2002, Agrawal, Kayal, and Saxena proved that $PRIMES$ is in **P**. In their article "Primes are in P," they demonstrate an algorithm that determines whether or not a number n is prime in fewer than l^{12} steps where l is the number of digits in n.

The **P** = **NP** question asks whether every **NP** problem, like $PRIMES$, is actually in **P**. This is among the most important unanswered questions of mathematics.

The big questions:

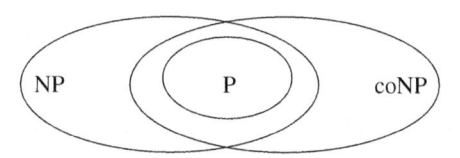

NP   P   coNP

Does coNP ∩ NP = P?
Does coNP = NP?
Does P = NP?

If **P** = **NP**, then **coNP** = **NP**. This is because any polynomial-time algorithm that determines whether an element is in a set A can also be used to determine whether an element is not in A. That is, if we define the class **coP** analogously to **coNP**, then **coP** necessarily equals **P**. This is not true for nondeterministic algorithms. A nondeterministic algorithm that determines whether an element is in A may be of no use in determining whether an element is not in A. We shall say much about **P** = **NP** and other important questions in Sections 7.8, 10.4, and 10.5. We presently present more examples.

The next examples are from graph theory. Given a finite graph, we ask whether or not the graph has certain properties. We said that every decision problem corresponds to a relation on the non-negative integers. To view the following examples in this manner, we code each graph as a sequence of 1s and 0s. There is a natural way to do this. If G is a graph having vertices $\{v_1, \dots, v_k\}$, then we define a $k \times k$ matrix as follows. The entry in row i and column j is 1 if v_i and v_k share an edge. Otherwise, this entry is 0. The resulting matrix is called the *adjacency matrix* of G. To input the graph G into a T-machine, we input the adjacency matrix as a k^2-tuple.

The graph problem (GRAPH)

The graph problem asks whether or not a given finite string of natural numbers is the adjacency matrix for some graph. We describe a polynomial-time algorithm for deciding this problem. First, we read through the string of numbers and check that each entry is either 1 or 0. At the same time, we count the entries to determine the length n of the sequence. We then determine whether or not n is a perfect square. We do this by checking whether $k^2 = n$ for each $k \leq \sqrt{n}$. Since n is the length of the input, this can be done in polynomial-time. (In contrast, recall the situation for PRIMES above. There, we could not check each $k \leq \sqrt{n}$ in polynomial-time since n was the input having length $log(n)$.) The matrix represents a graph if and only if it is symmetric and has 0s along the diagonal. This is because the edge relation in a graph is, by definition, symmetric and irreflexive. Checking these two properties is clearly a feasible task. The input is an adjacency matrix for a graph if and only if each of the above conditions is verified.

Since $GRAPH$ is in **P**, we may consider decision problems that take finite graphs as input. By this we mean that the decision problem takes a finite string of numbers as input and verifies that the input represents a graph before proceeding.

The connectivity problem (CON)

The connectivity problem asks whether or not a given finite graph is connected. Recall that a graph is connected if, given any two vertices x and y, there exists a path from x to y. We show that CON is in **P**. First, we demonstrate a polynomial-time algorithm for the problem $PATH$. This decision problem takes a finite graph and two vertices of the graph as input and determines whether or not there exists a path between the given vertices.

Given: a graph with n vertices and two particular vertices x and y.
let $S_1 := \{x\}$, let $T_1 := \{x\}$, let $i := 1$
for $i \geq 1$
  if $y \in T_i$ then
    output "x is connected to y" and halt
  else
    let $S_{i+1} = \{v : v \notin S_i$ and v shares an edge with some $z \in T_i\}$
    let $T_{i+1} = T_i \cup S_{i+1}$
  if S_{i+1} is empty then
    output "no path" and halt
  else let $i := i + 1$
end for
halt

We leave it to the reader to verify that this algorithm determines whether or not x is connected to y in polynomial-time. Now consider CON. A graph with n vertices is in CON if and only if the vertex corresponding to the first column of the adjacency matrix is connected to each of the other $n - 1$ vertices. So we can use the above algorithm $n - 1$ times to determine whether or not a graph with n vertices is in CON. It follows that CON, like $PATH$, is in **P**.

The clique problem (CLIQUE)

The clique problem corresponds to the set of all pairs (G, k), where G is a finite graph and k is a natural number such that G has the k-clique as a subgraph. This problem is in **NP** as the following algorithm demonstrates.

Given: a graph G with n vertices and a natural number $k \leq n$.
choose a subgraph H of G of size k
if every pair of vertices from H shares an edge
then output "clique present"
halt

This algorithm is clearly nondeterministic. There are many ways to choose the subgraph H in the first step. Prior to halting, this algorithm checks whether there exists an edge between every pair of vertices in H. Since $|H| \leq n$, there are at most $n(n-1)/2 < n^2$ pairs of vertices to consider. So this nondeterministic algorithm halts in polynomial time and $CLIQUE \in$ **NP**. The problem of determining whether a given finite graph does not contain a k-clique is **coNP**.

The max-clique problem (MAXCLIQUE)

The max-clique problem corresponds to the set of all pairs (G, k) where G is a finite graph and k is a natural number such that the largest clique in G has size k. This problem is decidable. Given a finite graph G and natural number k, we could check every subgraph of G of size k or larger. Not only is this algorithm not feasible, it is not polynomial time. It is unknown whether or not MAXCLIQUE is in **NP**.

The satisfiability problem for propositional logic (PSAT) and related problems

Recall from the first section of the first chapter that formulas of propositional logic contain the symbols $\neg$, $\wedge$, $\vee$, $\rightarrow$, and $\leftrightarrow$, "(", and ")" as well as any finite number of atomic formulas which we denote A_1, A_2, A_3, and so forth. We now consider decision problems that take as input finite sequences of these symbols. To conform to our formal definition of a *decision problem*, we code each string of symbols as a natural number. For the present informal discussion, it is unnecessary to delve into the details of the coding.

The Satisfiability Problem (PSAT) corresponds to the set of satisfiable formulas of propositional logic. This problem is decidable since, given any formula F of propositional logic, we can compute a truth table to determine whether or not F is satisfiable. Recall that the truth table has 2^n rows where n is the number of atomic formulas occurring in F. However, to show that F is satisfiable, we need only compute one row of the table. This provides the following nondeterministic polynomial-time algorithm for $PSAT$:

Given: a formula F of propositional logic
compute one row of the truth table for F
if F has truth value 1 in this row
then output "satisfiable"
halt

So $PSAT$ is in **NP**. It follows that the decision problem $P - UNSAT$ of determining whether F is unsatisfiable is **coNP**.

It is not known whether $PSAT$ is in **P**. In fact, $PSAT$ is in **P** if and only if $\mathbf{P} = \mathbf{NP}$. This decision problem is **NP**-complete. We define and discuss this phenomenon in the next section. We prove that $PSAT$ is **NP**-complete in Section 10.4.

7.7.2 Time and space. We said that an algorithm is feasible if it can be executed in a reasonable amount of time and space. We now specify how time and space are to be measured.

First, we define the concept of an "algorithm." That is, we provide this notion with a definition that suffices for the complexity classes we are to consider. To do this, we modify the notion of a T^{++} program. The purpose of T^{++} was to serve as a simplified programming language to ease the coding process of Section 7.4. Because of their simplicity, T^{++} programs are terribly inefficient. They cannot even add in polynomial-time. Consider a T^{++} program that outputs $n + n$ given input n in bin B_1. The command Add 1 must be repeated n times in such a program. Since the length of n is measured in terms of $log(n)$, this program takes exponentially long.

We add commands to T^{++} to obtain the more efficient programming language $T^{\sharp}$. The $T^{\sharp}$ programs are executed by an upgraded version of the T-machine that recognizes the new commands. We view the bins as program variables that take on non-negative integer values. The main feature of $T^{\sharp}$ is that it allows us to work with the *decimal presentation* of a number. Suppose that B_1 equals 8472. The decimal presentation of B_1 assigns the values 8, 4, 7, and 2 to B_1, B_2, B_3, and B_4, respectively.

- The $T^{\sharp}$ command *Dec* converts the value v of B_1 into its decimal presentation and stores the result and in bins $B_1, B_2, \ldots, B_l$ where l is the length of v.

- The $T^\sharp$ command $iDec(j)$ is the inverse of the Dec command. This command regards the values of $B_1, B_2, \ldots, B_j$ as a decimal presentation of a number. The command $iDec(j)$ computes the value of this number and sets B_1 equal to the result.

So $iDec(5)$ computes $B_1 + 10 * B_2 + 100 * B_3 + 1000 * B_4 + 10000 * B_5$ and sets B_1 equal to the result. Writing a T^{++} program to do this would not be pleasant, but it certainly could be done. The $T^\sharp$ commands Dec and $iDec(5)$ are convenient names for T^{++} subroutines. The same is true for the $T^\sharp$ command $Length(i, j)$.

- $Length(i, j)$ sets B_j equal to the length of B_i.

Finally, $T^\sharp$ also includes a variety of ways to move data.

- $Copy(i, j)$ sets B_j equal to B_i.
- $Copy(i, B_j)$ sets B_v equal to B_i where v is the value of B_j.
- $Copy(B_i, j)$ sets B_j equal to B_v where v is the value of B_i.

The commands for $T^\sharp$ include the T^{++} commands (Add i, Rmv i, RmvP i, and GOTO i) and also, for each i and j in $\mathbb{N}$, the commands:

$Dec, iDec(i), Length(i, j), Copy(i, j), Copy(i, B_j),$ and $Copy(B_i, j)$.

A $T^\sharp$ *program* is a finite enumerated list of $T^\sharp$ commands.

Definition 7.59 An *algorithm* is a $T^\sharp$ program.

In previous sections, we tacitly defined an algorithm to be a T^{++} program. For computability theory, the two definitions are equivalent. The additional commands increase the efficiency, but not the computing power, of the programs. By Corollary 7.34, the functions computable by $T^\sharp$ programs are precisely the T-computable functions.

Definition 7.60 Each executed $T^\sharp$ command constitutes one *step* of the computation of an algorithm.

This definition of "step" makes precise the earlier definition of "polynomial-time." We claim that this definition captures our intuitive notion of what can and cannot be accomplished in polynomial-time. Basic operations such as addition, multiplication, and division are polynomial-time as they should be. If I were to add two 7-digit numbers together without electronic (or mechanical) assistance, then I would rely on an algorithm I learned long ago and add the numbers digit-by-digit. Using the command Dec, we can mimic this familiar algorithm with a

$T^\sharp$ program. Likewise, we can write $T^\sharp$ programs that carry out the polynomial-time procedures taught in grammar school for subtraction, multiplication, or long division. It follows that $T^\sharp$ programs (unlike T^{++} programs) can compute polynomials in polynomial-time. We now turn from *time* to another measure of computational complexity.

Definition 7.61 The *space* of a computation is the number of bins that are used.

To be more precise, we must state what it means to "use" a bin.

Definition 7.62 A computation *uses* bin B_i if the value of B_i is altered at some step of the computation.

Definition 7.63 An algorithm is *polynomial-space* if there exists a $k \in \mathbb{N}$ such that, given any input of length $n > 1$, the algorithm uses fewer than n^k bins.

Space complexity also considers *sublinear* classes. These are classes where the number of bins used is less than the length of the input. We do not consider such classes in time complexity since we cannot even read the input in sublinear time.

Definition 7.64 An algorithm is *logarithmic-space* if, given any input of length n (for sufficiently large n), the algorithm uses fewer than $\log n$ bins.

There exist nonfeasible algorithms that use only a small fixed number of bins (see Exercise 7.8). To make space a useful measure of complexity, we consider algorithms that bound the size of the bins. We say that an algorithm is *bounded* if each bin has value less than 10 at each step of the computation (provided this is true of the input).

Definition 7.65 A decision problem is *polynomial-space* if there exists a bounded polynomial-space algorithm that decides the problem. The set of all polynomial-space decision problems is denoted **PSPACE**.

Definition 7.66 A decision problem is *logarithmic-space* if there exists a bounded logarithmic-space algorithm that decides the problem. The set of all logarithmic-space decision problems is denoted **L**.

We state without proof two facts regarding the relationship between these complexity classes. For proofs, we refer the reader to either [36] or [47].

Proposition 7.67 $L \neq PSPACE$.

Proposition 7.68 $L \subset P \subset PSPACE$.

By the former proposition, at least one of the two inclusions in the latter proposition must be a proper inclusion. It is not known, however, which of these is proper.

Computability and complexity 347

7.7.3 Nondeterministic polynomial-time. An algorithm is a $T^\sharp$ program. By this definition, any algorithm is *deterministic* in the sense that, once the input is given, each step of the computation is completely determined. If we repeatedly execute an algorithm with the same input, then we will repeatedly observe the same computation and the same outcome. We now provide a formal definition for the notion of a *nondeterministic* algorithm.

Let $T^\sharp_{ND}$ be the programming language obtained by adding to $T^\sharp$ the commands "GOTO i OR j" for each i and j in $\mathbb{N}$. The T-machine, upon reading this command, proceeds either to line (i) or line (j) of the program. A $T^\sharp_{ND}$ *program* is a finite enumerated list of $T^\sharp_{ND}$ commands.

Definition 7.69 An *nondeterministic algorithm* is a $T^\sharp_{ND}$ program.

The GOTO i OR j command allows $T^\sharp$ programs to nondeterministically jump to one of any number of lines. For example, the following commands cause the T-machine to go to line (5), (6), (7), or (8) of the program.

(1) GOTO 2 OR 5
(2) GOTO 3 OR 6
(3) GOTO 7 OR 8

The GOTO i OR j command can also be used to choose an arbitrary value for a bin. For example, the following commands cause the T-machine to assign to bin B_2 an arbitrary number from 1 to 9.

(1) Zero B_2
(2) Add B_2
(3) Compute $B_2 \stackrel{\cdot}{-} 8$ and store result in B_3
(4) GOTO 6
(5) GOTO 8
(6) RmvP B_3
(7) GOTO 2 OR 8
(8) ...

The reader may verify that the nondeterministic algorithms described earlier in this section can be written as $T^\sharp_{ND}$ programs.

The definition of a *nondeterministic algorithm* completes the definition of **NP** (Definition 7.58). The class **NL** ("nondeterministic log-space") is defined analogously. A decision problem is in this class if any correct "yes" output can be obtained by a nondeterministic algorithm using logarithmic-space. Replacing "yes" with "no" yields the class **coNL**. In contrast to the open question **NP** = **coNP**, we have the following fact.

Theorem 7.70 NL = coNL

For a proof of this, we refer the reader to section 8.6 of [47]. This theorem represents one of only a few known facts regarding the relationship between these complexity classes. We also have the following extension of Proposition 7.67.

Proposition 7.71 $L \subset NL \subset P \subset NP \subset PSPACE$.

By Proposition 7.68, at least one of the above inclusions must be proper. This is essentially all that is known. In particular the possibility that $L = \mathbf{NP}$ remains open.

7.8 NP-completeness

We define the concept of an **NP**-complete problem and provide examples. Informally, the **NP**-*complete* decision problems are the most difficult problems in **NP**. Cook's theorem states that $PSAT$ is **NP**-complete. We prove this theorem as a consequence of Fagin's theorem in Section 10.3. In the present section, we take the **NP**-completeness of $PSAT$ as fact and use $PSAT$ to find other examples **NP**-complete sets.

Definition 7.72 A function $f : (\mathbb{N} \cup \{0\})^m \to (\mathbb{N} \cup \{0\})^n$ is a *polynomial-time function* if there exists a polynomial-time algorithm that computes f.

Definition 7.73 For $A \subset (\mathbb{N} \cup \{0\})^m$ and $B \subset (\mathbb{N} \cup \{0\})^n$, we say that A is *polynomial-time reducible* to B, denoted $A \leq_p B$, if there exists a polynomial-time function $f : A \to B$ so that $x \in A$ if and only if $f(x) \in B$.

If $A \leq_p B$, then the decision problem associated with A is at least as hard as the decision problem for B. The relation $\leq_p$ is a refinement of recursively reducible relation $\leq_r$ from Section 7.6. The relation $\leq_p$ distinguishes between recursive sets whereas $\leq_r$ does not.

Definition 7.74 A set A is **NP**-*complete* if it is in **NP** and for every set B in **NP**, $B \leq_p A$.

Let $PROB$ be an arbitrary decision problem. We refer to $PROB$ as **NP**-complete if the corresponding relation on the non-negative integers is **NP**-complete. Likewise, we view $\leq_p$ as a relation on decision problems. To show that $PROP$ is **NP**-complete, it suffices to show that $PROP$ is in **NP** and $PSAT \leq_p PROB$. This follows from the **NP**-completeness of $PSAT$ (which we are currently taking on faith).

Proposition 7.75 Let $CNF - SAT$ be the problem of determining whether or not a given formula of propositional logic in CNF is satisfiable. This problem is **NP**-complete.

Proof Clearly, $CNF - SAT \leq_p PSAT$ (via the identity function). Since $PSAT$ is in **NP** so is $CNF - SAT$. Recall the CNF algorithm from Section 1.6. This algorithm produces a CNF formula equivalent to a given formula F. This algorithm is polynomial-time. So $PSAT \leq_p CNF - SAT$. Since $PSAT$ is **NP**-complete, so is $CNF - SAT$. □

Proposition 7.76 $CLIQUE$ is **NP**-complete.

Proof The algorithm for $CLIQUE$ from Section 7.7.1 demonstrates that $CLIQUE$ is in **NP**. To show that $CLIQUE$ is **NP**-complete, we show that $CNF - SAT \leq_p CLIQUE$. We sketch a proof of this. Given a formula F in CNF, we define a graph G_F. Let $F = \{C_1, \ldots, C_k\}$ where each C_i is a set of literals. Let $\{L_1, \ldots, L_m\}$ be the set of all literals occurring in some clause of F. We must define the vertices and edge relation for G_F. The set of all pairs (C_i, L_j) where L_j is in C_i serves as the set of vertices. There is an edge between two vertices (C_i, L_j) and (C_s, L_t) if and only if $C_i \neq C_s$ and $L_j \not\equiv \neg L_t$. The process of producing the adjacency matrix for G_F given F is feasible and therefore polynomial-time. The following claim proves the proposition.

Claim F is satisfiable if and only if G_F has a subgraph isomorphic to the k-clique (where k is the number of clauses in F).

Proof Suppose F is satisfiable. Then $\mathcal{A} \models F$ for some assignment $\mathcal{A}$. Each clause of F contains a literal to which $\mathcal{A}$ assigns the truth value 1. For $i = 1, \ldots, k$, let $f(i)$ be such that $L_{f(i)}$ is in C_i and $\mathcal{A} \models L_{f(i)}$. Consider the set of vertices $V_k = \{(C_1, L_{f(1)}), (C_2, L_{f(2)}), \ldots, (C_k, L_{f(k)})\}$. Since $\mathcal{A} \models L_{f(1)} \wedge L_{f(2)}$, it cannot be the case that $L_{f(1)} \equiv \neg L_{f(2)}$. By the definition of the edge relation in G_F, vertices $(C_1, L_{f(1)})$ and $(C_2, L_{f(2)})$ share an edge. Likewise, every pair of vertices in V_k share an edge and so G_F has the k-clique as a subgraph. Conversely if some subset of vertices $\{(C_1, L_{f(1)}), (C_2, L_{f(2)}), \ldots, (C_k, L_{f(k)})\}$ form a k-clique, then we can find an assignment $\mathcal{A}$ such that $\mathcal{A} \models L_{f(i)}$ for each i. □

By verifying that $PSAT$ is **NP**-complete in theorem 10.20, we conclude that $CNF - SAT$, $CLIQUE$, and many other decision problems are also **NP**-complete. This is analogous to the situation in Section 7.6.1 where we demonstrated that the set J is nonrecursive and then reduced countless other sets to J in Rice's theorem. In this way, we were able to demonstrate the nonrecursiveness of many sets by explicitly verifying the nonrecursiveness of one set. For **NP**-completeness, we do not have an analogue for Rice's theorem. To obtain many examples of **NP**-complete problems, we must consider the problems one-by-one. The proof of **NP**-completeness usually involves a construction of some sort. For example, we "constructed" the graph G_F to prove Proposition 7.76. The construction relates two ostensibly different problems and can be quite convoluted. We describe various known **NP**-complete problems. Instead of proofs of **NP**-completeness, we provide references at the end of the section.

The Sum k problem ($SUMk$)

Recall the $SUM10$ Problem from the Preliminaries. Given a finite set of integers, we are asked whether or not there exists a subset that adds up to 10. Similarly, we define the $SUMk$ for any integer k. It is easy to see that these problems are in **NP**: choose an arbitrary subset and check to see if it sums up to k. Moreover, $SUMk$ is **NP**-complete. Given an arbitrary formula F, there is a process for constructing a finite set of integers X_F such that X_F has a subset that sums to k if and only if F is satisfiable.

The Hamilton problem

Let G be a finite graph. A *Hamilton path* for G is a path that includes each vertex of G once and only once. The Hamilton problem is to determine whether or not a given graph has a Hamilton path. This problem is closely related to the Traveling Salesman problem. Given a set of cities, the Traveling Salesman problem is to find the shortest route that visits each city. There is no known algorithm that efficiently solves this basic problem. The Traveling Salesman problem is not a decision problem and we do not refer to it as being **NP**-complete. The associated Hamilton problem, however, is **NP**-complete.

The k-Colorability problem ($kCOLOR$)

A graph is said to be *k-colorable* if the vertices can be colored with k different colors in such a way that no two vertices of the same color share an edge. The k-Colorability problem is to determine whether or not a given finite graph is k-colorable. For $k > 2$, this problem is **NP**-complete. For $k = 2$, it is not (see Exercise 7.8).

Minesweeper

Most Microsoft operating systems are equipped with the game of Minesweeper. It is played on a grid that is, for our purposes, infinite. There are bombs behind some of the squares of the grid. The player does not know the location of these bombs. The goal is to uncover squares that do not have bombs. When a square is uncovered, the player either sees a bomb and the game is over, or there is a number that tells the player how many bombs are under squares adjacent to the uncovered square.

Now suppose we are playing this game and are considering a certain covered square. We may have previously uncovered nearby squares that give us some information. There are three possibilities. Either we can deduce from the known information that there is a bomb under the square we are considering, or we can conclude that there is no bomb, or there is not enough information. To be a good Minesweeper player, we want to be able to determine which of these three situations is confronting us. This problem turns out to be **NP**-complete.

More precisely, if we were to uncover many squares simultaneously, then we would see an arrangement of bombs and natural numbers. Let $\sum$ be the set of all finite arrangements consisting of bombs and natural numbers, including arrangements that do not follow the rules of Minesweeper. Let MS be the set of those configurations that can occur in the Minesweeper game. For example, a configuration that has a square containing the number 3 surrounded by 5 bombs is not in MS. The decision problem corresponding to the set MS is **NP**-complete. This problem is equivalent to the problem in the previous paragraph. If we can determine whether a given configuration is consistent with the rules of Minesweeper, then we can play the game effectively. Given a covered square, if we can determine whether the existence of a bomb (or the lack thereof) is contradictory, then we can determine which of the three possible scenarios is confronting us.

Since they are each **NP**-complete, the above problems are equivalent. If we have an algorithm for solving one of these problems, then we can use that algorithm to solve all of these problems. The proof of Proposition 7.76 demonstrates how an algorithm for $CLIQUE$ can be used for $PSAT$. Likewise, if you are really good at playing Minesweeper, then you can use your skill to determine whether a given graph has a Hamilton path, or is 3-colorable, or whether a given formula of propositional logic is satisfiable. More importantly, if you have a polynomial-time algorithm for solving the Minesweeper Problem, then you can win mathematical fame and the Clay Institute fortune by verifying that $\mathbf{P} = \mathbf{NP}$.

Proposition 7.77 Some **NP**-complete problem is in **P** if and only if $\mathbf{P} = \mathbf{NP}$.

Proposition 7.78 Some **NP**-complete problem is in **coNP** if and only if $\mathbf{NP} = \mathbf{coNP}$.

Each of these propositions follow immediately from the definition of **NP**-complete.

Exercises

7.1 Let $\{f_1, f_2, f_3, \ldots\}$ be an enumeration of the set of unary primitive recursive functions. Let U_{pr} be the binary function defined by $U_{pr}(x,y) = f_y(x)$.

(a) Show that the set of unary primitive recursive functions can be enumerated in such a way that $U_{pr}(x,y)$ is computable.

(b) Show that the unary function defined by $g(x) = U_{pr}(x,x) + 1$ is not primitive recursive.

(c) Can part (a) and (b) be repeated with the set of recursive functions in place of the set of primitive recursive functions? Why or why not?

7.2 Show that the function $G(x) = \underbrace{x^{x^{\cdot^{\cdot^{\cdot^{x}}}}}}_{x \text{ times}}$ is primitive recursive.

7.3 Show that the function $F(x) = x!(x-1)!(x-2)! \cdots 3!2!1!$ is primitive recursive.

7.4 Show that the function $E(x,n) = n!(1 + x + \frac{x^2}{2} + \frac{x^3}{3!} + \cdots + \frac{x^n}{n!})$ is primitive recursive.

7.5 Show that the function $P(x,y) = x!/(x \mathbin{\dot{-}} y)!$ is primitive recursive.

7.6 Let $f(x,y)$ and $g(x)$ be primitive recursive functions.

(a) Show that the function $h_a(x,y) = \sum_{z < g(y)} f(x,z)$ is primitive recursive.

(b) Let $h_b(x)$ be the least value of y less than $g(x)$ such that $f(x,y) = 0$. If no such y exists, then $h_b(x) = g(x)$. Show that the function $h_b(x)$ is primitive recursive.

7.7 Let n be a natural number. If we take the definition of "recursive function" and replace the zero function $Z(x) = 0$ with the constant function $c_n(x) = n$, then we obtain the set of *n-recursive functions*. That is, the set of n-recursive functions is the smallest set

- containing the basic functions $s(x)$, $p_i^k(\overline{x})$, and the constant function $c_n(x)$, and

- closed under composition, primitive recursion, and unbounded search.

For each $n \in \mathbb{N}$, let $\mathbb{N}_{\geq n} = \{n, n+1, n+2, \ldots\}$. Prove that a function on $\mathbb{N}_{\geq n}$ is recursive if and only if it is n-recursive.

7.8 Write T^{++} programs that perform the following tasks on a T-machine:
   (a) Swap the contents of B_1 and B_2.
   (b) Set B_2 equal to B_1 and leave B_1 unchanged.
   (c) Add B_1 and B_2 and store the result in B_2.
   (d) Multiply B_1 and B_2 and store the result in B_2.

7.9 Show that the set of T-computable functions is closed under compositions and primitive recursion.

7.10 Let $f(x)$ be a function on the non-negative integers.
   The *history function* of $f(x)$, denoted $f^H(x)$, is defined inductively as

$$f^H(0) = 1 \quad \text{and} \quad f^H(n+1) = 2^{f(0)} 3^{f(1)} \cdots p_n^{f(n)},$$

where p_i denotes the ith prime.
   Suppose that $f(x) = g(f^H(x))$ for some primitive recursive function $g(x)$.
   (a) Show that $f^H(x)$ is primitive recursive.
   (b) Show that $f(x)$ is primitive recursive.

7.11 Let $h_1(\bar{x})$ and $h_2(\bar{x})$ be k-ary primitive recursive functions. Let $g_1(\bar{x}, y, z)$ and $g_2(\bar{x}, y, z)$ be $(k+2)$-ary primitive recursive functions. Suppose that $(k+1)$-ary functions $f_1(\bar{x}, y)$ and $f_2(\bar{x}, y)$ are defined by *simultaneous recursion* as follows.

$$f_1(0, \bar{x}) = h_1(\bar{x}),$$
$$f_2(0, \bar{x}) = h_2(\bar{x}),$$
$$f_1(n+1, \bar{x}) = g_1(\bar{x}, f_1(n, \bar{x}), f_2(n, \bar{x})), \text{ and}$$
$$f_2(n+1, \bar{x}) = g_2(\bar{x}, f_1(n, \bar{x}), f_2(n, \bar{x})).$$

Show that both f_1 and f_2 are primitive recursive functions. (Hint: Consider history functions for f_1 and f_2 as defined in the previous exercise.)

7.12 Give an explicit definition of the Ackermann function in terms of unbounded search.

7.13 Let f be a k-ary function. The *graph* of f is the $k+1$-ary relation consisting of all $(\bar{x}, y)$ such that $f(\bar{x}) = y$.
   (a) Show that f is a primitive recursive function if and only if the graph of f is a primitive recursive set.

(b) Show that f is a recursive function if and only if the graph of f is a recursively enumerable set.

(c) Show that f is a total recursive function if and only if the graph of f is a recursive set.

7.14 Let $f(x, y) = \lfloor x/y \rfloor$ where $\lfloor x/y \rfloor$ denotes the greatest integer less than x/y. Let $g(x)$ be a primitive recursive function and let

$$h(x, y) = \begin{cases} f(x, y) & y \neq 0 \\ g(x) & y = 0. \end{cases}$$

Show that $h(x, y)$ is primitive recursive. (Use the previous exercise and the fact that a relation is primitive recursive if and only if it is definable by a Δ_0 formula.)

7.15 Let $f(x)$, $g(x)$, and $h(x)$ be primitive recursive functions. Let

$$e(x) = \begin{cases} f(x)^{g(x)} & \text{if } f(x) + g(x) > 0 \\ h(x) & \text{if } f(x) + g(x) = 0. \end{cases}$$

Show that $e(x)$ is primitive recursive.

7.16 Let $\varphi(x, y)$ be a Δ_0 formula and let $f(x)$ be a primitive recursive function. Show that the formula $\exists y(y < f(x) \wedge \varphi(x, y))$ is Δ_0, where $y < f(x)$ is an abbreviation for the V_{ar}-formula $\exists z(y + z = f(x))$.

7.17 Assuming that every primitive recursive set is Δ_0, prove the following.
(a) Every recursively enumerable set is Σ_1.
(b) Every recursive set is both Σ_1 and Π_1.

7.18 Let A be an infinite set. Prove that A is recursive if and only if it is the range of an increasing recursive function.

7.19 Show that every infinite recursively enumerable set has an infinite recursive subset.

7.20 Let A and B be recursively enumerable sets. Show that there exist recursively enumerable subsets $A_1 \subset A$ and $B_1 \subset B$ such that $A_1 \cap B_1 = \emptyset$ and $A_1 \cup B_1 = A \cup B$.

7.21 Show that the union of two recursively enumerable sets is recursive enumerable. Moreover, show that the function $f(x, y)$ defined by $W_x \cup W_y = W_{f(x,y)}$ is a recursive function.

7.22 Repeat the previous exercise with intersections instead of unions.

7.23 Show that there exists a partial recursive function that cannot be extended to a total recursive function.

7.24 Let A and B be subsets of $\mathbb{N} \cup \{0\}$. We say that A and B are *recursively separable* if $A \subset R$ and $B \subset \bar{R}$ for some recursive subset R of $\mathbb{N}$. Otherwise, A and B are said to be *recursively inseparable*. Show that there exist recursively enumerable A and B that are recursively inseparable.

7.25 Let FIN, INF, COF, TOT, and REC be as defined in Section 7.6.1.
 (a) Show that $FIN \leq_r COF$.
 (b) Show that $INF \leq_r COF$.
 (c) Show that $TOT \leq_r COF$.
 (d) Show that $COF \leq_r REC$.

7.26 (a) Show that TOT is Π_2.
 (b) Show that FIN is Σ_2.
 (c) Show that COF is Σ_3.

7.27 Classify the set $SQR = \{e | \varphi_e^1(x) = x^2\}$ as either Π_n or Σ_n for some n.

7.28 Let Γ be either Π_n or Σ_n for some n. We say that $U \subset \mathbb{N}^2$ is Γ-*complete* if U is Γ and for any set A that is Γ, $A \leq_r U$. Show that $REC = \{e : W_e$ is recursive$\}$ is Σ_3-complete.

7.29 Let Γ be either Π_n or Σ_n for some n. We say that $U \subset \mathbb{N}^2$ is Γ-*universal* if for any set A that is Γ, there exists e such that $A = \{x | (e, x) \in U\}$.
 (a) Show that it U is Γ-universal, then U is Γ-complete (see Exercise 7.28).
 (b) Show that if U is Σ_n-universal, then $\bar{U}$ is Π_n-universal.
 (c) Show that if U is Σ_n-universal, then U is not Δ_n.
 (Hint: consider the set $\{x | (x, x) \in U\}$.)

7.30 Refer to the previous exercise.
 (a) Show that the set H_1 from Section 7.6.1 is Σ_1 universal.
 (b) Let $f : (\mathbb{N} \cup \{0\})^2 \to (\mathbb{N} \cup \{0\})$ be a recursive one-to-one correspondence.
  Show that if U_n is Π_n-universal, then the set

$$\{(e, n) | \exists y (e, f(x, y)) \in U_n\}$$

  is Σ_n-universal.
 (c) Using the previous exercise, conclude that there exist Σ_n sets and Π_n sets that are not Δ_n for each $n \in \mathbb{N}$.

7.31 Show that constant functions cannot by computed by a T^{++} algorithm in polynomial-time. Which functions can be computed by T^{++} in polynomial-time?

7.32 Prove that every decision problem in **NP** is decidable.

7.33 Recall the k-Colorability problem from Section 7.8.

(a) Describe a polynomial-time nondeterministic algorithm for the k-colorability problem for $k > 2$.

(b) Describe a polynomial-time algorithm for the 2-colorability problem.

7.34 Describe an algorithm that computes $f(n) = 2^n$ and uses only k bins for some $k \in \mathbb{N}$.

7.35 For each $k \in \mathbb{N}$, let $kPSAT$ correspond to the set of satisfiable formulas of propositional logic that contain at most k atomic subformulas.

(a) Show that $2PSAT$ is in **P**.

(b) Using the fact that $PSAT$ is **NP**-complete, show that $3PSAT$ is **NP**-complete.

7.36 Let $PVAL$ correspond to the set of valid sentences of propositional logic. To which complexity classes does this problem belong? Is $PVAL$ complete for any of these classes? (Use the fact that $PSAT$ is **NP**-complete.)

7.37 Let φ be a sentence in the vocabulary $\mathcal{V}_R$ of graphs. The φ-Graph problem is to determine whether or not a given finite graph models φ. Show that the φ-Graph problem is in **P** for any first-order sentence φ.

8 The incompleteness theorems

In this chapter we prove that the structure $\mathbf{N} = (\mathbb{N}|+,\cdot,1)$ has a first-order theory that is undecidable. This is a special case of Gödel's First Incompleteness theorem. This theorem implies that any theory (not necessarily first-order) that describes elementary arithmetic on the natural numbers is necessarily undecidable. So there is no algorithm to determine whether or not a given sentence is true in the structure $\mathbf{N}$. As we shall show, the existence of such an algorithm leads to a contradiction. Gödel's Second Incompleteness theorem states that any decidable theory (not necessarily first-order) that can express elementary arithmetic cannot prove its own consistency. We shall make this idea precise and discuss the Second Incompleteness theorem in Section 8.5. Gödel's First Incompleteness theorem is proved in Section 8.3.

Although they are purely mathematical results, Gödel's Incompleteness theorems have had undeniable philosophical implications. Gödel's theorems dispelled commonly held misconceptions regarding the nature of mathematics. A century ago, some of the most prominent mathematicians and logicians viewed mathematics as a branch of logic instead of the other way around. It was thought that mathematics could be completely formalized. It was believed that mathematical reasoning could, at least in principle, be mechanized. Alfred North Whitehead and Bertrand Russell envisioned a single system that could be used to derive and enumerate all mathematical truths. In their three-volume *Principia Mathematica*, Russell and Whitehead rigorously define a system and use it to derive numerous known statements of mathematics. Gödel's theorems imply that any such system is doomed to be incomplete. If the system is consistent (which cannot be proved within the system by Gödel's Second theorem), then there necessarily exist true statements formulated within the system that the system cannot prove (by Gödel's First theorem). This explains why the name "incompleteness" is attributed to these theorems and why the title of Gödel's 1931 paper translates (from the original German) to "On Formally Undecidable Propositions of *Principia Mathematica* and Related Systems" (translated versions appear in both [13] and [14]).

Depending on one's point of view, it may or may not be surprising that there is no algorithm to determine whether or not a given sentence is true in $\mathbf{N}$. More surprising is the fact that we can prove that there is no such algorithm. The proof itself is truly remarkable. Gödel's proof introduced the notions of primitive recursive sets and relations discussed in the previous chapter. Gödel's proof gave impetus to the subject of computability theory years before the advent of the computer. Moreover, Gödel deduced from his proof that any decidable system that can perform arithmetic on the natural numbers cannot prove its own consistency. This is a gem of mathematical reasoning and is, by any measure, among the great results of twentieth century mathematics.

8.1 Axioms for first-order number theory

We discuss some consequences of Gödel's First Incompleteness theorem. We prove this theorem in Section 8.3. In the present section we accept this theorem as a fact.

We distinguish two related first-order theories. The *theory of arithmetic*, denoted T_A is the theory of the structure $\mathbf{A} = \{\mathbb{Z}|+,\cdot,0,1\}$. The theory T_N of the structure $\mathbf{N} = (\mathbb{N}|+,\cdot,1)$ is *first-order number theory*. We focus on the latter of these theories. By Gödel's First Incompleteness theorem, not only T_N, but also related theories (such as T_A) are undecidable.

We claim that elementary number theory is contained in T_N. Peruse any book on the subject and you will find that the statements of interest (the theorems, lemmas, conjectures, and so forth) can be formulated as first-order sentences in the vocabulary $\mathcal{V}_N = \{1, +, \cdot\}$. For example, the following $\mathcal{V}_N$-formulas represent the concepts of divisibility and prime number:

let $div(x, y)$ be $\exists z(x \cdot z = y)$
and $prime(x)$ be $\forall z(div(z, x) \to z = 1 \lor z = x) \land \neg(x = 1)$.

By definition, $\mathbf{N} \models div(a, b)$ if and only if a divides b and $\mathbf{N} \models prime(a)$ if and only if a is prime. We can express that there are infinitely many primes with the $\mathcal{V}_N$-sentence

$$\forall x(prime(x) \to \exists y(prime(y) \land y > x),$$

where $y > x$ is the $\mathcal{V}_N$-formula $\exists z(x + z = y)$. Less obvious is the fact that we can express the *Fundamental Theorem of Arithmetic* as a $\mathcal{V}_N$-sentence. There are also $\mathcal{V}_N$-sentences that express the Chinese Remainder theorem, Fermat's Last theorem, Goldbach's Conjecture, Gauss' Law of Quadratic Reciprocity, and Riemann's hypothesis.

The reader does not need to be familiar with the all of the theorems of number theory listed in the previous paragraph. In the next section we show that for every recursive subset A of $\mathbb{N}^n$ (for some n), there exists a $\mathcal{V}_N$-formula $\varphi_A(x_1, \ldots, x_n)$ such that $\bar{a} \in A$ if and only if $\mathbf{N} \models \varphi_A(\bar{a})$. This justifies our admittedly vague claim that elementary number theory is contained in first-order number theory.

The reader does need to be familiar with some basic number theory. The reader should recall the Fundamental Theorem of Arithmetic and the fact that there are infinitely many primes from the previous chapter (Sections 7.1 and 7.3). In the present chapter, we shall state and use the Chinese Remainder theorem. The reader should also be aware that the unanswered question of whether Riemann's hypothesis is true is the most important open question of modern mathematics.

If T_N had a decidable axiomatization, then, in principle, we could use the methods of Chapter 3 to answer every open question of number theory. By Gödel's First Incompleteness theorem, this is not the case. Any deductive system (such as formal proofs or resolution) has inherent limitations. Since it is undecidable, T_N does not have a decidable first-order axiomatization (see Proposition 5.10). However, we now demonstrate a second-order theory containing T_N that does have a decidable axiomatization.

Let us axiomatize T_N. We begin with the following two axioms:

$$\forall x \neg (x + 1 = 1),$$

$$\forall x \forall y (x + 1 = y + 1 \rightarrow x = y).$$

Any model of these two sentences is necessarily infinite. Together they say that the function defined by $(x + 1)$ is one-to-one but not onto. The following axioms describe addition:

$$\forall x \forall y (x + y = y + x),$$

$$\forall x \forall y (x + (y + 1) = (x + y) + 1).$$

Multiplication is described in a similar manner:

$$\forall x \forall y (x \cdot y = y \cdot x),$$

$$\forall x (x \cdot 1 = x),$$

$$\forall x \forall y (x \cdot (y + 1) = (x \cdot y) + x).$$

Each of these axioms is first-order. The final axiom, called the Induction Axiom, is second-order. Second-order logic is a topic of the next chapter. Presently, it suffices to say that, in second-order logic, one can quantify over relations. In the following second-order sentence, $S(x)$ represents an arbitrary unary relation.

$$\forall^1 S (S(1) \wedge \forall x (S(x) \rightarrow S(x + 1))) \rightarrow \forall x S(x).$$

In this sentence, "$\forall^1 S$" is read "for all unary relations $S\ldots$" In effect, the Induction Axiom says that any subset of the universe that contains 1 and is closed under the function $x+1$ necessarily contains every element in the universe.

These axioms describe **N** completely and categorically. The first several axioms inductively define addition and multiplication and the final axiom states that induction works.

Proposition 8.1 Let M be a $\mathcal{V}_N$-structure. If each of the above axioms holds in M, then M is isomorphic to **N**.

Proof As a $\mathcal{V}_N$-structure, M contains an element that interprets the constant 1. We inductively construct an isomorphism $f: \mathbf{N} \to M$ by sending 1 to this element and $(n+1)$ to the successor $f(n) + 1$ of $f(n)$ in M (for each $n \in \mathbb{N}$). If M satisfies the Induction Axiom, then the range of f must be all of the underlying set of M. It follows that f is an isomorphism. □

It follows from this proposition that second-order logic does not have completeness. That is, we cannot hope to define second-order analogues for first-order resolution and formal proofs. Any such formal system would provide a systematic way to determine whether or not a sentence is a consequence of the above axioms and, hence, whether or not the sentence is true in **N**. By Gödel's First Incompleteness theorem (which we are presently taking on faith), there is no such algorithm.

It also follows from Proposition 8.1 that no first-order sentence nor set of first-order sentences can express the Induction Axiom. First-order theories are subject to the Löwenheim–Skolem theorems and are incapable of describing an infinite structure such as **N** up to isomorphism. Although we cannot say "for all subsets $S \dots$" in first-order logic, we can say "for all definable subsets $S \dots$" For each $\mathcal{V}_N$-formula $\varphi(x)$ in one free variable, let ψ_φ be the $\mathcal{V}_N$-sentence

$$(\varphi(1) \wedge \forall x(\varphi(x) \to \varphi(x+1))) \to \forall x \varphi(x).$$

Let Γ_{ind} be the set of all such sentences ψ_φ. This set of sentences is a first-order "approximation" of the Induction Axiom.

Definition 8.2 Let Γ_N be the set of first-order sentences obtained from the above axiomatization by replacing the Induction Axiom with the set Γ_{ind}.

By Gödel's First Incompleteness theorem, Γ_N is incomplete. So these sentences do not describe the structure **N** up to elementary equivalence. However, the set Γ_N is a natural fragment of T_N to consider. This fragment is often referred to as *Peano Arithmetic*. We describe the difference between T_N and Peano arithmetic in terms of both $\mathcal{V}_N$-structures and $\mathcal{V}_N$-sentences.

Let M be a $\mathcal{V}_N$-structure. Then M interprets the $\mathcal{V}_N$-terms $1, 1+1, 1+1+1$, and so forth. Denote the elements of M that interpret these terms as a_1, a_2, a_3, and so forth. If $M \models \Gamma_N$, then the function $e: \mathbf{N} \to M$ defined by $e(n) = a_n$ is an embedding. So any model of Γ_N contains a substructure that is isomorphic to **N**. In this sense, Γ_N fully defines multiplication and addition on the natural numbers. Since Γ_N is incomplete, some of its models contain elements beyond the natural numbers which exhibit behavior not witnessed in **N**. Whereas **N** can be embedded into every model of Γ_N, it cannot be elementarily embedded into every model.

Since it is incomplete, there necessarily exists a $\mathcal{V}$-sentence φ in T_N that cannot be derived from Γ_N. It should be clear from the previous paragraph that such a sentence φ cannot be atomic. In fact, it cannot be a Σ_1 sentence (as defined in Section 7.6.2).

Proposition 8.3 If $\varphi(x_1,\ldots,x_k)$ is equivalent to a Σ_1 formula, then $\Gamma_N \vdash \varphi(t_{n_1},\ldots,t_{n_k})$ if and only if $\mathbf{N} \models \varphi(n_1,\ldots,n_k)$, where, for each $n \in \mathbb{N}$, t_n denotes the $\mathcal{V}_N$-term $\underbrace{1 + 1 + \cdots + 1}_{n \text{ times}}$.

Proof Since $\Gamma_N \subset T_N$, if $\Gamma_N \vdash \varphi(t_{n_1},\ldots,t_{n_k})$ then $\mathbf{N} \models \varphi(n_1,\ldots,n_k)$. We must prove the converse of this. Suppose that $\mathbf{N}$ models $\varphi(n_1,\ldots,n_k)$. By completeness, it suffices to show that every model M of Γ_N models $\varphi(t_{n_1},\ldots,t_{n_k})$. If this is true for some formula φ, then it is also true for any formula that is equivalent to φ. So we may assume that φ is in prenex normal form.

Case 1 φ is a Δ_0 formula (as defined in Section 7.2).

We proceed by induction on the number of quantifiers in φ. Since it is Δ_0, each quantifier is bounded. Let M be an arbitrary model of Γ_N. If φ is quantifier-free, then $M \models \varphi(t_{n_1},\ldots,t_{n_k})$ since $e : \mathbf{N} \to M$ (as defined above) is an embedding. Now suppose that φ has $n + 1$ quantifiers. Our induction hypothesis is that the proposition holds for any Δ_0 formula having at most n quantifiers. To ease notation, suppose that φ has only one free variable x. If the first quantifier in φ is $\forall$, then there exists a Δ_0 formula $\psi(x,y)$ having n quantifiers such that φ is equivalent to the formula $\forall y(y < x \to \psi(x,y))$, where $y < x$ is an abbreviation for the $\mathcal{V}_N$-formula $\exists z(y + z = x)$.

If $\mathbf{N} \models \forall y(y < n \to \psi(n,y))$ for some $n \in \mathbb{N}$, then
$\mathbf{N} \models \psi(n,m)$ for each $m < n$, in which case
$M \models \psi(t_n, t_m)$ for each $m < n$ by our induction hypothesis.

It follows from the axioms in Γ_N that $M \models y < t_n$ if and only if $y = t_m$ for some $m < n$. It follows that $M \models \forall y(y < t_n \to \psi(t_n, y))$ as we wanted to show.

Now suppose that the first quantifier in φ is $\exists$. Then φ has the form $\exists y \psi(x,y)$ for some Δ_0 formula $\psi(x,y)$ having n quantifiers.

If $\mathbf{N} \models \exists y \psi(n, y)$ for some $n \in \mathbb{N}$,
then $\mathbf{N} \models \psi(n, m)$ for some $m \in \mathbb{N}$, in which case
$M \models \psi(t_n, t_m)$ by our induction hypothesis, and so $M \models \exists y \psi(t_n, y)$.
By induction, the proposition holds for any Δ_0 formula φ.

Case 2 φ is a Σ_1 formula.

By definition, φ is a Σ_1 formula if it has the form $\exists y \psi$ for some Δ_0 formula ψ. We have already considered such formulas in case 1. In the proof of case 1, we did not use the fact that the quantifier $\exists$ is bounded in Δ_0 formulas. So our proof of case 1 also proves case 2. □

In the final Section 8.5, we explicitly demonstrate a $\mathcal{V}_N$-sentence φ_0 that is true in $\mathbf{N}$ but cannot be derived from Γ_N. If we augment Γ_N by adding φ_0 as an axiom, then the result is still incomplete by Gödel's First Incompleteness theorem. There must exist a $\mathcal{V}_N$-sentence φ_1 that is true in $\mathbf{N}$ that cannot be derived from $\Gamma_N \cup \{\varphi_0\}$. Likewise, continuing in this manner, there exists a $\mathcal{V}_N$-sentence φ_{n+1} that is true in $\mathbf{N}$ that cannot be derived from $\Gamma_N \cup \{\varphi_0, \varphi_1, \ldots, \varphi_n\} \subset T_N$. Any decidable set of sentences in T_N is necessarily incomplete. Contrapositively, any axiomatization of T_N, such as T_N itself, is necessarily undecidable.

8.2 The expressive power of first-order number theory

The reason that $T_N = Th(\mathbf{N})$ is undecidable is that so many subsets of $\mathbf{N}$ are $\mathcal{V}_N$-definable. In this section we prove that every recursive subset of $\mathbf{N}^n$ is a definable subset of $\mathbf{N}$. Moreover, these sets are definable by a Σ_1 formulas. The key to this immense expressive power is the fact that we can quantify over sequences of variable length. We demonstrate this idea with an example.

Example 8.4 Let n be a natural number. Let S_n be the subset of $\mathbb{N}$ consisting of those natural numbers that can be written as the sum of the squares of n primes. That is, $x \in S_n$ if and only if there exists prime numbers $q_1, \ldots, q_n$ (not necessarily distinct) such that

$$x = (q_1)^2 + (q_2)^2 + \cdots + (q_n)^2.$$

The set S_n is a definable subset of $\mathbf{N}$. Recall the $\mathcal{V}_N$-formula $prime(x)$ from the previous section. Let $\varphi_n(x)$ be the $\mathcal{V}_N$-formula

$$\exists z_1 \cdots \exists z_n \left(\bigwedge_{i=1}^{n} prime(z_i) \wedge (z_1 \cdot z_1 + \cdots + z_n \cdot z_n = x) \right).$$

Clearly, $\varphi_n(x)$ defines the set S_n.

The incompleteness theorems 363

Now consider the subset $\mathcal{S}$ of $\mathbb{N}^2$ consisting of the pairs of natural numbers (x,n) such that $x \in S_n$. That is (x,n) is in $\mathcal{S}$ if and only if x can be written as the sum of the squares of n primes. Let us attempt to find a $\mathcal{V}_N$-formula $\varphi_\mathcal{S}(x,y)$ that defines this set. The formula $\varphi_\mathcal{S}(x,y)$ must say that there exist prime numbers $q_1, \ldots, q_y$ such that $x = (q_1)^2 + (q_2)^2 + \cdots + (q_y)^2$. The obstacle to writing such a formula is the phrase "there exist $q_1, \ldots, q_y$." It seems that the number y of existential quantifiers in this formula is determined by a free-variable of the formula itself! Later in this section, we show that we can overcome this obstacle and define the first-order $\mathcal{V}_N$-formula $\varphi_\mathcal{S}(x,y)$.

To define the set $\mathcal{S}$ from the previous example, the formula $\varphi_\mathcal{S}(x,y)$ must express that there exists a sequence of length y having certain properties. This is an example of what we mean by "quantifying over sequences of variable length." Using $\mathcal{S}$ as an example, we demonstrate a technique for expressing the formula $\varphi_\mathcal{S}(x,y)$. We then use this technique to show that any recursive subset of $\mathbf{N}$ is definable.

To quantify over sequences of variable length, we encode the sequence. There are many ways to do this. The method we use encodes a given sequence as a triple of numbers l, m, and k. Any finite sequence of natural numbers uniquely defines its "code" $[l, m, k]$. Conversely, given l, m, and k, we can "decode" $[l, m, k]$ to recover the original sequence. The coding and decoding process takes place within first-order number theory. This allows us to replace "there exists a sequence of length y such that..." with an expression of first-order number theory that begins "$\exists l \exists m \exists k \ldots$."

Given a finite sequence $\bar{a}$ of natural numbers, the code for $\bar{a}$ describes the sequence completely and categorically. The number m is the maximum number in the sequence and the number l represents the length of the sequence. To fully describe a particular sequence, we must provide more information. The number k completes the description. This number is more complicated than the maximum or the length of the sequence. We demonstrate k with some examples.

First we show how to decode a sequence.

Example 8.5 Suppose we are given the code $[3, 5, 590]$. This code represents a unique sequence of natural numbers a_1, a_2, a_3. The first number of the code is the length of the sequence. The second number in the code tells us that one of the three numbers in the sequence is 5 and no number in the sequence is larger than 5. The third number $k = 590$ is the "key" to decoding the sequence a_1, a_2, a_3. This key works as follows.

- Let d be the least number bigger than m that is divisible by each number less than $l + 1$.

In our example, $d = l! = 3 \cdot 2 \cdot 1 = 6$.

- The first number of the sequence is the remainder when k is divided by $d+1$.

In our example, 590 has a remainder of 2 when divided by $6 + 1 = 7$ ($590 = 84 \cdot 7 + 2$). So the first number of the sequence is $a_1 = 2$.

- The second number of the sequence is the remainder when k is divided by $2d + 1$.

In our example, $2d + 1 = 13$. Since $590 = 45 \cdot 13 + 5$, the second number is $a_2 = 5$.

- The third number of the sequence is the remainder when k is divided by $3d + 1$.

In our example, $3d + 1 = 19$. Since $31 \cdot 19 = 589$, the remainder $a_3 = 1$. So $[3, 5, 590]$ is the code for the sequence $(2, 5, 1)$.

Next, we demonstrate how to encode a given sequence.

Example 8.6 Suppose we wish to encode the sequence $(2, 8, 1)$. Clearly $l = 3$ and $m = 8$. We must find a key k for the encryption. To find k, we must think about how $[3, 8, k]$ will be decoded. We want k to be such that

the remainder when k is divided by $d + 1$ is 2,
the remainder when k is divided by $2d + 1$ is 8, and
the remainder when k is divided by $3d + 1$ is 1.

Recall from the previous example that d is the least number exceeding m divisible by each number less than $l + 1$. Since $3! = 6$ and $m = 8$, d is equal to 12. We must find k such that the remainders when k is divided by 13, 25, and 37 are respectively 2, 8, and 1.

There is a systematic way to find such a number k. The first step is to get a calculator.

Let $d_1 = d+1$, $d_2 = 2d+1$, and $d_3 = 3d+1$. Then $k = a_1 b_1 d_2 d_3 + a_2 b_2 d_1 d_3 + a_3 b_3 d_1 d_2$, where the b_is are defined as follows:

Let b_1 be least such that $b_1 d_2 d_3$ has remainder 1 when divided by d_1.

In our example, $d_2 d_3 = 25 \cdot 37 = 925$. We must find the least multiple of 925 that has a remainder of 1 when divided by $d_1 = 13$. By checking each multiple $925, 2 \cdot 925, \ldots, 12 \cdot 925$, we find that $7 \cdot 925 = 6475 = 13 \cdot 488 + 1$ (the Euclidean algorithm works for this too). So $b_1 = 7$.

Likewise, let b_2 and b_3 be least such that $b_2 d_1 d_3$ has remainder 1 when divided by d_2 and $b_3 d_1 d_2$ has remainder 1 when divided by d_3.

In our example, we must find the least multiple of $481 = d_1 d_3$ that has remainder 1 when divided by $25 = d_2$ and also the least multiple of $325 = d_1 d_2$ that has remainder 1 when divided by $37 = d_3$. Clearly, $b_2 = 21$ and $b_3 = 23$ since and $21 \cdot 481 = 10101 = 404 \cdot 25 + 1$. and $23 \cdot 325 = 202 \cdot 37 + 1$. (I confess, I am using a computer algebra system.)

Now that we know b_1, b_2, and b_3, we can compute

$$k = a_1 b_1 d_2 d_3 + a_2 b_2 d_1 d_3 + a_3 b_3 d_1 d_2$$
$$= 2 \cdot 7 \cdot 25 \cdot 37 + 8 \cdot 21 \cdot 13 \cdot 37 + 1 \cdot 23 \cdot 13 \cdot 25 = 101233.$$

There is more than one value for k that works as a code for the sequence $(2, 8, 1)$. What really matters is not the number k, but the remainder when k is divided by $d_1 d_2 d_3$. In our example, since 101233 has a remainder of 5033 when divided by $d_1 d_2 d_3 = 12025$, we can take k to be 5033, 101233, or any other number that has remainder 5033 when divided by 12025.

Whereas there exist more than one code for a sequence, the opposite is not true. There exists exactly one sequence for each code. The sequence $(2, 8, 1)$ can be recovered by decoding either of the codes $[3, 8, 5033]$ or $[3, 8, 101233]$.

In the previous examples we have successfully coded sequences of length three using only three numbers l, m, and k. This is not too impressive. However, any finite sequence can be coded in a similar manner. Regardless of the length of the sequence, the code requires only the three numbers l, m, and k. In the remainder of this section, we first show that the technique described in these examples works and, secondly, use this technique to write $\mathcal{V}_N$-formulas to define the set S from Example 8.4 and other subsets of $\mathbb{N}^n$.

We view the coding process as a computable function. Given any finite sequence of natural numbers as input, this function outputs the code $[l, m, k]$. As was pointed out in Example 8.6, there exists more than one code for each sequence. To make it a function, we restrict the output to the code that has the smallest value for k. To verify that our coding technique works, we must show that this function is defined for all finite sequences of natural numbers and that it is one-to-one. That is, we must show that every finite sequence has a code and no two sequences have the same code.

Let us again walk through the coding process. Given a sequence $(a_1, \ldots, a_l)$, let l be the length of the sequence and let m be the largest a_i in the sequence. Let d be the least number exceeding m divisible by each number less than $l + 1$. Clearly, such d is uniquely defined. We must find k such that k has remainder a_i when divided by $d_i = i \cdot d + 1$ (for each $i = 1, \ldots, l$). To obtain the smallest value for k, let k_1 be the remainder when k is divided by $d_1 \cdot d_2 \cdots \cdot d_n$. Output $[l, m, k_1]$ as the code for $(a_1, \ldots, a_l)$. The only step of this process that requires verification is the existence of k. For this we use the Chinese Remainder theorem.

Definition 8.7 A set of natural numbers $\{d_1, d_2, \ldots, d_l\}$ is said to be *relatively prime* if, for $1 \leq i < j \leq n$, there is no number that divides both d_i and d_j other than 1.

The Chinese Remainder theorem Let $\{d_1, \ldots, d_l\}$ be a relatively prime set of natural numbers. Let $(a_1, \ldots, a_l)$ be a sequence of natural numbers with $a_i \leq d_i$ for $i = 1, \ldots, l$. There exists $k \in \mathbb{N}$ such that when k is divided by d_i, the remainder is a_i for each $i = 1, \ldots, l$. Moreover, there exists a unique such k that is less than $d_1 \cdot d_2 \cdots d_l$.

Proof idea The idea behind the proof was demonstrated in Example 8.6. The following formula computes k:

$$k = a_1 b_1 D_1 + a_2 b_2 D_2 + \cdots + a_n b_n D_n,$$

where $D_i = (d_1 \cdot d_2 \cdots d_n)/d_i$ and b_i is least such that $b_i D_i$ has remainder 1 when divided by d_i. The existence of b_i follows from the fact that the set $\{d_1, \ldots, d_n\}$ is relatively prime. Note that d_i divides D_j for $i \neq j$. So k has the same remainder as $a_i b_i D_i$ when divided by d_i. By design, $b_i D_i$ has remainder 1. So $a_i b_i D_i$ has remainder $a_i \cdot 1$ when divided by d_i.

For a detailed proof, the reader is referred to any book on elementary number theory such as [19]. □

Proposition 8.8 For any l and d in $\mathbb{N}$, if d is divisible by each number less than $l+1$, then $\{d+1, 2 \cdot d+1, \ldots, l \cdot d+1\}$ is a relatively prime set of natural numbers.

Proof Consider $d_i = i \cdot d + 1$ and $d_j = j \cdot d + 1$ for $1 \leq i < j \leq l$. Let p divide both d_i and d_j. We want to show that $p = 1$.

Claim p divides d.

Proof Since p divides both d_i and d_j, p divides $d_j - d_i = j \cdot d - i \cdot d = (j - i)d$. It follows that either p divides d or p divides $(j - i)$. Note that $(j - i) < l$. By assumption, $(j - i)$ divides d. This proves the claim.

Since p divides d, it also divides $i \cdot d$. Since p divides d_i, it also divides $d_i - i \cdot d = 1$. It follows that $p = 1$. □

Corollary 8.9 Every finite sequence of natural numbers has a code $[l, m, k]$. Moreover, no two sequences have the same code.

Proof Any finite sequence has a length l and a maximal entry m. The existence of k follows from the previous proposition, which allows us to apply the Chinese Remainder theorem.

To see that no two sequences have the same code, consider the decoding process described in Example 8.5. To decode $[l, m, k]$ we first find d, the least number greater than m divisible by each number less than $l + 1$. Clearly, such a number exists and is unique. We then divide k by $d + 1, 2 \cdot d + 1, \ldots, l \cdot d + 1$

and take the remainders as the sequence $a_1, \ldots, a_l$. Since these remainders are uniquely determined, so is the sequence $a_1, \ldots, a_l$. □

So the process we have described for coding finite sequences works. We next show that this process can be translated into the language of first-order number theory. The following table translates some key phrases:

To say	Use the $\mathcal{V}_N$-formula	Denoted by
x is smaller than y	$\exists z(x + z = y)$	$x < y$
x is not bigger than y	$x < (y + 1)$	$x \leq y$
x divides y	$\exists z(x \cdot z = y)$	$div(x, y)$
y is divisible by each number less than $x+1$	$\forall z(z \leq x \to div(z, y))$	$D(x, y)$
y divided by x has remainder z	$(z < x) \land$ $\exists q(y = q \cdot x + z)$	$rem(x, y, z)$

The $\mathcal{V}_N$-formula $rem(x, y, z)$ is the graph of a function. Given any x and y, there is at most one z that satisfies $rem(x, y, z)$. Let $Rem(x, y)$ represent this function. That is

$$z = Rem(x, y) \text{ is defined by } \exists w(rem(x, y, w) \land z = w).$$

Definition 8.10 Let $f : \mathbb{N}^k \to \mathbb{N}$ be a k-ary function on the natural numbers. We say that f is *definable* if there exists a $\mathcal{V}_N$-formula $\varphi_f(\bar{x}, y)$ such that

$$f(\bar{a}) = b \text{ if and only if } \mathbb{N} \models \varphi_f(\bar{a}, b).$$

We say that f is *defined by* $\varphi_f(\bar{x}, y)$.

We now show that we can decode sequences within first-order number theory.

Proposition 8.11 Let $code\,(l, m, k, i)$ be the ith element in the sequence obtained by decoding $[l, m, k]$. This function is definable.

Proof Let $\varphi(x, l, m)$ be the $\mathcal{V}_N$-formula $x > m \land D(l, x)$. Let $\delta(x, l, m)$ be the $\mathcal{V}_N$-formula

$$\varphi(x, l, m) \land \forall z(z < x \to \neg \varphi(z, l.m))$$

saying that x is least such that $\varphi(x, l, m)$ holds. That is, $\delta(x, l, m)$ says that $x = d$. The function $code(l, m, k, i) = z$ is defined by

$$(i \leq l) \land \exists d(\delta(d, l, m) \land z = Rem(k, i \cdot d + 1)). \quad \square$$

Next, we define a $\mathcal{V}_N$-formula $seq(l, m, k)$ that says $[l, m, k]$ is the code for a finite sequence. If we randomly choose the numbers, l, m, and k, then $[l, m, k]$ will probably not be a code. For example, consider $[2, 9, 24]$. Since $l = 2$ and $m = 9$, $d = 10$. We have $a_1 = Rem(24, 10+1) = 2$ and $a_2 = Rem(24, 20+1) = 3$. Since the maximum entry of the sequence (a_1, a_2) is 3 and not 9, $[2, 9, 12]$ is not a code for this sequence. Since $[2, 9, 12]$ is not the code for $(2, 3)$, it is not the code for any sequence. To say that $[l, m, k]$ is the code for a sequence, we must say that m is the maximum number in the sequence obtained by decoding $[l, m, k]$.

Proposition 8.12 There exists an $\mathcal{V}_N$-formula $seq(x, y, z)$ such that $\mathbf{N} \models seq(l, m, k)$ if and only if $[l, m, k]$ is the code for some sequence.

Proof Let $seq(l, m, k)$ be the formula

$$\forall i(i \leq l \rightarrow code(l, m, k, i) \leq m) \wedge \exists i(i \leq l \wedge code(l, m, k, i) = m)). \qquad \square$$

Example 8.13 We now define the formula $\varphi_S(x, y)$ from Example 8.4. We want to say there exist primes $q_1, \ldots, q_y$ such that $q_1^2 + q_2^2 + \cdots + q_y^2 = x$. To do this, we say there exists the sequence $(a_1, \ldots, a_y)$ defined as the partial sums

$$a_1 = q_1^2, a_2 = q_1^2 + q_2^2, \ldots, a_y = q_1^2 + q_2^2 + \cdots + q_y^2$$

and that $x = a_y$. Let $ps(x)$ be the $\mathcal{V}_N$-formula

$$\exists z(prime(z) \wedge z \cdot z = x)$$

saying that x is a prime number squared and let $psd(x, y)$ be the $\mathcal{V}_N$-formula

$$\exists z(x + z = y \wedge ps(z))$$

saying that the difference $y - x$ is the square of a prime number.
Let $\varphi_S(x, y)$ is the $\mathcal{V}_N$-formula

$$\exists m \exists k \exists l(seq(m, k, l) \wedge ps(a_1) \wedge \forall i(i < y \rightarrow psd(a_i, a_{i+1})) \wedge x = a_y,$$

where a_n is an abbreviation for $code(m, k, l, n)$.

We are now able to show that every recursive subset of $\mathbf{N}$ is Σ_1. We first show that every recursive function on the natural numbers is definable by a Σ_1 formula.

Proposition 8.14 Let $f: \mathbb{N}^n \to \mathbb{N}$ be an n-ary function on the natural numbers. If f is recursive, then f is definable by a Σ_1 formula.

Proof By Exercise 7.8, it suffices to prove this for the 1-recursive functions on $\mathbb{N}$. These are the function generated by composition, primitive recursion, and unbounded search from the constant function $c_1(x)$ and the basic functions $s(x)$

and $p_i^n(x_1, \ldots, x_n)$ from Section 7.1. By the definitions of these functions:

$s(x) = y$ is defined by the $\mathcal{V}_N$-formula $y = x + 1$,
$c_1(x) = y$ is defined by the $\mathcal{V}_N$-formula $y = 1$, and
$p_i^n(x_1, x_2, \ldots, x_n) = y$ is defined by the $\mathcal{V}_N$-formula $y = x_i$.

Now suppose that f is a composition of functions:

$$f(x_1, \ldots, x_n) = h(g_1(x_1, \ldots, x_n), \ldots, g_m(x_1, \ldots, x_n)).$$

Suppose further that the m-ary function h is defined by a Σ_1 formula $\varphi_h(x_1, \ldots, x_m, y)$ and, for each $i = 1, \ldots m$, the n-ary functions g_i is defined by a Σ_1 formula $\varphi_{g_i}(x_1, \ldots, x_n, y)$. Then f is defined by the Σ_1 formula

$$\exists z_1 \cdots \exists z_m \left(\bigwedge_{i=1}^m \varphi_{g_i}(x_1, \ldots, x_n, z_i) \wedge \varphi_h(z_1, \ldots, z_m, y) \right).$$

Next suppose that f is obtained from functions h and g by primitive recursion:

$$f(1, x_2, \ldots, x_n) = g(x_2, \ldots, x_n),$$
$$f(x_1 + 1, x_2, \ldots, x_n) = h(x_1, \ldots, x_n, f(x_1, \ldots, x_n)).$$

We must show that if both g and h are definable by Σ_1 formulas, then so is f. For this, we must quantify over sequences of variable length.

Suppose that g and h are defined by $\varphi_g(x_2, \ldots, x_n, y)$ and $\varphi_h(x_1, \ldots, x_n, x_{n+1}, y)$. To find $\varphi_f(x_1, \ldots, x_n, y)$, consider how to compute $f(z, x_2, \ldots, x_n)$ for given z and tuple $x_2, \ldots, x_n$. We first compute

$$f(1, x_2, \ldots, x_n) = g(x_2, \ldots, x_n) = a_1.$$

We then use a_1 to compute

$$f(2, x_2, \ldots, x_n) = h(1, x_2, \ldots, x_n, a_1) = a_2.$$

We then use a_2 to compute

$$f(3, x_2, \ldots, x_n) = h(2, x_2, \ldots, x_n, a_2) = a_3,$$

and so forth. We generate the sequence $a_1, a_2, \ldots$ where each a_i equals $f(i, x_2, \ldots, x_n)$. The formula $\varphi_f(z, x_2, \ldots, x_n, y)$ says that there exists such a sequence of length z and y is the last number in the sequence. That is, $y = a_z$. More explicitly, let $\varphi_f(z, x_2, \ldots, x_n, y)$ be the formula

$$\exists l \exists m \exists k (seq(l, m, k) \wedge z = l \wedge \varphi_g(x_2, \ldots, x_n, a_1) \wedge$$
$$\forall i (i < y \rightarrow \varphi_h(i, x_2, \ldots, x_n, a_i, a_{i+1})) \wedge y = a_z),$$

where each a_i abbreviates the $\mathcal{V}_N$-term $code(l,m,k,i)$. Then $\varphi_f(\bar{x},y)$ defines the function f. Moreover, if both φ_g and φ_h are Σ_1 formulas, then, by its definition, so is φ_f. We have shown that the proposition holds for every primitive recursive function f.

To show that every recursive function is definable by a Σ_1 formula, we must consider unbounded search. Suppose that $f(x_1,\ldots,x_n)$ is defined as the least natural number y such that both $h(x_1,\ldots,x_n,y) = 0$ and $h(x_1,\ldots,x_n,z)$ is defined for all $z < y$. Then f is defined by the $\mathcal{V}_f$-formula

$$\forall z(z < y \to \exists w(\varphi_h(x_1,\ldots,x_n,z,w) \wedge \neg(w=0))) \wedge \varphi_h(x_1,\ldots,x_n,y,0).$$

Since the universal quantifier is bounded, this formula is Σ_1 provided that φ_h is. □

Corollary 8.15 For any $A \subset \mathbb{N}^k$, A is recursive if and only if it is both Σ_1 and Π_1.

Proof If A is both Σ_1 and Π_1, then it is recursive by Corollary 7.55. Conversely, suppose that A is recursive. Then the function $f(\bar{x}) = s(\chi_A(\bar{x}))$ is recursive where χ_A is the characteristic function of A. By the previous proposition, f is defined by a Σ_1 formula $\varphi_f(\bar{x},y)$. Since $f(\bar{a}) = 2$ if and only if $\bar{a} \in A$, the Σ_1 formula $\varphi_f(\bar{x},(1+1))$ defines the set A. Moreover, the compliment of A is defined by the Σ_1 formula $\varphi_f(\bar{x},1)$. So A is Π_1 as well as Σ_1. □

Corollary 8.16 If $M \models \Gamma_N$ and $\mathbf{N} \subset M$, then any recursive subset of $\mathbb{N}^n$ is a definable subset of M.

Proof This follows immediately from Proposition 8.3 and Corollary 8.15. □

8.3 Gödel's First Incompleteness theorem

We now prove that T_N is undecidable. We show that there is no algorithm that determines whether or not a given $\mathcal{V}_N$-sentence is true in T_N. Suppose that ALG is an algorithm that takes $\mathcal{V}_N$-sentences as input and outputs either "yes" or "no." Further suppose that an output of "yes" means that $T_N \vdash \varphi$. We show that the converse cannot be true. We show that there necessarily exists a $\mathcal{V}_N$-sentence β in T_N for which ALG does not give the correct affirmative output of "yes."

The key to our proof is on the first page of this book where we find the following English sentence:

> Let n be the smallest natural number that cannot be defined in fewer than 20 words.

The incompleteness theorems

To make this sentence more precise, let us only consider the words defined in the *Oxford English Dictionary*. Let A denote the sentence:

> There exists a least natural number that cannot be defined using fewer than twenty words from the Oxford English Dictionary.

The "A" is for "antinomy." Since there are only finitely many words in the *Oxford English Dictionary* (contained in a mere finitely many volumes), there are only finitely many possible sentences having at most 20 words. It follows that the negation of A cannot be true. This negation entails that every number can be defined in 20 words or less. On the other hand, A cannot be true. If A were true, then it would contradict itself since it contains fewer than 20 words.

So, in English, we can write sentences such as A so that neither A nor the negation of A is true. What prevents us from writing such sentences in first-order logic? Of course, a $\mathcal{V}_N$-sentence cannot refer to the *Oxford English Dictionary*. Let us modify the sentence A to obtain a sentence of first-order logic. Instead of "words" let us count the number of symbols occuring in the $\mathcal{V}_N$-sentence. The symbols include variables, the fixed symbols of first-order logic, and the set $\mathcal{V}_N$. The *length* of a $\mathcal{V}_N$-formula is the number of symbols in the formula. For example, $\exists x(y+y=x)$ has length 9 (counting parentheses). Consider now the sentence:

> The number n is definable by a $\mathcal{V}_N$-formula $\varphi(x)$ having length at most l.

We view this sentence as a formula, call it $A'(n,l)$, having free variables n and l. When we say that "n is definable by $\varphi(x)$" we mean that

$$\mathbf{N} \models \varphi(a) \text{ if and only if } a = n.$$

We now prove that $A'(n,l)$ cannot be expressed in first-order logic.

Proposition 8.17 There does not exist a $\mathcal{V}_N$-formula $\psi(x,y)$ such that:

$$\mathbf{N} \models \psi(n,l) \text{ if and only if}$$

n is definable by a $\mathcal{V}_N$-formula having length at most l.

Proof This follows from the elementary fact that T_N is complete and consistent (by Theorem 2.86). Given any $\mathcal{V}_N$-sentence φ, either φ or $\neg\varphi$ is in T_N. If there exists a formula $\psi(x,y)$ as described in the proposition, then we can write a $\mathcal{V}_N$-sentence Ψ_A that is neither true nor false in $\mathbf{N}$. We now describe this sentence.

If $\psi(x,y)$ is a $\mathcal{V}_N$-formula, then so is $\neg\psi(x,y) \wedge \forall z(\neg\psi(z,y) \rightarrow \exists w(x+w=z))$. Let $\theta(x,y)$ denote this formula. This formula says that x is the least number not definable by a $\mathcal{V}_N$-formula having length at most y.

Let l be the length of $\theta(x,y)$. Then $\theta(x,(1+1))$ has length $l+m\cdot 4$, where m is the number of occurences of the variable y in $\theta(x,y)$. Of course, $m \leq l$. So the length of $\theta(x,(1+1))$ is at most $5l$. In general, for any $\mathcal{V}_N$-term t, the length of $\theta(x,(t))$ is at most $(h+2)l$ where h is the length of t (the "+ 2" is for the two parentheses around t).

Let t_n be the $\mathcal{V}_N$-term $\underbrace{1+1+1+\cdots+1}_{n \text{ times}}$. This term has length $2n-1$.

Therefore, the term $(t_n)\cdot(t_n)$ representing n^2 has length $2(2n-1)+5 = 4n+3$. Let s_n (the "square" of n) denote this $\mathcal{V}_N$-term. It follows that the formula $\theta(x,(s_n))$ has length at most $(4n+5)l$. Let N be such that $N^2 > (4N+5)l$. That is, let N be any integer greater than $2l+\sqrt{5l+4l^2}$. Let Ψ_A be the sentence $\exists x \theta(x,s_N)$.

Claim N does not model the sentence $\neg\Psi_A$.

Proof If $\mathbf{N} \models \neg\Psi_A$, then $\mathbf{N} \models \forall x \neg\theta(x,s_N)$. By the definition of θ, this means that there is no least number x such that $\neg\psi(x,s_N)$ holds. By induction, $\psi(x,s_N)$ holds for every non-negative integer x. This means that every non-negative integer is definable by a formula having length at most N^2. This is impossible since, up to equivalence, there are only finitely many $\mathcal{V}_N$-formulas of length less than N^2. We must include the phrase "up to equivalence" because there are infinitely many variables we may use. However, our choice of variables does not matter. Every $\mathcal{V}_N$-formula having length at most N^2 is equivalent to a formula that uses only the following symbols:

$$x_1, x_2, \ldots, x_{N^2}, \wedge, \vee, \neg, \rightarrow, \leftrightarrow, (,), \exists, \forall, =, 0, 1, \cdot, \text{ and } +.$$

Since this is a finite list of symbols, only finitely many formulas of length N^2 comprise these symbols.

Claim N does not model the sentence Ψ_A.

Proof The sentence Ψ_A asserts the existence of a least number x not definable by a formula of length less than N^2. But $\theta(x,s_N)$ is a formula of length at most $(4N+5)l < N^2$ that defines x. So Ψ_A is contradictory.

Since T_N is consistent, each of the two claims hold. Since T_N is complete the proposition holds. □

It may seem that our discussion regarding the sentence A has not accomplished much. We have established that this sentence is an antinomy and as such cannot be expressed as a $\mathcal{V}_N$-sentence. Thus, we have provided no information beyond that contained in the first paragraph of this book. However, we are now prepared to prove the Gödel's First Incompleteness theorem.

Theorem 8.18 T_N is undecidable.

Proof Let ALG be an algorithm that halts in a finite number of steps and outputs either "yes" or "no" given any $\mathcal{V}_N$-sentence as input. Suppose that, for any $\mathcal{V}_N$-sentence φ, if ALG outputs "yes," then $T_N \vdash \varphi$. We demonstrate a $\mathcal{V}_N$-sentence β in T_N for which ALG produces the incorrect output of "no."

Let n be an non-negative integer. We say that n is ALG-definable by a $\mathcal{V}_N$-formula $\varphi(x)$ if ALG outputs "yes" when given the sentence

$$\forall x(\varphi(x) \leftrightarrow x = t_n)$$

as input (where t_n is the $\mathcal{V}_N$-term representing n). Let R be the set of all $(n, l) \in \mathbb{N}^2$ such that n is ALG-definable by a formula $\varphi(x)$ having length at most l.

Claim R is recursive.

Proof We show that the characteristic function $\chi_R(x, y)$ is computable and appeal to the Church–Turing thesis. To compute $\chi_R(n, l)$, given n and l, first list every $\mathcal{V}_N$-formula in one free variable having length at most l. If we only include formulas having variables among $x, x_1, x_2, \ldots, x_l$, then this list is finite. For each formula $\varphi(x)$ in the list, run the algorithm ALG with the sentence $\forall x(\varphi(x) \leftrightarrow x = t_n)$ as input. If the algorithm halts with output "yes" for some $\varphi(x)$ in our list, then $\chi_R(n, l) = 1$. Otherwise, if "no" is the output for each $\varphi(x)$ in the list, $\chi_R(n, l) = 0$.

We have described an algorithm that computes $\chi_R(x, y)$. We conclude that this function is recursive.

By Corollary 8.15, R is a definable subset of $\mathbf{N}$. Let $\psi_{ALG}(x, y)$ be a $\mathcal{V}_N$-formula that defines R. We now define a $\mathcal{V}_N$-sentence Ψ_{ALG} in a manner analogous to the definition of Ψ_A in the proof of Proposition 8.17. Let $\theta_{ALG}(x, y)$ be the formula

$$\neg \psi_{ALG}(x, y) \land \forall z(\neg \psi_{ALG}(z, y) \to \exists w(x + w = z))$$

saying that x is the least number not ALG-definable by a $\mathcal{V}_N$-formula having length at most y. Let N be greater than $2l + \sqrt{5l + 4l^2}$, where l is the length of $\theta_{ALG}(x, y)$. Then N^2 is greater than the length of $\theta_{ALG}(x, s_N)$, where s_N is the $\mathcal{V}_N$-term representing N^2 as $(t_N) \cdot (t_N)$.

Let Ψ_{ALG} be the sentence $\exists x \theta_{ALG}(x, s_N)$. Since only finitely many numbers are ALG-definable by some $\mathcal{V}_N$-formula having length at most N^2,

$$T_N \vdash \Psi_{ALG}.$$

So $\mathbf{N} \models \theta_{ALG}(a, s_N)$ for some $a \in \mathbb{N} \cup \{0\}$. Since it asserts that x is the least number such that $\psi_{ALG}(x, s_N)$ does not hold, the formula $\theta_{ALG}(x, s_N)$ uniquely defines a. So

$$T_N \vdash \forall x(\theta_{ALG}(x, s_N) \leftrightarrow x = t_a).$$

Let β be the sentence $\forall x(\theta_{ALG}(x, s_N) \leftrightarrow x = t_a)$. Suppose we execute ALG with input β. If the output is "yes," then a is ALG-definable by the formula $\theta_{ALG}(x, s_N)$ (by the definition of "ALG-definable"). Since the length of this formula is less than N^2 and a is not ALG-definable by such a formula, we conclude that the output must be "no."

Thus we have demonstrated a formula $\beta \in T_N$ for which ALG returns the output "no." Since ALG is an arbitrary algorithm, we conclude that T_N is undecidable. □

We make two comments regarding the above proof. First, note that this proof relies on the Church–Turing thesis. This is convenient, but by no means necessary. Since ALG is an arbitrary algorithm, it would be quite tedious to prove directly that $R(x, y)$ is recursive. In the next two sections, we prove that T_N is undecidable by an alternative proof that avoids the Church–Turing thesis. In fact, our "alternative proof" follows the original proof given by Kurt Gödel. The above proof is due to George Boolos [5].

Secondly, note that the proof of Theorem 8.18 proves more than the statement of Theorem 8.18.

Theorem 8.19 (Gödel's First Incompleteness theorem) If T is a decidable theory containing Γ_N, then T is incomplete.

Proof Repeat the proof of Theorem 8.18 using Corollary 8.16 where needed. □

The proof of Theorem 8.18 also extends to certain theories that do not contain Γ_N (for example, see Exercise 8.11).

8.4 Gödel codes

In this section, we set the stage for the proof of Gödel's Second Incompleteness theorem. To each $\mathcal{V}$-formula φ, we assign a natural number called the *Gödel code* of φ. We let $[\varphi]$ denote this natural number. Gödel codes for formulas are analogous to the codes of T^{++} programs defined in Section 7.4 and are defined in a similar manner.

We assign the odd natural numbers to symbols as follows:

(	)	¬	∧	∨	→	↔	=	∃	∀	+	·	0	1	x_1	x_2	...
1	3	5	7	9	11	13	15	17	19	21	23	25	27	29	31	...

Any finite string of symbols corresponds to a finite sequence of odd numbers. Suppose that s is the string of symbols corresponding to the numbers $a_1, \ldots, a_k$.

We assign to this string the number $[s] = 2^{a_1} \cdot 3^{a_2} \cdot 5^{a_3} \cdots p_k^{a_k}$ where p_k denotes the $(k)^{th}$ prime number. This is the *Gödel code* for s.

Given a natural number n, there are many decision problems we may consider.

- Is n the Gödel code for a symbol?
- Is n the Gödel code for a variable?
- Is n the Gödel code for a $\mathcal{V}_N$-term?
- Is n the Gödel code for a $\mathcal{V}_N$-formula?
- Is n the Gödel code for a $\mathcal{V}_N$-sentence in Skolem normal form?

We can decide the answers to these and other decision problems by factoring n. If n is odd, then it is the code for some symbol. The number n codes a string of symbols if and only if n is divisible by each of the first k prime numbers (for some k) and is divisible by no other prime numbers. Such n must be even since it is divisible by the first prime $p_1 = 2$. If n codes a sequence $a_1, \ldots, a_k$, then we can recover this sequence from the prime factorization of n. Since the factorization process is recursive, each of the above decision problems is decidable. By Corollary 8.15, the sets corresponding to these problems are definable subsets of $\mathbf{N}$.

So there exist $\mathcal{V}_N$-formulas $var(x)$, $term(x)$, and $form(x)$ that define the set of Gödel codes for variables, terms, and formulas. That is, $\mathbf{N} \models form(n)$ if and only if $n = [\varphi]$ for some $\mathcal{V}_N$-formula φ. Moreover, we claim that these formulas are Δ_0 (see Exercise 8.7). Likewise, there exist Δ_0 $\mathcal{V}_N$-formulas corresponding to rules of derivation from Section 3.1.

Example 8.20 Consider the (,)-Introduction rule from Table 3.1. This rule states that if $\Gamma \vdash \varphi$, then $\Gamma \vdash (\psi)$. We define a $\mathcal{V}$-formula $\varphi_{PI}(x,y)$ such that

$$\mathbf{N} \models \varphi_{PI}(a,b) \quad \text{if and only if } a = [\varphi] \quad \text{and} \quad b = [(\varphi)]$$

for some $\mathcal{V}_N$-formula φ. Let $\varphi_0(x,y)$ be the $\mathcal{V}_N$-formula that says there exists m such that $x = 2^{a_1} 3^{a_2} \cdots p_m^{a_m}$ and $y = 2^1 \cdot 3^{a_1} \cdots p_m^{a_{m-1}} p_{m+1}^{a_m} p_{m+2}^3$ for some $a_1, \ldots, a_m$ in $\mathbb{N}$. Then

$$\mathbf{N} \models \varphi_0(a,b) \quad \text{if and only if } a = [s] \quad \text{and} \quad b = [(s)]$$

for some string of symbols s. Let $\varphi_{PI}(x,y)$ be the formula $form(x) \wedge \varphi_0(x,y)$. By Exercise 8.7 $form(x)$ is a Δ_0 formula. It follows that $\varphi_{PI}(x,y)$ is also Δ_0.

Example 8.21 Consider the $\wedge$-Introduction rule. We demonstrate a formula $\varphi_{\wedge I}(x,y,z)$ so that

$$\mathbf{N} \models \varphi_{\wedge I}(a,b,c) \quad \text{if and only if } a = [\varphi],\ b = [\psi],\ \text{and } c = [(\varphi) \wedge (\psi)]$$

for some $\mathcal{V}_N$-formulas φ, ψ, and θ. Let $\varphi_1(a,b,c)$ be the $\mathcal{V}_N$-formula expressing that, for some m and n,

$$a = 2^{a_1} 3^{a_2} \cdots p_m^{a_m}, b = 2^{b_1} 3^{b_2} \cdots p_n^{b_n}, \quad \text{and}$$

$$c = 2^{a_1} 3^{a_2} \cdots p_m^{a_m} p_{m+1}^7 p_{m+2}^{b_1} \cdots p_{n+m+1}^{b_n}.$$

This formula says that c is the Gödel code for the string of symbols formed by putting a $\wedge$ (having code 7) between the symbols coded by a and those coded by b. When using $\wedge$-Introduction, we must be careful to include parentheses where needed. Let $\varphi_{\wedge I}(x,y,z)$ be the $\mathcal{V}_N$ formula

$$\exists u \exists v (\varphi_{PI}(x,u) \wedge \varphi_{PI}(y,v) \wedge \varphi_0(u,v,z)).$$

We claim that this formula, although it is not Δ_0, is equivalent to a Δ_0 formula. By the definition of $\varphi_{PI}(x,y)$, there is a primitive recursive function $f(x)$ such that $\varphi_{PI}(x,y)$ implies $y < f(x)$. So $\varphi_{\wedge I}(x,y,z)$ is equivalent to the Δ_0 formula

$$\exists u \exists v ((u < f(x)) \wedge (v < f(y)) \wedge \varphi_{PI}(x,u) \wedge \varphi_{PI}(y,v) \wedge \varphi_0(u,v,z)),$$

where, as usual, $x < y$ abbreviates the $\mathcal{V}_N$-formula $\exists z(x + z = y)$.

In a similar manner, many of the rules for derivations from Chapter 3 can be expressed with Δ_0 formulas. Let Generic rule be any rule from Table 3.1 other than $\vee$-Elimination and $\rightarrow$-Introduction. There is a Δ_0 formula $\varphi_{rule}(x_1, x_2, y)$ such that $\mathbf{N} \models \varphi_{rule}(a_1, a_2, b)$ if and only if both

- $a_1 = [\varphi_1], a_2 = [\varphi_2]$, and $b = [\psi]$ for some $\mathcal{V}_N$-formulas φ_1, φ_2, and ψ, and
- the Generic rule states that if $\Gamma \vdash \varphi_1$ and $\Gamma \vdash \varphi_2$, then $\Gamma \vdash \psi$.

This applies not only to rules in Table 3.1, but also to the rules of Tables 3.2 and 3.3 as well as DeMorgan's rules, the Contradiction rule, the Cut rule, Modus Ponens, $\wedge$-Distributivity, and so forth.

Since $\mathcal{V}_N$ can express rules for derivations, $\mathcal{V}_N$ can express formal proofs. There are two rules from Table 3.1 that we have not considered. Since it contains $\Gamma \cup \{\varphi\} \vdash \psi$ in its premise, $\rightarrow$-Introduction does not have the same format as our Generic rule. The formula representing $\rightarrow$-Introduction is more complicated than the formulas for the other rules. Rather than dealing with this rule, we simply ignore it. For the same reason, we ignore $\vee$-Elimination. These rules are redundant. By Exercise 1.16, the rules mentioned in the previous paragraph form a complete set of rules for first-order logic.

So we have a complete set of rules of derivation for first-order logic and each rule is represented by a Δ_0 formula. Since there are only finitely many rules in this set, there exists a $\mathcal{V}_N$-formula $Der(x,y,z)$ that says x, y, and z each code formulas and the formula coded by z can be derived from the formulas coded by x and y by one of the rules in our set. This formula is the disjunction of the formulas corresponding to each of the rules. We have

$\mathbf{N} \models Der(a,b,c)$ if and only if

- a, b, and c are the Gödel codes for $\mathcal{V}_N$-formulas φ, ψ, and θ, and
- θ follows from φ and ψ by $\wedge$-Introduction or $\vee$-Distributivity or $\forall$-Introduction... and so forth.

Since it is a disjunction of Δ_0 formulas, $Der(x,y,z)$ is also a Δ_0 formula. For the remainder of this section, we exploit this fact and show that, for certain $\mathcal{V}_N$-theories T, there exists a $\mathcal{V}_N$-formula $Pr_T(x)$ that says "x is the Gödel code of a formula that can be formally derived from T."

Definition 8.22 A set of $\mathcal{V}_N$-sentences Γ is *recursive* if the set of Gödel codes $C_\Gamma = \{[\varphi] | \varphi \in \Gamma\}$ is a recursive subset of $\mathbb{N}$. Likewise, Γ is said to be *primitive recursive* if C_Γ is.

So a theory is recursive if and only if it is decidable. The set Γ_N from Section 8.1 is clearly recursive. In fact, Γ_N is primitive recursive.

Proposition 8.23 For any recursive set of $\mathcal{V}_N$-sentences T, there exists a primitive recursive set of $\mathcal{V}_N$-sentences T_0 such that $T \vdash \varphi$ if and only if $T_0 \vdash \varphi$ for any $\mathcal{V}_N$-formula φ.

Proof The proof is known as Craig's trick.

Let $D_T = \{[\varphi] | T \vdash \varphi\}$. If T is recursive, then D_T is recursively enumerable. Let $f(x)$ be a recursive function having D_T as its range. Then the set of consequences of T can be enumerated as $\varphi_1, \varphi_2, \varphi_3, \ldots$, where, for each $n \in \mathbb{N}$, $f(n) = [\varphi_n]$.

$$ \text{Let } T_0 = \{ \underbrace{(\varphi_n) \wedge (\varphi_n) \wedge \ldots \wedge (\varphi_n)}_{n \text{ times}} | n \in \mathbb{N} \}. $$

This is Craig's trick. Clearly, $T \vdash \varphi$ if and only if $T_0 \vdash \varphi$. Moreover, we can determine whether or not a sentence φ is in T_0 in a primitive recursive manner: first count the number m of conjunctions in φ, and then compute $f(1), \ldots, f(m+1)$. The formula φ is in T_0 if and only if it is the conjunction of a formula having Gödel code $f(i)$ for some $i = 1, \ldots, m+1$. (Exercise 8.15 offers a more explicit proof.) □

We now define the formula $Pr_T(x)$ for any recursive $\mathcal{V}$-theory T.

Proposition 8.24 For any recursive set of $\mathcal{V}_N$-sentences T, there exists a $\mathcal{V}_N$-formula $Pr_T(x)$ such that $\mathbf{N} \models Pr_T([\varphi])$ if and only if $T \vdash \varphi$.

Proof By the previous proposition, it suffices to prove this for primitive recursive T. Let $\psi_T(x)$ be a Δ_0 formula such that $\mathbf{N} \models \psi_T([\varphi])$ if and only if $\varphi \in T$. We define a subset R of $\mathbb{N}^2$ inductively as follows:

$(x, 1) \in R$ if and only if $\mathbf{N} \models \psi_T(x)$,
$(x, n+1) \in R$ if and only if there exist y and z
such that $(y, n) \in R$, $(z, n) \in R$, and $\mathbf{N} \models Der(y, z, x)$.

The set R is primitive recursive by its definition. By Exercise 8.6 (which can be extracted from our proof of Proposition 8.14), there exists a Δ_0 $\mathcal{V}_N$-formula $Dr_T(x, n)$ that defines this set. So $\mathbf{N} \models Dr_T([\varphi], n)$ if and only if φ is a sentence that can be derived from T in at most n steps. Let $Pr_T(x)$ be the formula $\exists n \, Dr_T(x, n)$. □

Now suppose that T is a recursive set of $\mathcal{V}_N$ sentences that contains Γ_N as a subset. In this case, we claim that $T \vdash \varphi$ if and only if $T \vdash Pr_T(t_\varphi)$, where t_φ is a $\mathcal{V}_N$-term such that $\mathbf{N} \models t_\varphi = [\varphi]$. Essentially, this follows from the fact that Γ_N is strong enough to express the coding process described in the previous section. In particular, we use Proposition 8.3 to show that $T \vdash \varphi$ implies $T \vdash Pr_T(t_\varphi)$ in the following proposition. The converse is left as Exercise 8.17.

Notation 1 As in the proof of Proposition 8.17, we let t_n denote the $\mathcal{V}_N$-term $\underbrace{1 + 1 + 1 + \cdots + 1}_{n \text{ times}}$ for each $n \in \mathbb{N}$. For any $\mathcal{V}_N$-formula φ, we let t_φ denote the $\mathcal{V}_N$-term t_n where n is the Gödel code $[\varphi]$ of φ.

Proposition 8.25 Let T be a recursive set of $\mathcal{V}_N$-sentences such that $\Gamma_N \subset T$. For any $\mathcal{V}_N$-sentence φ, if $T \vdash \varphi$, then $T \vdash Pr_T(t_\varphi)$.

Proof Suppose that $T \vdash \varphi$. Then $\mathbf{N} \models Pr_T([\varphi])$ by Proposition 8.24. By the definition of $Pr_T(x)$, $\mathbf{N} \models Dr_T([\varphi], n)$ for some $n \in \mathbb{N}$. It follows that $\mathbf{N} \models Dr_T(t_\varphi, t_n)$. Since $Dr_T(x, y)$ is Δ_0, we have $T \vdash Dr_T(t_\varphi, t_n)$ by Proposition 8.3. Finally, $T \vdash Pr_T(t_\varphi)$ by $\exists$-Introduction. □

Although $T \vdash \varphi$ implies $T \vdash Pr_T(t_\varphi)$, it is NOT ALWAYS TRUE that

(†) $\quad T \vdash \varphi \rightarrow Pr_T(t_\varphi)$.

This is the crux of incompleteness. Just because something is true does not mean that we can prove it is so. As we shall show in the next section, there necessarily exists a sentence γ such that the sentence $\gamma \rightarrow Pr_T(t_\gamma)$ cannot be derived from T. By the definition of Pr_T in Proposition 8.24, $\gamma \rightarrow Pr_T(t_\gamma)$ is a consequence of T. So $\gamma \rightarrow Pr_T(t_\gamma)$, like the sentence β from the previous section, is a true sentence that cannot be derived from T. Thus, the failure of (†) reasserts Gödel's First Incompleteness theorem. Moreover, this failure has implications regarding the nature of provability from which we deduce Gödel's Second Incompleteness theorem.

Although it is not true in general, (†) does hold in certain cases. In particular, it holds whenever φ has the form $Pr_T(t_\psi)$ for some $\mathcal{V}_N$-sentence ψ.

Lemma 8.26 Let T be a recursive set of $\mathcal{V}_N$-sentences such that $\Gamma_N \subset T$. Let φ be any $\mathcal{V}_N$-sentence and let ρ be the sentence $Pr_T(t_\varphi)$. Then $T \vdash Pr_T(t_\varphi) \to Pr_T(t_\rho)$.

The proof of Lemma 8.26 is tedious and we omit it. We make some remarks regarding this proof at the end of this section.

Lemma 8.27 Let T be a recursive set of $\mathcal{V}_N$-sentences such that $\Gamma_N \subset T$. If $\mathbf{N} \models Der(a,b,c)$, then $T \vdash (Pr_T(t_a) \land Pr_T(t_b)) \to Pr_T(t_c)$.

Proof Since $\mathbf{N} \models Der(a,b,c)$, a, b, and c are the Gödel codes for some $\mathcal{V}$-formulas φ, ψ, and θ. Moreover, θ follows from φ and ψ by one of our rules of derivation. We see that

$$\mathbf{N} \models (Pr_T(t_a) \land Pr_T(t_b)) \to Pr_T(t_c).$$

The sentence $(Pr_T(t_a) \land Pr_T(t_b)) \to Pr_T(t_c)$ is a $\forall_2$ sentence in disguise, and so we cannot directly apply Proposition 8.3.

Instead, consider the formula

$$Dr_T(t_a, x) \land Dr_T(t_b, y) \to Dr_T(t_c, x+y+1).$$

If we can derive this formula from T, then we can obtain the desired result by $\exists$-Distribution. We must show that

$$M \models Dr_T(t_a, x) \land Dr_T(t_b, y) \to Dr_T(t_c, x+y+1).$$

for each model M of T. This is certainly true if M happens to be $\mathbf{N}$. However, M may contain "infinite" elements (in the sense of Exercise 4.10). Moreover, since M may not be elementarily equivalent to $\mathbf{N}$, these infinite elements may behave differently than the natural numbers. By compactness, $M \models Dr_T(\zeta, \eta)$ for infinite ζ and η in some model M of T. We must show that, even for these strange infinite numbers, if $M \models Dr_T(t_a, \zeta)$ and $M \models Dr_T(t_b, \eta)$, then $M \models Dr_T(t_c, \zeta + \eta + 1)$.

To see that this is the case, let us unravel the formula $Dr_T(t_a, x)$. In the proof of Proposition 8.24, this formula was defined by primitive recursion from the formulas $\psi_T(x)$ and $Der(x,y,z)$. Recall that, in $\mathbf{N}$, the formula $\psi_T(x)$ defines the set of Gödel codes for sentences in T. In M, this formula may hold for an element ζ which is infinite and therefore is not the Gödel code for any sentence.

Even so, the formula $Dr_T(t_a, x)$ asserts that there exists a sequence $a_1, \ldots, a_x$ so that $a_x = t_a$ and, for each $k = 1, \ldots, x$, either $\psi_T(a_k)$ or

$Der(a_i, a_j, a_k)$ for some i and j less than k. Since this formula is Σ_1 (in fact, it is Δ_0), this assertion is true in any model of T by Proposition 8.3.

Since $\mathbf{N} \models Der(a, b, c)$, we have $T \vdash Der(t_a, t_b, t_c)$ by Proposition 8.3. From this *and the definition of Dr_T in the previous paragraph* it follows that for any $M \models T$, if $M \models Dr_T(t_a, \zeta)$ and $M \models Dr_T(t_b, \eta)$, then $M \models Dr_T(t_c, \zeta + \eta + 1)$. □

Corollary 8.28 Let T be a recursive set of $\mathcal{V}_N$-sentences such that $\Gamma_N \subset T$. Let $a = [\varphi]$, $b = [\psi]$, and $c = [(\varphi) \wedge (\psi)]$ for some $\mathcal{V}_N$-sentences φ and ψ. Then

$$T \vdash (Pr_T(t_a) \wedge Pr_T(t_b)) \to Pr_T(t_c).$$

Corollary 8.29 Let T be a recursive set of $\mathcal{V}_N$-sentences such that $\Gamma_N \subset T$. Let $a = [\varphi]$, $b = [\psi]$, and $c = [(\varphi) \to (\psi)]$ for some $\mathcal{V}_N$-sentences φ and ψ. Then $T \vdash (Pr_T(t_a) \wedge Pr_T(t_c)) \to Pr_T(t_b)$.

The proof of Lemma 8.26 is considerably more difficult than the proof of Lemma 8.27. To prove Lemma 8.26, we must unravel the sentence $Pr_T(t_\rho)$. By its definition $\mathbf{N} \models Pr_T(t_\rho) \leftrightarrow Pr_T([Pr_T([\varphi])])$. So,

$$\mathbf{N} \models Pr_T(t_\rho) \text{ if and only if } T \vdash Pr_T([\varphi]).$$

From this observation, we see that

$$\mathbf{N} \models Pr_T(t_\varphi) \to Pr_T(t_\rho).$$

But, to prove Lemma 8.26, we must show that this is true for all models M of T. We must show that if $M \models Dr_T(t_\varphi, \eta)$ for some η, then $M \models Dr_T(t_\rho, \zeta)$ for some ζ. To see that this is the case, note that, given a formal derivation of φ from T, one can construct, in a primitive recursive manner, a formal derivation of $Pr_T(t_\varphi)$ from T. It is the proof of this intuitive fact that is quite tedious.

Typically, I have provided references where proofs have been omitted. In the case of Lemma 8.26, I know of no such reference. Even Gödel's original paper omits such intuitive but tedious details. Rather than a reference, I recommend to the reader Exercise 8.19. (That is, do it yourself reader!) Whereas a formal proof of Lemma 8.26 makes for horrible prose, it makes a good exercise for understanding the subtle concepts of the present section in preparation for the next section.

8.5 Gödel's Second Incompleteness theorem

Let T be a recursive $\mathcal{V}_N$-theory that contains Γ_N. In the previous section, we showed that there exists a $\mathcal{V}_N$-formula $Pr_T(x)$ such that $T \vdash Pr_T(t_\varphi)$ if and only if $T \vdash \varphi$. This holds for any $\mathcal{V}_N$-sentence φ. Suppose now that φ is a contradiction. For definiteness, let φ be the sentence $\neg(1 = 1)$. The Gödel code for this sentence is

$$2^5 \cdot 3^1 \cdot 5^{27} \cdot 7^{17} \cdot 11^{27} \cdot 13^3 =$$

4792483061057586040882874334500072421275090224027633666992187500000.

For our sanity, let c denote this number. In this section, we consider the sentence $\neg Pr_T(t_c)$. Gödel's Second Incompleteness theorem states that this sentence cannot be derived from T.

For any theory T, $T \vdash \neg(1 = 1)$ if and only if T is inconsistent. Since we are assuming that T is a theory, it is not the case that $T \vdash \neg(1 = 1)$. By Proposition 8.24, $\mathbf{N} \models \neg Pr_T(t_c)$. And so, $\neg Pr_T(t_c)$ is a true sentence that cannot be derived from T. In this sense, Gödel's Second Incompleteness is a special case of the First Incompleteness theorem. The Second Incompleteness theorem says more than the First Incompleteness theorem since it asserts that the sentence $\neg Pr_T(t_c)$ in particular cannot be derived from T.

To prove Gödel's Second Incompleteness theorem, we use the following Fixed Point lemma (often referred to as the *Diagonalization lemma*). The key to this lemma and its proof is a certain $\mathcal{V}_N$-formula $D(x, y)$ that we now define. For any formula φ, let δ_φ denote the formula $\exists x(x = t_\varphi \wedge \varphi)$. If φ happens to have x as its only free variable, then δ_φ is equivalent to the sentence $\varphi(t_\varphi)$ asserting that φ holds of its own Gödel number. Given the Gödel number for φ, we can compute the Gödel number for δ_φ in a primitive recursive manner. By Exercise 8.6, the relation between the Gödel number of φ and that of δ_φ is definable by a Δ_0-formula. Let $D(x, y)$ be a Δ_0 formula that says $x = [\varphi]$ and $y = [\delta_\varphi]$ for some $\mathcal{V}_N$-formula φ.

Lemma 8.30 (Fixed Point lemma) Let T be a recursive set of $\mathcal{V}_N$-sentences such that $\Gamma_N \subset T$. Let $\varphi(x)$ be a $\mathcal{V}_N$-formula having one free variable. There exists a $\mathcal{V}_N$-sentence ψ such that $T \vdash \psi \leftrightarrow \varphi(t_\psi)$.

Proof Let ψ be the sentence δ_θ where $\theta(x)$ is the formula $\exists y(D(x, y) \wedge \varphi(y))$.
That is, ψ is the formula $\exists x(x = t_\theta \wedge \theta(x))$.
We see that $\psi \equiv \theta(t_\theta) \equiv \exists y(D(t_\theta, y) \wedge \varphi(y))$.
By Completeness, we have (*) $T \vdash \psi \leftrightarrow \exists y(D(t_\theta, y) \wedge \varphi(y))$.
Since $D(x, y)$ is Δ_0 and $\mathbf{N} \models D(t_\theta, t_\psi)$, we have $T \vdash D(t_\theta, t_\psi)$.
The primitive recursive function that takes $[\varphi]$ to $[\delta_\varphi]$ is one-to-one.
It follows that, for any model M of T,
$M \models \forall y(D(t_\theta, y) \leftrightarrow y = t_\psi)$, and so $T \vdash \forall y(D(t_\theta, y) \leftrightarrow y = t_\psi)$.
Substituting this into (*) yields:
$T \vdash \psi \leftrightarrow \exists y(y = t_\psi \wedge \varphi(y))$, and so $T \vdash \psi \leftrightarrow \varphi(t_\psi)$ as we wanted to show. □

Corollary 8.31 Let T be a decidable $\mathcal{V}_N$-theory containing Γ_N. There exists a $\mathcal{V}_N$-sentence γ such that $T \vdash \gamma \leftrightarrow \neg Pr_T(t_\gamma)$.

Proof Apply the Fixed Point lemma to the formula $\neg Pr_T(x)$. □

Proposition 8.32 If T and γ are as in the previous corollary, then $T_N \vdash \neg Pr_T(t_\gamma)$ and $T \nvdash \gamma$.

Proof The sentence γ asserts that it is not provable from T. If $T \vdash \gamma$, then $T \vdash \neg Pr_T(t_\gamma)$. By Proposition 8.25, if $T \vdash \gamma$, then $T \vdash Pr_T(\gamma)$. Since T is consistent, it must be the case that $T \nvdash \gamma$. Since γ cannot be derived from T, $\neg Pr_T(t_\gamma) \in T_N$ by Proposition 8.24. □

The previous proposition provides an alternative proof for Gödel's First Incompleteness theorem. For any recursive subset T of T_N, the sentence γ that assert "I am not provable from T" must be both true and not provable from T. This is the proof Gödel originally gave for the First Incompleteness theorem. The Second Incompleteness theorem is deduced by showing that γ and $\neg Pr_T(t_c)$ are T-equivalent.

Theorem 8.33 (Gödel's Second Incompleteness theorem) If T is a decidable $\mathcal{V}_N$-theory that contains Γ_N, then $T \nvdash \neg Pr_T(t_c)$.

Proof Let γ be as in Corollary 8.31. We show that γ and $\neg Pr_T(t_c)$ are T-equivalent.

Since $\neg(1 = 1)$ is contradictory, $T \vdash \neg(1 = 1) \to \gamma$.
Let $b = [\neg(1 = 1) \to \gamma]$. By Proposition 8.25, $T \vdash Pr_T(t_b)$.
Corollary 8.29 states that $T \vdash (Pr_T(t_c) \wedge Pr_T(t_b)) \to Pr_T(t_\gamma)$.
Since $T \vdash Pr_T(t_b)$, we have $T \vdash Pr_T(t_c) \to Pr_T(t_\gamma)$.
By contraposition, $T \vdash \neg Pr_T(t_\gamma) \to \neg Pr_T(t_c)$.
This establishes $T \vdash \gamma \to \neg Pr_T(t_c)$ by the definition of γ.

We now derive the converse from T.

Let $p = [Pr_T([\gamma])]$. Then $T \vdash Pr_T(t_\gamma) \to Pr_T(t_p)$ by Proposition 8.26.
Since $\neg\gamma \equiv Pr_T(t_\gamma)$, $T \vdash Pr_T(t_{\neg\gamma}) \leftrightarrow Pr_T(t_p)$.
By the previous two lines, $T \vdash Pr_T(t_\gamma) \to Pr_T(t_{\neg\gamma})$.
Let $d = [\gamma \wedge \neg\gamma]$. By Corollary 8.28, $T \vdash (Pr_T(t_\gamma) \wedge Pr_T(t_{\neg\gamma})) \to Pr_T(t_d)$.
Since $T \vdash Pr_T(t_\gamma) \to Pr_T(t_{\neg\gamma})$, $T \vdash Pr_T(t_\gamma) \to Pr_T(t_d)$.
Since c and d are both Gödel codes for contradictions,
$T \vdash Pr_T(t_c) \leftrightarrow Pr_T(t_d)$.
By the previous two lines, $T \vdash Pr_T(t_\gamma) \to Pr_T(t_c)$.
By contraposition, $T \vdash \neg Pr_T(t_c) \to \neg Pr_T(t_\gamma)$.
By the definition of γ, we have $T \vdash \neg Pr_T(t_c) \to \gamma$.

We have successfully shown that $T \vdash \neg Pr_T(t_\gamma) \leftrightarrow \gamma$. Gödel's Second Incompleteness theorem then follows from Proposition 8.32. □

The sentence $\neg Pr_T(t_c)$ is commonly denoted $Con(T)$. This notation reflects the fact that $\mathbf{N} \models Con(T)$ if and only if T is consistent. Now, suppose that we have a recursive set of $\mathcal{V}$-sentences Γ and we want to determine whether or not $T = \Gamma \cup \Gamma_N$ is consistent. Attempting to derive $Con(T)$ from T would be an extremely naive approach. The reason is that, if T happens to be inconsistent, then any $\mathcal{V}_N$-sentence can be derived from T. So if we are successful in deriving $Con(T)$ from T, then it is possible that T is inconsistent. By the Second Incompleteness theorem, it is not only possible, but necessary that T be inconsistent:

$T \vdash Con(T)$ if and only if T is inconsistent.

Whereas $Con(T)$ defines consistency semantically in $\mathbf{N}$, it means precisely the opposite from the syntactic perspective of T.

8.6 Goodstein sequences

In this section, we describe a true statement regarding the natural numbers that can be formulated as a $\mathcal{V}$-sentence but cannot be derived from Γ_N. Gödel's First Incompleteness theorem guarantees the existence of such sentences, but does not provide an explicit example. Inexplicit examples are provided by the sentences γ from the previous section and β from the proof of Theorem 8.18. Since they are $\mathcal{V}_N$-sentences in T_N, both γ and β are statements regarding the natural numbers. However, we do not know what these sentences express. The Fixed Point lemma implies the existence of γ such that $\Gamma_N \vdash \gamma \leftrightarrow \neg Pr_{\Gamma_N}(t_\gamma)$. This is all we know about of the sentence γ. Likewise, we know that the $\mathcal{V}$-sentence β exists, but we do not know what it says about the natural numbers.

We consider sequences of non-negative integers known as *Goodstein sequences*. Given any natural number n, there is a unique Goodstein sequence that begins with n as its first term. Let us denote this sequence s_n. The best way to describe these sequences is to provide an example. Suppose $n = 14$. Let $a_1, a_2, a_3, \ldots$ denote the terms of the sequence s_{14}. Then $a_1 = 14$.

To find a_2, first express the number 14 *totally in base 2*. That is, write 14 as a sum of powers of 2: $14 = 2^3 + 2^2 + 2$. Moreover, write the exponents as sums of powers of two. Repeat this until every number occuring in the expression is a power of 2. In this case, write 14 as $2^{(2+1)} + 2^2 + 2$. To find a_2, change each 2 in this expression to a 3 and then subtract 1:

$$a_2 = (3^{(3+1)} + 3^3 + 3) - 1 = (81 + 9 + 3) - 1 = 93 - 1 = 92.$$

To find a_3, first express the number a_2 totally in base 3: $92 = 3^{(3+1)} + 3^3 + 2$. The number 2 (which is not a power of 3) represents the sum $3^0 + 3^0$. In general, when writing a number totally in base n, we allow numbers less than n as coefficients.

Now change each 3 to a 4 and subtract 1:

$$a_3 = (4^{(4+1)} + 4^4 + 2) - 1 = (1024 + 256 + 2) - 1 = 1282 - 1 = 1281,$$

and so forth. To find a_m, first express a_{m-1} totally in base m, then change each m to $m+1$, and then subtract 1 from the result. This rule generates each Goodstein sequence. If a_m equals 0 for some m, then the Goodstein sequence terminates.

Continuing with the sequence s_{14}, we have

$$a_4 = (5^{(5+1)} + 5^5 + 1) - 1 = (15625 + 3125 + 1) - 1 = 18751 - 1 = 18750,$$
$$a_5 = (6^{(6+1)} + 6^6) - 1 = (279936 + 46656) - 1 = 326592 - 1 = 326591.$$
Now $326591 = 5 \cdot 6^{(6)} + 5 \cdot 6^5 + 5 \cdot 6^4 + 5 \cdot 6^3 + 5 \cdot 6^2 + 5 \cdot 6 + 5.$
So $a_6 = (5 \cdot 7^{(7)} + 5 \cdot 7^5 + 5 \cdot 7^4 + 5 \cdot 7^3 + 5 \cdot 7^2 + 5 \cdot 7 + 5) - 1 = 4215754.$

So the sequence s_{14} begins $14, 92, 1281, 18750, 326591, 4215754, \ldots$. Clearly, the next few terms of this sequence get larger and larger. In this respect, there is nothing special about the number 14. The sequence s_n gets quite large for most choices of n. This is not true for $n = 1, 2$, or 3 but it is true for $n = 4$ (try it). For some values of n, s_n grows very rapidly. For example, let $n = 18$. Then, written in base 2, $18 = 2^5 + 2$. Written *totally* in base 2, we have

$$a_1 = 18 = 2^{(2^2+1)} + 2, \quad \text{and so}$$
$$a_2 = (3^{(3^3+1)} + 3) - 1 = 3^{28} + 2 = 22876792454963.$$

Computing a Goodstein sequence without a computer would be unpleasant. Even with a computer, this computation may not be feasible. Each step of the computation consists of two parts: we must increase the base and subtract 1. Whereas the first part may increase our number greatly, the second part decreases the result slightly.

Although they may appear cumbersome, Goodstein sequences possess the following charming property.

Theorem 8.34 (Goodstein) *Every Goodstein sequence converges to zero.*

Proof Let $s_n = (a_1, a_2, a_3, \ldots)$ be an arbitrary Goodstein sequence.
We define a sequence $b_1, b_2, b_3, \ldots$ of ordinals as follows. For each $m \in \mathbb{N}$, let b_m be the ordinal obtained by replacing each occurrence of $(m+1)$ with ω

in the total base $(m+1)$ representation of a_m. For example, if s_n is s_{14}, then:

$$a_1 = 2^{(2+1)} + 2^2 + 2 \quad \text{implies} \quad b_1 = \omega^{(\omega+1)} + \omega^\omega + \omega$$
$$a_2 = 3^{(3+1)} + 3^3 + 2 \quad \text{implies} \quad b_2 = \omega^{(\omega+1)} + \omega^\omega + 2$$
$$a_3 = 4^{(4+1)} + 4^4 + 1 \quad \text{implies} \quad b_3 = \omega^{(\omega+1)} + \omega^\omega + 1$$
$$a_4 = 5^{(5+1)} + 5^5 \quad\quad\quad \text{implies} \quad b_4 = \omega^{(\omega+1)} + \omega^\omega.$$

Note that the sequence of b_is is decreasing. Continuing, we see that

$$b_5 = 5 \cdot \omega^{(\omega)} + 5 \cdot \omega^5 + 5 \cdot \omega^4 + 5 \cdot \omega^3 + 5 \cdot \omega^2 + 5 \cdot \omega + 5,$$
$$b_6 = 5 \cdot \omega^{(\omega)} + 5 \cdot \omega^5 + 5 \cdot \omega^4 + 5 \cdot \omega^3 + 5 \cdot \omega^2 + 5 \cdot \omega + 4, \text{ and so forth.}$$

Increasing the base in the sequence of a_is has no effect on the sequence of b_is. Because we subtract 1 at each stage, b_{i+1} is necessarily smaller than b_i. This observation proves the theorem.

For any Goodstein sequence $a_1, a_2, a_3 \cdots$, the corresponding sequence of ordinals $b_1 > b_2 > b_3 \cdots$ is decreasing. By Exercise 4.22, this latter sequence must be finite. This is easily proved by induction on the ordinals. We conclude that the sequence $a_1, a_2, a_3, \ldots$ must be finite. This only happens if $a_m = 0$ for some m. □

Although each sequence eventually reaches zero, it may take a very long time for this to happen. For example, we know that the mth term of the sequence s_{18} is zero for some m. Since each step the sequence decreases by at most 1, the number m must be at least 22876792454963 (since this is the value of a_2 for this sequence).

Note that the statement of Goodstein's theorem can be formulated as a $\mathcal{V}_N$-sentence. Using the techniques of Section 8.2 we can define a $\mathcal{V}_N$-formula $\text{Good}(l, m, k)$ that holds if and only if $[l, m, k]$ codes the initial l nonzero terms of a Goodstein sequence. We can express that every Goodstein sequence is finite by saying that every such initial segment is contained in an initial segment having 1 as its final term. Clearly $a_m = 1$ if and only if $a_{m+1} = 0$ and the sequence terminates. (So the fact that 0 is not in our vocabulary is not a problem.) Let Φ_{Good} denote this sentence. In 1982, Kirby and Paris proved the following.

Theorem 8.35 $\Gamma_N \not\vdash \Phi_{Good}$.

Clearly, our proof of Goodstein's theorem, since it refers to the infinite ordinal ω, cannot be carried out in Γ_N. Kirby and Paris's theorem shows that no formal proof in Γ_N can prove this theorem. Kirby and Paris show that Goodstein's theorem is equivalent to an induction axiom that allows us to prove the consistency of Γ_N. In this way, Theorem 8.35 can be deduced from Gödel's Second Incompleteness theorem. We refer the reader to [23] for the details of this proof.

Exercises

8.1. Explain why Gödel's Incompleteness theorems do not contradict the Completeness theorem (also proved by Kurt Gödel).

8.2. Encode the following finite sequences as a triple $[l, m, k]$ using the method described in Section 8.2: (a) $(1, 1, 1)$; (b) $(3, 3, 3, 3, 3)$; (c) $(1, 2, 3)$.

8.3. What finite sequence is coded by the triple $[4, 5, 373777]$?

8.4. The Fibonacci sequence is $1, 1, 2, 3, 5, 8, \ldots$ (each number in the sequence is the sum of the previous two.) A number is called a *Fibonacci number* if it is one of the numbers in this sequence. Write a $\mathcal{V}_N$-formula $\phi(x)$ such that $\mathbf{N} \models \phi(a)$ if and only if a is a Fibonacci number.

8.5. (a) Express the formula $1 + 2 + \cdots + x = \frac{x(x+1)}{2}$ as a $\mathcal{V}_N$-formula $\varphi(x)$.

(b) Show that $\Gamma_N \vdash \forall x \varphi(x)$ where Γ_N is the set of axioms from Section 8.1.

8.6. Prove that a definable subset D of $\mathbf{N}$ is definable by a Δ_0 $\mathcal{V}_N$-formula if and only if D is primitive recursive.

8.7. Show that the following sets of natural numbers are primitive recursive by describing a Δ_0 formula that defines the set:
(a) $\mathcal{T} = \{n|n \text{ is the Gödel code for a } \mathcal{V}_N\text{-term }\}$
(b) $\mathcal{F} = \{n|n \text{ is the Gödel code for a } \mathcal{V}_N\text{-formula }\}$
(c) $\mathcal{S} = \{n|n \text{ is the Gödel code for a } \mathcal{V}_N\text{-sentence }\}$.

8.8. Let T be recursive $\mathcal{V}_N$-theory containing Γ_N. Show that the set $\{n|\mathbf{N} \models Pr_T(n)\}$ is not primitive recursive.

8.9. Show that the decision problems corresponding to each of the four sets defined in the previous two problems are in **NP**. If $\mathbf{P} \neq \mathbf{NP}$, then which these problems are in **P**?

8.10. Consider the structure $\mathbf{R} = \{\mathbb{R}|+, \cdot, 0, 1\}$. The theory of $\mathbf{R}$ is decidable. For each $n \in \mathbb{N}$ the set $\{1, 2, 3, \ldots, n\}$ is a definable subset of $\mathbf{R}$. Let $\mathbf{R}_e$ be an expansion of $\mathbf{R}$ in which the natural numbers is a definable subset. Show that the theory of $\mathbf{R}_e$ is undecidable.

8.11. Let $\mathcal{V}$ be a finite vocabulary and let T be a $\mathcal{V}$-theory. Let $D = \{t_1, t_2, t_3, \ldots\}$ be a set of $\mathcal{V}$-terms. A subset B of D^2 is *recursive* if $B = \{(t_i, t_j)|(i, j) \in I\}$ for some recursive subset I of $\mathbb{N}^2$. Suppose that
- for some $M \models T$, each recursive subset of D^2 is a definable subset of M.
- for each $m \in \mathbb{N}$ there exists a term $t_n \in D$ such that n is more than m times the length of t_n. (i.e., there exist terms $t_n \in D$ that are arbitrarily short relative to n.) Prove that T is undecidable.

The incompleteness theorems

8.12. **(Löb)** Let T be a recursive subset of T_N. Show that there exists a V_N-sentence φ so that $T \vdash \varphi \leftrightarrow Pr_T(t_\varphi)$. Show that, unlike the sentence γ that asserts its own unprovability, φ can be derived from T.

8.13. **(Tarski)** Let $V = \{[\varphi] | \mathbf{N} \models \varphi\}$. Use the Fixed Point lemma to show that V is not a definable subset of $\mathbf{N}$.

8.14. Show that for any two V_N-formulas $\varphi_1(x)$ and $\varphi_2(x)$ in one free variable there exist V_N-sentences ψ_1 and ψ_2 such that
$$T \vdash \psi_1 \leftrightarrow \varphi_1(t_{\psi_2}) \text{ and } T \vdash \psi_2 \leftrightarrow \varphi_2(t_{\psi_1}).$$

8.15. Let T be a recursive set of V-sentences.
    Let $T_0 = \{\underbrace{(\varphi) \wedge (\varphi) \wedge \ldots \wedge (\varphi)}_{n_\varphi \text{ times}} | T \vdash \varphi\}$ and let $C_0 = \{[\varphi] | \varphi \in T_0\}$.
    Recall the primitive recursive function $bin(e, x, n, 1)$ from the proof of Theorem 7.33. Show that there exists a T^{++} program P_e such that $bin(e, x, x, 1)$ is the characteristic function of C_0.

8.16. Let T be a deductively closed V_N-theory and let $C_T \subset \mathbf{N}$ be the set of Gödel codes of sentences in T. Show that the following are equivalent.
    (i) T is decidable.
    (ii) C_T is recursively enumerable.
    (ii) T is axiomatized by a primitive recursive set of sentences.

8.17. Let T be a recursive set of V_N-sentences that contains Γ_N. Show that if $T \vdash Pr_T(t_\varphi)$ then $T \vdash \varphi$.

8.18. Let T be a recursive set of V_N-sentences such that $\Gamma_N \subset T$. Let $f(\bar{x})$ be a primitive recursive function on $\mathbb{N}$ and let $\varphi_f(\bar{x}, y)$ be a V_N-formula expressing that $f(\bar{x}) = y$. Show that $T \vdash \varphi_f(\bar{x}, y) \to Pr_T(t_d)$, where d is the Gödel code for $\varphi_f(\bar{x}, y)$. (Proceed by induction on the primitive recursive function f.)

8.19. Sketch a proof for Lemma 8.26.

8.20. Let T_n be the set of Σ_n sentences in T_N. Let $S_n = \{[\varphi] | \varphi \in T_n\}$. Show that S_n is a Σ_n set that is not Π_n (see Exercise 7.29).

8.21. Let T_n be the set of Σ_n sentences in T_N. Show that, for any $n \in \mathbb{N}$, T_n is incomplete.

8.22. **(Rosser)** Let T be a recursive V_N-theory that contains Γ_N.
    Let $Y = \{[\varphi] | T \vdash \varphi\}$ and $N = \{[\varphi] | T \vdash \neg\varphi\}$. Show that Y and N are recursively inseparable (as defined in Exercise 7.24).

9 Beyond first-order logic

We consider various extensions of first-order logic. Informally, a logic $\mathcal{L}$ is an *extension* of first-order logic if every sentence of first-order logic is also a sentence of $\mathcal{L}$. We also require that $\mathcal{L}$ is closed under conjunction and negation and has other basic properties of a logic. In Section 9.4, we list the properties that formally define the notion of an *extension of first-order logic*. Prior to Section 9.4, we provide various natural examples of such extensions. In Sections 9.1–9.3, we consider, respectively, second-order logic, infinitary logics, and logics with fixed-point operators.

We do not provide a thorough treatment of any one of these logics. Indeed, we could easily devote an entire chapter to each. Rather, we define each logic and provide examples that demonstrate the expressive power of the logics. In particular, we show that none of these logics has compactness.

In the final Section 9.4, we prove that if a proper extension of first-order logic has compactness, then the Downward Löwenhiem–Skolem theorem must fail for that logic. This is Lindström's theorem. The Compactness theorem and Downward Löwenheim–Skolem theorem are two crucial results for model theory. Every property of first-order logic from Chapter 4 is a consequence of these two theorems. Lindström's theorem implies that the only extension of first-order logic possessing these properties is first-order logic itself.

9.1 Second-order logic

Second-order logic is the extension of first-order logic that allows quantification of relations. The symbols of second-order logic are the same symbols used in first-order logic. The syntax of second-order logic is defined by adding one rule to the syntax of first-order logic. The additional rule makes second-order logic far more expressive than first-order logic. Specifically, the syntax of second-order logic is defined as follows. Any atomic first-order formula is a formula of second-order logic. Moreover, we have the following four rules:

(R1) If φ is a formula then so is $\neg\varphi$.
(R2) If φ and ψ are formulas then so is $\varphi \wedge \psi$.
(R3) If φ is a formula, then so is $\exists x \varphi$ for any variable x.
(R4) If φ is a formula, then so is $\exists R^n \varphi$ for any n and n-ary relation R.

Beyond first-order logic

Definition 9.1 A string of symbols is a second-order formula if and only if it can be built up from atomic formulas using these four rules.

Recall that rules (R1), (R2), and (R3) define the syntax for first order logic. These rules regard the primitive symbols $\neg$, $\wedge$, and $\exists$. We allow the same abbreviations $\forall$, $\vee$, $\rightarrow$, and $\leftrightarrow$ from first-order logic. For any formula φ, we naturally define $\forall R^n \varphi$ to be the formula $\neg \exists R^n \varphi$. We define a second-order *sentence* in the same manner that we defined the notion of a first-order sentence.

Definition 9.2 A second-order sentence is a formula having no free variables.

This definition does not refer to relations. In second-order logic, relations like variables may have free or bound occurences within a formula. A second-order sentence may have free relations but not free variables.

Let $\mathcal{V}$ be a vocabulary. A second-order sentence φ is a $\mathcal{V}$-*sentence* if each constant and function occurring in φ is in $\mathcal{V}$ and each relation having free occurrence in φ is in $\mathcal{V}$. A second-order $\mathcal{V}$-sentence may contain relations that are not in $\mathcal{V}$ provided that they are bound by a quantifier.

Example 9.3 Let $\mathcal{V} = \{P, Q\}$ be the vocabulary consisting of unary relations P and Q. Consider the following sentence:

$$\forall x \forall y (R(x,y) \rightarrow (P(x) \wedge Q(y))) \wedge$$
$$\forall x (P(x) \rightarrow \exists y R(x,y)) \wedge$$
$$\forall x \forall y \forall z (R(x,y) \wedge R(x,z) \rightarrow y = z).$$

Call this sentence φ_0. This is a first-order sentence, but it is not a $\mathcal{V}$-sentence since the relation R is not in $\mathcal{V}$. Now let φ be the second-order sentence $\exists R^2 \varphi_0$. This is a $\mathcal{V}$-sentence since the free relations are in $\mathcal{V}$.

We now define the semantics of second-order logic. Let M be a $\mathcal{V}$-structure and let φ be a second-order $\mathcal{V}$-sentence. We must say what it means for M to be a model of φ. We use the notation $M \models \varphi$ to express that M is a model of φ. We define this concept by induction on the complexity of φ. If φ is first-order, then $M \models \varphi$ is as defined in Section 2.3. Now, suppose $M \models \varphi_0$ has been defined and let φ have the form $\exists R^n \varphi_0$. Then $M \models \varphi$ if and only if there exists an interpretation of R on the universe of M which makes φ_0 true. Put another way, $M \models \varphi$ if and only if there exists an expansion M' of M to the vocabulary $\mathcal{V} \cup \{R\}$ such that $M' \models \varphi_0$.

Example 9.4 Let $\mathcal{V} = \{P, Q\}$ and let M be a $\mathcal{V}$-structure. Let φ_0 and φ be as in Example 9.3. It makes no sense to ask whether M models φ_0 since M is an $\mathcal{V}$-structure and φ_0 contains the relation R which is not in $\mathcal{V}$. Whether a structure models φ_0 depends on how R is interpreted. The sentence φ asserts that φ_0 holds for some interpretation of R. As was pointed out in Example 9.3, the second-order sentence φ is a $\mathcal{V}$-sentence. So for every $\mathcal{V}$-structure M, either $M \models \varphi$ or $M \models \neg\varphi$. To determine which is the case, let us consider what φ_0 and φ say.

Let $P(M) = \{a \in U | M \models P(a)\}$ and let $Q(M) = \{a \in U | M \models Q(a)\}$.

The sentence φ_0 says that for each x in $P(M)$ there exists a unique y in $Q(M)$ such that $R(x,y)$ holds. Note that if this is true, then there must be at least as many elements in $Q(M)$ as in $P(M)$. The sentence φ says that φ_0 holds for some interpretation of R. This sentence is true in M if and only if $|P(M)| \leq |Q(M)|$.

Recall Examples 4.73 and 4.74 from Section 4.7. In Example 4.73 it was shown that no set of first-order sentences can express $|P(M)| = |Q(M)|$. This is a consequence of the Downward Löwenhiem–Skolem theorem. Clearly, we can modify φ in the previous example to obtain a second-order sentence that holds in a structure M if and only if $P(M)$ and $Q(M)$ have the same size. In Example 4.74, it was shown that the graph-theoretic property of connectedness cannot be expressed in first-order logic. We now show that this, too, can be expressed in second-order logic.

Example 9.5 Let $\mathcal{V}_R = \{R\}$ be the vocabulary of graphs. We write a second-order $\mathcal{V}_R$-sentence φ_{con} that holds in a graph G if and only if G is connected. This sentence asserts that there exists a linear order with certain properties. Recall that a binary relation $L(x,y)$ is a linear order if and only if the following three sentences hold:

$$\forall x \forall y (L(x,y) \vee L(y,x) \vee y = x)$$

$$\forall x \forall y (L(x,y) \rightarrow \neg L(y,x))$$

$$\forall x \forall y \forall z ((L(x,y) \wedge L(y,z)) \rightarrow L(x,z)).$$

Let $\varphi_{lo}(L)$ be the conjunction of these sentences.

To define the sentence φ_{con} we make the following observation: G is connected if and only if the vertices of G can be linearly ordered so that for each vertex v, if v is not the first vertex, then there exists a previous vertex in the order adjacent to v. We express this as follows:

$$\exists L^2 (\varphi_{lo}(L) \wedge \forall x (\exists y L(y,x) \rightarrow \exists y (L(y,x) \wedge R(y,x)))).$$

Let φ_{con} be this sentence.

Example 9.4 implies that the Downward Löwenhiem–Skolem theorem fails in second-order logic. The previous example implies the failure of compactness. The next two examples demonstrate these failures in a direct way. We show that there exist second-order sentences expressing that "the universe is infinite" and "the universe is uncountable."

Example 9.6 We demonstrate a second-order sentence φ_{inf} that holds in a structure if and only if the structure is infinite. Let φ_0 be the conjunction of the following first-order sentences:

1. $\forall x \exists y R(x, y)$.
2. $\forall x \forall y \forall z (R(x, y) \land R(x, z) \rightarrow y = z)$.
3. $\forall x \forall y \forall z (R(x, y) \land R(z, y) \rightarrow x = z)$.
4. $\exists y \forall x \neg R(x, y)$.

Let $M \models \varphi_0$. By sentences 1 and 2, we can view $R(x, y)$ as a function on the universe of M. Given any x in the universe, this function outputs the unique y for which $R(x, y)$ holds. Sentence 3 asserts that this function is one-to-one. By sentence 4, this function is not onto. This is only possible if M is infinite (any one to one function from a finite set to itself must be onto). So, if $M \models \varphi_0$, then M must be infinite. Let φ_{inf} be the sentence $\exists R^2 \varphi_0$. Infinite structures and only infinite structures model this sentence.

Example 9.7 We demonstrate a second-order sentence $\varphi_{uncount}$ that holds in a structure if and only if the structure is uncountable. Let $Q(x)$ be a unary relation. In the manner demonstrated in the previous example, we can write a second-order sentence $\varphi(Q)$ that holds if and only if $Q(x)$ defines a finite set. Let $\varphi_{lo}(L)$ be the sentence from Example 9.5 asserting that the binary relation L defines a linear order. Now let φ_{count} be the sentence:

$$\exists L^2(\varphi_{lo}(L)) \land \forall x \forall Q^1 \forall y ((Q(y) \rightarrow L(y, x)) \rightarrow \varphi(Q)).$$

Suppose $M \models \varphi_{count}$. The sentence φ_{count} says that we can linearly order the elements of the universe of M in such a way that each element has only finitely many predecessors. This is possible if and only if the universe is at most countable. So $M \models \varphi_{count}$ if and only if M is countable. Let $\varphi_{uncount}$ be $\neg \varphi_{count}$.

We see that second-order logic does not share the properties of first-order logic discussed in Chapter 4. The previous examples show that the two main results of Chapter 4, the Compactness theorem and the Downward Löwenheim–Skolem theorem, are not true in second-order logic. From the failure of compactness, we can deduce the failure of completeness (this was also shown in Section 8.1). There is no algorithmic way to determine whether or not a given second-order sentence is a consequence of a given set of second-order sentences. Likewise, there is no method for determining whether or not a structure models a certain second-order sentence, or whether or not two given structures model the same second-order sentences, and so forth. In short, second-order logic is too expressive to admit a useful model theory.

Because second-order logic is too powerful, it is natural to consider various fragments of second-order logics. *Monadic second-order logic* is the fragment that only allows second-order quantification over unary relations. So in monadic second-order logic, one can consider subsets of the universe U of a structure, but not subsets of U^n for $n > 1$. In *weak monadic second-order logic*, one can consider only finite subsets of U. We now turn our attention to other extensions of first-order logic.

9.2 Infinitary logics

The logic $\mathcal{L}_{\omega_1\omega}$ is the extension of first-order logic which allows countable conjunctions. That is, we have the following rule for forming formulas. (R2) If $\{\varphi_1, \varphi_2, \varphi_3, \ldots\}$ is a countable set of formulas, then $\bigwedge_i \varphi_i$ is also a formula.

This is in addition to the rules of first-order logic which state that any atomic formula is a formula and

(R1) If φ is a formula then so is $\neg\varphi$, and
(R3) If φ is a formula, then so is $\exists v\varphi$ for any variable v.

Note that countable disjunctions are also allowed since $\neg \bigwedge \varphi_i \equiv \bigvee \neg \varphi_i$.

Let M be a first-order structure. If the vocabulary of M is countable, then there is a single sentence of $\mathcal{L}_{\omega_1\omega}$ that describes M up to first-order elementary equivalence. Namely, the conjunction of the sentences in $Th(M)$ is a $\mathcal{L}_{\omega_1\omega}$ sentence. Moreover, $\mathcal{L}_{\omega_1\omega}$ sentences can state precisely which types are realized in M. For each k-type $p \in S(Th(M))$, there exists a $\mathcal{L}_{\omega_1\omega}$ sentence of the form $\exists x_1 \cdots \exists x_k p(x_1, \ldots, x_k)$. It follows immediately from Proposition 6.27 that the logic $\mathcal{L}_{\omega_1\omega}$ describes countable homogeneous structures up to isomorphism. As we shall show, $\mathcal{L}_{\omega_1\omega}$ describes any countable structure, whether homogeneous or not, up to isomorphism.

Definition 9.8 Structures M and N are said to be $\mathcal{L}_{\omega_1\omega}$-elementarily equivalent, denoted $M \equiv^{\mathcal{L}_{\omega_1\omega}} N$, if M and N model the same $\mathcal{L}_{\omega_1\omega}$ sentences.

We show that countable structures are $\mathcal{L}_{\omega_1\omega}$-elementarily equivalent if and only if they are isomorphic. To show this, we consider pebble games. Pebble games provide a method for determining whether or not two given structures in the same relational vocabulary are equivalent with respect to various logics including first-order logic and $\mathcal{L}_{\omega_1\omega}$. Pebble games also serve as a useful tool for the finite-variable logics of Section 10.2.

Let M and N be structures in the relational vocabulary $\mathcal{V}$. There are various pebbles games that can be played on M and N. Each pebble game is played by two players named Spoiler and Duplicator. The disjoint union of the underlying sets of M and N serves as the game board for the pebble games. Let A and B denote the underlying sets of M and N, respectively. Since we may change the names of elements (using subscripts, for example) there is no loss of generality in assuming that A and B are disjoint. In each game, Spoiler and Duplicator alternately place pebbles on elements of A and B. Spoiler's goal is to show that the two structures are somehow different. In contrast, Duplicator's objective is to show that M and N are partially isomorphic.

Definition 9.9 Let M and N be structures in the same relational vocabulary $\mathcal{V}$. Let $f: M \to N$ be a function that has as its domain a subset of U_M (the underlying set of M). This function is called a *partial isomorphism* if it preserves literals. That is, $M \models \varphi(a_1, \ldots, a_k)$ if and only if $M \models \varphi(f(a_1), \ldots, f(a_k))$ for all k-tuples from the domain of f and all atomic $\mathcal{V}$-formulas. (If the domain of f happens to be all of U_M, then f is an isomorphism.)

Each pebble game is played with a specified number of pairs of pebbles. Each pair has a distinct color. A specified number of rounds comprises each game. In each round, Spoiler first chooses a color; mauve, say. Spoiler places a mauve pebble on an element of one of the structures. Duplicator completes the round by taking the other mauve pebble and placing it on an element of the opposite structure. The color of the pebbles determines a one-to-one correspondence between those elements of A and those of B which have pebbles on them. After any round, if this one-to-one correspondence is not a partial isomorphism, then Spoiler wins the game. Duplicator's goal is defensive; to prevent Spoiler from winning.

The *Ehrenfeucht–Fraisse game* of length m played on structures M and N is denoted $EF_m(M, N)$. It is played with m pairs of pebbles and comprises m rounds. Spoiler places different colored pebbles in each round. After m rounds, all of the pebbles have been placed and the game is over.

Proposition 9.10 Let $\mathcal{V}$ be a relational vocabulary and let M and N be $\mathcal{V}$-structures. The following are equivalent:

(i) Duplicator can always prevent Spoiler from winning $EF_m(M, N)$.
(ii) For any $\mathcal{V}$-sentence φ in prenex normal form having at most m variables, $M \models \varphi$ if and only if $N \models \varphi$.

Proof idea Suppose that $M \models \exists x \varphi(x)$ and $N \models \forall x \neg \varphi(x)$. If φ is quantifier-free, then Spoiler can win $EF_1(M, N)$ by placing her pebble on an element of M such that $M \models \varphi(a)$. Since $N \models \forall x \neg \varphi(x)$ Duplicator cannot match this move.

The proposition can be proved by induction on m by extending this idea. We leave the details as Exercise 9.10. □

Corollary 9.11 Let $\mathcal{V}$ be a relational vocabulary and let M and N be $\mathcal{V}$-structures. Then $M \equiv N$ if and only if, for each $m \in \mathbb{N}$, Duplicator prevents Spoiler from winning $EF_m(M, N)$.

Proof This follows immediately from the previous proposition and the fact that every sentence of first-order logic is equivalent to a sentence in prenex normal form. □

In the definition of $EF_\omega(M, N)$, we allow the possibility that $m = \omega$, in which case play continues indefinitely. If at any point during the game the correspondence given by the color of the pebbles is not a partial isomorphism, Spoiler wins. This game provides the following characterization of $\mathcal{L}_{\omega_1 \omega}$-equivalence.

Proposition 9.12 Let $\mathcal{V}$ be a countable relational vocabulary and let M and N be $\mathcal{V}$-structures. Then $M \equiv^{\mathcal{L}_{\omega_1 \omega}} N$ if and only if Duplicator can always prevent Spoiler from winning $EF_\omega(M, N)$.

We do not prove this proposition. Intuitively, the proof of Proposition 9.12 is similar to the proof of Proposition 9.10. We use Proposition 9.12 to show that two countable structures are $\mathcal{L}_{\omega_1 \omega}$-equivalent if and only if they are isomorphic.

Proposition 9.13 Let $\mathcal{V}$ be a relational vocabulary and let M and N be countable $\mathcal{V}$-structures. Then $M \equiv^{\mathcal{L}_{\omega_1 \omega}} N$ if and only if $M \cong N$.

Proof Suppose that $M \equiv^{\mathcal{L}_{\omega_1 \omega}} N$. We prove that $M \cong N$ using a back-and-forth argument. Let U_M and U_N denote the underlying sets of M and N, respectively. Enumerate these sets as

$$U_M = \{a_1, a_2, a_3, \ldots\} \quad \text{and} \quad U_N = \{b_1, b_2, b_3, \ldots\}.$$

We construct an isomorphism $f : M \to N$ step-by-step. We use the fact that Duplicator can match Spoiler's moves to prevent her from winning $EF_\omega(M, N)$ (Proposition 9.12). In odd numbered rounds of the game (including the first round of play) Spoiler finds the least i such that a_i does not have a pebble on it, and then places a pebble on that element (so she chooses a_1 in round 1). Duplicator matches Spoiler's move by placing a pebble on some element of U_M. Likewise, in even numbered rounds, Spoiler finds the least i such that b_i does not have a pebble on it, and then places a pebble on that element. In choosing elements in this way, Spoiler guarantees that every element of U_N and U_M will eventually have a pebble. The color of the pebbles determine a function $f : M \to N$. Since Duplicator matches Spoiler, this function is a partial isomorphism. Since it is one-to-one and onto, it is an isomorphism. □

Theorem 9.14 (Scott) Let $\mathcal{V}$ be a countable vocabulary and let M be a countable $\mathcal{V}$-structure. There exists a single sentence of $\mathcal{L}_{\omega_1\omega}$ that describes M up to isomorphism.

To prove Scott's theorem, one describes a countable set of $\mathcal{L}_{\omega_1\omega}$-sentences T_{EF} that allow Duplicator to prevent Spoiler from winning $EF_\omega(M, N)$ for any model N of T_{EF}. For a full proof, refer to [16].

Example 9.15 By Scott's theorem, the first-order theory of the structure $\mathbf{N} = (\mathbb{N}|+,\cdot,1)$ is a consequence of a single sentence of $\mathcal{L}_{\omega_1\omega}$. We describe such a sentence Φ_{Scott}. Recall the axioms Γ_N from Section 8.1. Let Φ_{Scott} be the conjunction of the sentences in Γ_N together with the following sentence of $\mathcal{L}_{\omega_1\omega}$:

$$\forall x(x = 1 \vee x = (1+1) \vee x = ((1+1)+1) \vee x = (((1+1)+1)+1) \vee \cdots).$$

Since the sentences Γ_N define multiplication and addition on the natural numbers, any model of Φ_{Scott} is isomorphic to $\mathbf{N}$.

From this example and Gödel's First Incompleteness theorem, it follows that $\mathcal{L}_{\omega_1\omega}$, like second-order logic, does not have completeness. That is, there is no formal system of deduction that is both sound and complete for $\mathcal{L}_{\omega_1\omega}$ (this also follows from the failure of compactness). Unlike second-order logic, the Downward Löwenhiem–Skolem theorem, the Tarski–Vaught criterion, and preservation theorems are true for $\mathcal{L}_{\omega_1\omega}$ (see Exercise 9.7).

As the title of this section suggests, there are infinitary logics other than $\mathcal{L}_{\omega_1\omega}$. For any infinite ordinals α and β, the logic $\mathcal{L}_{\alpha\beta}$ is defined as follows. Any formula of first-order logic is a formula of $\mathcal{L}_{\alpha\beta}$. Moreover, we have the following rules:

(R1) If φ is a formula then so is $\neg\varphi$.
(R2) If $\{\varphi_i | i < \beta\}$ is a set of formulas, then $\bigwedge_{i<\beta} \varphi_i$ is a formula.
(R3) If φ is a formula and $(x_i | i < \alpha)$ is a (possibly infinite) tuple of elements, then $\exists(x_i | i < \alpha)\varphi$ is a formula.

So $\mathcal{L}_{\omega\omega}$ is another name for first-order logic.

The logic $\mathcal{L}_{\omega_1\omega}$ holds a unique place among infinitary logics since it shares some of the properties of first-order logic (such as the Downward Löwenhiem–Skolem theorem). In particular, pebble games provide a useful characterization of $\mathcal{L}_{\omega_1\omega}$-equivalence. Other infinitary logics are not so nice. Since they can quantify over infinite sets, the logics $\mathcal{L}_{\alpha\beta}$ for $\alpha > \omega$ have an expressive power comparable to second-order logic.

9.3 Fixed-point logics

We consider expansions of first-order logic that allow for inductive definitions. Inductive definitions are common in mathematics and computer science. We have used inductive definitions in this book to define *primitive recursive* functions, *formulas* of propositional and first-order logic, and other notions.

Example 9.16 Consider the notion of a connected component of a graph. We define this concept inductively as follows. Let v be a vertex of a graph G. Let $C_0(v) = \{v\}$. For each $n \in \mathbb{N}$, $C_n(v) = \{x | G \models R(x,y) \text{ for some } y \in C_{n-1}(v)\}$. If G is a finite graph, then $C_m(v) = C_{m+1}(v)$ for some m. If this is the case, then $C_m(v)$ is the connected component of v in G. In any case, the connected component of v in G is defined as $\bigcup_{n \in \mathbb{N}} C_n(v)$.

Although first-order logic can define the sets $C_m(x)$ for each m, it cannot define the notion of a connected component (see Example 4.74). In this sense, first-order logic is not *closed under inductive definitions*. Second-order logic and infinitary logics are closed in this manner (see Exercise 9.15). We now consider logics that include various *fixed-point operators*. Intuitively, these logics are minimal expansions of first-order logic that are closed under inductive definitions. There is more than one way to make the notion of "inductive definition" precise. Each corresponds to a different fixed-point operator.

Inflationary fixed-point logic (IFP)

An *operator* is similar to a function. A function from set A to set B outputs an element $f(a) \in B$ given an element $a \in A$ as input. The definition of *function* requires that A and B are *sets*. The notion of an *operator* extends this notion to classes of objects other than sets. We consider certain operators defined on the class of first-order structures.

Let $\varphi(x_1, \ldots, x_k)$ be a first-order formula in the vocabulary $\mathcal{V} \cup \{P\}$ containing the k-ary relation P. We define an operator $\mathcal{O}_{\varphi,P}$ on $(\mathcal{V} \cup \{P\})$-structures. Given a $(\mathcal{V} \cup \{P\})$-structure M as input, the operator $\mathcal{O}_{\varphi,P}$ outputs the $(\mathcal{V} \cup \{P\})$-structure $\mathcal{O}_{\varphi,P}(M)$ defined as follows. The underlying set of $\mathcal{O}_{\varphi,P}(M)$ is the same as M and $\mathcal{O}_{\varphi,P}(M)$ interprets $\mathcal{V}$ the same way as M. So as $\mathcal{V}$-structures, $\mathcal{O}_{\varphi,P}(M)$ is identical to M. The interpretation of P may not be the same.

The structure $\mathcal{O}_{\varphi,P}(M)$ interprets P as $P(M) \cup \varphi(M)$, where $P(M)$ and $\varphi(M)$ denote the subsets of M defined by $P(\bar{x})$ and $\varphi(\bar{x})$, respectively.

Now let N be a $\mathcal{V}$-structure. We may view N as a $(\mathcal{V} \cup \{P\})$-structure that interprets P as the empty set. That is, let N_0 be the expansion of N to $\mathcal{V} \cup \{P\}$ such that $N_0 \models \forall \bar{x} \neg P(\bar{x})$. Then N and N_0 are essentially the same structure. The operator $\mathcal{O}_{\varphi,P}$ generates a sequence of structures. For each $i \in \mathbb{N}$, let $N_{i+1} = \mathcal{O}_{\varphi,P}(N_i)$. Consider the sequence $N_0, N_1, N_2, N_3, \ldots$. If N is a finite structure, then $N_{m+1} = N_m$ for some $m \in \mathbb{N}$. This is because, if N_{m+1}

and N_m are not the same, then the set defined by $P(\bar{x})$ in N_{m+1} is larger than the set defined in N_m. This can happen for only finitely many m if N is finite. If N is infinite, then we continue the sequence. For each ordinal α, let N_α interpret P as $\bigcup_{\beta<\alpha} P(N_\beta)$. Eventually, $N_{\alpha+1}$ must equal N_α for some α. We refer to such a structure as the *fixed-point* for the operator $\mathcal{O}_{\varphi,P}$ on N. We let N_f denote this fixed-point structure.

Example 9.17 Let G be a graph. Then G is a structure in the vocabulary $\mathcal{V}_R = \{R\}$. Let P be a binary relation and let $\varphi(x)$ be the $(\mathcal{V}_R \cup \{P\})$-formula

$$x = y \lor \exists z (R(x,z) \land P(z,y)).$$

Let $G_0, G_1, G_2, \ldots$ be the sequence of $(\mathcal{V}_R \cup \{P\})$-structures generated by the operator $\mathcal{O}_{\varphi,P}$. Then

G_0 interprets $P(x,y)$ as the empty relation,
G_1 interprets $P(x,y)$ as the relation $x = y$,
G_2 interprets $P(x,y)$ as the relation $x = y \lor R(x,y)$,

and so forth. For each $i \in \mathbb{N}$, $G_i \models P(a,b)$ if and only if there exists a path from a to b in G of length at most $i - 1$. Let G_f denote the fixed-point of this sequence. Then $G_f \models P(a,b)$ if and only if a is in the connected component of b in the graph G.

Whereas G is bi-definable with each G_i, it may not be bi-definable with G_f. As was demonstrated in Example 4.74, first-order logic can express that there exists a path between x and y of length $i \in \mathbb{N}$, but it cannot express that there exists a path. So the fixed-point structure may contain a definable subset that is not definable in the original structure G.

We now define *inflationary fixed-point logic* (denoted *IFP*). Let N be a first-order $\mathcal{V}$-structure. In the logic *IFP*, every subset of N that is definable in some fixed-point structure N_f is definable in N. The logic *IFP* is the least extension of first-order logic with this property.

More precisely, the syntax of *IFP* is defined by the following rule together with the rules (R1), (R2), and (R3) from first-order logic:

(R_{IFP}) For any k-ary relation P, any $(\mathcal{V} \cup \{P\})$-formula $\varphi(x_1, \ldots, x_k)$ in k free variables, and any $\mathcal{V}$-terms $t_1, \ldots, t_k$, $[ifp_{\varphi,P}](t_1, \ldots, t_k)$ is a $\mathcal{V}$-formula of *IFP*.

For any $\mathcal{V}$-structure N, $N \models [ifp_{\varphi,P}](t_1, \ldots, t_k)$ means the tuple $(t_1, \ldots, t_k)$ is in the set defined by $P(x_1, \ldots, x_k)$ in the fixed-point structure N_f of the operator $\mathcal{O}_{\varphi,P}$ on N. This defines the semantics of *IFP*.

Partial fixed-point logic (PFP)

We obtain variations of *IFP* by varying the operator $\mathcal{O}_{\varphi,P}$. By definition, the structure $\mathcal{O}_{\varphi,P}(M)$ interprets P as $P(M) \cup \varphi(M)$. It follows that the set defined by P is increasing in the sequence $N_1, N_2, N_3, \dots$ defined above. That is, for each i, $P(N_i) \subset P(N_{i+1})$. The word "inflationary" refers to this fact.

Now suppose we modify the operator $\mathcal{O}_{\varphi,P}$. Let $\mathcal{O}^{pfp}_{\varphi,P}$ be such that $\mathcal{O}^{pfp}_{\varphi,P}(M)$ interprets P as $\varphi(M)$ instead of $P(M) \cup \varphi(M)$. Again consider the chain of structures $N_1, N_2, N_3, \dots$ generated by $\mathcal{O}^{pfp}_{\varphi,P}(N_i) = N_{i+1}$. Unlike the inflationary operator, it is not necessarily true that $P(N_i) \subset P(N_{i+1})$. Because of this, there is no guarantee that a fixed-point exists for this operator.

Example 9.18 Let $\mathcal{V} = \{\leq, S, 1\}$ and let M be the structure $(\mathbb{N}|\leq, S, 1)$ that interprets the binary relation S as the successor relation and interprets $\leq$ and 1 in the obvious way. Let N be any model of $Th(M)$. Let P be a unary relation and let $\varphi(x)$ be the following $(\mathcal{V} \cup \{P\})$-sentence:

$$(x = 1) \vee \exists y (P(y) \wedge S(y,x) \wedge \forall z (P(z) \rightarrow (x \leq y))).$$

This formula says that either $x = 1$ or x is the successor of the greatest element y for which $P(y)$ holds. Let N_0 the expansion of N that interprets P as the empty relation. Let $N_i = \mathcal{O}^{pfp}_{\varphi,P}(N_{i-1})$ for each $i \in \mathbb{N}$.

Then $P(N_1) = \{1\}$, $P(N_2) = \{1,2\}$, $P(N_3) = \{1,3\}$, $P(N_4) = \{1,4\}$, and so forth. We see that there is no fixed-point structure for this sequence. In contrast, if the sequence $N_1, N_2, N_3, \dots$ is instead generated by the inflationary operator, then $P(N_m) = \{1, 2, 3, \dots, m\}$ for each $m \in \mathbb{N}$. The inflationary fixed-point structure interprets $P(x)$ as "x is the nth successor of 1 for some n."

Partial fixed-point logic, denoted *PFP*, is defined the same way as *IFP* using $\mathcal{O}^{pfp}_{\varphi,P}$ in place of $\mathcal{O}_{\varphi,P}$. The logic *PFP* can express that a term is in the fixed-point N_f of this operator *provided this fixed point exists*. The syntax of *PFP* is defined by the following rule together with the rules (R1), (R2), and (R3) from first-order logic.

(R_{PFP}) For any k-ary relation P, any $(\mathcal{V} \cup \{P\})$-formula $\varphi(x_1, \dots, x_k)$ in k free variables, and any $\mathcal{V}$-terms $t_1, \dots, t_k$,
$[pfp_{\varphi,P}](t_1, \dots, t_k)$ is a $\mathcal{V}$-formula of *PFP*.

For any $\mathcal{V}$-structure N, $N \models [pfp_{\varphi,P}](t_1, \dots, t_k)$ means that the fixed-point structure N_f of the operator $\mathcal{O}_{\varphi,P}$ on N exists and the tuple $(t_1, \dots, t_k)$ is in the set defined by $P(x_1, \dots, x_k)$ in N_f.

Example 9.19 We show that *PFP*, like *IFP*, can express that there exists a path between vertices of a graph. Recall P and φ from Example 9.17. Since $\varphi(x,y)$ is the formula $x = y \lor \exists z(R(x,z) \land P(z,y))$ we see that $P(M) \subset \varphi(M)$ for any $\{P, R\}$-structure M. For this reason, the operators $\mathcal{O}_{\varphi,P}$ and $\mathcal{O}_{\varphi,P}^{pfp}$ are identical.

Least fixed-point logic (LFP)

Let φ be a formula in a vocabulary containing the relation P. The relation P is said to have a *negative* occurrence in φ if φ is equivalent to a formula in conjunctive prenex normal form in which the literal $\neg P(\bar{t})$ occurs as a subformula for some tuple of terms $\bar{t}$. We say that φ is *positive* in P if P has no negative occurences in φ.

Example 9.20 The formula $(x = 1) \lor \exists y(P(y) \land S(y,x) \land \forall z(P(z) \to (x \leq y)))$ from Example 9.18 is not positive in P. The subformula $P(z) \to (x \leq y)$ is equivalent to $\neg P(z) \lor (x \leq y)$.

Let N_1, N_2, N_3,... be the sequence generated by $\mathcal{O}_{\varphi,P}^{pfp}$. If φ is positive in P, then we have $P(N_1) \subset P(N_2) \subset P(N_3) \subset \ldots$ and the fixed-point structure exists. *Least Fixed-Point Logic*, denoted *LFP*, is the variation of *PFP* that allows $[pfp_{\varphi,P}](t_1,\ldots,t_k)$ as a formula only if φ is positive in P. This formula is interpreted the same way in *LFP* as in *PFP*. Clearly, every formula of *LFP* is also a formula of *PFP*. Whether or not the converse is true is an open question. The following theorem relates this open question to a question from complexity theory.

Theorem 9.21 (Abiteboul–Vianu) *LFP* is equivalent to *PFP* on finite structures if and only if **PSPACE** = **P**.

This theorem indicates a close relationship between fixed-point logics and complexity classes. Indeed, the development of fixed-point logics over the last two decades has been primarily motivated by complexity theory. A theorem of Immerman and Vardi states that, in some sense, *LFP* is equivalent to the class **P**. We shall state this theorem precisely in Section 10.3 where we discuss the subject of descriptive complexity.

Note the phrase "on finite structures" in the previous theorem. This means that for any sentence φ of *LFP* there exists a sentence ψ of *PFP* such that $M \models \varphi$ if and only if $M \models \psi$ for any finite structure M. Likewise, Shelah and Gurevich showed in 1986 that *LFP* and *IFP* are equivalent on finite structures. This result was improved in 2003 by Stephan Kreutzer who proved the following remarkable fact.

Theorem 9.22 (Kreutzer) *LFP* is equivalent to *IFP*.

9.4 Lindström's theorem

In this section, we define the concept of an arbitrary *extension of first-order logic*. Each of the logics defined in this chapter are examples of such extensions. The semantics of each of these logics is defined in terms of the structures defined in Chapter 2. Given any $\mathcal{V}$-structure M and any $\mathcal{V}$-sentence φ of one of these logics, either $M \models \varphi$ or $M \models \neg\varphi$. Two sentences in the same vocabulary, but not necessarily in the same logic, are said to be *equivalent* if they hold in the same structures.

A logic is a formal language. The logics we are considering are designed to describe structures and only structures (as defined in Chapter 2). Moreover, extensions of first-order logic behave like first-order logic in certain fundamental ways that we shall describe. In defining "extensions of first-order logic," we are by no means providing a definition for the notion of a "logic." There are important logics (such as fuzzy logic, intuitionistic logic, and modal logic) that do not fall under the scope of what we are about to describe.

For a logic $\mathcal{L}$ to be an extension of first-order logic, it must possess certain basic properties. For any vocabulary $\mathcal{V}$, we let $\mathcal{L}(\mathcal{V})$ denote the set of satisfiable $\mathcal{V}$-sentences of the logic $\mathcal{L}$. We assume not only that the set $\mathcal{L}(\mathcal{V})$ exists, but also that it behaves as expected. In particular, if a $\mathcal{V}$-structure models some sentence φ of $\mathcal{L}$, then $\varphi \in \mathcal{L}(\mathcal{V})$. Furthermore, to be an *extension of first-order logic*, $\mathcal{L}$ must possess each of the following properties:

- *The Extension Property:* Any satisfiable first-order $\mathcal{V}$-sentence is equivalent to a sentence in $\mathcal{L}(\mathcal{V})$. (That is, an extension of first-order logic must contain first-order logic.)
- *The Expansion Property:* Let M be a $\mathcal{V}$-structure and let M' be an expansion of M. For any sentence φ in $\mathcal{L}(\mathcal{V})$, $M \models \varphi$ if and only if $M' \models \varphi$. (In particular, if $\mathcal{V}_1 \subset \mathcal{V}_2$, then $\mathcal{L}(\mathcal{V}_1) \subset \mathcal{L}(\mathcal{V}_2)$.)
- *The Relational property:* Let M be a structure in a vocabulary $\mathcal{V} \cup \{f\}$ containing a k-ary function f. Let M_R be the structure obtained by replacing f with a $(k+1)$-ary relation R. That is, $M_R \models R(\bar{x}, y)$ if and only if $M \models f(\bar{x}) = y$. As $\mathcal{V}$-structures, M_R and M are identical.

  For any $\varphi \in \mathcal{L}(\mathcal{V} \cup \{f\})$ there exists $\psi \in \mathcal{L}(\mathcal{V} \cup \{R\})$ such that, for any structures M and M_R as described above, $M \models \varphi$ if and only if $M_R \models \psi$. (This allows us to assume that vocabularies are relational with no loss of generality.)
- *The Relativization Property:* Let M be a $\mathcal{V}$-structure. Let $D \subset M$ be a substructure of M such that the universe of D is defined by a first-order $\mathcal{V}$-formula $\varphi(x)$ in one free variable. For any $\psi \in \mathcal{L}(\mathcal{V})$ there exists $\psi_\varphi \in \mathcal{L}(\mathcal{V})$ such that $M \models \psi_\varphi$ if and only if $D \models \psi$.

- *The Small Vocabulary Property:* If a sentence φ of $\mathcal{L}$ is satisfied in a model of size κ, then it is satisfied in a structure having vocabulary of size κ.
- *The Closure Property:* if φ and ψ are sentences of $\mathcal{L}$, then so are $\varphi \wedge \psi$, $\neg \varphi$, and $\exists x \varphi$.
- *The Isomorphism Property:* if $M \cong N$, then for any sentence φ of $\mathcal{L}$, $M \models \varphi$ if and only if $N \models \varphi$.

Definition 9.23 An *extension of first-order logic* is any formal language that satisfies each of the above properties.

Each of the logics we have considered is an extension of first-order logic. For the fixed-point logics of the previous section, this is a nontrivial fact (it is far from obvious that they are closed under negation). Whereas each extension of first-order logic must possess each of the above properties, the following are the properties of first-order logic that may or may not be shared by an extension of first-order logic:

- *The Compactness property:* for any set Γ of sentences of $\mathcal{L}$, if every finite subset of Γ has a model, then Γ has a model.
- *The Downward Löwenhiem–Skolem property:* for any sentence φ of $\mathcal{L}$, if $M \models \varphi$, then there exists $N \subset M$ such that $N \models \varphi$ and $|N| = \aleph_0$.

Let $\mathcal{L}_1$ and $\mathcal{L}_2$ be two extensions of first-order logic. We say that $\mathcal{L}_2$ is at least as strong as $\mathcal{L}_1$, and write $\mathcal{L}_1 \leq \mathcal{L}_2$ if every sentence of $\mathcal{L}_1$ is equivalent to a sentence of $\mathcal{L}_2$. If both $\mathcal{L}_1 \leq \mathcal{L}_2$ and $\mathcal{L}_2 \leq \mathcal{L}_1$, then we say that $\mathcal{L}_1$ and $\mathcal{L}_2$ have the same expressive power. Lindström's theorem implies that first-order logic is the strongest logic possessing both the Downward Löwenhiem–Skolem property and the Compactness property. The following proposition is somewhat of a "warm-up" exercise for Lindström's theorem.

Proposition 9.24 Let $\mathcal{L}$ be an extension of first-order logic that has the Downward Löwenhiem–Skolem property. Suppose that $\mathcal{L}_{\omega_1 \omega} \leq \mathcal{L}$. For any structures M and N, if there exists an $\mathcal{L}$ sentence that holds in M but not N, then there exists such a sentence in $\mathcal{L}_{\omega_1 \omega}$.

Proof Suppose that $M \equiv^{\mathcal{L}_{\omega_1 \omega}} N$. Suppose for a contradiction that $M \models \varphi$, and $N \models \neg \varphi$ for some $\mathcal{L}$ sentence φ. By the Downward Löwenhiem–Skolem property, φ has a model of size $\aleph_0$. By the Small Vocabulary Property, we may assume that $\varphi \in \mathcal{L}(\mathcal{V})$ for a countable vocabulary $\mathcal{V}$. By the Expansion property (which may just as well be called the "Reduct property"), we may assume that M and N are $\mathcal{V}$-structures. By the Relational property, we may assume that $\mathcal{V}$ is relational. By the Downward Löwenhiem–Skolem property, there exist countable $M_0 \subset M$ and

$N_0 \subset N$ such that $M_0 \equiv^{\mathcal{L}_{\omega_1\omega}} N_0$, $M_0 \models \varphi$, and $N_0 \models \neg\varphi$. By Proposition 9.13, $M_0 \cong N_0$. This contradicts the Isomorphism Property. □

This proposition states that, in some sense, $\mathcal{L}_{\omega_1\omega}$ is the strongest extension of first-order logic possessing the Downward Löwenhiem–Skolem property. Toward Lindström's theorem, let us strengthen this proposition. Lindström's Theorem regards first-order logic rather than $\mathcal{L}_{\omega_1\omega}$. Prior to stating this theorem, we strengthen Proposition 9.24 in a different direction. We replace the Downward Löwenhiem–Skolem property with a weaker property.

Definition 9.25 Let κ be a cardinal. An extension of first-order logic is said to have *Löwenhiem–Skolem number κ* if for every set Γ of sentences of $\mathcal{L}$, if Γ has a model and $|\Gamma| \leq \kappa$, then Γ has a model of size at most κ.

If $\mathcal{L}$ has the Downward Löwenhiem–Skolem property, then $\mathcal{L}$ has Löwenhiem–Skolem number $\aleph_0$. The converse is not necessarily true. If a sentence φ has a countable model, then it is not necessarily true that every model of φ has a countable substructure that models φ as the Downward Löwenhiem–Skolem property asserts.

Proposition 9.26 Let $\mathcal{L}$ be an extension of first-order logic that has Löwenhiem–Skolem number $\aleph_0$. Suppose that $\mathcal{L}_{\omega_1\omega} \leq \mathcal{L}$. For any structures M and N, if there exists an $\mathcal{L}$ sentence that holds in M but not N, then there exists such a sentence in $\mathcal{L}_{\omega_1\omega}$.

Proof Suppose that $M \equiv^{\mathcal{L}_{\omega_1\omega}} N$. Suppose for a contradiction that $M \models \varphi$, and $N \models \neg\varphi$ for some $\mathcal{L}$ sentence φ. Let U be the underlying set of M and let V be the underlying set of N. By the Isomorphism property, we may assume that U and V are disjoint. By the Relational property, we may assume that the vocabulary $\mathcal{V}$ of M and N is relational. Since $\mathcal{L}$ has Löwenhiem–Skolem number $\aleph_0$ and φ is satisfiable (M is a model), φ has a countable model. By the Small Vocabulary property, we may assume $\mathcal{V}$ is countable.

We describe a structure $\mathbb{M}$ having $U \cup V$ as an underlying set. The vocabulary for $\mathbb{M}$ is $\mathcal{V}_{EF} = \mathcal{V} \cup \{P_U, P_V, R_1, R_2, \dots\}$. The unary relation P_U defines the set U and the unary relation P_V defines the set V in $\mathbb{M}$. So M is isomorphic to the substructure $P_U(\mathbb{M})$ of $\mathbb{M}$ and N is isomorphic to the substructure $P_V(\mathbb{M})$ of $\mathbb{M}$. By the Relativization property, there exists a sentence φ_U of $\mathcal{L}$ such that, for any $\mathcal{V}_{EF}$-structure $\mathbb{S}$, $\mathbb{S} \models \varphi_U$ if and only if $P_U(\mathbb{S}) \models \varphi$. Likewise, there is a sentence $\neg\varphi_V$ of $\mathcal{L}$ such that $\mathbb{S} \models \neg\varphi_V$ if and only if $P_V(\mathbb{S}) \models \neg\varphi$.

Each R_n in the vocabulary of $\mathbb{M}$ is a $2n$-ary relation. We now list countably many first-order sentences that describe the interpretation of these relations in

M. For each $n \in \mathbb{N}$

$$\mathbb{M} \models R_n(x_1, \ldots, x_n, y_1, \ldots, y_n) \to \bigwedge_{i=1}^{n} P_U(x_i) \wedge \bigwedge_{i=1}^{n} P_V(y_i).$$

For each $n \in \mathbb{N}$ and each quantifier-free first-order $\mathcal{V}$-formula θ in n free variables,

$$\mathbb{M} \models R_n(x_1, \ldots, x_n, y_1, \ldots, y_n) \to (\theta(x_1, \ldots, x_n) \leftrightarrow \theta(y_1, \ldots, y_n)).$$

That is if $R_n(a_1, \ldots, a_n, b_1, \ldots, b_n)$ holds then the function $f(a_i) = b_i$ is a partial isomorphism from $P_U(\mathbb{M})$ to $P_V(\mathbb{M})$. Moreover, for each $n \in \mathbb{N}$,

$$\mathbb{M} \models R_n(x_1, \ldots, x_n, y_1, \ldots, y_n) \to \forall x \exists y R_{n+1}(x_1, \ldots, x_n, x, y_1, \ldots, y_n, y),$$

and

$$\mathbb{M} \models R_n(x_1, \ldots, x_n, y_1, \ldots, y_n) \to \forall y \exists x R_{n+1}(x_1, \ldots, x_n, x, y_1, \ldots, y_n, y).$$

Finally, $\mathbb{M} \models \forall x \exists y R_1(x, y) \wedge \forall y \exists x R_1(x, y)$.

We have listed countably many first-order sentences that hold in $\mathbb{M}$. Let Φ_{EF} be the conjunction of these sentences. We claim that, for any $\mathcal{V}_{EF}$-structure $\mathbb{S}$, $\mathbb{S} \models \Phi_{EF}$ if and only if $P_U(\mathbb{S}) \equiv^{\mathcal{L}_{\omega_1 \omega}} P_V(\mathbb{S})$. That is, Φ_{EF} expresses that Duplicator wins $EF_\omega(P_U(\mathbb{S}), P_V(\mathbb{S}))$. The final sentence in the above list states that Duplicator can match Spoiler's first move. The previous sentences express that Duplicator can continue to match Spoiler indefinitely.

By the Closure property, $\Phi_{EF} \wedge \varphi_U \wedge \neg \varphi_V$ is equivalent to a sentence of $\mathcal{L}$. This sentence is satisfiable since $\mathbb{M}$ is a model. Since $\mathcal{L}$ has Löwenhiem–Skolem number $\aleph_0$, there exists a countable model $\mathbb{C}$ of this sentence. Since $\mathbb{C} \models \Phi_{EF}$, $P_U(\mathbb{C}) \equiv^{\mathcal{L}_{\omega_1 \omega}} P_V(\mathbb{C})$. By Proposition 9.13, $P_U(\mathbb{C}) \cong P_V(\mathbb{C})$. Since $\mathbb{C} \models \varphi_U \wedge \neg \varphi_V$, $P_U(\mathbb{C}) \models \varphi$ and $P_V(\mathbb{C}) \models \neg \varphi$. This contradicts the isomorphism property. □

To state Lindström's theorem in its most general form, we consider a weak version of the Compactness property.

Definition 9.27 Let κ be a cardinal. An extension of first-order logic is said to have *compactness number* κ if for any set Γ of sentences of the logic, if $|\Gamma| \leq \kappa$ and every finite subset of Γ has a model, then Γ has a model.

Theorem 9.28 (Lindström) If an extension $\mathcal{L}$ of first-order logic has Löwenhiem–Skolem number $\aleph_0$ and compactness number $\aleph_0$, then $\mathcal{L}$ has the same expressive power as first-order logic.

Proof Let $\mathcal{L}_{fo}$ denote first-order logic. We must show that $\mathcal{L} \leq \mathcal{L}_{fo}$. That is, each sentence φ of $\mathcal{L}$ is equivalent to some first-order sentence ψ_φ.

Claim 1 If $M \equiv N$, then $M \models \varphi$ if and only if $N \models \varphi$.

Proof Suppose for a contradiction that $M \equiv N$, $M \models \varphi$, and $N \models \neg\varphi$. Recall the sentence $\Phi_{EF} \wedge \varphi_U \wedge \neg\varphi_V$ from the proof of Proposition 9.26. The sentence Φ_{EF} is the conjunction of countably many first-order sentences. Let us now regard Φ_{EF} as a countable set (since we can no longer take infinite conjunctions). The Relativization property guarantees the existence of the $\mathcal{L}$ sentences φ_U and $\neg\varphi_V$. We claim that $\Phi_{EF} \cup \{\varphi_U, \neg\varphi_V\}$ has a countable model $\mathbb{M}$. Since $M \equiv N$, Duplicator wins $EF_k(M, N)$ for each $k \in \mathbb{N}$. It follows from the definition of Φ_{EF} that every finite subset of $\Phi_{EF} \cup \{\varphi_U, \neg\varphi_V\}$ is satisfiable. Since $\mathcal{L}$ has compactness number $\aleph_0$, this set has a model. Since $\mathcal{L}$ has Löwenheim–Skolem number $\aleph_0$, this countable set of sentences has a countable model. This leads to the same contradiction as in the proof of Proposition 9.26. This contradiction proves the claim.

It remains to be shown that Lindström's theorem follows from the claim. We must show that φ is equivalent to some first-order sentence. Let $M \models \varphi$. Since $\mathcal{L}$ has Löwenheim–Skolem number $\aleph_0$, we may assume that M is countable. Since $\mathcal{L}$ has compactness number $\aleph_0$, we can use the following compactness argument that was used repeatedly in Chapter 4.

Let T be the first-order theory of M. Let C be the set of all $\psi \in T$ such that each model of φ also models ψ (C is the set of "consequences" of φ in T).

Claim 2 Every model of Δ models φ for some finite subset Δ of C.

Proof Suppose not. Then, by compactness, $C \cup \{\neg\varphi\}$ has a model N_1. Let T_1 be the first-order theory of N_1. Consider the set $T_1 \cup \{\varphi\}$. If this set does not have a model, then some finite subset does not have a model (again by compactness). So, for some $\theta \in T_1$, $\varphi \cup \{\theta\}$ has no model and $\neg\theta \in C$. This contradicts the facts that $C \subset T_1$ and T_1 is consistent. This contradiction proves that $T_1 \cup \{\varphi\}$ does have a model N_2. But now $N_1 \equiv N_2$ (since T_1 is complete), $N_2 \models \varphi$ and $N_1 \models \neg\varphi$. This directly contradicts Claim 1 and proves Claim 2.

So Δ and φ have the same models and φ is equivalent to the conjunction of the finitely many first-order sentences in Δ. □

Exercises

9.1. Let G be a graph. A *Hamilton circuit* in G is a path that begins and ends at the same vertex and includes each of the other vertices once and only once.
  (a) Write a second-order sentence that holds in G if and only if G has a Hamilton circuit.
  (b) Write a $\mathcal{L}_{\omega_1\omega}$ sentence that holds in G if and only if G has a Hamilton circuit.

9.2. Write a second-order sentence that holds in a structure M if and only if M is finite and the universe of M contains an odd number of elements.

9.3. Let K_n denote the n-clique. Show that for any $k \in \mathbb{N}$, Duplicator can prevent Spoiler from winning $EF_k(K_n, K_m)$ for sufficiently large n and m.

9.4. Let $\mathcal{V}_< = \{<\}$. Let L_1 and L_2 be finite $\mathcal{V}_<$-structures that interpret the binary relation $<$ as a linear order. Show that if both $|L_1|$ and $|L_2|$ are larger than 2^k, then Duplicator can prevent Spoiler from winning $EF_k(L_1, L_2)$.

9.5. Let G be a finite graph. The *degree* of a vertex v in G is the number of vertices adjacent to v. An *Euler path* is a path that includes each edge once and only once. Leonhard Euler proved that a finite graph has an Euler path if and only if there are at most two vertices of odd degree.
  (a) Show that there is no first-order sentence Φ_{Euler} such that $G \models \Phi_{Euler}$ if and only if G has an Euler path. (Use Exercise 9.3.)
  (b) Write a second-order sentence that holds in G if and only if G has a Euler circuit. (Use the formula from Exercise 9.2.)
  (c) Write a $\mathcal{L}_{\omega_1\omega}$ sentence that holds in G if and only if G has a Euler circuit.

9.6. A graph is said to be *k-colorable* if the vertices of the graph can be colored with k colors in such a way that no two vertices of the same color share an edge. Write a second-order existential sentence that holds in a graph G if and only if G is k-colorable.

9.7. Each of the following are proved in Chapter 4 for first-order logic. Show that each remains true when first-order logic is replaced with $\mathcal{L}_{\omega_1\omega}$.
  (a) Proposition 4.31 (the Tarski–Vaught criterion)
  (b) Theorem 4.33 (the Downward Löwenhiem–Skolem theorem)
  (c) Theorem 4.47 (the Los–Tarski theorem)

9.8. Let $\mathcal{V}_< = \{<\}$. Describe a $\mathcal{V}_<$-sentence in the logic $\mathcal{L}_{\omega_1\omega_1}$ that says $<$ is a well-ordering of the underlying set.

9.9. Suppose that Duplicator prevents Spoiler from winning $EF_\omega(M, N)$. Show that $M \models \varphi$ if and only if $N \models \varphi$ for each $\mathcal{V}$-formula φ of $\mathcal{L}_{\omega_1\omega}$. (Use induction on the complexity of φ.)

9.10. Prove Proposition 9.10.

9.11. Let $\mathcal{S}$ be the set of all finite strings of symbols from the set

$$\{A, B, C, \ldots, X, Y, Z,), (, \wedge, \neg\}.$$

We describe a structure M having $\mathcal{S}$ as an underlying set. The vocabulary of M contains:
- a unary relation $a(s)$ interpreted as $s = A$ or $s = B,\ldots$, or $s = Z$,
- a unary function $n(x)$ interpreted as $n(s) = \neg s$,
- a unary function $p(x)$ interpreted as $p(s) = (s)$, and
- a binary function $c(x, y)$ interpreted as $p(s, t) = s \wedge t$,

where s and t denote arbitrary elements of $\mathcal{S}$. Let $form(s)$ be a formula that says the string s is a formula of propositional logic. Show that $form(s)$ is not a first-order formula. Show that $form(s)$ can be expressed in any of the fixed-point logics from Section 9.3.

9.12. Let $\mathcal{L}_Q$ be the extension of first-order logic that contains a new quantifier Q. The syntax of $\mathcal{L}_Q$ is defined by adding the following rule to the rules that define the syntax of first-order logic:

(RQ) If φ is a formula of $\mathcal{L}_Q$, then so is $Qx\varphi$.

The formula $Qx\varphi(x)$ is to be interpreted as "there exist uncountably many x such that $\varphi(x)$ holds." This describes the semantics of $\mathcal{L}_Q$. Show that $\mathcal{L}_Q$ is a logic that possesses compactness but not the Downward Löwenhiem–Skolem Property.

9.13. Show that every formula of second-order logic can be put into prenex normal form. That is, show that each formula is equivalent to a formula of the form $Q_1 R_1^{i_1} Q_2 R_2^{i_2} \ldots Q_k R_k^{i_k} \varphi$ where each Q_j is either $\forall$ or $\exists$ and φ is a first-order formula.

9.14. Let $MonSO$ denote monadic second-order logic as defined at the end of Section 10.1. Using the fact that $MonSO$ has Löwenhiem–Skolem number $\aleph_0$ and is compact, show that
   (a) there is no sentence of $MonSO$ that holds in connected graphs and only connected graphs;
   (b) there does exist a sentence of $MonSO$ that holds in nonconnected graphs and only nonconnected graphs;
   (c) Why does $MonSO$ not contradict Lindström's theorem?

9.15. Show that $LFP \leq \mathcal{L}_{\omega_1\omega}$.

9.16. Show that $\mathcal{L}_{\omega_1\omega} \leq SOL$.

9.17. Show that $LFP \leq IFP$.

9.18. Let T be a first-order $\aleph_0$-categorical $\mathcal{V}$-theory. Show that every $\mathcal{V}$-formula of IFP is T-equivalent to a first-order $\mathcal{V}$-formula.

9.19. Let $\mathcal{L}$ be an extension of first-order logic. The *Hanf number* of $\mathcal{L}$ is the least cardinal κ such that every sentence of $\mathcal{L}$ that has a model of size κ has arbitrarily large models.
   (a) Show that $\mathcal{L}$ is equivalent to first-order logic if and only if $\mathcal{L}$ has Löwenhiem–Skolem number $\aleph_0$ and Hanf number $\aleph_0$.
   (b) Show that every extension of first-order logic has a Hanf number.

9.20. **(Lindström)** Let $\mathcal{L}$ be an extension of first-order logic. Show that $\mathcal{L}$ is equivalent to first-order logic if and only if $\mathcal{L}$ has Löwenhiem–Skolem number $\aleph_0$ and the set of sentences of $\mathcal{L}$ that hold in every model is a recursively enumerable set.

10 Finite model theory

This final chapter unites ideas from both model theory and complexity theory. Finite model theory is the part of model theory that disregards infinite structures. Examples of finite structures naturally arise in computer science in the form of databases, models of computations, and graphs. Instead of satisfiability and validity, finite model theory considers the following finite versions of these properties.

- A first-order sentence is *finitely satisfiable* if it has a finite model.
- A first-order sentence is *finitely valid* if every finite structure is a model.

Finite model theory developed separately from the "classical" model theory of previous chapters. Distinct methods and logics are used to analyze finite structures. In Section 10.1, we consider various finite-variable logics that serve as useful languages for finite model theory. We define variations of the pebble games introduced in Section 9.2 to analyze the expressive power of these logics.

Pebble games are one of the few tools from classical model theory that is useful for investigating finite structures. In Section 10.2, it is shown that many of the theorems from Chapter 4 are no longer true when restricted to finite models. There is no analog for the Completeness and Compactness theorems in finite model theory. Moreover, we prove Trakhtenbrot's theorem which states that the set of finitely valid first-order sentences is not recursively enumerable.

Descriptive complexity is the subject of 10.3. This subject describes the complexity classes discussed in Chapter 7 in terms of the logics introduced in Chapter 9. We prove Fagin's theorem relating the class **NP** to existential second-order logic. We prove the Cook–Levin theorem as a consequence of Fagin's Theorem. This theorem states that the Satisfiability Problem for Propositional Logic is **NP**-complete. We conclude this chapter (and this book) with a section describing the close connection between logic and the $\mathbf{P} = \mathbf{NP}$ problem.

10.1 Finite-variable logics

In this section, we discuss appropriate logics for the study of finite models. First-order logic, since it describes each finite model up to isomorphism, is too strong. For this reason, we must weaken the logic. It may seem counter-intuitive that we should gain knowledge by weakening our language. Recall that, for infinite structures, first-order logic is quite weak (compared to logics from the previous chapter). This weakness is demonstrated in the Compactness theorem, the Löwenhiem–Skolem theorems, and the other theorems of Chapter 4. It is precisely these properties that make first-order logic a productive logic for the study of infinite structures. The weakness of first-order logic gives rise to model theory. With this in mind, we consider the following logics.

Definition 10.1 For $k \in \mathbb{N}$, let $\mathcal{L}^k$ be the fragment of first-order logic obtained by restricting to the k variables $x_1, x_2, \ldots, x_k$ (or any other set of k variables).

There are two reasons that first-order logic is not appropriate for finite model theory. One reason is that it is too strong. The other reason is that it is too weak. This is the *Fundamental Joke of Finite Model Theory*. It is too strong for the reasons we have mentioned. First-order logic is too weak because it is incapable of defining basic properties of finite structures. For example, there is no first-order sentence expressing that a finite structure is a connected graph. Also, there is no first-order sentence expressing that a finite structure has an even number of elements. For this reason, we consider various strengthenings of $\mathcal{L}^k$.

Definition 10.2 Let $\mathcal{C}^k$ be k-variable logic with counting quantifiers. That is, $\mathcal{C}^k$ contains all $\mathcal{L}^k$ formulas and is closed under negation, conjunction, and quantification by the counting quantifiers $\exists^{\leq n}$ for each $n \in \mathbb{N}$.

Since $\exists^{\leq n}$ is first-order expressible, we regard $\mathcal{C}^k$ as a fragment of first-order logic. So although $\mathcal{C}^k$ is stronger than $\mathcal{L}^k$, it cannot express properties such as connectedness that cannot be expressed in first-order logic.

Definition 10.3 For $k \in \mathbb{N}$, let $\mathcal{L}^k_{\omega_1\omega}$ be the fragment of $\mathcal{L}_{\omega_1\omega}$ obtained by restricting to the k variables $x_1, x_2, \ldots, x_k$.

Finite model theory also considers finite-variable logics with a fixed-point operator and other extensions of $\mathcal{L}^k$. However, we restrict our attention to the above logics. We demonstrate what can and cannot be expressed in these logics by providing some examples and stating without proof some basic facts.

Consider the logic $\mathcal{L}^3$. Since this logic only allows three variables, it is a restrictive language. However, by using the variables in an economical way, we can express more than may be apparent. We can repeatedly use (and re-use) each of the three variables any number of times. For example, let $\mathcal{V}_S$ be the vocabulary consisting of a single binary relation S. Let $Z_S = (\mathbb{Z}|S)$ be the $\mathcal{V}_S$-structure that interprets S as the successor relation on the integers. That is, $Z_S \models S(a,b)$ if and only if $b = a + 1$. The following $\mathcal{V}_S$-formula says that y is the $(n+1)^{th}$

successor of x in Z_S:

$$\exists z_1 \exists z_2 \cdots \exists z_n \left(S(x, z_1) \wedge \bigwedge_{i=1}^{n-1} S(z_i, z_{i+1}) \wedge S(z_n, y) \right).$$

If n is large, then this is not a formula of $\mathcal{L}^3$. However, we claim that it is equivalent to a $\mathcal{L}^3$ formula. We inductively define formulas $\psi_i(x, y)$, for $i \in \mathbb{N}$, that say y is the ith successor of x. We use the following convenient notation that allows us to keep track of bound variables as well as free variables.

Notation 1 We let $\varphi(x_1, \ldots, x_n \,|\, y_1, \ldots, y_m)$ denote an arbitrary formula having free variables $x_1, \ldots, x_n$ and bound variables $y_1, \ldots, y_m$. If a variable has both free and bound occurences within φ, then this variable is listed only as a free variable among the x_is.

We now define the formulas $\psi_i(x, y)$ for $i \in \mathbb{N}$.

Let $\psi_1(x, y)$ be the formula $S(x, y)$.
Let $\psi_2(x, y|z)$ be the formula $\exists z(S(x, z) \wedge S(z, y))$.
For $i \geq 2$, let $\psi_{i+1}(x, y)$ be $\exists z(S(x, z) \wedge \psi_i(z, y|x))$.

Note that x occurs in $\psi_i(x, y|z)$ as both a free variable and a bound variable for $i > 2$.

Notation 2 For any formula φ, let $V(\varphi)$ denote the number of variables occurring in φ.

Notation 3 Let $qd(\varphi)$ denote the *quantifier depth of* φ defined inductively as follows: $qd(\varphi) = 0$ for atomic φ; $qd(\neg \varphi) = qd(\varphi)$; $qd(\varphi \wedge \psi) = qd(\varphi \vee \psi) = max\{qd(\varphi), qd(\psi)\}$; and $qd(\exists x \varphi(x)) = qd(\forall x \varphi(x)) = qd(\varphi(x)) + 1$.

Example 10.4 For the formulas $\psi_i(x, y|z)$ as defined above, $V(\psi_i) = 3$ and $qd(\varphi_i) = i - 1$.

Notation 4 Let M and N be structures in the same vocabulary $\mathcal{V}$.

For $k \in \mathbb{N}$, $M \equiv^k N$ means that $M \models \varphi$ if and only if $N \models \varphi$ for any $\mathcal{V}$-sentence φ with $V(\varphi) \leq k$. That is $M \equiv^k N$ means that M and N cannot be distinguished by sentences of the logic $\mathcal{L}^k$.

For k and m in $\mathbb{N}$, $M \equiv^k_m N$ means $M \models \varphi$ if and only if $N \models \varphi$ for any $\mathcal{V}$-sentence φ with $V(\varphi) \leq k$ and $qd(\varphi) \leq m$.

We describe pebble games that determine whether or not two structures in the same relational vocabulary are equivalent with respect to various finite–variable logics. Recall the *Ehrenfeucht–Fraisse game* of length m from Section 9.2. This game is played with m pairs of pebbles and comprises m rounds.

Spoiler places different colored pebbles in each round. After m rounds, all of the pebbles have been placed and the game is over.

Proposition 10.5 Let $\mathcal{V}$ be relational and let M and N be $\mathcal{V}$-structures. Duplicator prevents Spoiler from winning $EF_m(M, N)$ if and only if $M \equiv_m^m N$.

This is a restatement of Proposition 9.10.

We now define the k-pebble game of length $m > k$. Let M and N be structures in the relational vocabulary $\mathcal{V}$. The k-pebble game of length m is denoted $P_m^k(M, N)$. It is played just like $EF_k(M, N)$ for the first k rounds. So after the first k rounds, there are k pebbles on each structure which are in one-to-one correspondence by color. But whereas $EF_k(M, N)$ was finished at this point, $P_m^k(M, N)$ is not. Spoiler continues the game by choosing one of the pebbles that has been placed and moving it to another element within the same structure. Duplicator then takes the pebble of the same color in the other structure and moves it. Play continues in this manner for $m - k$ moves. We allow the possibility that $m = \omega$, in which case play continues indefinitely. We omit the subscript and write $P^k(M, N)$ to denote this version of the game. If at any point during the game the correspondence given by the color of the pebbles is not a partial isomorphism, Spoiler wins.

Proposition 10.6 Let M and N be structures in the same relational vocabulary. Duplicator prevents Spoiler from winning $P_m^k(M, N)$ if and only if $M \equiv_m^k N$.

Proof This can be proved by extending the proof of Proposition 9.10. □

Proposition 10.7 Let M and N be finite structures in the same relational vocabulary.
The following are equivalent:

(i) M and N satisfy the same sentences of $\mathcal{L}_{\omega_1\omega}^k$.

(ii) $M \equiv^k N$.

(iii) Duplicator prevents Spoiler from winning $P_m^k(M, N)$ for each m.

(iv) Duplicator prevents Spoiler from winning $P^k(M, N)$.

Sketch of proof It is clear that (i) implies (ii). By the previous proposition, (ii) implies (iii). That (ii) implies (iv) follows from the assumption that M and N are finite. Finally, (iv) implies (i) can be proved in the same manner as the previous proposition (using the idea of the proof of Proposition 9.10). □

Example 10.8 Let M be a connected graph (not necessarily finite). Suppose that $N \equiv^3 M$. We claim that N is also a connected graph. That N is a graph follows from the fact that the axioms for the theory of graphs each use at most three variables. By compactness, there is no sentence that expresses connectivity. However, we claim that there does exist a set of sentences of $\mathcal{L}^3$ that expresses this. To see this, play the game $P^3(M,N)$. If N is not connected, then Spoiler can place pebbles on two vertices a and b of N such that there is no path from a to b in N. Duplicator must place his corresponding two pebbles on vertices of M. No matter which vertices c and d Duplicator chooses, there exists a path from c to d in M. Spoiler places her third pebble adjacent to c in M along the path toward d. Spoiler then removes the pebble from c and places it adjacent to the previous pebble. In this manner Spoiler "walks" the two pebbles toward the pebble on d. Eventually, these pebble will reach d at which point Spoiler wins the game. So if N is not connected, then Spoiler has a strategy for winning $P^3(M,N)$.

We can modify the k-pebble game of length $m > k$ to obtain a game $C_m^k(M,N)$ which captures the notion of $\mathcal{C}^k$-equivalence. This game is again played by two players with k pairs of colored pebbles. But each round of $C_m^k(M,N)$ consists of two steps. First, Spoiler selects a color, hazel say, and a finite subset X_1 of either M or N. Duplicator then selects a subset X_2 of the opposite structure so that $|X_2| = |X_1|$. Spoiler then places a hazel pebble onto an element of X_i for $i = 1$ or 2. Duplicator then places the other hazel pebble in X_{3-i}. This is repeated k times after which all of the pebbles have been placed. As with $P_m^k(M,N)$, $C_m^k(M,N)$ continues for $m - k$ rounds beyond the initial placing of the pebbles. Again, we allow for the possibility that $m = \omega$ in which case the game continues indefinitely. If at any time during the game, the color correspondence of pebbles does not define a partial isomorphism between M and N, Spoiler wins.

Proposition 10.9 Duplicator can always prevent Spoiler from winning $C_m^k(M,N)$ if and only if M and N model the same sentences of $\mathcal{C}^k$ of quantifier depth at most m.

The proof of this proposition is similar to the proof of Proposition 9.10 in Section 9.2 (as are the proofs of Propositions 10.6 and 10.7). Detailed proofs of these propositions can be found in [11].

10.2 Classical failures

Many of the theorems of classical model theory fail when restricted to finite structures. We show that the Completeness and Compactness theorems become false statements when "satisfiable" is replaced by "finitely satisfiable." Moreover, the set of finitely satisfiable sentences is not recursive. This is Trakhtenbrot's theorem. We leave it as an exercise to show that certain preservation theorems and Beth's Definability theorem no longer hold in the finite setting (Exercises 10.6–10.8). The Downward Löwenheim–Skolem theorem, as stated in Theorem 4.33, remains true but becomes trivial. The Upward Löwenhiem–Skolem theorem cannot be formulated for finite structures. In short, many of the essential methods and tools of classical model theory cannot be applied in finite model theory.

A key result of Chapter 4 states that every model has a theory and every theory has a model (Theorem 4.27). In finite model theory, this is only half true. It is true that every finite model has a theory. In fact, every finite model has a finitely axiomatizable theory (by Proposition 2.81). It is not true that every theory has a finite model. Most theories discussed in Chapters 5 and 6 have no finite model. We show that the consequences of Theorem 4.27 also do not hold in the finite.

Proposition 10.10 (Failure of Compactness) There exists a set of first-order sentences Γ such that every finite subset of Γ is finitely satisfiable, but Γ is not.

Proof Let $\Gamma = \{\exists^{\geq n} x(x = x) | n \in \mathbb{N}\}$. □

In Chapter 4, we deduced the compactness of first-order logic from completeness. Following this same argument, we deduce the failure of completeness from the failure of compactness. The argument from Chapter 4 goes as follows. By the finite nature of proofs:

$$\Gamma \vdash \varphi \text{ implies } \Delta \vdash \varphi$$

for some finite subset Δ of Γ. By the definition of "$\models$,"

$$\Delta \models \varphi \text{ implies } \Gamma \models \varphi.$$

Completeness provides the vertical arrows in the diagram:

$$\begin{array}{ccc} \Gamma \models \varphi & \Rightarrow & \Delta \models \varphi \\ \updownarrow & & \updownarrow \\ \Gamma \vdash \varphi & \Leftarrow & \Delta \vdash \varphi. \end{array}$$

So $\Gamma \models \varphi$ if and only if $\Delta \models \varphi$ for some finite subset Δ of Γ. If we take φ to be a contradiction, then this is precisely compactness.

Proposition 10.11 (Failure of Completeness) There does not exist a formal proof system for first-order logic that is both sound and complete with respect to finite satisfiability.

Proof Replace $\models$ with $\models_{fin}$ in the above argument and use $\Gamma = \{\exists^{\geq n} x(x = x) | n \in \mathbb{N}\}$ to derive a contradiction. □

These are failures not only of first-order logic, but also the finite-variable logics of the previous section. Since the set Γ used in the proofs of Propositions 10.10 and 10.11 is a set of sentences of $\mathcal{C}^1$, these propositions hold for the logics $\mathcal{C}^k$. To see that the same is true for the logics $\mathcal{L}^k$, replace Γ with the following set of $\mathcal{L}^1$ sentences in a vocabulary containing unary function f:

$$\{\forall x \neg (f(x) = x), \forall x \neg (f(f(x)) = x), \forall x \neg (f(f(f(x))) = x), \ldots\}.$$

We next prove a result that is stronger than Proposition 10.11. Consider the following decision problems:

- *The Finite Satisfiability Problem:* Given a first-order sentence, determine whether or not it is finitely satisfiable.
- *The Finite Validity Problem:* Given a first-order sentence, determine whether or not it is finitely valid.

We show that the Finite Satisfiability Problem is not decidable. Equivalently, we show that the set of finitely satisfiable sentences is not recursive. As a corollary to this, we show that the Finite Validity Problem is not semi-decidable. In contrast, the validity problem for first-order logic is semi-decidable. By the Completeness theorem, we can derive every valid sentence from the basic rules for deduction in Chapter 3. In the finite setting, not only does the Completeness theorem fail, but there is no way to recursively enumerate all of the finitely valid sentences. This is Trakhtenbrot's theorem.

Theorem 10.12 (Trakhtenbrot) The set of finitely satisfiable sentences is not recursive.

Proof Given T^{++} program P_e, we write a first-order sentence φ_e such that φ_e is finitely satisfiable if and only if P_e halts on input 0. This reduces the Finite Satisfiability Problem to the Halting Problem. The Halting Problem is undecidable by Proposition 7.48. If we successfully define φ_e, then we can conclude that the Finite Satisfiability Problem, too, is undecidable.

So it suffices to define the sentence φ_e. This sentence has the form

$$\varphi_< \land \varphi_s \land \varphi_p \land \varphi_c \land \theta_{init} \land \theta_{halt} \land \psi_1 \land \psi_2 \land \cdots \psi_L,$$

where L is the number of lines in P_e. To describe the sentence φ_e we describe each of the sentences in this conjunction one-by-one.

The sentence $\varphi_<$ is a sentence in the vocabulary $\{<\}$ saying that $<$ is a discrete linear order of the underlying set and there exists a smallest element and a largest element in this order. By a *discrete* linear order, we mean that for each element x other than the maximal element, there exists a least element greater than x. That is, every element other than the maximal element has a successor. The sentence φ_s says that the unary function s is the successor function. To make this a total function, we say that, if x is maximal, then $s(x) = x$. Likewise, φ_p says that the unary function p is the predecessor function and define $p(x) = x$ for the smallest element. Since both s and p are definable in terms of $<$, the inclusion of these functions is not necessary but convenient.

So $\varphi_<$, φ_s, and φ_p are sentences in the vocabulary $\{<, s, p\}$. We expand this vocabulary by adding constants $C = \{0, 1, 2, 3, \ldots, L, L+1\}$ where L is the number of lines in P_e. The sentence φ_c describes these constants. It says 0 is the smallest element of the order and that $i + 1$ is the successor of i for $i = 0, \ldots, L$:

$$p(0) = 0 \wedge s(0) = 1 \wedge s(1) = 0 \wedge s(1) = 2 \wedge \cdots \wedge s(L) = L + 1.$$

To describe the sentences ψ_i, we expand the vocabulary again. Recall the coding process from Section 7.4. In particular, note that if $j > e$, then the program P_e cannot possibly mention bin B_j and its contents remain empty throughout the computation. To see that this is correct, suppose that Add B_j occurs as line (i) of P_e. Then p_i^{4j} occurs as a factor of e and so e must be bigger than j. The vocabulary of ψ_i includes unary functions l and b_i for $i = 1, \ldots, e$. Let $\mathcal{V}_e$ denote this vocabulary: $\mathcal{V}_e = \{<, s, p, l, b_1, \ldots, b_e, 0, 1, \ldots, L, L+1\}$.

The sentences ψ_i have an intended interpretation. Suppose we run the program P_e on a T-machine beginning with all of the bins empty. The T-machine will execute the commands of P_e one-by-one in the order dictated by P_e. Each element of our underlying set is intended to represent a "step" of this computation. The function $b_i(x)$ represents the value of B_i after step x of the computation and $l(x)$ represents the line of the program to be read next. The sentence θ_{init} describes the initial configuration of the T-machine:

$$l(0) = 1 \wedge \bigwedge_{i=1}^{e} b_i(0) = 0.$$

Each ψ_i for $i > 0$ corresponds to a line of the program. The lines of the program are not to be confused with the steps of the computation. The lines of the program are $(1), (2), \ldots, (L)$. The computation of the program may take any number of steps. The computation begins with line (1), but it then may jump to any other line (if the GOTO command occurs) and it may execute the same line several times within a loop. The T-machine continues to execute until it is told to read a line of the program that does not exist. If this happens, then the machine halts. The sentence θ_{halt} expresses this as

$$\forall x((l(x) = 0 \vee l(x) = L + 1) \rightarrow s(x) = x).$$

This sentence says that if the next line to be read after step x is either line (0) or line $(L + 1)$, then x is the last step of the computation (it is the maximal element in the order).

The sentence ψ_i describes line (i) of P_e. We demonstrate this sentences with a few examples.

- If "(9) Add 4" occurs as the ninth line of P_e, then ψ_9 is

$$l(x) = 9 \to \left(l(s(x)) = 10 \land b_4(s(x)) = s(b_4(x)) \land \bigwedge_{i \neq 4} b_i(s(x)) = b_i(x) \right).$$

- If line (4) is GOTO 12, then there are two possibilities. If there is no line (12) in the program, then ψ_4 is

$$l(x) = 4 \to \left(l(s(x)) = 0 \land \bigwedge_{i=1}^{e} b_i(s(x)) = b_i(x) \right).$$

Note that $l(s(x)) = 0$ implies that P_e halts at step $x+1$ (by θ_{halt}). Otherwise, if line (12) does exist, then replace $l(s(x)) = 0$ in this sentence with $l(s(x)) = 12$.

- If line (1) is RmvP B_6, then ψ_1 is the sentence

$$l(x) = 1 \to \left(b_6(x) = p(x) \land \bigwedge_{i \neg 6} b_i(s(x)) = b_i(x) \right)$$
$$\land (b_6(x) = 0 \to l(s(x)) = s(l(x))) \land (b_6(x) > 0 \to l(s(x)) = p(l(x))).$$

In this manner, each line of the program can be expressed as a $\mathcal{V}_e$-sentence. This completes the description of the $\mathcal{V}_e$-sentence φ_e. The computation of P_e on input 0 can be viewed as a model of this sentence. Moreover, any finite model describes such a computation. By design, φ_e has a finite model if and only if P_e halts on input 0. Since the Halting Problem is undecidable, so is the Finite Satisfiability Problem. □

Corollary 10.13 (Trakhtenbrot) The set of finitely valid sentences is not recursively enumerable.

Proof We claim that the set of finitely satisfiable sentences is recursively enumerable. Given a sentence φ, we can consider each of the finitely many structures in the vocabulary of φ having size n for $n = 1, 2, 3, \ldots$. If φ has a finite model, then (theoretically if not practically) we can find such a model in a finite number of steps. Since the set of finitely satisfiable sentences is not recursive by the previous theorem, the complement of this set cannot be recursively enumerable. The compliment is the set of sentences that are not satisfiable in a finite structure. A sentence φ is not finitely satisfiable if and only if $\neg \varphi$ is finitely valid. It follows that the set of finitely valid sentences too is not recursively enumerable. □

10.3 Descriptive complexity

This section provides an introduction to the branch of finite model theory known as *descriptive complexity*. The goal of this subject is to describe complexity classes in terms of various logics. The seminal result of this subject was proved by Ronald Fagin in 1974. Fagin's theorem states that, in some precise sense, the class **NP** is equivalent to existential second-order logic. Grädel's theorem describes the class **P** in a similar manner. In this section, we prove both Fagin's theorem and Grädel's theorem.

Descriptive complexity provides definitions for complexity classes that are independent of our choice of computing machine or programming language. In Section 7.7, we formally defined the class **P** in terms of the programming language $T^\sharp$. In the same section, we defined **NP** in terms of $T^\sharp_{NP}$. Grädel's theorem and Fagin's theorem allow us to avoid these contrived programming languages. These and other results of descriptive complexity show that certain classes of decision problems, defined in terms of natural constraints on space and time, are robust notions that can be analyzed using the tools of logic.

In Chapter 7, we regarded a decision problem as a set of natural numbers. Every subset A of $\mathbb{N}$ corresponds to the decision problem that asks whether or not a given natural number is in A. Conversely, every decision problem corresponds to a problem of this form. In descriptive complexity, we change this point of view. Instead of viewing decision problems as sets of natural numbers, we view each decision problem as a set of finite structures.

Let $\mathcal{V}$ be a finite vocabulary. Let $\mathcal{S}$ be a set of finite $\mathcal{V}$-structures. We assume that we have a fixed method for coding each structure in $\mathcal{S}$ as a tuple of natural numbers. For example, if $\mathcal{V} = \{R\}$ and $\mathcal{S}$ is the set of finite graphs, then each $\mathcal{V}$-structure has an associated adjacency matrix as described in Section 7.7. Moreover, we assume that, given a finite $\mathcal{V}$-structure M, we can determine whether or not M is in $\mathcal{S}$ in polynomial time (as is the case for graphs).

Each $\mathcal{V}$-sentence φ corresponds to a decision problem on $\mathcal{S}$. Given a finite $\mathcal{V}$-structure M as input, the φ-$\mathcal{S}$ Problem asks whether or not M models the sentence φ. The sentence φ is not necessarily first-order.

Notation 5 We let $\exists SO$ denote the set of second-order sentences of the form

$$\exists_1 R_1^{i_1} \exists_2 R_2^{i_2} \cdots \exists_k R_k^{i_k} \psi,$$

where ψ is a first-order $\mathcal{V}$-sentence.

We let $\exists SO$-*Horn* denote the set of sentences in $\exists SO$ for which the first-order part ψ is a Horn sentence as defined in Exercise 3.25.

Fagin's theorem states that the φ-$\mathcal{S}$ problem is in **NP** if and only if φ is equivalent to a sentence in $\exists SO$. If $\mathcal{V}$ contains a binary relation $<$ that is interpreted by each structure in $\mathcal{S}(\mathcal{V})$ as a linear order, then the φ-$\mathcal{S}$ problem is in **P** if and only if φ is equivalent to a sentence in $\exists SO$-$Horn$. This is Grädel's theorem.

The following proposition supplies one direction of Fagin's theorem.

Proposition 10.14 If φ is equivalent to a sentence in $\exists SO$, then the φ-$\mathcal{S}$ problem is in **NP**.

Proof Suppose that φ is equivalent to a sentence in $\exists SO$ having k existential second-order quantifiers. We proceed by induction on k. For the base step, we take $k = 0$. In this case, φ is first-order and the proposition is a restatement of Exercise 7.37. (This exercise is easily verified by induction on the complexity of φ.) Now suppose φ has the form:

$$\exists_1 R_1^{i_1} \exists_2 R_2^{i_2} \cdots \exists_{k+1} R_k^{i_{k+1}} \psi,$$

where ψ is first-order. Given a structure M in $\mathcal{S}$ as input, choose an arbitrary subset of U^{i_1} where U is the underlying set of M. Let M_1 be the expansion of M to the vocabulary $\mathcal{V} \cup \{R_1\}$ that interprets the i_1-ary relation R_1 as the arbitrarily chosen set. By induction, there is a nondeterministic algorithm that determines whether or not the second-order $(\mathcal{V} \cup \{R_1\})$-sentence $\exists_2 R_2^{i_2} \cdots \exists_{k+1} R_k^{i_{k+1}} \psi$ holds in M_1. If M_1 models this sentence, then M models φ. Thus, we have a nondeterministic polynomial-time algorithm for the φ-$\mathcal{S}$ Problem. $\square$

The following proposition supplies one direction of Grädel's theorem.

Proposition 10.15 If φ is equivalent to a sentence in $\exists SO$-$Horn$, then the φ-$\mathcal{S}$ Problem is in **P**.

Proof The proof we shall give works regardless of the number of second-order variables. For convenience, suppose that φ has the form $\exists R^k \psi_0$ for some Horn sentence ψ_0. So ψ_0 has the form $\forall x_1 \cdots \forall x_t \psi(x_1, \ldots, x_t)$, where $\psi(\bar{x})$ is a disjunction of conjunction of literals in the vocabulary $\mathcal{V} \cup \{R\}$. Now suppose we are given a finite $\mathcal{V}$-structure M. We describe a procedure for determining whether or not $M \models \varphi$.

There are $|M|^t$ t-tuples of elements in the underlying set of M. For each t-tuple $\bar{a}$, we consider the sentence $\psi(\bar{a})$. For each $\mathcal{V}$-literal in this sentence we check to see whether or not the literal holds for $\bar{a}$ in M. If it is false, we delete that literal from the sentence $\psi(\bar{a})$, and if it is true, we delete each clause containing that literal. In this way, we reduce the above sentence to a sentence $\theta_{\bar{a}}$ in the vocabulary $\{R, a_1, \ldots, a_t\}$. Let σ denote the conjunction of the sentences $\theta_{\bar{a}}$ taken over all t-tuples $\bar{a}$ in M.

The sentence σ is satisfiable if and only if there exists some interpretation of R on M that makes $\psi(\bar{a})$ true for each $\bar{a}$ in M. That is, σ is satisfiable if and only if $M \models \varphi$. Since ψ is a Horn formula, σ is a Horn sentence (the conjunction of Horn sentences is Horn). Since it is a quantifier-free sentence, we may regard σ as a formula of propositional logic and use the Horn algorithm from Section 1.7. This is a polynomial-time algorithm. □

Prior to proving the converses to these propositions, we prove a far weaker statement. Suppose that a φ-$\mathcal{S}$ problem is decided by a polynomial-time T^{++} program. Since the programming language T^{++} is extremely inefficient, this is a strong hypothesis. We show that, if this hypothesis holds, then φ is equivalent to a sentence of $\exists SO$. We prove this by analyzing the proof of Trakhtenbrot's theorem. Grädel's theorem and Fagin's theorem are proved by extending this analysis.

Proposition 10.16 Let $\mathcal{S}$ be a set of finite $\mathcal{V}$-structures and let φ be a $\mathcal{V}$-sentence such that the φ-$\mathcal{S}$ problem can be decided by a T^{++} program that runs in polynomial time. Then φ is equivalent to a sentence in $\exists SO$.

Proof For convenience, suppose that $\mathcal{V} = \{R\}$ and $\mathcal{S}$ is the set of finite graphs. We describe a $\exists SO$ sentence Ψ such that $G \models \Psi$ if and only if $G \models \varphi$ for any finite graph G.

Let P_e be a T^{++} program that decides whether or not a given graph models φ. Recall the sentence φ_e from the proof of Trakhtenbrot's theorem. This is defined as the sentence $\varphi_< \wedge \varphi_s \wedge \varphi_p \wedge \varphi_c \wedge \theta_{init} \wedge \theta_{halt} \wedge \psi_1 \wedge \psi_2 \wedge \cdots \psi_L$ in the vocabulary $\mathcal{V}_e = \{<, s, p, l, b_1, \ldots, b_e, 0, 1, \ldots, L, L+1\}$. Let us simplify this vocabulary. Note that the constants are not necessary since each of the elements represented by $0, \ldots, L+1$ is definable (they are the first $L+2$ elements of the order). So we may replace each constant in φ_e with a subformula that defines the same element. We may also replace each function in φ_e with a relation. That is, replace the unary successor function s with the binary successor relation S, the unary function l with a binary relation $LINE$, b_1 with BIN_1 and so forth. Let φ_E be the result of these changes. Consider the $\exists SO$ sentence

$$\exists <^2 \exists S^2 \exists P^2 \exists LINE^2 \exists BIN_1^2 \cdots \exists BIN_e^2 \varphi_E.$$

Let us abbreviate this sentence $\exists RELATIONS \varphi_E$.

If a structure M models $\exists RELATIONS \varphi_E$, then P_e must halt on input 0. In particular, given a finite graph G, if $G \models \exists RELATIONS \varphi_E$, then P_e halts on input 0. We want a sentence that tells us whether or not P_e halts given input G. For this, we must alter the subformula θ_{init} of φ_e. Recall that θ_{init} describes the initial configuration of the T-machine as

$$l(0) = 1 \wedge \bigwedge_{i=1}^{e} b_i(0) = 0.$$

We change this initial configuration to encode G. Let n be the number of vertices in G. We code G as a sequence of 0s and 1s of length n^2 (the adjacency matrix for G). The order $<$ imposes an order on the set of ordered pairs of vertices. We change θ_{init} to the sentence

$$l(0) = 1 \wedge \bigwedge_{i=1}^{e} b_i(0) = \delta_i,$$

where $\delta_i = 1$ if the ith ordered pair of vertices share an edge and otherwise $\delta_i = 0$. Again, we can replace 0 and 1 with formulas defining the first and second elements of the order. Note that it is possible that $|G| = n > e$ in which case not all of G will be encoded. This is one of the major deficiencies of the programming language T^{++}.

Let Ψ_0 be the sentence obtained by changing the subformula θ_{init} in $\exists RELATIONS\varphi_E$ as we have described. Now if $G \models \Psi_0$, then G can be viewed as the computation of P_e given the code of G as input. That is, each step of the computation is represented by a vertex of the graph. However, this computation may take more than $n = |G|$ steps. The key to the proof, and the key to Fagin's theorem as well, is the following elementary observation. The set of k-tuples in G has size n^k. Since P_e is polynomial-time, there exists k such that this program halts in fewer than n^k steps given input of size n. We modify Ψ_0 so that it refers to k-tuples of vertices. Let Ψ be the $\exists SO$ sentence that is the result of this modification. □

Theorem 10.17 (Grädel) Let $\mathcal{V}$ be a relational vocabulary containing the binary relation $<$. Let $\mathcal{S}$ be a set of finite $\mathcal{V}$-structures each of which interpret $<$ as a linear order. A $\mathcal{V}$-sentence φ is equivalent to a sentence of $\exists SO$-$Horn$ if and only if the φ-s problem is in **P**.

Proof One direction of this theorem is proved as Proposition 10.15. For the other direction, suppose that the φ-S Problem is in **P**. Then there exists a $T^{\sharp}$ program P_e that decides this problem in polynomial time. We can code programs of $T^{\sharp}$ in the same manner that we coded T^{++} programs in Section 7.4. So we can write a sentence φ_e as in the proof of Trakhtenbrot's theorem. Following the proof of the previous proposition, we find a sentence Ψ in $\exists SO$ that works. However, we want a sentence that is Horn. For this we make two observations regarding the sentence φ_e from Trakhtenbrot's theorem. Since the structures in $\mathcal{S}$ are ordered, we may drop the subformula $\varphi_< \wedge \varphi_s \wedge \varphi_p \wedge \varphi_c$ from φ_e. The second observation is that the sentence that remains, namely

$$\theta_{init} \wedge \theta_{halt} \wedge \psi_1 \wedge \psi_2 \wedge \cdots \psi_L$$

is equivalent to a conjunction of Horn sentences. For example, θ_{halt} is the sentence:

$$\forall x((l(x) = 0 \vee l(x) = L+1) \to s(x) = x).$$

This is equivalent to the following conjunction of Horn sentences:

$$\forall x(l(x) = 0 \to s(x) = x) \wedge \forall x(l(x) = L+1 \to s(x) = x).$$

Likewise, the sentences ψ_i are equivalent to conjunctions of Horn sentences. For example, if "(9) Add 4" occurs as the ninth line of P_e, then ψ_9 is

$$l(x) = 9 \to \left(l(s(x)) = 10 \wedge b_4(s(x)) = s(b_4(x)) \wedge \bigwedge_{i \neq 4} b_i(s(x)) = b_i(x) \right).$$

This sentence is equivalent to

$$(l(x) = 9 \to l(s(x)) = 10) \wedge (l(x) = 9 \to b_4(s(x))$$
$$= s(b_4(x))) \wedge \bigwedge_{i \neq 4} (l(x) = 9 \to b_i(s(x)) = b_i(x)).$$

The theorem then follows from the fact that a conjunction of Horn sentences is equivalent to a Horn sentence. $\square$

Grädel's theorem is can be paraphrased as $\exists SO\text{-}Horn = \mathbf{P}$. We say that $\exists SO\text{-}Horn$ *captures* the complexity class $\mathbf{P}$ on ordered structures. Least fixed-point logic also captures $\mathbf{P}$. This was proved independently by Immerman and Vardi.

Theorem 10.18 (Immerman–Vardi) Let $\mathcal{V}$ be a relational vocabulary containing the binary relation $<$. Let $\mathcal{S}$ be a set of finite $\mathcal{V}$-structures each of which interpret $<$ as a linear order. A $\mathcal{V}$-sentence φ is equivalent to a sentence of *LFP* if and only if the φ-$\mathcal{S}$ problem is in $\mathbf{P}$.

This theorem can be proved analogously to the proof of Grädel's theorem. We omit this proof (a proof is contained in [17]). Note that both of these theorems are restricted to ordered structures. This is a common restriction in complexity theory. Suppose we are given a graph having n vertices. There are $n!$ ways to arrange these vertices into a linear order. So there are $n!$ ways to input the same graph into a T-machine. Moreover, there is no known polynomial-time algorithm to determine whether or not two given graphs are the same or not. So if a program P is polynomial-time, it may produce different outputs when given two different presentations of the same graph as input. If we restrict to ordered graphs, then we avoid this problem. Fagin's theorem is one of the few results in descriptive complexity that is not restricted to ordered structures. This is because we can assert that there exists a linear order in $\exists SO$.

Theorem 10.19 (Fagin) Let $\mathcal{V}$ be a relational vocabulary and let $\mathcal{S}$ be a set of finite $\mathcal{V}$-structures. A $\mathcal{V}$-sentence φ is equivalent to a sentence of $\exists SO$ if and only if the φ-$\mathcal{S}$ problem is in **NP**. Moreover, we may further require that the first-order part of the $\exists SO$ sentence is universal.

Proof One direction of this theorem is proved as Proposition 10.14. For the other direction, suppose that the φ-$\mathcal{S}$ Problem is in **NP**. Then there exists a $T_N^\sharp P$ program P_e that decides this problem in polynomial time. We can code programs of $T_N^\sharp$ in the same manner that we coded T^{++} programs. Following the proof of Proposition 10.16, we can find a $\exists SO$ sentence Ψ as desired.

The moreover clause is verified by inspecting the sentence Ψ. It is interesting to note that, in the nondeterministic case, the sentences ψ_i are not necessarily $\exists SO$-$Horn$. For example, suppose that line (9) of P_e is the $T_N^\sharp P$ command GOTO 2 or 3. Then ψ_9 must express $l(x) = 9 \rightarrow (l(x+1) = 2 \vee l(x+1) = 3)$ which is not a Horn sentence. □

Finally, we show that $PSAT$ is **NP**-complete. This was first proved by Stephen Cook in 1971.

Theorem 10.20 (Cook) $PSAT$ is **NP**-complete.

Proof Suppose we have an algorithm that determines $PSAT$ in polynomial-time. We show that, using this algorithm, we can determine any decision problem is in **NP** in polynomial-time. Suppose we are given a φ-$\mathcal{S}$ problem for some $\mathcal{V}$-sentence φ and some set of finite $\mathcal{V}$-structures $\mathcal{S}$. If this problem is in **NP**, then by Fagin's theorem, φ is equivalent to a formula of the form

$$\exists R_1^{i_1} \cdots \exists R_k^{i_k} \forall x_1 \cdots \forall x_n \psi(x_1, \ldots, x_n)$$

where $\psi(\bar{x})$ is a quantifier-free first-order formula in conjunctive prenex normal form. Suppose we are given a structure M in $\mathcal{S}$ an want to determine whether or not M models the above $\exists SO$ sentence. Consider the conjunction $\bigwedge_{\bar{a} \in M} \psi(a_1, \ldots, a_n)$ taken over all n-tuples of M. As in the proof of Proposition 10.15, this sentence reduces to one in which every atomic subformula has the form $R_j(a_1, \ldots, a_{i_j})$ for some $j = 1, \ldots, k$. Since this is a quantifier-free sentence, we may view this as a formula of propositional logic. If we can determine whether or not this formula is satisfiable, then we can determine whether or not M models the above $\exists SO$ sentence. In this way, the φ-$\mathcal{S}$ problem is reducible to $PSAT$. Since this problem is an arbritrary problem in **NP**, $PSAT$ is **NP**-complete. □

10.4 Logic and the P = NP problem

The question of whether or not **P** = **NP** is one of the most important unanswered questions of mathematics. In this final section, we reformulate this and related questions as questions of pure logic.

To show that **P** = **NP**, it suffices to find a polynomial-time algorithm for determining whether or not a given formula of propositional logic is satisfiable. This follows from Cook's theorem. Of course, the same is true for any **NP**-complete problem. Consider the 3-Color Problem. A graph is 3-colorable if and only if it can be divided into three subsets such that no two vertices in the same set shares an edge. This property can be expressed as a sentence of $\exists SO$. By Fagin's theorem, the 3-Color Problem is in **NP**. In fact, this problem can be shown to be **NP**-complete. So if this problem is in **P**, then so is $PSAT$ as is every **NP** problem. So to show that **P** = **NP**, it suffices to define 3-Colorability on ordered graphs using a sentence of $\exists SO\text{-}Horn$ (by Grädel's theorem). Likewise, to prove that **NP** = **coNP** it suffices to write one clever sentence. If there exists a sentence of $\exists SO$ that says that a graph is *not* 3-Colorable, then the 3-Color Problem is in **coNP** as well as **NP**. By the **NP**-completeness of this problem, this would imply that **NP** = **coNP**. Conversely, one can show that **NP** ≠ **coNP** by playing pebble games. One must construct a 3-colorable graph M with the property that, for any $k \in \mathbb{N}$ and any expansion M' of M to a finite relational vocabulary $\mathcal{V}$, there exists a $\mathcal{V}$-structure N such that N is not 3-colorable and $M \equiv^k N$. If one could achieve this, then one could further conclude that **P** ≠ **NP**.

The question of **NP** = **coNP** is related to the question of whether or not $\exists SO$ is an extension of first-order logic. Recall the definition of such an extension from Section 9.4. The point of difficulty is the Closure Property. Is $\exists SO$ closed under negations? Clearly, the negation of an $\exists SO$ sentence is equivalent to a $\forall SO$ sentence (where $\forall SO$ is defined analogously to $\exists SO$). By Fagin's theorem $\exists SO$ captures **NP**. As a corollary of this, we see that $\forall SO$ captures **coNP**. It follows that **NP** = **coNP** if and only if $\exists SO \equiv \forall SO$. Moreover, if $\exists SO \equiv \forall SO$, then every second-order sentence is equivalent to a sentence of $\exists SO$. This can be shown by induction on the complexity of the second-order quantifiers using $\exists SO \equiv \forall SO$ as the base step. So the following are equivalent:

(i) **NP** = **coNP**,
(ii) $\exists SO$ is an extension of first-order logic (as defined in Section 9.4), and

(iii) $\exists SO$ is equivalent to second-order logic.

Finally, the **NP** = **coNP** problem is related to the finite spectra of first-order sentences as defined in Exercise 2.3. Recall that the finite spectrum of a sentence φ is the set of $n \in \mathbb{N}$ such that φ has a model of size n. Asser's Problem asks whether or not the set of finite spectra is closed under complements. That is, if $A \subset \mathbb{N}$ is the spectrum for a sentence φ, then is there a sentence for which the complement of A is the spectrum? If not, then one can conclude that **NP** $\neq$ **coNP**.

We leave it to the reader to verify and expand upon the claims of this section and to resolve the problems of whether or not **P** = **NP** = **coNP**.

Exercises

10.1. Let M_4 be the structure in the vocabulary $\mathcal{V}_E = \{E\}$ that interprets the binary relation E as an equivalence relation having exactly four classes each containing exactly four elements.
   (a) Show that $M_4 \equiv^4 N$ for any $\mathcal{V}_E$ structure N that interprets E as an equivalence relation having more than four classes each containing more than four elements.
   (b) Show that M_4 is not $\mathcal{L}^3$-equivalent to any $\mathcal{V}_E$-structure that does not interpret E as an equivalence relation.

10.2. Let $\mathcal{V}_< = \{R\}$. Show that there exists a $\mathcal{V}_R$-sentence φ of $\mathcal{L}^3_{\omega_1\omega}$ such that $M \models \varphi$ if and only if M is a connected graph.

10.3. Let $\mathcal{V}_< = \{<\}$. Show that there exists a $\mathcal{V}_<$-sentence φ of $\mathcal{L}^3_{\omega_1\omega}$ such that $M \models \varphi$ if and only if M interprets $<$ as a linear order and $|M|$ is an odd natural number.

10.4. Let $\mathcal{V}$ be a relational vocabulary. Let T be a complete $\mathcal{V}$-theory that is axiomatized by a set of $\mathcal{L}^k$ sentences. Show that $V(\varphi) \leq k$ for each atomic $\mathcal{V}$-formula φ.

10.5. Let $\mathcal{V}$ be a finite relational vocabulary that contains the binary relation S. Let M be a $\mathcal{V}$-structure that has underlying set $\mathbb{N}$ and interprets S as the successor relation. Show that there exists $k \in \mathbb{N}$ such that $M \equiv^{\mathcal{L}^k_{\omega_1\omega}} N$ implies $M \cong N$.

10.6. This exercise demonstrates that the Beth Definability theorem fails when restricted to finite structures. Let $\mathcal{V} = \{<, P\}$ and let T be the incomplete $\mathcal{V}_<$ saying that $<$ is a linear order and the unary relation P holds for the odd elements in the order (the first element of the order, the third, the fifth, and so forth).

(a) Show that P is implicitly defined by T in terms of $\{<\}$ on finite structures. That is, show that any two expansions of a finite linear order to a model of T are isomorphic.

(b) Show that P is not explicitly defined by T in terms of $\{<\}$ on finite structures. That is, there is no formula $\varphi(x)$ in the vocabulary $\{<\}$ such that $M \models \varphi(x) \leftrightarrow P(x)$ for any finite model M of T. (Use the previous exercise.)

10.7. Show that Craig's theorem does not hold for finite structures. That is, demonstrate a $\mathcal{V}_1$-sentence φ_1 and a $\mathcal{V}_2$-sentence φ_2 so that $\varphi_1 \models_{fin} \varphi_2$ but there is no sentence in the vocabulary $\mathcal{V}_1 \cap \mathcal{V}_2$ for which both $\varphi_1 \models_{fin} \theta$ and $\theta \models_{fin} \varphi_2$.

10.8. Demonstrate a first-order sentence that is not equivalent to a universal sentence but is preserved under substructures of finite models.

10.9. Let V denote the set of first-order sentences that are valid. Let V_{fin} denote those sentences that are finitely valid. Trakhtenbrot proved that V_{fin} and the complement of V are recursively inseparable sets. To prove this strengthened version of Theorem 10.12, let A and B be any recursively inseparable pair of sets (Exercise 7.24). Let S be any set of first-order sentences such that $V \subset S \subset V_{fin}$. Show that if S is recursive, then A and B are not recursively inseparable.

Bibliography

[1] J. T. Baldwin, *Fundamentals of Stability Theory*, Springer-Verlag (Perspectives in Logic), 1989.
[2] J. T. Baldwin and A. H. Lachlan, On strongly minimal sets, *Journal of Symbolic Logic*, vol. 36, pp. 79–96, 1971.
[3] J. Barwise, ed., *Handbook of Mathematical Logic*, North-Holland (Studies in Logic), 1977.
[4] G. Boolos and R. Jeffery, *Computability and Logic*, Cambridge University Press, 1989.
[5] G. Boolos, New proof of the Gödel Incompleteness Theorem, *Notices of the American Mathematical Society*, vol. 36, pp. 388–390, 1989.
[6] S. Buechler, *Essential Stability Theory*, Springer-Verlag (Perspectives in Logic), 1996.
[7] C. C. Chang and H. J. Keisler, *Model Theory*, North-Holland (Studies in Logic), 1990.
[8] N. Cutland, *Computability*, Cambridge University Press, 1980.
[9] R. Diestel, *Graph Theory*, Springer-Verlag (Graduate Texts in Mathematics), 2000.
[10] L. van den Dries, *Tame Topology and o-Minimal Structures*, Cambridge University Press (LMS Lecture Notes), 1998.
[11] H. D. Ebbinghaus and J. Flum, *Finite Model Theory*, Springer-Verlag (Perspectives in Logic), 1995.
[12] H. D. Ebbinghaus, J. Flum, and W. Thomas, *Mathematical Logic*, Springer-Verlag (Undergraduate Texts in Mathematics), 1989.
[13] K. Gödel, *Collected Works,* vol. 1 (S. Feferman et al., eds.), Oxford University Press, 1986.
[14] J. van Heijenoort, *From Frege to Gödel: A Source Book in Mathematical Logic, 1879–1931*, Harvard University Press, 1967.
[15] I. N. Herstein, *Topics in Algebra*, John Wiley and Sons, 1975.
[16] W. A. Hodges, *Model Theory*, Cambridge University Press (Encyclopedia of Mathematics), 1993.
[17] N. Immerman, *Descriptive Complexity*, Springer-Verlag (Texts and Monographs in Computer Science), 1999.
[18] T. Jech, *Set Theory*, Academic Press, 1978.
[19] G. A. Jones and J. M. Jones, *Elementary Number Theory*, Springer-Verlag (Springer Undergraduate Mathematics), 1998.
[20] J. A. Kalman, *Automated Reasoning with Otter*, Rinton Press, 2001.
[21] I. Kaplansky, *Set Theory and Metric Spaces*, Chelsea Publishing Co., 1977.
[22] R. Kaye, Minesweeper is **NP**-complete, *Mathematical Intelligencer*, vol. 22(2), pp. 9–15, 2000.
[23] L. Kirby and J. Paris, Accessible independence results in Peano arithmetic, *Bulletin of the LMS*, vol. 14, pp. 285–293, 1982.

[24] F. Kirwan, *Complex Algebraic Curves*, Cambridge University Press (LMS Student Texts), 1992.
[25] K. Kunen, *Set Theory: An Introduction to Independence Proofs*, North-Holland (Studies in Logic), 1995.
[26] D. Lascar, *Stability in Model Theory*, Longman, 1988.
[27] P. Lindström, On model-completeness, *Theoria*, vol. 30, pp. 183–196, 1964.
[28] P. Lindström, On extensions of elementary logic, *Theoria*, vol. 35, pp. 1–11, 1969.
[29] D. Marker, *Model Theory: An Introduction*, Springer-Verlag (Graduate Texts in Mathematics), 2003.
[30] D. Marker, M. Messmer, and A. Pillay, *Model Theory of Fields*, Springer-Verlag (Lecture Notes in Logic), 1996.
[31] Y. Matiyasevich, *Hilbert's 10th Problem*, MIT Press, 1993.
[32] M. Morley, Categoricity in power, *Transactions of the American Mathematical Society*, vol. 114, pp. 514–538, 1965.
[33] A. Nerode and R. A. Shore, *Logic for Applications*, Springer-Verlag (Texts and Monographs in Computer Science), 1994.
[34] S. H. Nienhuys-Cheng and R. de Wolf, *Foundations of Inductive Logic Programming*, Springer-Verlag (Lecture Notes in Artificial Intelligence), 1997.
[35] M. Otto, *Bounded Variable Logics and Counting*, Springer-Verlag (Lecture Notes in Logic), 1997.
[36] C. H. Papadimitiou, *Computational Complexity*, Addison-Wesley, 1994.
[37] A. Pillay, *An Introduction to Stability Theory*, Oxford University Press (Logic Guides), 1983.
[38] A. Pillay, *Geometric Stability Theory*, Oxford University Press (Logic Guides), 1996.
[39] B. Poizat, *Cours de Theorie des Modeles*, Nur Al-Mantiz Wal-Ma'rifah, 1985.
[40] A. Robinson, *Selected Papers*, vol. 1, (H. J. Keisler et al., eds.), Yale University Press, 1979.
[41] J. A. Robinson, A machine oriented logic based on the resolution principle, *Journal of the ACM*, vol. 12, pp. 23–41, 1965.
[42] K. A. Ross, *Elementary Analysis: The Theory of Calculus*, Springer-Verlag (Undergraduate Texts), 1980.
[43] S. Shelah, *Classification Theory and the Number of Non-isomorphic Models*, North-Holland (Studies in Logic), 1990.
[44] S. Shelah, *Cardinal Arithmetic*, Oxford University Press (Logic Guides), 1994.
[45] J. Shoenfield, *Mathematical Logic*, Addison-Wesley, 1967.
[46] J. Shoenfield, *Recursion Theory*, Springer-Verlag (Lecture Notes in Logic), 1993.
[47] M. Sipser, *Introduction to the Theory of Computation*, PWS Publishing Co., 1997.
[48] R. I. Soare, *Recursively Enumerable Sets and Degrees*, Springer-Verlag (Perspectives in Logic), 1987.
[49] R. Vaught, Denumerable models of complete theories, In *Infinitistic Methods. Proceedings of the Symposium on Foundations of Mathematics*, pp. 303–321, Pergamon, 1961.
[50] F. O. Wagner, *Simple Theories*, Kluwer Academic Press, 2000.

Index

0–1 laws   217, 221
Δ_0 formula   314
Π_n set   337
$\sum_n$ set   337
$\sum_n$-complete   359
$\sum_n$-universal   359
$\exists_n$ formula   182
$\forall_n$ formula   182
$\mathcal{L}_{\omega_1\omega}$ 392

Abiteboul–Vianu theorem   399
Ackermann function   307, 327
Ackermann, Wilhelm   307
algebraic closure   241
algebraic formula   241
algebraically closed fields   253
algorithm   347
amalgamation
  property   238
  theorem   171
Aristotle   1
arithmetic hierarchy   336
Asser's problem   424
assignment   7
atomic formula
  first-order logic   55
  propositional logic   1
atomic structure   276
automorphism   297
Ax's theorem   266
Axiom of Choice   156
axiomatization   200
  finite   210
  quasi-finite   210

back-and-forth argument   214
Baldwin–Lachlan theorem   289, 294
basic functions   302
basis   242
Beth Definability theorem   52, 191
bi-definable   227
Boolos, George   374
bound variable   57
bounded search   306
Buechler, Steven   294

cardinal number   152
categoricity   206
  countably categorical   209
  totally categorical   209
  uncountably categorical   209
characteristic function   312
Chinese Remainder theorem   366
Church–Turing thesis   310, 316
clause   37
clique   68
CNF   28
CNF algorithm   30
Cohen, Paul   162
compactness   45
  of first-order logic   167
  of propositional logic   47
compactness number   403
complete theory   89
complete type   268
completeness   44
  of propositional logic   47
  of first-order logic   167
computable function   xx, 79, 301
conjunction   2
conjunctive normal form   28
conjunctive prenex normal form   109
coNL   351
connected graph   67
coNP   341
coNP = NP   342, 354
consequence   9, 12
Consequence Problem   10, 36, 64, 99
consistent   90
Continuum hypothesis   162
contradiction   8, 63
Contradiction rule   16
contrapositive   17
Cook's theorem   422
countable   76
counting quantifier   95
Craig Interpolation theorem   52, 189
Craig's trick   377
Cut rule   37

Davis, Martin   331
decidable theory   203
decision problem   8, 300

Index

Dedikind, Richard 161
deductive closure 199
definable closure 265
definable subset 62, 93, 239
DeMorgan's rules 11
dense linear order 71, 211
denumerable 76
diagram 86
dimension 244
Diophantine set 330
disjunction 5
disjunctive normal form 28
Distributivity rules 11
DNF 28
Duplicator 393

Easton's theorem 163
elementary
  diagram 86
  extension 85
  substructure 85
  chain 181
  class 90
  embedding 80
  equivalence 82
embedding 80
Engler, Erwin 215
Equivalence problem 36, 64
equivalent formulas 10
Erdös, Paul 216, 316
Euclid 16
Euler path 405
existential formula 81
existential quantifier 54
existentially closed 235
expansion 61
extension of first-order logic 401

Fagin's theorem 422
Fagin, Ronald 417
feasible decision problem 338
Fibonacci number 386
field 249
Finite Satisfiability problem 414
finite spectrum 260
Finite Validity problem 414
finite-variable logics 409
finitely satisfiable 408
finitely valid 408
Fixed Point lemma 381
formal proof 15, 102
formula
  first-order logic 55
  propositional logic 2
  second-order logic 389
Four Color theorem 197
free variable 56
function xviii

Fundamental Theorem
  of algebra 253
  of arithmetic 358, 316, 321
fuzzy logic 1

Gödel's First Incompleteness
  theorem 374
Gödel's Second Incompleteness
  theorem 382
Gödel, Kurt xx, 162, 333, 357, 358, 374, 382
General Continuum hypothesis 162
Goldbach's conjecture 93, 141, 358
Goodstein sequences 383
Goodstein's theorem 384
Grädel's theorem 420
graph 66
group 201
Gurevich, Yuri 399

Halting Problem 334
Hamilton circuit 404
Hanf number 407
Henkin construction 148, 165, 271
Herbrand
  structure 114
  universe 115
  vocabulary 115
Hilbert's Nullstellensatz 257
Hilbert's 10th problem 332
Hilbert, David 332
homogeneous structure 277, 285
Horn formula 32
Horn algorithm 33
Horn sentence 134
Hrushovski, Ehud 210, 256

IFP 397
Immerman, Neil 399
Immerman–Vardi theorem 421
inconsistent 51
independent set 242
index set 334
induction
  mathematical 23
  on complexity of formulas 25
  on ordinals 156
infinitary logics 395
Inflationary Fixed-point Logic 397
isolated type 271
isomorphism 82
isomorphism property 231

Joint Embedding Lemma 172

k-Colorability problem 353
k-colorable graph 353, 405

Kim, Byunghan 292, 294
Kirby–Paris theorem 385
Kleene Normal Form 325
Kleene's Recursion theorem 326
Kleene, Stephen 333
Knight, Robin 294
Kreutzer's theorem 400
Kreutzer, Stephan 399

L 348
Löb's theorem 387
Löwenhiem–Skolem number 402
Löwenhiem–Skolem theorems
    Downward 169, 405
    Upward 167
Lachlan's theorem 294
Least Fixed-Point Logic 399
length of input 337
LFP 399
limit ordinal 154
Lindström's theorem
    on extensions of first-order logic 403
    on model-completeness 237
linear order 71
linear resolution 129
linearly ordered set 153, 202
literal 28
LN-resolution 133
locally modular 252
logarithmic-space 348
logic xiii
Los–Tarski Theorem 178, 405
Lyndon's Interpolation theorem 199
Lyndon's Preservation theorem 199

Matiyasevich's theorem 331
Matiyasevich, Yuri 331, 333
Minesweeper 353
minimal structure 240
minimal unsatisfiable set 51, 131
model 89
model companion 264
model-complete 233
modular 252
Monadic second-order logic 392
monster model 286
Morley's theorem 209, 210, 291, 299
Morley, Michael 210, 291
most general unifier 121

N-resolution 128
negation 2
NL 351
nondeterministic algorithm 339, 349
NP 341
    complete 351
    =coNP 423

o-minimal 246
Omitting Types theorem 272
operator 396
order property 291
ordinal number 153
Otter 137

P 35
P-resolution 146
P=NP 36, 342, 354, 423
Partial Fixed-point Logic 398
partial isomorphism 393
partial order 197
Peano Arithmetic 360
pebble games 392
    Ehrenfeucht–Fraisse game 393
    k-pebble game 410
PFP 398
PNF 109
polynomial
    space 348
    time xiv, 35, 338
    time reduction 351
Post, Emile 333
power set 77
prenex normal form 109
prime model 279
primitive recursion 303
primitive recursive functions 304
primitive recursive set 312
primitive symbols 2
projective geometry 202
Prolog 137
provably equivalent 20
PSPACE 348
Putnam, Hilary 331

quantifier elimination 222
quantifier-free formula 81

random graph 220
realizable 268
recursive
    functions 310
    set 312
recursively enumerable set 328
recursively inseparable 358
recursively reducible 334
reduct 61
relation xvii
relational
    database 69
    vocabulary 221
resolvent 38, 124
Rice's theorem 335
Robinson's Joint Consistency 186
Robinson, Abraham 186
Robinson, John Allen 120

Robinson, Julia   331
Rosser's theorem   387
Russell, Bertrand   357
Ryll-Nardzewski, Czeslaw   215

Satisfiability Problem   63
   for first-order logic   299
   for propositional logic   8, 35, 344
satisfiable   8, 63
saturated structure   281, 285
Scott's theorem   395
second-order logic   360, 388
semi-decidable   328
sentence   56
   second-order logic   389
set   xvi
Shelah, Saharon   163, 289, 292, 399
Silver's theorem   163
simple theories   292
size of a structure   71
Skolem normal form   111
SLD-resolution   135
small theory   275
SNF   111
spectrum   289
   finite   92
   uncountable spectrum   289
Spoiler   393
stable theory   291
strict order property   292
strongly minimal   240
structure   xvi, 59
subformula
   first-order logic   56
   propositional logic   3
substructure   83
successor ordinal   154
Svenonius, Lars   215

T-computable function   318
T-resolution   146

T^{++} Program   317
Tarski's Undefinability of Truth   387
Tarski, Alfred   247, 332
Tarski–Vaught criterion   168, 405
tautology   8, 63
Tautology rule   16
theory   89
Trakhtenbrot's theorem   414, 416
transfinite induction   157
Traveling Salesman problem   353
truth table   4
Turing, Alan   316, 333
type   268

unbounded search   310
uncountable   76
unification   121
Unification algorithm   122
universal formula   85
universal model   283, 285
universal quantifier   54

valid   8, 63
Validity Problem   8, 35, 63
van den Dries, Lou   247
Vardi, Moshe   399
Vaught's conjecture   294
Vaught, Robert   290
vector space   250
vocabulary   59

weak Monadic second-order logic   392
well ordered set   153
Well Ordering Principle   156
Whitehead, Alfred North   357
wild theories   293
Wilkie, Alex   247

Zermeleo–Frankel set theory   162
ZFC   162
Zil'ber's Theorem   210